T0251316

SHAPE CLASSIFICATION AND ANALYSIS

THEORY AND PRACTICE

SECOND EDITION

IMAGE PROCESSING SERIES

Series Editor: Phillip A. Laplante, Pennsylvania State University

Published Titles

SHAPE CLASSIFICATION AND ANALYSIS

THEORY AND PRACTICE

SECOND EDITION

Luciano da Fontoura Costa
Roberto Marcondes Cesar, Jr.

CRC Press
Taylor & Francis Group
Boca Raton   London   New York

CRC Press is an imprint of the
Taylor & Francis Group, an **informa** business

CRC Press
Taylor & Francis Group
6000 Broken Sound Parkway NW, Suite 300
Boca Raton, FL 33487-2742

© 2009 by Taylor & Francis Group, LLC
CRC Press is an imprint of Taylor & Francis Group, an Informa business

No claim to original U.S. Government works
Printed in the United States of America on acid-free paper
10 9 8 7 6 5 4 3 2 1

International Standard Book Number-13: 978-0-8493-7929-1 (Hardcover)

Library of Congress Cataloging-in-Publication Data

Costa, Luciano da Fontoura.
    Shape classification and analysis : theory and practice / Luciano da Fontoura Costa, Roberto Marcondes Cesar, Jr. -- 2nd ed.
        p. cm. -- (Image processing series)
    Rev. ed. of: Shape analysis and classification. 2001.
    Includes bibliographical references and index.
    ISBN 978-0-8493-7929-1 (hardcover : alk. paper)
    1. Image processing--Digital techniques. 2. Image analysis. 3. Form perception.
I. Cesar, Roberto Marcondes. II. Costa, Luciano da Fontoura. Shape analysis and classification. III. Title. IV. Series.

TA1637.C67 2009
621.36'7--dc22   2009005501

Visit the Taylor & Francis Web site at
http://www.taylorandfrancis.com

and the CRC Press Web site at
http://www.crcpress.com

To the memory of my grandmother Clorinda, my brightest teacher.
To my wife, Soraya, my ray of sunlight.

To the memory of my grandmother Nair, who showed me how great is being alive.
To my continuously growing family and, especially, to Adriana and Tainá.

Contents

Preface - Second Edition

THE SECOND EDITION of *Shape Analysis and Classification: Theory and Practice* incorporates important enhancements from the last reprint version. The book has been graphically reformulated and we hope the reader will like the new design (fonts, equations, figures, and references). We have implemented a number of corrections and improvements, many of which suggested by readers of the first edition, to whom we are very grateful. New material has been added, in particular new sections on topics like graph and complex networks, dimensionality reduction as well as a new chapter on structural pattern recognition and shape representation using graphs. The reference list has been updated generally.

The book now includes a new type of Box called *Additional Resources* indicating useful resources on the World Wide Web (WWW) such as software, databases and videos. In particular, some open-source software projects related to the subjects covered in this book have been initiated since its first edition was issued. Links to these open-source projects are now provided throughout the book (also available from `http://ccsl.ime.usp.br/`). The user is invited to know, use and help to develop all these projects! The second edition of the book has a new URL: `http://www.vision.ime.usp.br/~cesar/shape/`. In addition, slides that may be useful in courses and talks based on the book are now available at: `http://www.vision.ime.usp.br/~cesar/shape/slides`.

In addition to all friends that helped us in the first edition, we are grateful to those colleagues, students and readers that provided important input regarding the book. Additionally to the people already mentioned in the preface of the first edition, we are grateful to: Alejandro Frery, Alexandre Crivellaro, Alexandre Noma, Ana Beatriz, Anselmo Kumazawa, Arnaldo Lara, Carlos H. Morimoto, Celina Takemura, David Martins Junior, Emerson Tozette, Fabrício Lopes, Fátima N. Sombra, Flávia Ost, Giovane Cesar, Heitor Nicoliello, Herbert Jelinek, Isabelle Bloch, Jeferson Silva, Jesus Mena, Jishu Ashimine, João Eduardo Ferreira, João Soares, Jorge Leandro, José Mario Martinez, Junior Barrera, Junior Barrera, Luis Augusto Consularo, Luis Gustavo Nonato, Luiz Henrique de Figueiredo, Luiz Velho, Marcel Claro, Marcel Jackowski, Marcelo Hashimoto, Marcos Moreti, Michael Cree, Nina S. T. Hirata, Paulo José da Silva e Silva, Pete Lestrel, Reinhard Klette, Roberto Andreani, Roberto Hirata Jr., Rogério Féris, Ronaldo F. Hashimoto, Routo Terada, Silvia Pinto, Teófilo Campos, Thiago Paixão, and Yossi Zana, as well as to the many students of undergrad and grad courses taught by the authors. In particular, the second edition would not be possible without the valuable work of David da Silva Pires. David, thank you very, very much.

As always, our friends at Taylor & Francis have provided a fundamental support to our work, and we are especially grateful to Amber Donley and Nora Konopka. We would also like to acknowledge the support provided by the following research funding agencies: FAPESP, CNPq, CAPES, FINEP, COFECUB and NSF.

We have worked hard to provide an improved text about the exciting areas of shape analysis and classification. We hope you will enjoy reading and using this new edition.

Preface - First Edition

MOST HUMAN ACTIVITIES—including the reading of this text—and interactions with the environment are performed by visual means, more specifically through the capture of electromagnetic waves and subsequent processing by the human visual system. Given the redundancy underlying most that is visual in nature, evolution has produced visual systems that pay great attention to abrupt variations along the objects, such as their boundaries delimiting uniform regions. Indeed, it was very likely the varied geometry of the outlines of uniform objects in the world that led to the human concept of *shape* and the evolution of visual processing abilities dedicated, or at least related, to the capture, processing, analysis and classification of shapes. Besides, it should be observed that many other natural processes also involve geometric objects, normally interacting through fields and waves. For instance, most biochemical reactions involve the matching between two or more molecules, a process that is governed by their respective shapes and force fields. Similarly, interactions between living beings also involve shape geometry and interactions. For example, the mouth of a whale must be particularly effective for collecting plankton and the wings of birds have to be effective for flying. In brief, the importance of shapes is not only essential to humans, but seems to provide a basic principle underlying most natural processes [Costa et al., 1999].

While the importance of the visual information is now fully acknowledged, defining a scientific area on its own, it was only recently that computer vision came to establish itself as an off-the-shelf tool for both researchers and practitioners. That is to say, it has been the continuous advances in computer technology, including higher processing speed, storage capabilities, better acquisition devices (e.g., cameras and scanners), along with more powerful concepts and algorithms and a progressive cost reduction, that paved the way to the dissemination of imaging techniques through a large number of practical applications in the most diverse areas. Indeed, it is the opinion of the authors of this book that we are undergoing an unprecedented *technological opportunity* for the effective application of image and shape analysis resources to many areas. Yet, while this tendency has been duly reflected in an ever-increasing number of books on imaging, shape processing and analysis, biological and computer vision and pattern recognition, it is felt that only a few (if any) existing textbooks provide an introductory, modern, relatively comprehensive and integrated coverage of shape representation, analysis *and* recognition, which are closely intertwined. In light of the above reasons, we cannot avoid the commonplace of observing that the current book will fill the gap not only between the areas of shape analysis and classification, including one of the most comprehensive lists of practical shape features, but also between theory and practice. While concentrating on 2D shapes, the current book also provides the basis for the treatment of higher dimensional objects.

This book is aimed at serving as a largely self-contained introductory textbook on shape analysis and recognition, including several of the most modern and promising tendencies such as scale-space, analyzing wavelets, fractals, computational geometry and so on. An important trend that has characterized the evolution of shape analysis is its inter- and multidisciplinary nature, involving an increasing variety of concepts from both mathematics and computer science, including differential geometry, Fourier analysis, signal processing, probability and multivariate calculus, to name a few. This implies that students and researchers alike often experience difficulties while trying to understand many approaches to shape analysis. On the other hand, it is worthy emphasizing that shape analysis is such an important task in so many areas, from biology to material sciences, that it would be highly desirable that experts in all those different areas could understand and apply the techniques explored in this book. As a matter of fact, even those practitioners intending simply to use imaging software should also be acquainted with the basic concepts and techniques in image and shape analysis, in order not only to properly apply the several tools usually bundled into such software, but also interpret the respectively obtained results. As the authors had in mind to create a didactic book that would be accessible to a broad range of readers, a comprehensive and mostly self-contained introduction/review of the involved basic mathematical concepts has been especially prepared and included. In addition, whenever possible the mathematical detail is always preceded by the respective conceptual characterization and discussion and several examples have been included in order to help the assimilation of the presented concepts and techniques. Several special boxes have also been included in the book, being organized according to the following types:

Examples: step-by-step examples of the introduced concepts and methods. Such examples can also be treated as exercises, in the sense that the reader should try to solve the presented problems before following up their respective solutions;

Algorithms: high-level algorithms that can be straightforwardly implemented in most programming languages, such as C, Java, Delphi®, Pascal and MATLAB® scripts;

Notes: additional information on relatively secondary or more advanced topics;

"To Probe Further" Pointers: references to more advanced complementary scientific literature, so that the reader can probe further on the state-of-the-art of each specific topic.

In order to increase its readability, this book has been generously illustrated with graphics and images. In addition, a *WWW homepage* (see the box at the end of this preface) has been created as a complement, which will be continuously updated with additional examples, data, programs, errata and URL pointers to related areas. Given the introductory and multidisciplinary nature of the present book, as well as space limitations, several important related works and bibliographical references have certainly been overlooked, for which we apologize in advance. In addition, please observe that although great care has been taken in preparing this book, the authors can accept no responsibility for the direct or indirect consequences of the use of algorithms or any other information in this book.

The book starts by discussing the main operations often involved in shape analysis. Chapter 2 provides a self-contained review of the main mathematical concepts that are fundamental to proper understanding of shape analysis and recognition. Since shape analysis by computers involves acquiring and processing digital images of the objects of interest, an overview of the most typical image processing techniques is provided in Chapter 3. It should be observed that while most students and practitioners from other fields will find Chapters 2 and 3 helpful, computer vision and pattern recognition experts may prefer to directly proceed to Chapter 4, which discusses and formalizes the main concepts in 2D shapes. Chapter 5 covers some of the main techniques for shape representation by digital means (i.e., computers), while Chapters 6 and 7 present several important techniques for extracting information from shapes, the latter concentrating on multiscale techniques, including curvature and wavelets. Chapter 8 presents the main techniques for classifying shapes, with emphasis on the more traditional and sound statistical approaches. Finally, some brief considerations about the future perspectives in shape processing and analysis are presented in Chapter 8.

We would like to express our gratitude to a number of people and institutions. First, we thank Prof. Philip Laplante for the kind invitation to contribute to the *CRC Image Processing Series*. Whatever the merits of the present book, Phil deserves the full credit for the initiative. Also fundamental was the kind support of all CRC staff, especially Ms. Elizabeth Spangenberger, Nora Konopka, Michele Berman, Sara Seltzer and Suzanne Lassandro. We thank Prof. Héctor Terenzi (USP) and Prof. Hernan Chaimovich (USP), through the *Pró-Reitorias* of the University of São Paulo, who made the whole project possible by kindly providing financial support. In addition, many of the reported results and techniques have been developed by the authors under sponsorship of several Brazilian funding agencies, including FAPESP, CNPq, CAPES and CCInt, having also involved several students. We also thank all those researchers who so kindly read and commented on preliminary versions of the chapters in this book, including Andréa G. Campos, Carlos P. Dionísio, Cristian Montagnoli, Jorge J. G. Leandro, Prof. Maria Cristina F. de Oliveira, Prof. Nelson D. A. Mascarenhas, Luiz A. Consularo, Luiz G. Rios, Renato Carmo, Rogério S. Feris and Vital Cruvinel Ferreira. Students who have also contributed in the production of some of the data and results presented in this book are Alexandre Crivellaro, Andréa G. Campos, Franklin C. Flores, Heraldo Madeira, Luiz A. Consularo, Marcel Brun, Nina T. Hirata, Regina C. Coelho, Roberto Hirata and Teófilo Campos. Important discussions and suggestions have been provided by many colleagues, especially Carlos Hitoshi, Damien Barache, Herbert Jelinek, Isabelle Bloch, J.-P. Antoine, João Kogler, Junior Barrera, Pierre Vandergheynst and Roberto Lotufo. A good deal of the diagrams in this book were drawn by Samuel Alvarez and additional technical support was provided by Adenilza Alves, Alexandre Silva, Cláudia Tofanelli, Edna Torres, Marcos Gonçalves and Rita de Cássia, to whom we also extend our acknowledgements. Luciano Costa is especially grateful to his wife, Soraya, for her understanding and help with the revision of this book.

♦♦♦♦♦♦♦♦♦   BOOK HOMEPAGE   ♦♦♦♦♦♦♦♦♦

The reader should not miss the book's homepage, containing a series of additional related information, including related pointers, data files, and errata, to be found at the following address:

http://www.ime.usp.br/~cesar/shape_crc/

The authors also appreciate receiving comments through e-mail, which should be addressed to luciano@if.sc.usp.br or cesar@ime.usp.br.

List of Figures

List of Tables

Start by doing what's necessary, then do what's possible, and suddenly you are doing the impossible.

Saint Francis of Assisi

Introduction

Chapter Overview

THIS CHAPTER PRESENTS an introduction to computational shape analysis. Starting with some considerations about computer vision and 2D shapes, it proceeds by illustrating some typical applications and discussing the main problems normally involved in shape analysis, and concludes by presenting the organization of the related topics in the book chapters.

1.1 Introduction to Shape Analysis

There is little doubt that one of the most stimulating research fields, from both the scientific and technological perspectives, are those related to the main human sense: *vision*. The acquisition and analysis of the visual information produced by the interaction between light and the world objects have represented powerful means through which humans and animals can quickly and efficiently learn about their surrounding environments. The advantages of this ability for survival can be immediately recognized, accounting for all efforts nature has taken in developing such flexible visual systems. As far as humans are concerned, more than 50% of their brains are somehow involved in visual information analysis, a task that underlies the majority of human daily activities. In fact, it is hard to identify which of these activities do *not* involve, either directly or indirectly, vision. Therefore, whenever necessary to automate (e.g., in case of dangerous or tedious situations) or to improve (e.g., to increase precision and repetition) human activities, effective computer vision systems become essential.

The origin of computer vision is intimately intertwined with computer history, having been motivated by a wide spectrum of important applications such as in robotics, biology, medicine, industry, security and physics, to name but a few. Such a great deal of applications is not surprising if we consider the aforementioned importance of the human vision sense. Nevertheless, though "seeing" seems to us to be *simple*, *natural* and *straightforward*, in practice the design of versatile and robust computational vision systems has proven to be difficult, and most of the flexible computer vision systems created thus far have met with limited success. In fact, vision requires real-time processing of a very large and heterogeneous data set (including shape, spatial orientation, color, texture, motion, etc.) as well as interactions with other equally important cognitive abilities, such as memory, feelings and language. Additional difficulties with the analysis of visual information derive from noise, occlusion and distortions, as well as from the fact that image formation involves mapping from a three-dimensional space (the scene) onto a two-dimensional support (the retina or the image plane in a camera, for instance), thus implying information to be lost. Notwithstanding these difficulties, there is no doubt that robust vision systems are viable, for nature has created highly adapted and efficient vision systems in so many animals. In this context, vision science has developed as an interdisciplinary research field, frequently involving concepts and tools from computer science, image processing, mathematics, physics, artificial intelligence, machine learning, pattern recognition, computer graphics, biology, medicine, neuroscience, neurophysiology, psychology and cognitive sciences. Although not always recognized, such areas have already provided computer vision with important insights. For instance, several important imaging concepts and techniques can be closely related to biologic principles, including the edge detection approach described in [Marr, 1982], the two-dimensional (2D) Gabor filter models developed in [Daugman, 1980], the artificial neural networks introduced by McCullogh and Pitts [Anderson, 1995], and the importance of high curvature points in shape perception described in [Attneave, 1954], to name but a few.

To probe further: *Shape Theories in Biology and Psychology*

A related and interesting study topic are the theories of human shape perception. The reader is referred to [Biederman, 1985; Edelman, 1999; Hubel & Wiesel, 2005; Leyton, 1988, 1992; Perret & Oram, 1993; Poggio & Edelman, 1990; Rosin, 1993; Siddiqi & Kimia, 1995; Zeki, 2000] for further reading on this issue.

Among all different aspects underlying visual information, the *shape* of the objects certainly plays a special role, a fact that can be experienced while reading the characters on this page, which are essentially characterized by their shapes. In a sense, shapes can be thought as being the *words* of the visual language. Indeed, the prominent role of vision and shape (or its synonymous *form*) to humans has implied several visually-related terms to be incorporated into the common vocabulary, including the words trans*form*ation, in*sight* and *imag*ination, and expressions such as *lick into shape*, *take shape*, *shape up*, and *in any shape or form*. As far

Figure 1.1: *Image containing a 3D object (a cat) and respective representation in terms of its 2D silhouette.*

as the pictorial information is concerned, the particular issue of 2D shapes, i.e., shapes defined on the plane, is of paramount importance. As mentioned above, image formation often involves mapping objects from the three-dimensional (3D) space onto 2D structures, such as a retina or a CCD. It is worth noting that even the 2D-object silhouette often conveys enough information allowing the recognition of the original object, as illustrated in Figure 1.1.

This fact indicates that the 2D shape analysis methods described in this book can often be applied for the analysis of 3D objects. While there are many approaches for obtaining the full 3D representation of objects in computer vision, be it by reconstruction (from stereo, from motion, from shading, etc.) or by using special devices (e.g., 3D scanners), dealing with 3D models still is computationally expensive, frequently to a prohibitive degree, hence the importance of 2D approaches for treating such situations. Of course, there are several objects, such as characters, which are defined in terms of 2D shapes and should therefore be represented, characterized and processed as such.

In a more general situation, 2D shapes are often the archetypes of objects belonging to the same pattern class, which is illustrated in Figure 1.2.

In spite of the lack of additional important pictorial information, such as color, texture, depth and motion, the objects represented by each of the silhouettes in this image can be promptly recognized. Some of these 2D shapes are abstractions of complex 3D objects, which are represented by simple connected sets of black points on the plane (see Chapter 4 for additional discussion on the issue of shapes).

This book is precisely about obtaining, processing and analyzing shape images in automated, or at least semi-automated, fashion by using digital computers. In a

Figure 1.2: *Some typical and easily recognizable 2D shapes.*

typical application, the image of a shape is digitized, yielding a digital shape that can be pre-processed, analyzed and (eventually) classified. As mentioned above, these techniques have been successfully applied to a wide range of practical problems, some of which are exemplified in the following table. In order to gain a deeper insight about computational shape analysis, two representative applications illustrating typical situations in practical shape analysis are outlined and discussed in the next section. An introductory overview of the several tasks involved in shape analysis is presented in the remainder of this chapter.

Research Field	Examples of Applications
Neuroscience	Morphological taxonomy of neural cells, investigations about the interplay between form and function, comparisons between cells of different cortical areas and between cells of different species, modeling of biologically realistic cells, and simulation of neural structures.
Document analysis	WWW, OCR (optical character recognition), multimedia databases, and historical documents.
Visual arts	Video restoration, special effects, video tracking, games, computer graphics, visualizations, and image synthesis.
Internet	Content-based information retrieval, watermarking, graphic design, and usability.

Continued on next page.

Continuation.

Research Field	Examples of Applications
Medicine	Tumor recognition, quantification of change and/or deformation of anatomical structures (e.g., endocardial contour of left ventricle of heart, corpus callosum), morphometric analysis for diagnosis (e.g., multiple sclerosis and Alzheimer's disease), numerical analysis of chromosomes, identification of genetic pathologies, laparoscopy, and genetic studies of dentofacial morphology.
Biology	Morphometric-based evolution comparison, taxonomy, interplay between form and function, comparative anatomy, cytology, identification and counting of cells (e.g., white blood cells), characterization of cells and nuclear shapes, growth and shape modifications, analysis of human gait, analysis of electrophoretic gels, and microscopy.
Physics	Analysis of particle trajectories, crystal growth, polymers, characterization of star clusters in astronomy, and several types of microscopy.
Engineering	Semiconductors, quality control, danger detection, machine interpretation of line drawings, computer-aided design of mechanical parts and buildings, automation, robotics, remote sensing, image and video format standards, and spatial exploration.
Security	Fingerprint/face/iris detection, biometrics, human gait, and signature verification.
Agriculture	Harvest control, seed counting and quality control, species identification, and fruit maturation analysis.

To probe further: *Shape Analysis*

The multidisciplinarity of image analysis, with respect to both techniques and applications, has motivated a rich and impressive set of information resources represented by conferences, books, WWW URLs and journals. Some of the more important of these are listed in the book Web page at: `http://www.ime.usp.br/~cesar/shape_crc/chap1.html`.

1.2 Case Studies

1.2.1 Case Study: Morphology of Plant Leaves

A special problem where shape analysis usually comes into play is the classification of biological entities based on respective morphological information, as illustrated in the following (see [Bruno et al., 2008a]). Figure 1.3 shows a series of 12 images of leaves obtained from four different species of plants.

Figure 1.3: *Set of plant leaves belonging to four classes.*

Observe that in this situation the classes of leaves are clearly defined in terms of the respective plant species. Now, suppose that we want to classify an unknown leaf, i.e., to assign it to one of the four plant *classes* presented in Figure 1.3. A typical *pattern recognition* approach to solve this problem is to measure a series of *features*, or *attributes*, from each leaf image in Figure 1.3, say a feature related to the brightness distribution of each leaf and a feature related to its size or extension.

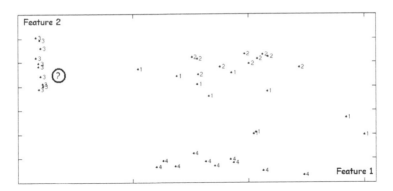

Figure 1.4: *Feature space obtained for plant leaves of the types shown in Figure 1.3.*

An example of the former feature type is the histogram entropy, denoted by f_1, which can be informally understood as a number indicating the degree of disorder of the gray levels inside each leaf (see Chapter 3 for additional discussion). The other feature could be characterized by the perimeter of each leaf, denoted by f_2. Therefore, each leaf is represented in terms of a pair (f_1, f_2), known as the *feature vector* associated with each pattern. The *feature space* of our pattern recognition problem, which is shown in Figure 1.4, is the 2D space defined by $f_1 \times f_2$ for all initially considered leaves.

In Figure 1.4, each point is labeled with its class number, i.e., 1, 2, 3 or 4, to the leaf classes of Figure 1.3. It is interesting to note that each pattern class has defined a respective *cluster* (informally speaking, a localized and separated cloud of points) in the feature space. Back to the initial problem of classifying an unknown leaf based on its image, it would be useful to have a *pattern classifier* that could produce the correct class for each supplied feature vector (f_1, f_2) corresponding to a new leaf not in the original database. For instance, in case the measured features (f_1, f_2) is that indicated by a "?" in Figure 1.4, then it would be reasonable to assume that the unknown leaf belongs to class 3, for the feature vector is much closer to that cluster than to all other remaining clusters in the feature space. This simple approach to automated classification is called, for obvious reasons, *nearest neighbor*.

To probe further: *Applications in Agriculture*

The page www.ee.surrey.ac.uk/Research/VSSP/demos/leaf/index.html presents an interesting application of the leaf classification problem to agriculture.

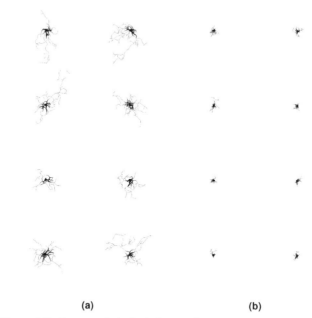

(a)   (b)

Figure 1.5: *Two morphological classes of prototypical cat ganglion cells:*
α-cells (a) and β-cells (b). The cells have been artificially
generated by using formal grammars [Costa et al., 1999].

1.2.2 Example: Morphometric Classification of Ganglion Cells

The second example of a practical shape analysis application concerns the morphological analysis of neural cells. The morphology of neurons has been recognized to be an extremely important property of such cells since the pioneering work of the neuroscientist [Ramon y Cajal, 1989]. The morphology has been often related to specific physiological properties of the cell (e.g., [Costa, 2005; Costa & Velte, 1999]). A specific example of this situation is defined by the two morphological classes of retinal ganglion cells in the cat, i.e., the α- and β-cells. Several studies by neuroscientists over decades have produced results clearly indicating an interesting relationship between the form and function of these cells. The morphological classes for α- and β-cells have been defined by taking into account especially the neural dendritic branching pattern, with α-cells presenting dendrites spread throughout a larger area, while the β-cells are more concentrated. Figures 1.5 (a) and (b) present two sets of α- and β-cell prototypes, respectively, clearly illustrating their shape differences.

Many are the motivations for developing objective morphological classification schemes for such neural cells. First, such objective parameters would help the creation of taxonomic classification with respect to neural morphology, as well as to review previously proposed classes typically developed in a subjective fashion (i.e., through human inspection). Furthermore, such studies can lead to advances regarding the characterization of the relationship between neural form and function, which is a particularly important problem in neuroscience. In addition, objective parameters about neural morphology are essential for paving the way towards more realistic models and computer simulations of neural structures [Ahnert & Costa, 2008; Costa et al., 1999]. Research has been carried out in order to develop quantitative measures about the geometry of neural cells (a research area called *neuromorphology*) that could properly reflect the differences between the different types of neural cells.

In this context, neuroscientists have started applying mathematical tools, such as the fractal dimension and the bending energy, in order to devise automatic tools allowing the effective classification of neurons with respect to their morphology. Nevertheless, the development of such features has met interesting and difficult problems that must be resolved by the researchers. For instance, there is no agreement among neuroscientists with respect to the number of morphological classes of neurons. In fact, the morphological classes can be defined and redefined by neuroscientists as new methods are developed. Therefore, an interesting shape analysis problem arises involving the following difficult questions:

① How many morphological classes of neurons are there?

② How can we assign cells to morphological classes?

③ What features (not only morphological, but also characterizing the respective neural activity) should we adopt in order to characterize the neural cells?

④ How reliable are the shape features and the classification?

⑤ What classification methods should we use in order to classify the neural cells with respect to the adopted features?

It is worth emphasizing that these questions are by no means restricted to the problem of morphological characterization of neural cells. In fact, many different practical situations in a wide diversity of fields face similar doubts. They are representative of both *supervised* and *unsupervised* classification schemes, as will soon become clear, and pattern recognition theory provides a set of mathematical tools that help scientists in answering (at least partially) the above questions.

1.3 Computational Shape Analysis

There are many problems usually addressed in the context of shape analysis and recognition by using computers, upon which this book is organized. In fact, computational shape analysis involves several important tasks, from image acquisition to shape classification. Figure 1.6 illustrates the shape processing tasks frequently

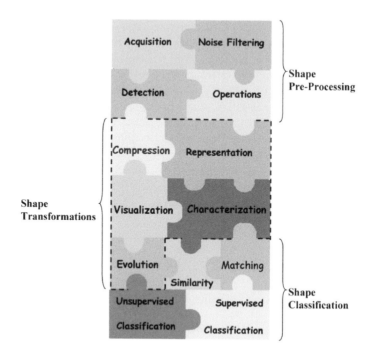

Figure 1.6: *Typical shape analysis tasks and their organization into three main classes.*

required for shape analysis, which can be broadly divided into three classes, namely *shape preprocessing*, *shape transformations* and *shape classification*.

The following sections address each of these classes of shape analysis operation:

1.3.1 Shape Pre-Processing

The first step toward the computational morphological analysis of a given object involves acquiring and storing an image of it and separating the object of interest from other non-important image structures. Furthermore, digital images are usually corrupted by noise and other undesirable effects (such as occlusion and distortions), therefore requiring the application of special procedures. The following subsections present a brief introduction to each of these problems, which are grouped together into the *shape pre-processing* category. The issues of shape acquisition and pre-processing are addressed in more detail in Chapter 3.

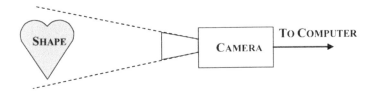

Figure 1.7: *Illustrative scheme of shape acquisition to be processed by a computer.*

Shape Acquisition

Shape acquisition involves acquiring an image (e.g., photograph) and digitizing it, so that it can be properly processed by a computer (Figure 1.7).

The type of image acquisition framework and devices to be used depend heavily on the application, e.g., a camera attached to a microscope can be used in histological applications, while a scanner can be used to acquire images of leaves, such as in the example in Section 1.2.1.

Shape Detection

One of the first steps generally required in shape analysis is to detect the shape, which involves locating the object of interest in the image so that it can be subsequently treated. For instance, Figure 1.8 shows an image including several shapes: in order to analyze them, it is necessary to locate each shape, which can have different visual properties (e.g., color and texture).

The most basic approach to shape detection is through image segmentation (e.g., by thresholding). When the image can be properly segmented so that the object of interest can be successfully isolated from other non-important image structures (including the background), then shape detection is a reasonably straightforward task. An interactive approach can also be adopted in a number of important practical situations. For instance, the object of interest can be detected by requesting a human operator to click inside the object of interest, which would be followed by a region-growing algorithm (see Chapter 3) in such a way that the resulting grown region corresponds to the detected shape. This procedure is usually implemented in most off-the-shelf image processing software, as it might have already been tried by the reader, which is usually represented as a magic wand that allows the selection of irregular image regions. However, there are many alternative approaches for object detection in digital images. For example, if the object of interest can be generally represented by a template, template-matching techniques could be applied in order to locate the object instances in the image. On the other hand, if the problem to be solved involves shape analysis in video sequences, then motion-based techniques can be used for detecting and locating the object in the image. Nevertheless, it is worth emphasizing that image segmentation can frequently become a difficult problem, mainly if the image acquisition conditions (such as illumination, camera position, focus, etc.) cannot be properly controlled.

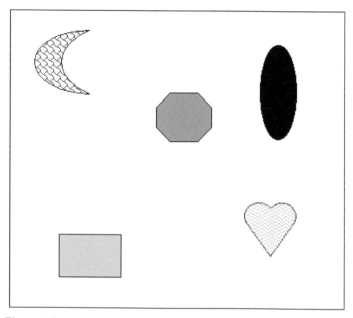

Figure 1.8: *Shape detection involves locating the objects of interest in the image. The shapes can present different visual properties, such as color and texture.*

Noise Filtering

Digital image processing systems generally have to cope with noisy images in nearly all practical situations, and shape analysis is by no means an exception: noisy shapes occur ordinarily, independently of the application (see Figure 1.9).

It is worth noting that, besides the perturbations inherent to digital images (consequences of the spatial and intensity quantizations involved in acquiring digital images), noise in shapes can also arise from the imaging operations that are typically applied in order to detect the shape of interest. Frequently the shape detection process is preceded by a series of image processing procedures, such as diverse filtering operations, data fusion and segmentation, which can introduce perturbations at nearly every processing stage. Furthermore, shape quantization and sampling, which are necessary for obtaining digital shapes, are usually a source of critical noise. All these noisy alterations are generally reflected as small modifications on the obtained shape, which can affect subsequent shape analysis procedures. Consequently, approaches to shape representation and description often attempt to be robust to noise or to incorporate some noise filtering mechanism. For instance, multiscale techniques such as curvature estimation (Chapter 7) adopt filtering as an inherent means to reduce or to eliminate quantization and other types of noise.

Figure 1.9: *If noise is present in the image, proper noise filtering procedures should be applied before and/or during shape analysis.*

Noise filtering is discussed in Chapter 3 and related techniques, such as multiscale
curvature estimation, are discussed in Chapters 6 and 7.

Shape Operations

There are many important operations that can be applied to shapes. For instance, if
the problem to be solved involves comparing two or more shapes, then they should
be normalized so that the comparison makes sense (Figure 1.10).

Normalization processes usually involve parameters such as scale, rotation and
translation. Shape warping, registration and morphing are also examples of shape
operations that can be applied to normalization and comparison. Typically, such
operations are based on defining a mapping between a set of points (*landmarks*)
along two or more shapes, which allows the generation, by interpolating, of a series of intermediate shapes that would possibly be obtained while *transforming*
one of the shapes into the other. Shape manipulation/handling can also include
interactive edition (e.g., elimination of portions of the shape) and operations aid-
ing visualization, as well as operations involving more than one shape (e.g., shape
addition and intersection). The most important shape operations are discussed in
Chapter 4.

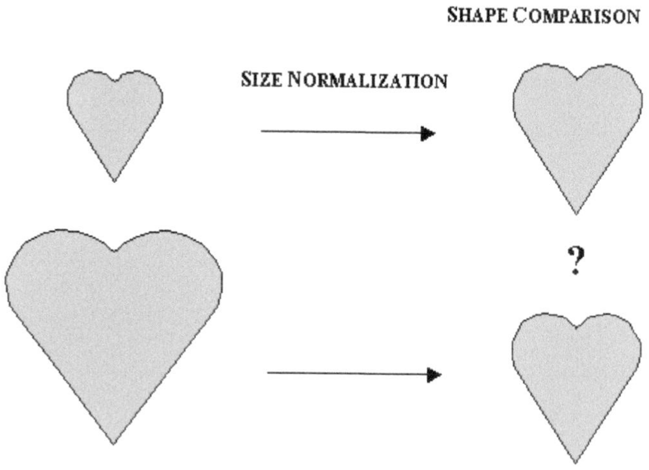

SHAPE COMPARISON

SIZE NORMALIZATION

Figure 1.10: *Shape operations can be involved while normalizing some visual properties, e.g., before comparing shapes. It is important to note that, in some applications, the differences in size are actually an important shape parameter, which would make size normalization inadequate in such situations. Similar comments apply to other visual properties, such as orientation and translation normalization.*

1.3.2 Shape Transformations

Once the shape of interest has been acquired and processed (for instance, noise has been substantially reduced), a set of techniques can be applied in order to extract information from the shape, so that it can be analyzed. Such information is normally extracted by applying suitable *shape transformations*. Such transformations are mappings that allow both representation of the shape in a more appropriate manner (with respect to a specific task) and extraction of measures that are used by classification schemes. The concept of shape transformation is covered in Chapter 4, while Chapters 5, 6, and 7 present computational techniques for feature extraction.

Shape Evolution

It is often important to study the properties of a sequence of shapes corresponding to an object that has evolved along a certain time period (an example is shown in Figure 1.11).

For instance, it is important to establish the correspondences between different points of the ventricular contour as the heart beats or to analyze development of neurons and other cells as they grow. All these problems can be treated in terms of shape transformations as shape evolution.

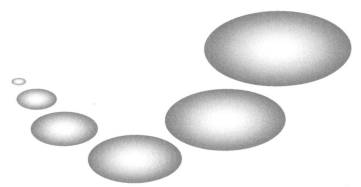

Figure 1.11: *Shape evolution involves analyzing an object that has modified its shape along a certain time period. This is often carried out by observing a series of shapes that the object has
assumed during the period.*

Shape Representation

Once the object of interest has been located in the image (through shape detection
and segmentation, see Section 1.3.1), its shape is understood as being formed by
the set of points found to belong to that object. In this sense, the first representation
of the object shape is the set of points identified in the original image. It is often
the case that such representation, though naturally produced by the shape detection
procedure, is not particularly useful, being even considered cumbersome for some
purposes because of the high number of data that is required (i.e., all the points of
the segmented shape have to be somehow stored). Therefore, the next problem to
be tackled is how to properly *represent* the shape, implying a suitable *shape representation* scheme to be defined with basis on specific tasks. Such schemes may
or may not allow the reconstruction of the original shape. In fact, this criterion
seems to have been first suggested by [Pavlidis, 1977] with respect to *information
preserving* (allow the reconstruction of the original shape) and *information nonpreserving techniques* (do not allow the reconstruction of the original shape). It
is worth emphasizing that information preserving representations are particularly
important due to the fact that different shapes are mapped onto different representations, whereas nonpreserving techniques can produce equal representations for
different shapes (which is called a degenerated or non-inverting mapping, as discussed with respect to functions in Chapter 2). Such nonpreserving techniques are
nevertheless usually adopted as shape measures that are useful for shape characterization and classification, as discussed in Chapter 4. Indeed, both approaches, which
have their advantages and shortcomings, are frequently applied to shape analysis
problems. In addition, it should be observed that some techniques only allow partial
reconstruction of the shape.

A more fundamental criterion for characterizing shape representation techniques
involves their classification as *boundary-based* and *region-based*. Boundary-based

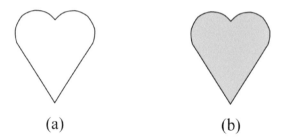

Figure 1.12: *Boundary-based* (a) *and region-based* (b) *shape representa-*
tions.

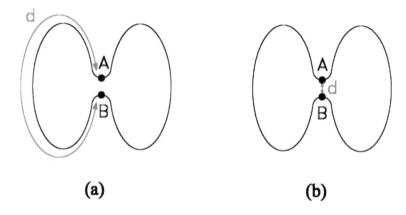

Figure 1.13: *Important difference implied by the boundary-based* (a) *and*
the region-based (b) *shape representations.*

(also known as *contour-based*) techniques represent the shape by its outline, while
region-based techniques treat the shape in terms of its respective 2D region (see
Figure 1.12).

By representing planar regions in terms of one-dimensional signals, contour-
based methods allow a simplified one-dimensional approach, while region-based
techniques involve 2D signals. This difference frequently implies that contour-
based methods are less computationally expensive than region-based methods,
though exceptions occur from time to time. Another important difference between
these two approaches can be better appreciated through the example in Figure 1.13.

In the contour-based approach, assuming that the contour is traversed counterclockwise, the distance between the points A and B *along the contour*, indicated
as d in Figure 1.13 (a), is larger than the distance in the region-based approach,
indicated as d in Figure 1.13 (b). Roughly speaking, if a contour-based local anal-

ysis is applied to point A, then point B does not imply a strong influence over the processing performed at A. The opposite situation is verified for the region-based approach, i.e., the point B can significantly affect processing done around A.

Shape Description or Characterization

Some of the most important problems involving shape analysis techniques require extracting information about objects in the real world. For example, one might want to investigate physical properties of biological entities by analyzing their shape, such as when studying spatial coverage, a concept that is also related to shape complexity. Such spatial coverage properties, particularly important in branching structures, can be related to the capacity of an object to interact with its surrounding environment, such as the capacity of roots of trees to extract water and food from the soil or of neural cells to interact with the extracellular medium, including other cells. In situations where relevant shape information is to be extracted, *shape description* or *characterization* techniques have to be applied (description and characterization are used as synonyms in this sense). Moreover, additional and equally important situations where shape description techniques are fundamental arise in shape recognition and shape classification. It should be observed that frequently some shape aspects are more important than others, depending on the task to be solved by the shape analysis system. For instance, many object recognition problems can be solved by first detecting some dominant points that usually occur in the shape (e.g., corners in polygonal figures). Clearly, the type of feature that should be detected depends on each specific problem as well as on the involved shapes, though some features have achieved special importance and popularity in shape analysis. For example, some of the most important aspects of a shape can be detected by analyzing the curvature of the shape boundary, especially in terms of corners and regions of constant curvature, such as circle segments (constant, non-null curvature) or straight lines (constant, null curvature). Furthermore, some shape features are also studied and usually considered as a consequence of biological facts (e.g., psychophysical results indicate that corners play a particularly important role in human shape analysis, see [Attneave, 1954]). To any extent, there are different approaches for extracting information about shapes, which can be classified as follows:

Shape measurements: One of the most common ways of describing shapes involves defining and measuring specific characteristics such as area, perimeter, number of corners, number of holes, curvature-based measures, preferential orientation of contour points, and so on (see Figure 1.14). The underlying idea of the description of a shape by a set of measures (i.e., numbers) is that the obtained measures are sufficient to reasonably represent the relevant information about that shape.

Shape transforms (signal processing-based): Transform techniques are popular in many different areas, from signal processing and telecommunications to

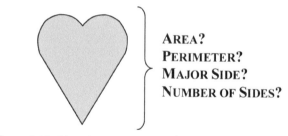

AREA?
PERIMETER?
MAJOR SIDE?
NUMBER OF SIDES?

Figure 1.14: *Many important approaches to shape analysis are based on measuring meaningful properties from the shape.*

ORIGINAL SHAPE   FOURIER REPRESENTATION

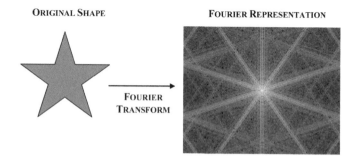

FOURIER
TRANSFORM

Figure 1.15: *Shape description by using its Fourier transform. Observe that the complete Fourier transform also includes phase information, in addition to the magnitude shown in this figure.*

optics and numerical solution of partial differential equations, also playing an important role in shape analysis. A signal transform is a mathematical tool that expresses the original signal in an alternative way, which is frequently more suitable for a specific task than the original one. For instance, the number two can be alternatively expressed in decimal Arabic ("2") or in Roman ("II"). While features can be obtained by measuring shape properties directly in their 2D or 3D space, a powerful and widely adopted alternative to such an approach consists in deriving features from transformed shapes. A simple example of obtaining a shape descriptor by using a transform technique is to calculate its Fourier transform (Figure 1.15) and to select some predefined coefficients (e.g., "select the first 5 Fourier coefficients"). There are many different transforms that can be used, though Fourier is one of the most popular. As a matter of fact, the Fourier transform is one of the most powerful and versatile linear transforms. Some other examples of important transforms are the wavelet, the Gabor and the Karhunen-Loève transform. It should be observed that invertible transforms could also be understood

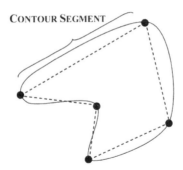

CONTOUR SEGMENT

Figure 1.16: *Original shape contour (solid line) and a possible representation by polygonal approximation (dashed line).*

as means for representation of shape. For instance, the original shape can be fully recovered from its Fourier transform representation by using the inverse Fourier transform (see Section 2.7).

Shape decomposition: The third class of shape description techniques presented herein is based on decomposing the shape into simpler parts, which are sometimes called *primitives*, as typically done in the context of structural and syntactical pattern recognition [Fu, 1982]. Since the meaning of "simpler parts" can vary widely in terms of each specific application and types of shapes, knowledge about the problem is usually decisive in this case. Nevertheless, there are some approaches that are versatile enough to be generally considered, being suitable for the most diverse applications. For instance, one of the most important problems in contour analysis involves fitting geometric primitives to contour portions, and the so-called *polygonal approximation* is an excellent example of this approach (Figure 1.16). In the polygonal approximation problem, also known as *piecewise linear approximation*, the original contour must be represented by a set of straight line segments, each line segment representing a portion of the original contour. It is important to note that such a representation can also be used to implement shape processing such as noise filtering (local noisy perturbations occurring in contour portions are eliminated when these portions are represented by line segments) and data compression (e.g., a digital straight contour segment involving hundreds of points can be almost exactly represented in terms of its two extremities). Other examples of shape decompositions are those based on circle segments and 2D polygonal regions, the latter being applied in region-based techniques. Proceeding a step further, the syntactic approach to pattern recognition problems associates abstract symbols to each geometric primitive in such a way that each shape can be represented by a sequence of such symbols. The subsequent shape recognition therefore involves parsing procedures operating over such symbol sequences (or strings; see Chapter 9).

Shape description through data structures: Several problems can be solved by representing aspects underlying the shape in terms of data structures. An illustrative example is the problem of representation of neural dendrites by the so-called dendrograms (typically binary trees), which suit several important applications in neurosciences [Cesar-Jr. & Costa, 1999; Costa et al., 2000] (see Figure 1.17). In addition to presenting a clear representation of

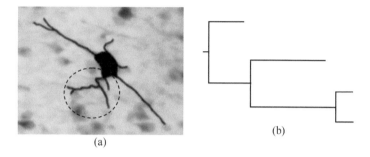

(a)   (b)

Figure 1.17: *A neural cell is presented in* (a) *while a dendrogram of one of its dendrites (that with terminations and branch points indicated) is shown in* (b).

the branching pattern, such hierarchical data structures can incorporate important additional shape measures such as size, width, local bending energy and angles in a particularly compact way. As a matter of fact, dendrograms have become important in neuroscience because they can be easily stored and handled by computer programs, thus allowing the standardization required for exchanging data among different laboratories, scientists and other professionals. It should be borne in mind that dendrograms are not only important in neuroscience, but also provide valuable descriptions of virtually any other branched structure, such as rivers, trees, vascular systems, etc.

Shape Visualization

Scientific visualization techniques are mainly concerned with the suitable presentation of large amounts of data to humans. As such, this area is particularly important both for supporting the development of shape analysis tools and as an aid for shape inspection by human operators. In the former situation, shape visualization can be used to effectively present the obtained results (e.g., features to be tested, intermediate results, filtered images), which can involve the superposition of such results over the original shapes or relating the several obtained data, in order to provide insights about the assets and shortcomings of the considered techniques. On the other hand, shape visualization is also important to aid human experts to solve specific problems, e.g., to help a physician decide how a broken bone should be treated.

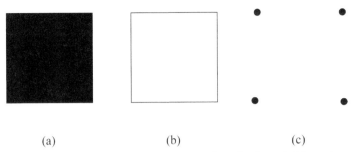

(a)   (b)   (c)

Figure 1.18: *An example of shape compression: the shape in* (a) *can be represented by its outline* (b), *which can then be represented by its corners* (c). *Therefore, the shape in* (a), *which could require hundreds or thousands of points in order to be properly represented in a discrete space, can now be conveniently represented by only the 4 points shown in* (c).

Shape Compression

Digital image applications generally involve processing a large amount of data, which can become prohibitive depending on the application, especially when real-time processing is required. Data compression is an issue often present in imaging applications, including shape analysis problems. For instance, applications that depend on large image databases (e.g., fingerprint recognition) usually require storing and computing very large sets of images. Some shape analysis approaches naturally offer good data compression solutions, e.g., contour-based approaches, which represent 2D shapes by 1D structures (Figure 1.18). Very high compression rates can be obtained by further compressing such contours. In fact, there are some approaches for image and video coding (for data compression) which make extensive use of contour shape representations (e.g., [Buhan et al., 1997]).

1.3.3 Shape Classification

Finally, after shape processing, representation and characterization (often involving feature extraction), classification algorithms are usually applied in order to assign each considered shape to a category. There are two particularly important aspects related to shape classification. The first is the problem of, given an input shape, deciding whether it belongs to some specific predefined class. This can also be thought of as a shape recognition problem, usually known as *supervised classification*. The second equally important aspect of shape classification is how to define or identify the involved classes in a population of previously unclassified shapes. This represents a difficult task, and expert knowledge acquisition problems are usually involved. The latter situation is known as *unsupervised classification* or *clustering*. Both supervised and unsupervised classification involve comparing shapes, i.e., deciding how *similar* two shapes are, which is done, in many situations, by *matching* specially important corresponding points of them (typically landmarks or

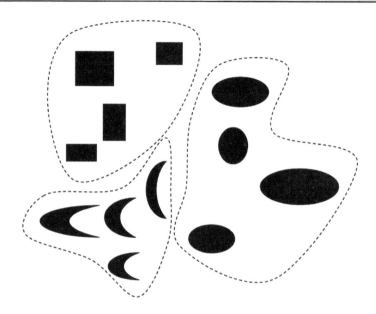

Figure 1.19: *In an unsupervised shape classification problem, the algorithm should discover shape classes from a given set of unclassified shapes.*

saliences). These four topics are outlined in the following sections. General shape
classification algorithms are covered in Chapter 8.

Unsupervised Shape Classification

As further explained in Chapter 8, classifying a shape can be understood as the
problem of assigning some class to it. Nevertheless, in many cases defining the
shape classes is itself a difficult problem. Therefore, it is important to devise methods that attempt to find shape classes based only on the unclassified pattern data,
an approach that is commonly known as *unsupervised learning*. The identification
of data clusters in the data sets is an ordinary way of defining shape classes, which
is carried out by *clustering algorithms*. For instance, for a given set of geometrical shapes, such as those shown in Figure 1.19, the expected output of a clustering
algorithm would be the three sets indicated by the dashed lines in that figure.

Supervised Shape Classification

When the shape classes are predefined, or examples are available for each class, it is
often desirable to create algorithms that take a shape as input and assign it to one of
the classes, i.e., that it *recognizes* the input shape (see Figure 1.20). For instance, an
important problem in medical imaging involves the recognition of mammographic
calcifications in order to verify the presence of tumors, the shapes of which are

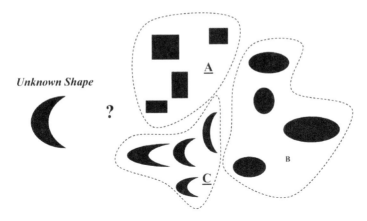

Figure 1.20: *Supervised shape classification: given a set of shape classes A, B and C, and an unknown shape, to which class does the unknown shape belong?*

related to the tumors being malignant or not. Observe that the terms *shape recognition* and *supervised shape classification* are often used interchangeably.

Shape Similarity

Shape similarity refers to establishing criteria that allow objective measures of how much two shapes are similar (or different) to each other, including issues such as when a given shape *A* can be considered more similar to another shape *B* than to *C*. An example is shown in Figure 1.21. It is worth observing that shape similarity criteria, which are fundamental to classifying shapes, are generally dependent on each specific problem. For instance, in a situation where size is an important parameter, two shapes with similar areas can be more similar to each other than two shapes with significantly different areas. Clearly, the shape features adopted for their characterization play a central role with respect to defining how similar two shapes are. Shape similarity is particularly important when trying to match two or more shapes.

Shape Matching

Shape matching is the process through which two or more shapes are associated, generally in a point-by-point fashion. There are many different applications for such techniques. For instance, images of the same region of the human body can be obtained using different acquisition modalities, such as tomography and magnetic resonance, and an important task in such situations is to *register* or *align* each image, in order to create a correspondence map between the several representations (see Figure 1.22). This task is a particular example of a more general problem known as *image registration*. One approach that can solve this problem

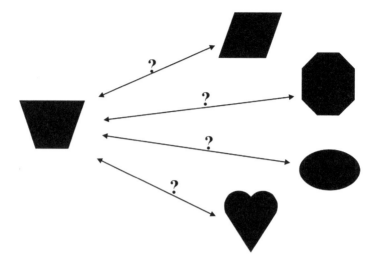

Figure 1.21: *Shape similarity: which shape is more similar? How can similarity be objectively measured?*

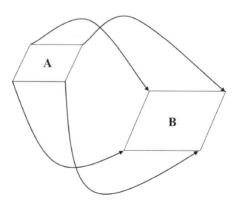

Figure 1.22: *Shape matching can involve finding the correct corresponding points between a given shape A and a target shape B.*

involves the detection of instances of homologous structures in both images. In addition, shape matching is important to different problems in data fusion and 3D reconstruction. In the former application, information about the same object is obtained by using different sensors, and the respective representations have to be merged. On the other hand, the latter involves establishing a correspondence between two-dimensional shapes obtained as slices of a three-dimensional structure: this correspondence allows reconstructing an approximation of the original 3D object. The reader interested in sensor fusion, image registration and shape matching is referred to [Bloch & Maître, 1997; Burr, 1981; Davis, 1979; Milios, 1989; Viola & Wells, 1997].

1.4 Additional Material

The book now includes a special box section called *Additional material*, which includes links to useful online material. In particular, the authors keep some open-source software projects online which are directly related to the theory and methods introduced in the book. The reader is invited to visit the projects' homepages, use the software and help to develop it in a open-source collaboration. The main projects, which are also referred in appropriated places in the book, are listed as an example in the box below.

Additional resources: *Slides, videos, software*

- Interactive image segmentation: `http://segmentacao.incubadora.fapesp.br/portal`

- Vessel segmentation: `http://retina.incubadora.fapesp.br/portal`

- Dimensionality reduction: `http://dimreduction.incubadora.fapesp.br/portal`

- Shape analysis: `http://code.google.com/p/imagerecognitionsystem/`

   Also, slides to help in courses based on the book are now available at the book homepage: `http://www.ime.usp.br/~cesar/shape/`

1.5 Organization of the Book

Chapter 2 presents an introduction to the mathematical background usually involved in image processing and shape analysis. Chapter 3 covers some of the most important techniques for image processing, and Chapter 4 presents a conceptual overview of the several concepts and issues involved in shape analysis. The problems related to shape representation are covered in Chapter 5, and Chapters 6 and 7 discuss several shape characterization features, with the latter concentrating on the multiscale paradigm. Chapter 8 covers the important issue of shape classification, concentrating on statistical approaches, while Chapter 8 concentrates on structural shape recognition issues. Appendix A concludes the book by presenting perspectives about the future development of shape analysis and classification.

Mathematics seems to endow one with something like a new sense.

Charles Darwin

2

Basic Mathematical Concepts

Chapter Overview

THIS CHAPTER PRESENTS the basic mathematics underlying shape analysis and classification. It starts by presenting some elementary concepts, including propositional logic, functions and complex numbers, and follows by covering important topics in linear algebra, such as vector spaces, linear transformations and metric spaces. Since several properties of shapes exhibit a differential nature, a review of the main related concepts from differential geometry and multivariate calculus is subsequently presented. Next, the key operations known as convolution and correlation are introduced and exemplified, which is followed by a review of probability and statistics, including probability distributions, autocorrelation and the Karhunen-Loève transform. The chapter concludes by presenting the main issues in Fourier analysis, from the Fourier series to discrete convolution performed in the frequency domain.

2.1   Basic Concepts

This chapter presents some basic mathematical concepts so as to provide the key to the full understanding and application of image and shape analysis, not only in the context of the present book but also of the most modern approaches covered elsewhere in the related literature. Every effort has been made to develop this chapter in an introductory fashion that should be accessible even to those with only elementary mathematical background. For those who are already familiar with the covered subjects, this chapter might still be read as a review. Although the topics are presented in a logical, progressive and integrated sequence throughout the several sections in this chapter, it is also possible to proceed directly to the specific topics treated in the respective sections. Readers can proceed directly to the chapters and sections they are most interested in, referring to this chapter only as needed. In order to help this approach, pointers are provided throughout the book indicating where related information can be found in the present chapter.

While every effort has been spent in providing a comprehensive and sound treatment for each of the considered mathematical concepts, it is observed that this should not be understood as a strictly mathematical textbook, in the sense that most proofs and formalizations have been deliberately omitted.

2.1.1 Propositional Logic

Before proceeding further, it is important to briefly revise some basic concepts in *propositional logic* (see, for instance, [James, 1996; Stolyar, 1984]) since they underly the proper and full understanding of all exact sciences. For instance, the simple but essential concepts of *necessary* and *sufficient* conditions are sometimes not properly understood. Perhaps the best way to understand such concepts is to pay attention to their own names. Thus, a *sufficient condition* is something that is *sufficient* or *enough* for the verification of some fact. This is symbolically represented as $A \Rightarrow B$, which reads A *implies* B, where the facts A and B are more formally known as *propositions*, i.e., statements that can only assume the logical values *true* (T) or *false* (F). Observe that $A \Rightarrow B$ is itself a proposition. In case we have a property or a theorem that says that it is enough to have A in order that B is verified, in such a way that A is clearly a sufficient condition, we can only be sure that A leads to B. However, it is important to observe that this does not mean that there is no other situation C also leading to B, i.e., $C \Rightarrow B$. For instance, it is sufficient that a number belongs to the set $\{4, 12, 20\}$ (fact A) in order to be divisible by 4 (fact B), but many other (actually an infinite amount) numbers are also divisible by 4.

Conversely, a *necessary condition* indicates a requirement that is necessary for some fact to be verified. This is represented as $A \Leftarrow B$, i.e., A *is implied by* B. It is interesting to note that this condition can be understood as the sufficient condition "read backward," i.e., $A \Leftarrow B$ is exactly the same as $B \Rightarrow A$. It is important to note that a condition A necessary to B does not grant the verification of B. For instance, the fact that a number is even (fact A) is a necessary condition in order to be divisible by four (fact B), but this does not imply that 2, which is even, is divisible by 4. Indeed, we can have a situation satisfying hundreds of necessary conditions required by B, but this will not necessarily imply B. On the other hand, it is enough to have a single necessary condition not satisfied in order to preclude B from being verified.

Another important fact to note is that the same statement A can be both *necessary and sufficient* for B, which is represented as $A \Leftrightarrow B$, meaning that B is verified *if and only if* A is true. In this case, A and B are said to be equivalent. For example, the fact that a number is a multiple of 4 is a necessary and sufficient condition for divisibility by 4. Observe that necessary and sufficient conditions provide a complete characterization of the stated property.

The three above *logical operations* are summarized in the following logical table:

A	B	$A \Rightarrow B$	$A \Leftarrow B$	$A \Leftrightarrow B$
T	**T**	T	T	T
T	**F**	F	T	F
F	**T**	T	F	F
F	**F**	T	T	T

It is clear from this table that $A \Leftrightarrow B$ is the same as $(A \Rightarrow B) \wedge (A \Leftarrow B)$, where $\wedge$ ("AND") is the widely known logical connective (see [James, 1996]). Perhaps the only amazing feature in this table is the fact that $F \Rightarrow T$. However, this is no more than an abbreviated form of saying that one can reach a true conclusion even when starting with some wrong hypothesis. For instance, in the past, people believed earth was stable because it was held on the back of an enormous tortoise. Although it is true that earth is stable, the reason for this has nothing to do with tortoises, but with classical physics.

2.1.2 Functions

A *function* f is a mapping that associates to each element x of a set U, the so-called *domain* of f, a single element y of a set W, which is denominated the *co-domain* of f. This is abbreviated as $f : U \to W$, $y = f(x)$, and y is said to be the *image* or *value* of the function at x, which is called the *argument* of f. It is frequently useful to represent functions graphically, as illustrated in Figure 2.1. The set $V \subseteq W$, called the *range* of f, corresponds to the values produced by function f when applied to all the elements in its domain U. This means that not all elements of W are necessarily covered by the function f, as illustrated in Figure 2.1 (a). When $V = W$, as illustrated in Figures 2.1 (b) and (c), the function is said to be *surjective* (U *onto* W); when every element of V is the image of a single element of U, as illustrated in Figure 2.1 (c), the function is said to be *injective* (or *one-to-one*). A function that is both surjective and injective is called *bijective* (or *one-to-one* and *onto*). An important property exhibited by bijective functions is the fact that they can be inverted: as they are surjective, there will be no point left out in W when inverting the function; and, as there is only one point of U being mapped into each point of W, the inverse will necessarily be a function. For instance, the function f in Figure 2.1 (c) can be inverted to produce the function f^{-1} illustrated in Figure 2.1 (d). The same is not true for the situations in Figures 2.1 (a) and (b), for different reasons. It should be observed that an injective but not surjective function is also invertible, but only on the restricted domain corresponding to the range of the original function. Invertible functions or maps are particularly important for representation of shapes, as they keep all the original geometric information.

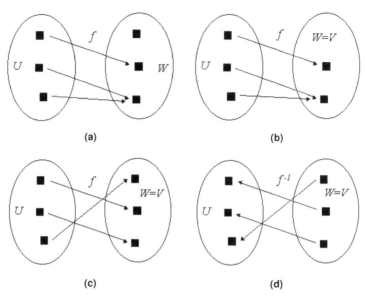

(a)   (b)

(c)   (d)

Figure 2.1: *Graphical representations of some types of functions (see text for explanation).*

A function where both U and W are real subsets (i.e., $U, W \subseteq \mathbb{R}$) is called a *real function*. Examples of real functions include:

$$f : \mathbb{R} \to \mathbb{R} \mid y = f(x) = x^2$$

and

$$g : \mathbb{R} \to \mathbb{R} \mid y = g(t) = \begin{cases} t & \text{if } 0 \leqslant t \leqslant w, \, w \in \mathbb{R}, \, w > 0, \\ 0 & \text{otherwise.} \end{cases}$$

The visual representation of real functions is always useful as a means to better understand their characteristics. Figure 2.2 depicts the graphical representations of the aforementioned functions.

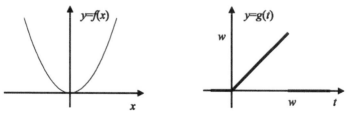

Figure 2.2: *Graphical representations of the functions $f(x)$ and $g(t)$.*

Some special types of functions include:

Differentiable (analytical, or smooth) functions: These are functions for which all derivatives exist. For instance, the function $f(t) = t^2$ is differentiable in $\mathbb{R}$, and its first derivative is $f'(t) = \frac{df}{dt} = 2t$. In case the derivatives exist only up to a maximum order k, the function is said to be differentiable of order k. Observe that this property can be verified at a specific value a (and we say the function is differentiable at a) or at sets of points, including the whole respective domain (in this case we simply say the function is differentiable).

Continuous functions: A function $f(t)$ is said to be *continuous* at a value a if and only if $\lim_{t \to a} f(t) = f(a)$. If the function is continuous for all points in a set S, it is said to be *continuous in S*. All differentiable functions are continuous, but not every continuous function is differentiable.

Periodic functions: Functions such as $f(t) = f(t + T)$, where T is a real value, are said to be periodic with period T. For example, the cosine function is periodical with period 2π. It can be easily verified that, if a function is periodic with period T, it is also periodic of period kT, where k is any integer value.

Even and odd functions: A function such that $f(t) = f(-t)$ is called *even*. A function such that $f(t) = -f(-t)$ is said to be *odd*. The cosine and sine functions are examples of even and odd functions, respectively. Every function can be decomposed as a sum of an even and an odd function.

2.1.3 Free Variable Transformations

Given a specific function $y = g(t)$, it is possible to obtain new functions by modifying the free variable t. Two common and useful transformations obtained in such a way involve the following:

① *Adding* or *subtracting* a constant $\tau > 0$ to the free variable, i.e., $y = g(t + \tau)$ and $y = g(t - \tau)$, which makes the function shift left and right, respectively, as illustrated in Figures 2.3 (a) and (b).

② *Multiplying* a constant value to the free variable, i.e., $y = g(at)$. If the constant a is positive and larger than 1, the function shrinks along the t-axis (observe that this shrinkage is centered towards the coordinate system origin, which does not move); if it is positive and smaller than 1 (i.e., $0 < a < 1$), the function is expanded around the coordinate origin (observe that the function becomes stationary at the origin). In case constant a is negative, the function is reversed with respect to the x-axis and will shrink for $a < -1$ and expand for $-1 < a < 0$. Figures 2.3 (c) through (f) illustrate such effects.

Special care must be taken when dealing with transformed functions of the type $y = g(at)$ where a is negative, since the function shifting is inverted, as illustrated in Figure 2.4 for $a = -1$.

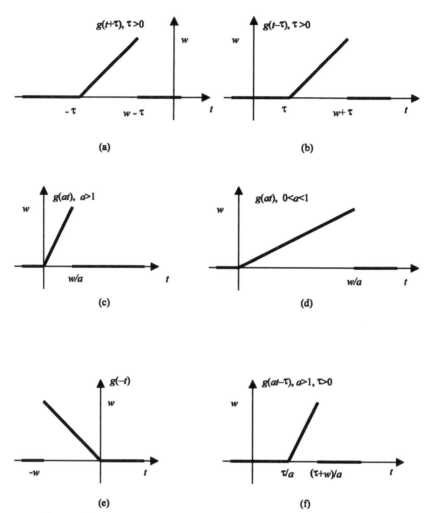

Figure 2.3: *Examples of new functions obtained from the function y = g (t) in Figure 2.2 by means of free variable transformations.*

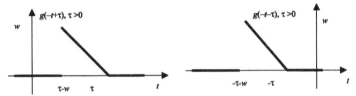

Figure 2.4: *Shift behavior for* $y = g(at)$, *where a is negative (here, a =* -1).

2.1.4 Some Special Real Functions

Some real functions which are especially relevant and useful in the context of this book are briefly presented and illustrated in the following:

Affine Mapping

Let a variable x be limited as $x_{min} \leqslant x \leqslant x_{max}$. In case we want to map x into a new variable y such that $y_{min} \leqslant y \leqslant y_{max}$, we can use the affine mapping function defined as:

$$y = mx + k, \tag{2.1}$$

where $m = \frac{y_{max} - y_{min}}{x_{max} - x_{min}}$ and $k = \frac{y_{min}x_{max} - x_{min}y_{max}}{x_{max} - x_{min}}$. The generic form of this function is illustrated in Figure 2.5.

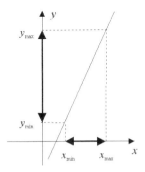

Figure 2.5: *The affine mapping function.*

Observe that the above presented strategy can also be used to map open intervals, such as from $x_{min} \leqslant x < x_{max}$ into $y_{min} \leqslant y < y_{max}$, from $x_{min} < x < x_{max}$ into $y_{min} < y < y_{max}$, or from $x_{min} < x \leqslant x_{max}$ into $y_{min} < y \leqslant y_{max}$.

Example: *Affine Mapping*

Consider that you want to normalize the function $g(x) = \cos(x)$, which is naturally constrained to the interval $-1 \leqslant g(x) \leqslant 1$, in order to obtain a new version of this function, $h(x)$, in such a way that $0 \leqslant h(x) \leqslant 255$.

We know that:

$$x_{min} = -1 \qquad\qquad x_{max} = 1$$
$$y_{min} = 0 \qquad\qquad y_{max} = 255$$

Therefore:

$$h(x) = \frac{(y_{max} - y_{min})\, g(x) + y_{min} x_{max} - x_{min} y_{max}}{x_{max} - x_{min}}$$
$$= \frac{255\, g(x) + 255}{2}$$
$$= \frac{255}{2}\left[g(x) + 1\right].$$

Exponential

The general form of this function is given by equation (2.2), where a is a real value (see the Box titled *Parameters*):

$$f(x) = e^{ax} = \exp(ax). \qquad (2.2)$$

Figure 2.6 depicts this function for several values of a. As is clear from this illustra-

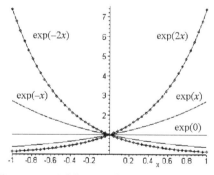

Figure 2.6: *The exponential function for several values of its parameter a.*

tion, the exponential function exhibits opposite behaviors for positive and negative values of the parameter a. More specifically, as x increases, the exponential function grows without bounds for positive values of a and decreases for negative values of a. For $a = 0$, the exponential function becomes the constant function $f(x) = 1$. Observe that the exponential is a strictly positive function, i.e., $\exp(ax) > 0$ for any real values a and x.

Note: *Parameters*

The exponential function, as well as many others, involves a constant value, a, that controls its specific behavior (see Figure 2.6). Such values are called *parameters*. A function can have several parameters, each defining a specific property of the function. Examples of other functions involving parameters are the *sigmoid* and *Gaussian*. The concept of parameter can also be extended to techniques, programs and models.

Sinc

This function, depicted in Figure 2.7, is defined as

$$\operatorname{sinc}(x) = \begin{cases} 1 & \text{if } x = 0, \\ \frac{\sin(\pi x)}{\pi x} & \text{otherwise.} \end{cases}$$

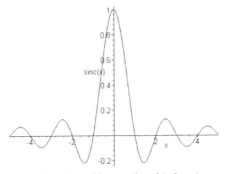

Figure 2.7: *The sinc function. Observe that this function crosses the x-axis at $x = \pm 1, \pm 2, \ldots$*

Sigmoid

This function, illustrated in Figure 2.8, is defined by equation (2.3), where a is the parameter that controls the smoothness of the transition of the function from -1 to 1.

$$\gamma(x) = \frac{1 - \exp(-ax)}{1 + \exp(-xa)}. \tag{2.3}$$

The higher the value of a, the steeper the transition. When $a \to \infty$, the sigmoid becomes more and more similar to the signum function (see Figure 2.8). The reader is invited to identify the behavior of the sigmoid function when $a < 0$ and $a = 0$. It should be observed that there are alternative versions for the sigmoid functions, such as those based on the arctangent function (see, for instance, [Fausett, 1994]).

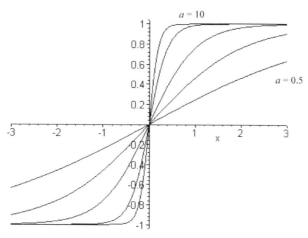

Figure 2.8: *The sigmoid function for a = 0.5, 1, 2, 5, and 10. The higher the value of a, the steeper the transition.*

The sigmoid function can be used to "equalize" the magnitude of a function or signal. This can be done by composing the function to be equalized, let us say $g(t)$, with the sigmoid function, i.e., $h(t) = \gamma(g(t)) = \frac{1-\exp(-ag(t))}{1+\exp(-ag(t))}$. Figure 2.9 depicts a graphical construction that is useful for visualizing the composition of functions, in this case showing how the sigmoid function can be used to equalize the magnitude of a function $g(t)$. Observe that the sigmoid also ensures that $-1 \leqslant \gamma(x) \leqslant 1$. Although the equalizing effect is more pronounced for higher values of the parameter a, this also implies higher distortion to the function.

Gaussian

This is one of the most important functions in science. It is given by equation (2.4), where μ and σ are two independent parameters that control the position of the function along the x-axis and the width of the function, and k is a real value.

$$g_\sigma(x) = k \, \exp\left\{-\frac{1}{2}\left(\frac{x-\mu}{\sigma}\right)^2\right\},$$ (2.4)

It can be immediately verified that the maximum value of the Gaussian function occurs for $x = \mu$, and is equal to the real value k. In order to better understand these

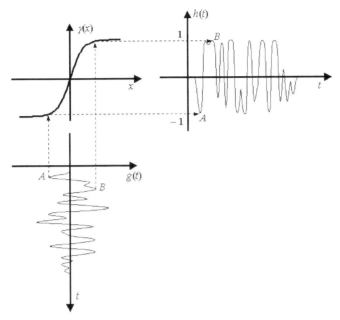

Figure 2.9: *Graphical construction for function composition. The composed function $h(t) = \gamma(g(t))$ can be graphically sketched by reflecting the values of $g(t)$ at the function $\gamma(x)$, as illustrated by the dashed lines for the points A and B. In this case, the composition with the sigmoid function provides equalization of the magnitude of the input function $g(t)$.*

parameters, observe that:

$$\frac{g_\sigma(\mu)}{g_\sigma(\mu \pm \sigma)} = \frac{\exp(0)}{\exp\left(-\frac{1}{2}\right)} = \sqrt{e} \Rightarrow g_\sigma(\mu \pm \sigma) = \frac{1}{\sqrt{e}} g_\sigma(\mu).$$

In other words, a displacement to the left or right of μ by σ implies the magnitude of the Gaussian to be reduced by a factor $\frac{1}{\sqrt{e}}$, as illustrated in Figure 2.10.

In case $k = \frac{1}{(\sigma\sqrt{2\pi})}$, the Gaussian is said to be *normalized*; its area becomes unitary, i.e., $\int_{-\infty}^{+\infty} g_\sigma(x)\,dx = 1$; and μ and σ correspond to the *mean* and the *standard deviation* of the Gaussian, respectively. Figure 2.11 illustrates three normalized Gaussian functions characterized by distinct means and standard deviations.

The first and second derivatives of the Gaussian function, which are particularly useful in shape analysis, are given by equations (2.5) and (2.6), respectively,

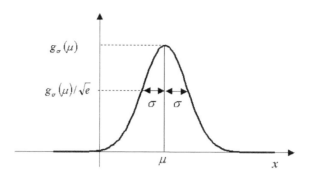

Figure 2.10: *Graphical interpretation of the parameter σ.*

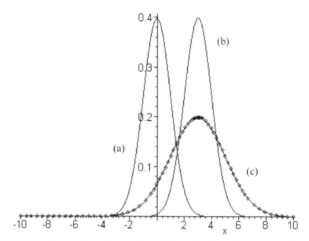

Figure 2.11: *Examples of normalized Gaussian functions:* $\mu = 0$ *and* $\sigma = 1$ *(a);* $\mu = 3$ *and* $\sigma = 1$ *(b); and* $\mu = 3$ *and* $\sigma = 2$ *(c).*

assuming zero mean.

$$\frac{\mathrm{d}g_\sigma(x)}{\mathrm{d}x} = -\frac{1}{\sqrt{2\pi}\,\sigma^3} x \exp\left\{-\frac{1}{2}\left(\frac{x}{\sigma}\right)^2\right\}. \tag{2.5}$$

$$\frac{\mathrm{d}^2 g_\sigma(x)}{\mathrm{d}x^2} = -\frac{1}{\sqrt{2\pi}\,\sigma^3}\exp\left\{-\frac{1}{2}\left(\frac{x}{\sigma}\right)^2\right\} + \frac{1}{\sqrt{2\pi}\,\sigma^5} x^2 \exp\left\{-\frac{1}{2}\left(\frac{x}{\sigma}\right)^2\right\}. \tag{2.6}$$

Figure 2.12 illustrates the first and second derivatives of the Gaussian function for $\sigma = 1$.

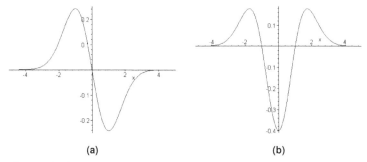

(a)   (b)

Figure 2.12: *The first* (a) *and second* (b) *derivatives of the Gaussian function for* $\sigma = 1$.

Step (or Heavyside)

This simple function, also called hard-limit, is defined as

$$v(x) = \begin{cases} 1 & \text{if } x \geqslant 0, \\ 0 & \text{otherwise,} \end{cases}$$

and illustrated in Figure 2.13.

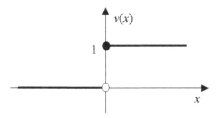

Figure 2.13: *The step (or heavyside) function.*

Signum

This function, hence sgn (x), returns the signal of x, i.e., sgn$(x) = -1$ iff $x < 0$ and sgn $(x) = 1$ for $x > 0$. Observe that sgn $(0) = 0$. The function signum is depicted in Figure 2.14.

Box (or pulse or rectangular)

Defined by equation (2.7) and illustrated in Figure 2.15, this function can be obtained through the product of two step functions, i.e., $u(x) = v_1(x - a) \, v_2(-x + b)$,

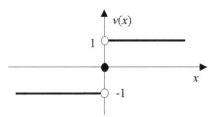

Figure 2.14: *The signum function.*

for $b > a$.

$$u(x) = \begin{cases} 1 & \text{if } a \leqslant x \leqslant b,\ a, b \in \mathbb{R}, \\ 0 & \text{otherwise.} \end{cases} \tag{2.7}$$

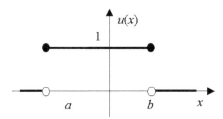

Figure 2.15: *A possible box function.*

Rem and Mod

The remainder function rem (x, a) produces the remainder of the division between x and a (observe that both x and a can be integer or real values), the latter acting as a parameter of the function. Figure 2.16 illustrates this function for a generic value of a. The *mod* function can be defined as $\text{mod}(x) = |\text{rem}(x)|$.

Kronecker Delta

This useful function is defined as

$$\kappa(x) = \begin{cases} 1 & \text{if } x = 0, \\ 0 & \text{otherwise.} \end{cases}$$

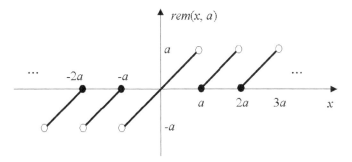

Figure 2.16: *The remainder function.*

For simplicity's sake, it is also possible to define the following:

$$\kappa_{x,w}(x) = \begin{cases} 1 & \text{if } x = w, w \in \mathbb{R}, \\ 0 & \text{otherwise.} \end{cases}$$

Dirac Delta

This is a particularly important function that is not strictly a function. Although it only makes sense under the theory of distributions [Morrison, 1994], it can nevertheless be easily understood in terms of the following properties:

- $\int_{-\infty}^{\infty} \delta(t) \, dt = 1$, i.e., the Dirac delta function has unit area; and

- $\delta(t) = \begin{cases} 0 & \text{if } t \neq 0, \\ \text{not defined} & \text{if } t = 0. \end{cases}$

Another important feature of the Dirac delta function is its ability to sample functions:

$$g(t) \, \delta(t - t_0) = g(t_0) \, \delta(t - t_0),$$

which implies that:

$$\int_{-\infty}^{\infty} g(t) \, \delta(t - t_0) \, dt = \int_{-\infty}^{\infty} g(t_0) \, \delta(t - t_0) \, dt = g(t_0) \int_{-\infty}^{\infty} \delta(t - t_0) \, dt = g(t_0).$$

The Dirac delta function is graphically represented by an arrow with length proportional to the area of the function, as illustrated in Figure 2.17. It should be observed that $\delta(at) = \frac{1}{|a|}\delta(t)$.

In case the reader is still uncomfortable with the Dirac delta nature, it is useful to consider it as the limit of a more traditional function. For instance, it can be

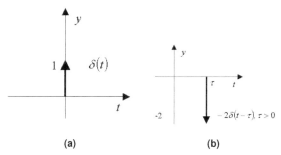

(a)   (b)

Figure 2.17: *The standard Dirac delta function* (a), *and a magnified and shifted version* (b).

proven that

$$\delta(t) = \lim_{\sigma \to 0} \frac{1}{\sigma \sqrt{2\pi}} \exp\left(-\frac{t^2}{2\sigma^2}\right),$$

i.e., the Dirac delta is the limit of the zero mean Gaussian function, which is known to have unitary area, as its standard deviation tends to zero.

Sampling Function (or Shah)

This function, henceforth represented as $\Psi_{\Delta t}(t)$, is of particular importance in digital signal processing (see Sections 2.7.5 and 2.7.6). It is defined as the sum of equally spaced (by Δt) Dirac deltas, i.e., $\Psi_{\Delta t}(t) = \sum_{n=-\infty}^{\infty} \delta(t - n\Delta t)$, as shown in Figure 2.18.

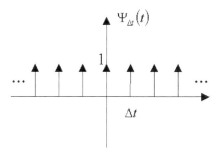

Figure 2.18: *The sampling function is an infinite sum of Dirac deltas.*

Floor

Given a real value x, the floor function produces the greatest integer value that is smaller than or equal to x. For instance, floor $(3.3) = 3$, floor $(2) = 2$, floor $(2.5) = 2$, floor $(8.9) = 8$, floor $(-3.3) = -4$, floor $(-2) = -2$, floor $(-2.5) = -3$, and floor $(-8.9) = -9$. This function is sometimes represented as floor $(x) = \lfloor x \rfloor$. Figure 2.19 graphically illustrates the floor function.

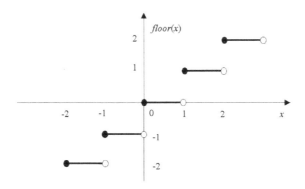

Figure 2.19: *The floor function.*

Ceil

This function acts over a real number x, producing as result the smallest integer value that is greater than or equal to x. For instance, ceil $(3.3) = 4$, ceil $(2) = 2$, ceil $(2.5) = 3$, ceil $(8.9) = 9$, ceil $(-3.3) = -3$, ceil $(-2) = -2$, ceil $(-2.5) = -2$, and ceil $(-8.9) = -8$. This function, which is sometimes represented as ceil $(x) = \lceil x \rceil$, can be related to the floor function as ceil $(x) = -$ floor $(-x)$. Figure 2.20 depicts the ceil function.

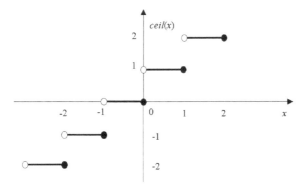

Figure 2.20: *The ceil function.*

Trunc (or Fix)

This function, sometimes also know as *floor*, produces as result the integer part of its input value. For instance, trunc $(3.3) = 3$, trunc $(2) = 2$, trunc $(2.5) = 2$, trunc (8.9) = 8, trunc (-3.3) = -3, trunc (-2) = -2, trunc (-2.5) = -2, and trunc $(-8.9) = -8$. Figure 2.21 illustrates this function.

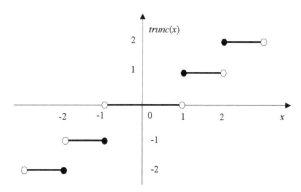

Figure 2.21: *The trunc function.*

Round

This function acts over a real number x, producing as result the integer value y = round (x) such that $y - 0.5 \leqslant x < y + 0.5$ if $x \geqslant 0$, and $y - 0.5 < x \leqslant y + 0.5$ if $x < 0$. For instance, round (3.3) = 3, round (2) = 2, round (2.5) = 3, round (8.9) = 9, round (-3.3) = -3, round (-2) = -2, round (-2.5) = -3, and round $(-8.9) = -9$. This function can be related to the floor function as round (x) = sgn (x) floor $(|x| + 0.5)$, and to the trunc function as round (x) = sgn (x) trunc $(|x| + 0.5)$. Figure 2.22 graphically illustrates the round function.

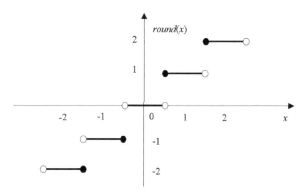

Figure 2.22: *The round function.*

2.1.5 Complex Functions

Complex functions correspond to the extension of real functions to include the situations where both the domain and co-domain are complex sets. In other words, complex functions take complex values into complex images. Of course, real functions are a particular case of complex functions. Before discussing complex functions in greater depth, it is convenient to review some basic concepts related to complex numbers.

Complex Numbers

Complex numbers, introduced by R. Bombelli in his treatise *L'Algebra* (1572), are not "complex," but quite simple. They complement the real numbers by including the imaginary number, which in this book is represented as j. Complex numbers have the general *Cartesian* form, where x is the *real part* of z, represented as Re $\{z\}$, and y is the *imaginary part*, represented as Im $\{z\}$. The infinite set of all complex numbers is denoted as $\mathbb{C}$. The *polar* form of a complex number, shown in Figure 2.23, is given as $z = \rho \exp(j\theta)$, where $|z| = \rho$ is the *magnitude* (or *modulus*) of z and $\phi(z) = \theta$ is its *argument* (or *phase*). Such a graphical representation of

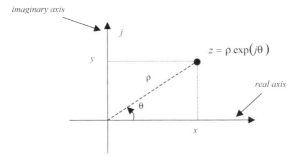

Figure 2.23: *A generic complex number z in the Argand plane.*

a general complex number takes place in the so-called *complex* or *Argand plane*. Thus, complex numbers can be informally understood as points in this plane.

It immediately follows from basic trigonometry (see also Figure 2.23) that the real and imaginary parts of a generic complex number z are $\rho \cos(\theta)$ and $\rho \sin(\theta)$, respectively, i.e., $z = \rho[\cos(\theta) + j\sin(\theta)] = x + jy$. It is also clear from Figure 2.23 that $\rho = \sqrt{x^2 + y^2}$ and $\theta = \arctan\left(\frac{y}{x}\right)$. The useful property $\exp(j\theta) = \cos(\theta) + j\sin(\theta)$ is known as *Euler's formula*.

The *addition* and the *product* between two complex numbers $z_1 = x_1 + jy_1 = \rho_1 \exp(j\theta_1)$ and $z_2 = x_2 + jy_2 = \rho_2 \exp(j\theta_2)$ are calculated as shown in the following:

$$z_1 + z_2 = (x_1 + x_2) + j(y_1 + y_2),$$
$$z_1 z_2 = (x_1 x_2 - y_1 y_2) + j(x_1 y_2 + x_2 y_1) = \rho_1 \rho_2 \exp[j(\theta_1 + \theta_2)].$$

The *complex conjugate* of a complex number $z = x + jy$, which is represented as z^*, is obtained by inverting the signal of its imaginary part, i.e., $z^* = x - jy$. Graphically, this operation corresponds to reflecting the point defined by the complex number with respect to the real axis, as illustrated in Figure 2.24. It can be

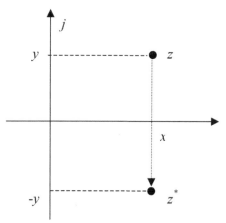

Figure 2.24: *The complex conjugation of a complex number is obtained by reflecting the number with respect to the real axis.*

easily verified that $|z|^2 = z^*z$ and $|z| = \sqrt{z^*z}$.

Complex Functions and Their Graphical Representation

Having presented the main concepts and properties of complex numbers, let us proceed to see how functions can be defined over these numbers. It is interesting to proceed through a progressive complexity sequence of possible types of functions. First, it is important to note that a generic complex function $w = g(z)$, defined on a domain U and taking values in a co-domain W, can be written as follows:

$$g: U \rightarrow W, w = g(z) = p(x, y) + jq(x, y),$$

where $z = x + jy$ is a generic complex number, and $p(x, y)$ and $q(x, y)$ are generic real functions.

The first situation to be considered is when U is real and W is complex. In this case, the function $g: U \rightarrow W, w = g(z)$ will associate a point in the Argand plane to each point in the real domain U. Consider, for instance, the following important function, known as *complex exponential*, which maps a real value t into a complex value $g(t)$:

$$g(t) = \rho \exp(jt) = \rho \cos(t) + j\rho \sin(t).$$

It is possible to represent such a function as a 3D curve where one of the axes is used to represent the free variable t. For the above example, and assuming $0 \leqslant t < 3.5\pi$, we have a helix, as shown in Figure 2.25.

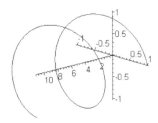

Figure 2.25: *The complex exponential function with unit radius defines a helix.*

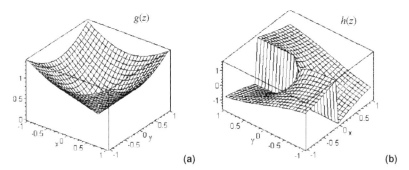

Figure 2.26: *The magnitude (a) and phase (b) functions.*

Another interesting situation is defined for complex U and real W. In this case, the function can be represented as a surface in a three-dimensional graph. For instance, the functions

$$g(z) = \sqrt{\text{Re}^2\{z\} + \text{Im}^2\{z\}} \quad \text{and}$$

$$h(z) = \arctan\left\{\frac{\text{Im}\{z\}}{\text{Re}\{z\}}\right\},$$

which associate to each complex value its respective magnitude and phase, can be graphically represented as shown in Figure 2.26.

The most general situation corresponding to both U and W complex, is slightly more cumbersome and is treated by considering two Argand planes and indicating mappings of points, lines and/or regions (e.g., circles, rectangles, etc.). Consider the following function, where $z = x + jy$:

$$g(z) = \exp(z)$$
$$= \exp(x + jy)$$
$$= \exp(x) \exp(jy).$$

First, it should be observed that

$$|z| \exp[j\phi(z)] = \exp(x) \exp(jy),$$

hence

$$|z| = \exp(x) \quad \text{and} \quad \phi(z) = y.$$

Thus, a vertical line $x = a$ is mapped by this function into a circle of radius $\exp(a)$, and a horizontal line $y = c$ is mapped into the ray making an angle y with the real axis. Observe that this ray does not reach the plane origin. The above mapping can be graphically represented as in Figure 2.27.

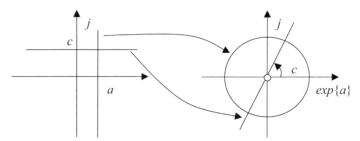

Figure 2.27: *The mapping of vertical and horizontal lines by* $g(z) = \exp(z)$.

Now, consider the rectangular region of the domain space defined as

$$a \leqslant x \leqslant b \quad \text{and} \quad c \leqslant y \leqslant d.$$

The reader should have no difficulty verifying that this rectangular region is mapped by $g(z)$ as illustrated in Figure 2.28.

Figure 2.29 presents a more comprehensive illustration of the mapping implemented by $g(z) = \exp(z)$ with respect to an orthogonal grid in the domain space.

As is clear from the above example, it is not always easy to identify the more interesting behavior to be illustrated for each specifically considered complex function.

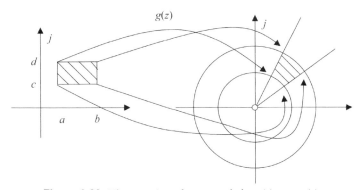

Figure 2.28: *The mapping of a rectangle by $g(z) = \exp(z)$.*

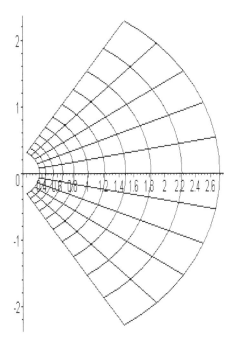

Figure 2.29: *The mapping of an orthogonal grid implemented by the function $g(z) = \exp(z)$.*

Analytical Functions and Conformal Mappings

The *limit L* of a complex function $g(z)$ at a point z_0, which is itself a complex value represented as:

$$L = \lim_{z \to z_0} g(z),$$

exists if and only if given an open ball B centered at L, with nonzero radius ε (i.e., the set of complex numbers z such that $0 < |z - L| < \varepsilon$), it is always possible to find an open ball centered at z_0, also with nonzero radius ($0 \neq |z - z_0| < \delta$), in such a way that the mapping of any point z inside this ball by the function g, i.e., $g(z)$, falls inside B.

In case

$$g(z_0) = \lim_{z \to z_0} g(z),$$

we say that the function is *continuous at* z_0. A function that is continuous at all the points of its domain is simply said to be *continuous*.

In case the limit:

$$g'(z_0) = \left. \frac{dg}{dz} \right|_{z_0} = \lim_{\Delta z \to 0} \frac{g(z + \Delta z) - g(z)}{\Delta z} = \lim_{z \to z_0} \frac{g(z) - g(z_0)}{z - z_0}$$

exists, the function $g(z)$ is said to be *differentiable* (of order 1) *at* z_0. In case $g(z)$ is differentiable in all points of a domain, it is simply said to be *differentiable* or *analytic* (of order 1) in that domain. The *Cauchy-Riemann conditions*, which provide a sufficient condition for a function $g(z) = p(x, y) + jq(x, y)$ to be analytic, are given by the following pair of equations:

$$\frac{dp}{dx} = \frac{dq}{dy} \quad \text{and} \quad \frac{dp}{dy} = -\frac{dq}{dx}.$$

Complex functions that preserve the senses and angles between curves, such as illustrated in Figure 2.30, are said to define a *conformal mapping*.

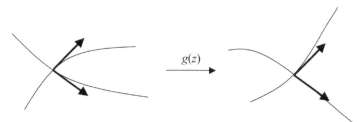

$g(z)$

Figure 2.30: *A conformal mapping does not change the senses and angles between curves.*

It can be proven [Churchill & Brown, 1989; Kreyszig, 1993] that every analytic function $g(z)$, except at the eventual points where $g'(z) = 0$, defines a conformal mapping. The magnification factor of such a mapping at a complex point z_0 is defined as:

$$M(x_0) = |g'(z_0)|. \tag{2.8}$$

This factor indicates how much a short line segment centered at z_0 is enlarged or reduced by the mapping. For instance, for the function $g(z) = \exp(z)$ considered in the conformal map example previously presented, we have $g'(z) = \exp(z)$ and $M(z) = |g'(z)| = \exp(z)$, indicating that the magnification factor increases exponentially with x, as clearly verified in Figure 2.29. Observe that this function is not conformal at $z = 0$.

The above condition $g'(z) \neq 0$ is required in order to have a conformal map, indicating that the small line segments will not collapse and, therefore, $g(z)$ will implement a one-to-one mapping.

2.2 Linear Algebra

Algebra, a word of Arabic origin ("al-jabr") meaning "transposition," became known in the West mainly as a consequence of the influential work by al-Khowarizmi, in the 9th century. In more recent times, algebra has been associated with arithmetic constructions involving not only explicit values (i.e., numbers), but also letters, corresponding to our modern concept of *variables*. Therefore, *linear algebra* can literally be understood as algebraic studies restricted to *linear* behavior, a concept that will be presented in this section.

Linear algebra is an important branch of mathematics, providing the bedrock not only for theoretical developments but also for many practical applications in physics and engineering, being intensively used in image processing, shape analysis and computer graphics. Generally speaking, it addresses the concepts of vector spaces and related issues such as linear operations, orthogonality, coordinate transformations, least square approximations, eigenvalues and eigenvectors. The current section provides a review of some of the most important and useful linear algebra concepts with respect to shape analysis and recognition.

2.2.1 Scalars, Vectors, and Matrices

A particularly useful aspect of algebra concerns its ability to represent and model entities from the real world. For instance, we can represent the perimeter of a shape in terms of a scalar p. The use of letters (i.e., variables) to represent values provides an interesting facility for *generalizing* developments and results. The basic representative concepts in algebra include *scalars*, *vectors*, *matrices* and *tensors*, which are described in the following (except for the relatively more advanced concept of tensor).

Scalars

Single numbers are usually referred to as *scalars*. Though typically corresponding to real numbers, scalars can also assume complex values. Examples of some physical properties represented by scalars are given in the left column of Table 2.1.

Property	Representation
Temperature	$T = 10°C$
Area of a region	$A = 25 \, \text{m}^2$
Electric current in an RC circuit	$I = 1.2 + 0.8j \, \text{mA}$ (complex value)

Table 2.1: *Representation of physical properties in terms of scalars.*

As illustrated in the right column of Table 2.1, all these properties can be precisely and conveniently described in terms of a single value (real or complex), which is the very essence of a scalar. Since they often correspond to physical properties, such values typically have *dimensional* units. However, we can also have *dimensionless* scalars, such as proportions. For instance, by adopting $T_0 = 100°C$ as a reference, we can represent any temperature in terms of the dimensionless scalar given by the proportion $\widetilde{T} = \frac{T}{T_0}$.

Vectors

Many situations arise in nature, such as those illustrated in the left column of Table 2.2, that cannot be effectively represented by a single scalar.

Property	Representation
Position of a point inside a square, with respect to its lower left corner	$p = \begin{bmatrix} 2.32 \\ 1.91 \end{bmatrix} \text{cm}$
Instantaneous velocity of a flying butterfly	$v = \begin{bmatrix} 3.4 \\ 1.1 \\ 2.0 \end{bmatrix} \text{m/s}$
The weights of all players in a basketball team	$w = (62, 80, 68, 79, 73) \, \text{kg}$

Table 2.2: *Representation examples in terms of vectors.*

In algebra, such quantities can be precisely represented in terms of *vectors*, which are composed of a number of scalar values arranged in a certain order. A vector differs from a set, which is a collection of scalars, by the fact that the *position* of each scalar is important. For instance, the butterfly velocity mentioned in the previous table would most probably become completely different if any pair of scalars in the respective vector were exchanged (except in the particular case where both values are equal).

A vector can be represented in several ways, including

$$
v = \vec{v} = \begin{bmatrix} a_1 \\ a_2 \\ \vdots \\ a_i \\ \vdots \\ a_N \end{bmatrix} = [a_1, a_2, \ldots, a_i, \ldots, a_N]^{\mathrm{T}} =
$$

$$
= (a_1, a_2, \ldots, a_i, \ldots, a_N) = a_1 \vec{e}_1 + a_2 \vec{e}_2 + \cdots + a_N \vec{e}_N
$$

where i is usually a strictly positive integer number (see next box), N is always a non-negative integer value, and the superscript T stands for the operation of *transposition*. The values a_i are the vector *components*, and $\vec{e}_i$ are the *elements of the basis* used to represent the vector space (see Section 2.2.2). The above vectors are said to have *dimension $N \times 1$*. All the above types of vector representation are completely equivalent, being alternatively used in the literature, and also in the present book, to suit specific requirements. For instance, the form $(a_1, a_2, \ldots, a_i, \ldots, a_N)$ is more suitable for inclusion in a text.

Figure 2.31 shows the graphical representation of the vector $v = (1, 2, 3)$, which "lies" in the three-dimensional space, i.e., $\mathbb{R}^3$. Actually, each distinct vector in any

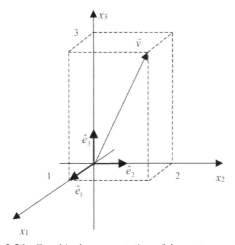

Figure 2.31: *Graphical representation of the vector $v = (1, 2, 3)$.*

space can be associated with a corresponding point in that space, and vice versa.

The vector representation of the entities in Table 2.2 are given in its respective right column. It is observed that scalars become a special case of vectors when $N = 1$.

Note: *Vector Indexing and Its Computational Implementation*

Although vectors and matrices are usually indexed by strictly positive values, such as $1 \leqslant i \leqslant 10$, it is also possible, and often particularly useful, to consider general integers for indexing, such as $0 \leqslant i \leqslant 20$, or even $-5 \leqslant i \leqslant 10$. However, since some programming languages and environments are restricted to strictly positive indexing values, a means for simulating general indexing should be devised.

We present two methods for coping with such a problem in the following example. Consider that, given a vector $\vec{v} = v(i)$, $i = a, a + 1, \ldots, b$ where $a \leqslant 0$ and $b \geqslant a$, we want to obtain a new vector $\vec{u} = u(i)$, $i = a, a + 1, \ldots, b$, such that each of its elements $u(i)$ corresponds to the multiplication of the respective element in $v(i)$ by i. Observe that the number of elements in both these vectors, also called their length or size, is given by $N = b - a + 1$. In case negative indexing is allowed, the solution is straightforward:

1.   **for** $i \leftarrow a$ **to** b
2.   **do**
3. $u(i) \leftarrow iv(i)$

In case only strictly positive indexes have to be used, we have little choice but to map the vectors $v(i)$ and $u(i)$ into respective auxiliary vectors $\tilde{v}(k)$ and $\tilde{u}(k)$, for $k = 1, 2, \ldots, N$, which can be obtained by making $k = i - a + 1$. Now, we have two alternative methods for implementing the above algorithm. The first is

1.   **for** $i \leftarrow a$ **to** b
2.   **do**
3. $\tilde{u}(i - a + 1) \leftarrow i\tilde{v}(i - a + 1)$

that consists in adding the value $(-a + 1)$ to every vector index. The second strategy is to use strictly positive loop variables:

1.   **for** $k \leftarrow 1$ **to** N
2.   **do**
3. $\tilde{u}(k) \leftarrow (k + a - 1)\tilde{v}(k)$

While the first approach allows a clearer understanding of the implemented method, the second can be faster because of the smaller number of arithmetic operations that are usually required. Once the vector $\tilde{u}(k)$ has been calculated by using any of the two above approaches, the sought result $u(i)$ can be obtained by making $u(i) = \tilde{u}(i - a + 1)$.

Matrices

Matrices can be understood as being generalized versions of vectors, which are extended to include more than a single column. Many situations in nature, such as tables of distances between the main world capitals, can be conveniently represented in terms of matrices, which have the general form

$$A = \begin{bmatrix} a_{1,1} & a_{1,2} & \cdots & a_{1,M} \\ a_{2,1} & a_{2,2} & \cdots & a_{2,M} \\ \vdots & \vdots & a_{i,j} & \vdots \\ a_{N,1} & a_{N,2} & \cdots & a_{N,M} \end{bmatrix} = \left[a_{i,j} \right]_{N,M},$$

where i, j, N and M are positive integer numbers. The above matrix, A, is said to have dimension $N \times M$.

In the example of the distances between the main world capitals, they can be represented as a matrix by assigning each capital city to an index, in such a way that $a_{i,j}$ will indicate the distance between the two respective cities, i.e., those indexed by i and j. For instance, if i = "*Athens*" and j = "*Washington D.C.*," then $a_{i,j}$ stands for the distance between Athens and Washington D.C. It should be observed that although the information in any matrix could, at least in principle, be also represented as a vector (obtained by stacking the columns as illustrated below), the latter representation is less explicit and meaningful even though it contains the same information as the former (matrix):

$$\begin{bmatrix} 3 & 0 & 1 \\ -1 & 3 & -2 \\ 2 & 1 & 2 \end{bmatrix} \rightarrow \begin{bmatrix} 3 \\ -1 \\ 2 \\ 0 \\ 3 \\ 1 \\ 1 \\ -2 \\ 2 \end{bmatrix}.$$

2.2.2 Vector Spaces

Vector space is a fundamental concept in linear algebra, since it provides the basic framework underlying not only the proper representation of elements known under the general name of *vectors* but also a series of useful operations between such elements. It should be observed that the term *vector* is herein considered at a substantially more comprehensive context than the standard vectors seen in the previous sections. As we will soon see, matrices, polynomials and even functions can be called vectors in linear algebra.

The two most important operations are addition between two vectors and multiplication between a scalar (real or complex) value and a vector. Informally speaking, a *vector space* S is a (possibly infinite) collection of vectors exhibiting the following properties, where $\vec{v}$, $\vec{p}$ and $\vec{q}$ are any two vectors in S:

☞ With respect to addition

① $\vec{v} + \vec{p} \in S$ (closure);
② $\vec{v} + \vec{p} = \vec{p} + \vec{v}$ (commutative law);
③ $\vec{v} + (\vec{p} + \vec{q}) = (\vec{v} + \vec{p}) + \vec{q}$ (associative law);
④ $\forall \vec{p} \in S, \exists \vec{0} \in S \mid \vec{0} + \vec{p} = \vec{p}$ (neutral element);
⑤ $\forall \vec{p} \in S, \exists (-\vec{p}) \in S \mid \vec{p} + (-\vec{p}) = \vec{0}$ (inverse element).

☞ With respect to scalar multiplication

⑥ $\alpha \vec{p} \in S$ (closure);
⑦ $\alpha(\beta \vec{p}) = (\alpha\beta)\vec{p}$ (associative law);
⑧ $\alpha(\vec{p} + \vec{q}) = \alpha\vec{p} + \alpha\vec{q}$ (distributive);
⑨ $(\alpha + \beta)\vec{p} = \alpha\vec{p} + \beta\vec{p}$ (distributive);
⑩ $1\vec{p} = \vec{p}$ (identity element).

The nature of the scalars involved in the multiplications (e.g., real or complex) defines the *field* onto which the vector space is defined. For instance, in case of real scalars, we say that the vector space is over the real field.

Examples of vector spaces include but are not limited to the following cases:

① The real numbers, $\mathbb{R}$, where each real number is understood as a vector.

② The complex numbers, $\mathbb{C}$, where each complex value is a vector.

③ The set $\mathbb{R}^2$, where each vector (x, y) is understood as a vector.

④ The set $\mathbb{R}^N$, where each vector $(x_1, x_2, \ldots, x_N)$ is a vector.

⑤ All polynomials of degree smaller or equal to n, P_n, plus the null polynomial. Each of such polynomials is called a vector.

⑥ All real (or complex) matrices with generic dimension $N \times M$.

⑦ The continuous real functions $f = f(t)$ in the real interval $[a, b]$. This vector space is usually represented as $C[a, b]$.

⑧ The complex functions f which are analytic on $|z| < 1$ and continuous on $|z| \leqslant 1$.

The reader should verify that all the above cases are vector spaces. A *counterexample*, i.e., a set that is *not* a vector space, is the set of all polynomials of degree *equal* to n. Let $n = 4$, $\vec{p} = x^4 - 2x^2$ and $\vec{q} = -x^4$, then the addition $\vec{p} + \vec{q} = -2x^2$ does not belong to the space. In other words, the result of an addition between two vectors in this space is not necessarily an element in this space (as it does not satisfy the addition *closure* property).

A non-empty subset X of a vector space S, such as the addition between any of its elements and the product of any of its elements by a scalar result a vector in X (i.e., closure with respect to addition and multiplication by a scalar), is called a *subspace of S*. It can be readily verified that the null vector $\vec{0}$ must be included in any subspace. For instance, the space $\mathbb{R}$ is a subspace of $\mathbb{R}^2$. In addition, observe that a subspace is also a vector space.

Given M vectors $\vec{p}_i$, for $i = 1, 2, \ldots, M$, in the vector space S, the *linear combination* of such vectors, resulting a vector $\vec{q}$ also in S, is defined as

$$\vec{q} = a_1 \vec{p}_1 + a_2 \vec{p}_2 + \cdots + a_M \vec{p}_M,$$

where a_i are any scalar values. The above M vectors are said to be *linearly independent (l. i.)* if and only

$$\vec{0} = a_1 \vec{p}_1 + a_2 \vec{p}_2 + \cdots + a_i \vec{p}_i + \cdots + a_M \vec{p}_M \Leftrightarrow a_1 = a_2 = \cdots = a_M = 0$$

In other words, it is not possible to express one of the vectors as a linear combination of the other vectors. Otherwise, the M vectors are said to be *linearly dependent (l. d.)*. A practical way to determine whether a set of vectors is l. i. can be obtained by using the determinants or rank of matrices, as described in the remainder of this section.

For any vector space S, it is always possible to identify a minimal set, in the sense of involving the minimum number of elements, of linearly independent vectors in S whose linear combinations produce (or *span*) all the possible vectors in S. Such a set of elementary vectors is called a *basis* of S. For instance, both

$$B_1 = \left\{ \begin{bmatrix} 0 \\ 1 \end{bmatrix}, \begin{bmatrix} 1 \\ 0 \end{bmatrix} \right\} \quad \text{and} \quad B_2 = \left\{ \begin{bmatrix} 0 \\ -1 \end{bmatrix}, \begin{bmatrix} 1 \\ 1 \end{bmatrix} \right\}$$

are valid bases for $\mathbb{R}^2$. The *dimension* of a vector space is defined as the number of vectors in any of the bases spanning this space. It should be observed that the vector space containing only the null vector has dimension zero and not one. The earlier examples ⑦ and ⑧ of vector spaces have infinite dimension.

Let S be an N-dimensional vector space and $B = \{\vec{b}_1, \vec{b}_2, \ldots, \vec{b}_N\}$ be one of its possible bases. Then any vector $\vec{p}$ in this space can be represented as a unique linear combination of the vectors in B, i.e., $\vec{p} = a_1 \vec{b}_1 + a_2 \vec{b}_2 + \cdots + a_N \vec{b}_N$, and scalars $a_1, a_2, \ldots, a_N$ are called *coordinates* of the vector $\vec{p}$ with respect to basis B, which are often represented as:

$$
\vec{p} = \begin{bmatrix} a_1 \\ a_2 \\ \vdots \\ a_N \end{bmatrix}_B = \begin{bmatrix} a_1 & a_2 & \cdots & a_N \end{bmatrix}_B^{\mathrm{T}}.
$$

Although a vector space can have an infinite number of alternative bases, it always has a special basis, in the sense of being the most elementary and simple one, which is called its respective *canonical basis*. For instance, the space $\mathbb{R}^N$ has the following canonical basis:

$$
C_N = \{\vec{e}_1, \vec{e}_2, \ldots, \vec{e}_N\} = \left\{ \begin{bmatrix} 1 \\ 0 \\ \vdots \\ 0 \end{bmatrix}, \begin{bmatrix} 0 \\ 1 \\ \vdots \\ 0 \end{bmatrix}, \ldots, \begin{bmatrix} 0 \\ 0 \\ \vdots \\ 1 \end{bmatrix} \right\}.
$$

These are the bases normally adopted as default for representing vectors. In such situations, the subscript indicating the basis is often omitted. For instance,

$$
\begin{bmatrix} -1 \\ 3 \\ 2 \end{bmatrix} = \begin{bmatrix} -1 \\ 3 \\ 2 \end{bmatrix}_{C_3} = -1 \begin{bmatrix} 1 \\ 0 \\ 0 \end{bmatrix} + 3 \begin{bmatrix} 0 \\ 1 \\ 0 \end{bmatrix} + 2 \begin{bmatrix} 0 \\ 0 \\ 1 \end{bmatrix}.
$$

It makes no sense to think about the orientations of these vectors with respect to some absolute reference, since this is not known to exist in the universe. What does matter are the intrinsical properties of the canonical basis, such as having unit magnitude and being orthogonal (see Section 2.2.4). Observe that all the thus far presented examples in this chapter have considered canonical bases.

Now, let $\vec{v}$ be a vector originally represented in terms of its coordinates with respect to the basis $A = \{\vec{a}_1, \vec{a}_2, \ldots, \vec{a}_N\}$. What will the new coordinates of this vector be when it is expressed with respect to the new basis $B = \{\vec{b}_1, \vec{b}_2, \ldots, \vec{b}_N\}$? This important problem, known as *change of coordinates*, can be addressed as follows. We start by expressing the vectors in the new basis B in terms of the coordinates relative to the original basis A:

$$
\vec{b}_1 = \alpha_{1,1}\vec{a}_1 + \alpha_{2,1}\vec{a}_2 + \cdots + \alpha_{N,1}\vec{a}_N,
$$
$$
\vec{b}_2 = \alpha_{1,2}\vec{a}_1 + \alpha_{2,2}\vec{a}_2 + \cdots + \alpha_{N,2}\vec{a}_N,
$$
$$
\vdots
$$
$$
\vec{b}_N = \alpha_{1,N}\vec{a}_1 + \alpha_{2,N}\vec{a}_2 + \cdots + \alpha_{N,N}\vec{a}_N.
$$

The above coefficients can be organized into the following matrix form:

$$C = \begin{bmatrix} \alpha_{1,1} & \alpha_{1,2} & \cdots & \alpha_{1,N} \\ \alpha_{2,1} & \alpha_{2,2} & \cdots & \alpha_{2,N} \\ \vdots & \vdots & \ddots & \vdots \\ \alpha_{N,1} & \alpha_{N,2} & \cdots & \alpha_{N,N} \end{bmatrix}.$$

The reader should notice that the coefficients appearing along each row of the previous system of equations are represented as columns in the above matrix. This is necessary in order to provide a more elegant matrix equation for coordinate exchanging, as will soon become clear.

Now, let $\vec{v}$ be any vector in our vector space, expressed in terms of its coordinates respectively to the basis B:

$$\vec{v}_B = v_1 \vec{b}_1 + v_2 \vec{b}_2 + \cdots + v_N \vec{b}_N.$$

Since every vector in basis B can be expressed in terms of the vectors in A (as in the above set of equations), we can write the following:

$$\begin{aligned} \vec{v}_A &= v_1 \left(\alpha_{1,1} \vec{a}_1 + \alpha_{2,1} \vec{a}_2 + \cdots + \alpha_{N,1} \vec{a}_N \right) \\ &+ v_2 \left(\alpha_{1,2} \vec{a}_1 + \alpha_{2,2} \vec{a}_2 + \cdots + \alpha_{N,2} \vec{a}_N \right) \\ &+ \cdots + v_N \left(\alpha_{1,N} \vec{a}_1 + \alpha_{2,N} \vec{a}_2 + \cdots + \alpha_{N,N} \vec{a}_N \right) \\ &= \left(\alpha_{1,1} v_1 + \alpha_{1,2} v_2 + \cdots + \alpha_{1,N} v_N \right) \vec{a}_1 \\ &+ \left(\alpha_{2,1} v_1 + \alpha_{2,2} v_2 + \cdots + \alpha_{2,N} v_N \right) \vec{a}_2 \\ &+ \cdots + \left(\alpha_{N,1} v_1 + \alpha_{N,2} v_2 + \cdots + \alpha_{N,N} v_N \right) \vec{a}_N \\ &= \hat{v}_1 \vec{a}_1 + \hat{v}_2 \vec{a}_2 + \cdots + \hat{v}_N \vec{a}_N. \end{aligned}$$

Or, in a more compact matrix form,

$$\vec{v}_A = \begin{bmatrix} \hat{v}_1 \\ \hat{v}_2 \\ \vdots \\ \hat{v}_N \end{bmatrix} = \begin{bmatrix} \alpha_{1,1} & \alpha_{1,2} & \cdots & \alpha_{1,N} \\ \alpha_{2,1} & \alpha_{2,2} & \cdots & \alpha_{2,N} \\ \vdots & \vdots & \ddots & \vdots \\ \alpha_{N,1} & \alpha_{N,2} & \cdots & \alpha_{N,N} \end{bmatrix} \begin{bmatrix} v_1 \\ v_2 \\ \vdots \\ v_N \end{bmatrix} \Leftrightarrow \vec{v}_A = C \vec{v}_B. \qquad (2.9)$$

Now, provided C is invertible (see Section 2.2.5), we have

$$\vec{v}_B = C^{-1} \vec{v}_A, \qquad (2.10)$$

which provides a practical method for changing coordinates. The above procedure is illustrated in the accompanying box.

Example: *Change of Coordinates*

Find the coordinates of the vector $\vec{v}_A = (-1, 2)^{\mathrm{T}}$ (represented with respect to the canonical basis) in the new basis defined by $B = \left\{\vec{b}_1 = (1, 1)^{\mathrm{T}}, \vec{b}_2 = (-2, 0)^{\mathrm{T}}\right\}$.

Solution:

Since matrix C is readily obtained as $C = \left[\begin{smallmatrix} 1 & -2 \\ 1 & 0 \end{smallmatrix}\right]$, we have:

$$\vec{v}_B = C^{-1}\vec{v}_A = \begin{bmatrix} 0 & 1 \\ -0.5 & 0.5 \end{bmatrix}\begin{bmatrix} -1 \\ 2 \end{bmatrix} = \begin{bmatrix} 2 \\ 1.5 \end{bmatrix}.$$

Figure 2.32 shows the representation of the above vector with respect to both considered bases.

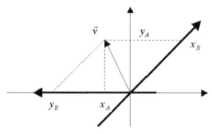

Figure 2.32: *The vector $\vec{v}$ represented with respect to both considered bases. The axes defined by bases A and B are represented by thin and thick arrows, respectively.*

The process of change of coordinates is particularly important since it allows us to obtain alternative, possibly simpler, representations of the same vector (as discrete signals are often represented). We will see in the remainder of this chapter that linear transformations (e.g., Fourier and Karhunen-Loève) provide interesting alternative representations of signals that are information preserving (in the sense that the inverse transform is always possible) and that allow information compaction, the latter being a consequence of the fact that the basis vectors in these transforms are similar to the signals typically found in nature (i.e., highly redundant or correlated).

2.2.3 | Linear Transformations

Given two vector spaces R and S, it is possible to define a mapping that assigns to every element of R a unique element in S. Such mappings, which are analogous to the concept of functions (Section 2.1.2), are called *transformations* and represented as $T: R \to S$. For instance, the transformation $T: \mathbb{R}^2 \to \mathbb{R}^2 \mid T(\vec{p}) = -\vec{p}$ maps each vector in $\mathbb{R}^2$ into the vector reflected with respect to the coordinate system origin. Many of the concepts discussed in Section 2.1.2 (i.e., domain, co-domain,

range, surjectivity, injectivity and bijectivity) can be immediately extended to transformations in vector spaces. A *linear transformation* is a transformation T such
that

$$T(\alpha \vec{p} + \beta \vec{q}) = \alpha T(\vec{p}) + \beta T(\vec{q}),$$

where α and β are generic scalars. It can be verified that *any linear transform can
be expressed in matrix form, and vice versa.* For instance, the above transform
$T_a: \mathbb{R}^2 \to \mathbb{R}^2 \mid T_a(\vec{p}) = -\vec{p}$ is linear and can be written as $T_a(\vec{p}) = A\vec{p}$, where

$$A = \begin{bmatrix} -1 & 0 \\ 0 & -1 \end{bmatrix}.$$

Its effect over the vector $(2, 3)^T$, therefore, is

$$T_a\left(\vec{p} = \begin{bmatrix} 2 \\ 3 \end{bmatrix}\right) = \begin{bmatrix} -1 & 0 \\ 0 & -1 \end{bmatrix}\begin{bmatrix} 2 \\ 3 \end{bmatrix} = \begin{bmatrix} -2 \\ -3 \end{bmatrix}.$$

Another example of linear transformation is given by $T_b: \mathbb{R}^3 \to \mathbb{R}^2, T_b(\vec{p}) = B\vec{p}$, where

$$B = \begin{bmatrix} 1 & 0 & 0 \\ 0 & 1 & 0 \end{bmatrix}.$$

It can be easily verified that this transformation maps a vector from the 3D
space $\mathbb{R}^3$ into its projection in the 2D space corresponding to the (x, y) plane. Such
situations are best characterized in terms of the *rank of the transformation*, which
corresponds to the dimension of the range of T (see Section 2.2.5). This value
can be verified to be equal to the number of linearly independent vectors defined
by each of the rows (or columns) of the transformation matrix. For instance, the
transformation defined by the following matrix:

$$D = \begin{bmatrix} 1 & 0 & 2 \\ 0 & 1 & 2 \\ 1 & 1 & 4 \end{bmatrix}$$

has *rank* 2, because the vector defined by the third row, i.e., $(1, 1, 4)^T$, is equal to
the sum of the vectors defined by the first two rows (i.e., it is a linear combination
of the other two rows), and will map vectors from $\mathbb{R}^3$ into vectors in $\mathbb{R}^2$. Such mappings, sometimes called *degenerated*, are not invertible. A practical procedure for
determining the rank of a matrix is presented in Section 2.2.5. A linear transformation such that the dimension of its domain is equal to the dimension of its range
is called a *full rank* transformation (in this case its rank is equal to the dimension
of the domain); otherwise the transformation is called *rank deficient*. In addition,
in case the rank of the transformation, which is equal to the rank of the respective
matrix, is equal to the dimension of the domain, the transformation also can be
verified to be full rank. While non-square matrices are necessarily rank-deficient, a
square matrix can be either rank deficient or full rank.

Given a linear transformation $T: R \rightarrow S \mid T(\vec{p}) = A\vec{p}$, its *inverse* can be defined as $T^{-1}: S \rightarrow R \mid \vec{p} = T^{-1}(T(\vec{p})) = A^{-1}T(\vec{p})$, where A^{-1} is the inverse of matrix A. It is clear that the inverse transformation is only possible if the inverse matrix A^{-1} exists, which means we can say that A is not a *singular* matrix. Only square matrices have respective inverses, and these can be proved to be unique. It can be verified that the inverse of a linear transformation is itself a linear transformation. In the two above examples, the inverse of T_b does not exist, but the inverse of T_a exists and is given in terms of the transformation matrix:

$$A^{-1} = \begin{bmatrix} -1 & 0 \\ 0 & -1 \end{bmatrix}.$$

Linear transforms taking vectors from an N-dimensional space into an M-dimensional space, $M < N$, (i.e., transformations which are not full rank) are said to be *degenerated*, and to find its inverse in this case is impossible. It should be observed at this early stage of the book that this type of transformation characterizes a large number of practical situations in shape analysis and vision. For instance, the 2D projections of the 3D world falling onto our retinas (or onto a digital camera) provide but a degenerate representation of the 3D imaged objects.

An important class of linear transformation is that implementing *rotations*. Figure 2.33 illustrates such a situation with respect to a vector $\vec{v}$ in the plane pointing at a point P, where the new and old coordinate systems are represented by full and dotted axes, respectively. It should be observed that rotating the old system by an angle θ (counterclockwise, with respect to the x-axis) corresponds to rotating vector $\vec{v}$, with respect to the coordinate system, by an angle $-\theta$. Consequently, both these problems can be treated in the same unified way.

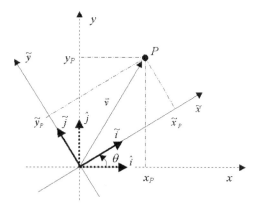

Figure 2.33: *Rotations of the coordinate system can be implemented by a specific class of linear transformations.*

The matrix representing the linear transformation, which rotates the coordinate system of the two-dimensional space $\mathbb{R}^2$ by an angle θ counterclockwise, is immediately obtained by using the coordinates exchange procedure discussed in Section 2.2.2. We start by expressing the basis vectors of the new space, i.e., $\widetilde{B} = \{\tilde{\imath}, \tilde{\jmath}\}$, in terms of the vectors of the old basis $\hat{B} = \{\hat{\imath}, \hat{\jmath}\}$:

$$\tilde{\imath} = \cos(\theta)\,\hat{\imath} + \sin(\theta)\,\hat{\jmath}$$

and

$$\tilde{\jmath} = -\sin(\theta)\,\hat{\imath} + \cos(\theta)\,\hat{\jmath},$$

which implies the following matrix:

$$C = \begin{bmatrix} \cos(\theta) & -\sin(\theta) \\ \sin(\theta) & \cos(\theta) \end{bmatrix}.$$

Thus, the sought transformation can be expressed as

$$P: \mathbb{R}^2 \to \mathbb{R}^2, T(\vec{v}) = C^{-1}\vec{v}.$$

Now, the inverse matrix C^{-1} can be easily obtained by taking into account the fact that matrices representing rotations are necessarily orthogonal (see Section 2.2.5). Since matrix A is orthogonal if $A^{-1} = A^{\mathrm{T}}$ (i.e., $AA^{\mathrm{T}} = I$), we have

$$P(\vec{v}) = C^{-1}\vec{v} = C^{\mathrm{T}}\vec{v} = \begin{bmatrix} \cos(\theta) & \sin(\theta) \\ -\sin(\theta) & \cos(\theta) \end{bmatrix}\vec{v}.$$

A generic transformation, in the sense of not being necessarily linear, taking vectors from a space S into vectors in this same space S, i.e., $T: S \to S$, is sometimes called an *operator*. A generic transformation taking vectors from a vector space S into scalars in a field F, i.e., $T: S \to F$, is called a *functional*.

2.2.4 Metric Spaces, Inner Products, and Orthogonality

So far, we have discussed vector spaces without worrying about the magnitude of vectors and how they can be measured, or defining distances between vectors. In this section we introduce the concepts of norm of a vector and inner products and distances between vectors, and consider some important vector spaces allowing such concepts.

Let S be a vector space. A transformation $T: S \to R, T(\vec{p}) = \|\vec{p}\|$ (i.e., a functional) obeying the following conditions is called a *norm* of S:

① $\|\vec{p}\| \geqslant 0$ for any $\vec{p} \in S$, and $\|\vec{p}\| = 0 \Leftrightarrow \vec{p} = \vec{0}$;

② $\|\alpha\vec{p}\| = |\alpha|\,\|\vec{p}\|$ for any $\vec{p} \in S$ and any real value α;

③ $\left\| \vec{p} + \vec{q} \right\| \leqslant \left\| \vec{p} \right\| + \left\| \vec{q} \right\|$ for any $\vec{p}, \vec{q} \in S$.

A vector space admitting a norm is called a *normed space*. A vector space can have many valid norms, and a vector with unit norm is said to be a *unit vector*. Every nonzero vector $\vec{p}$ defines the unique respective unit vector $\frac{\vec{p}}{\|\vec{p}\|}$. Normed spaces are more interesting than vector spaces without a norm because this concept allows us to measure the magnitude of the vectors in those spaces. Indeed, every normed space is also a *metric space*, in the sense that at least the distance $\text{dist}\{\vec{p}_1, \vec{p}_2\} = \|\vec{p}_1 - \vec{p}_2\| \geqslant 0$ can be defined. More formally, a metric space associated with a specific distance is a vector space where the following conditions are met:

① $\text{dist}\{\vec{p}_1, \vec{p}_2\} = 0 \Leftrightarrow \vec{p}_1 = \vec{p}_2$;

② $\text{dist}\{\vec{p}_1, \vec{p}_2\} = \text{dist}\{\vec{p}_2, \vec{p}_1\}$ (symmetry);

③ $\text{dist}\{\vec{p}_1, \vec{p}_3\} \leqslant \text{dist}\{\vec{p}_1, \vec{p}_2\} + \text{dist}\{\vec{p}_2, \vec{p}_3\}$ (triangular inequality).

The triangular inequality in $\mathbb{R}^2$ can be illustrated as shown in Figure 2.34.

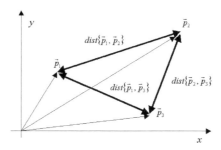

Figure 2.34: *Graphical illustration of the triangular inequality.*

It should be observed that not every metric space is a normed space. The *inner* (or *scalar*) *product* $\langle \vec{p}, \vec{q} \rangle$ (sometimes also represented as $\vec{p}.\vec{q}$) between two vectors $\vec{p}$ and $\vec{q}$, resulting a scalar (real or complex), is a transformation $T : S \times S \rightarrow F$, $T(\vec{p}, \vec{q}) = \langle \vec{p}, \vec{q} \rangle$ where the following conditions are verified:

① $\langle \vec{p}, \vec{q} \rangle = \langle \vec{q}, \vec{p} \rangle^*$ (commutative for real vectors);

② $\langle \vec{p} + \vec{q}, \vec{r} \rangle = \langle \vec{p}, \vec{r} \rangle + \langle \vec{q}, \vec{r} \rangle$;

③ $\langle \alpha \vec{p}, \vec{q} \rangle = \alpha^* \langle \vec{p}, \vec{q} \rangle$ (observe that $\langle \vec{p}, \alpha \vec{q} \rangle = \langle \alpha \vec{q}, \vec{p} \rangle^* = \alpha \langle \vec{p}, \vec{q} \rangle$);

④ $\langle \vec{p}, \vec{p} \rangle \geqslant 0$ and $\langle \vec{p}, \vec{p} \rangle = 0 \Leftrightarrow \vec{p} = \vec{0}$.

It is easy to see that a space allowing an inner product is also a normed and metric space, since we can always define a norm in terms of the inner product by making $\|\vec{p}\| = + \sqrt{\langle \vec{p}, \vec{p} \rangle}$.

Although not every vector space is normed, metric or allows inner products, most developments in the present book deal with concepts related to vector spaces with inner products. The box entitled *Metrics in* $\mathbb{C}^N$ exemplifies some of the valid norms, distances and inner products in those spaces.

Example: *Metrics in* $\mathbb{C}^N$

Let $\vec{p} = (p_1, p_2, \ldots, p_N)^{\mathrm{T}}$ and $\vec{q} = (q_1, q_2, \ldots, q_N)^{\mathrm{T}}$ be two generic vectors in the N-dimensional complex space $\mathbb{C}^N$.

☞ Norms of $\vec{p}$:

Euclidean: $\left\| \vec{p} \right\|_2 = \sqrt{p_1^2 + p_2^2 + \cdots + p_N^2}$;

City-block: $\left\| \vec{p} \right\|_1 = |p_1| + |p_2| + \cdots + |p_N|$;

Chessboard: $\left\| \vec{p} \right\|_\infty = \max\{|p_1|, |p_2|, \ldots, |p_N|\}$.

☞ Distances between $\vec{p}$ and $\vec{q}$:

Euclidean: $\left\| \vec{p} - \vec{q} \right\|_2 = \sqrt{(p_1 - q_1)^2 + (p_2 - q_2)^2 + \cdots + (p_N - q_N)^2}$;

City-block: $\left\| \vec{p} - \vec{q} \right\|_1 = |p_1 - q_1| + |p_2 - q_2| + \cdots + |p_N - q_N|$;

Chessboard: $\left\| \vec{p} - \vec{q} \right\|_\infty = \max\{|p_1 - q_1|, |p_2 - q_2|, \ldots, |p_N - q_N|\}$.

☞ Inner product between $\vec{p}$ and $\vec{q}$:

$$\langle \vec{p}, \vec{q} \rangle = \vec{p}.\vec{q} = \left(\vec{p}^{\mathrm{T}} \right)^* \vec{q} = p_1^* q_1 + p_2^* q_2 + \cdots + p_N^* q_N.$$

Example: *Norms, Distances and Inner Products in Function Spaces*

Let $f(t)$ and $g(t)$ be two generic functions in the space of the continuous functions in the interval $[a, b]$, i.e., $C[a, b]$.

☞ Norm of $f(t)$

$$\|f(t)\| = \sqrt{\int_a^b f^*(t)\, f(t)\, \mathrm{d}t} = \sqrt{\int_a^b |f(t)|^2\, \mathrm{d}t}.$$

☞ Distance between $f(t)$ and $g(t)$

$$\mathrm{dist}\{f(t), g(t)\} = \|f(t) - g(t)\| = \sqrt{\int_a^b |f(t) - g(t)|^2 \, dt}.$$

☞ Inner product between $f(t)$ and $g(t)$

$$\langle f, g \rangle = \int_a^b f^*(t) \, g(t) \, dt.$$

A particularly interesting interpretation of the inner product between two real vectors is in terms of *projections*. Consider the following illustration involving two vectors, $\vec{p}$ and $\vec{q}$, in $\mathbb{R}^2$.

Since $\langle \vec{p}, \vec{q} \rangle = \|\vec{p}\| \cdot \|\vec{q}\| \cos(\theta)$, the projection L (a scalar) of the vector $\vec{p}$ along the orientation defined by the vector $\vec{q}$ can be verified to be equal to the inner product between the vector $\vec{p}$ and the unitary vector $\frac{\vec{q}}{\|\vec{q}\|}$, i.e.

$$L = \|\vec{p}\| \cdot \left\| \frac{\vec{q}}{\|\vec{q}\|} \right\| \cos(\theta)$$

$$= \|\vec{p}\| \cos(\theta)$$

$$= \left\langle \vec{p}, \frac{\vec{q}}{\|\vec{q}\|} \right\rangle.$$

Figure 2.35 illustrates these concepts. An alternative, but equally interesting, interpretation of the inner product is as a measure of *similarity* between the orientations of the vectors $\vec{p}$ and $\vec{q}$. It can be verified that the maximum value of $\langle \vec{p}, \vec{q} \rangle$ is achieved when both vectors are parallel and have the same sense. On the other hand, the smallest value is achieved when these vectors are parallel but have inverse senses. The inner product is zero whenever $\theta = 90°$, and in this case the vectors are said to be *orthogonal*. More generally, given a set of M vectors $\vec{p}_i$, $i = 1, 2, \ldots, M$, in a vector space S, these vectors (and the set) are said to be *orthogonal* if and only

$$\left\langle \vec{p}_i, \vec{p}_j \right\rangle = 0, \forall i, j = 1, 2, \ldots, M; i \neq j.$$

In addition, if $\|\vec{p}_i\| = \langle \vec{p}_i, \vec{p}_i \rangle = 1, \forall i = 1, 2, \ldots, M$, the vectors (and the set) are said to be *orthonormal*. It should be observed that orthonormality implies orthogonality, but not the other way round. Important examples of orthonormal sets

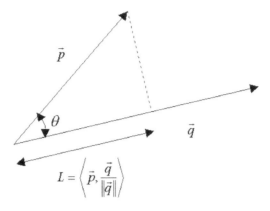

Figure 2.35: *The projection L of vector $\vec{p}$ over vector $\vec{q}$ can be expressed in terms of a scalar product.*

of vectors are the canonical vectors in $\mathbb{R}^N$ and the normalized complex exponential functions.

It is interesting to interpret a linear transformation involving finite dimensional spaces in terms of inner products. Indeed, each coordinate q_i of the resulting vector $\vec{q}$ in $\vec{q} = A\vec{p}$ can be understood as the inner product between the i-th row of the matrix A and the vector $\vec{p}$:

$$\vec{q} = A\vec{p} \Rightarrow \begin{bmatrix} q_1 \\ q_2 \\ \vdots \\ q_i \\ \vdots \\ q_N \end{bmatrix} = \begin{bmatrix} a_{1,1} & a_{1,2} & \cdots & a_{1,i} & \cdots & a_{1,N} \\ a_{2,1} & a_{2,2} & \cdots & a_{2,i} & \cdots & a_{2,N} \\ \vdots & \vdots & \ddots & \vdots & \vdots & \vdots \\ a_{i,1} & a_{i,2} & \cdots & a_{i,i} & \cdots & a_{i,N} \\ \vdots & \vdots & \vdots & \vdots & \ddots & \vdots \\ a_{N,1} & a_{N,2} & \cdots & a_{N,i} & \cdots & a_{N,N} \end{bmatrix} \begin{bmatrix} p_1 \\ p_2 \\ \vdots \\ p_i \\ \vdots \\ p_N \end{bmatrix}$$

$$\Rightarrow q_i = \begin{bmatrix} a_{i,1} & a_{i,2} & \cdots & a_{i,i} & \cdots & a_{i,N} \end{bmatrix} \begin{bmatrix} p_1 \\ p_2 \\ \vdots \\ p_i \\ \vdots \\ p_N \end{bmatrix}.$$

Consequently, the elements of the vector resulting from a linear transformation can be understood as a measure of similarity between the orientations of the input vector $\vec{p}$ and each of the respective vectors defined by the rows of the transformation matrix. This interpretation is essential for the full conceptual understanding of several properties in signal and image transforms, including the Fourier and Karhunen-Loève transforms.

2.2.5 More about Vectors and Matrices

We have thus far limited our discussion of vectors and matrices as elements of vector spaces, and as representations of linear transforms. This section provides additional concepts and properties including the more general cases of complex vectors and matrices, i.e., those vectors and matrices having complex numbers as elements, such as

$$\vec{v} = \begin{bmatrix} 2 \\ -j \\ 1+j \\ 3 \end{bmatrix}, \qquad A = \begin{bmatrix} 1 & 0 & 3j \\ -j & 3 & -1-j \\ 0 & 2+j & 2 \end{bmatrix}.$$

Some Basic Concepts

The *null* $N \times N$ matrix, henceforth represented as Φ, is a matrix having all elements equal to zero. A matrix A having dimension $N \times N$ is said to be a *square matrix*. Its *main diagonal* corresponds to the elements $a_{i,i}$, $i = 1, 2, \ldots, N$. A square matrix having all the elements below (above) its main diagonal equal to zero is said to be an *upper (lower) triangular matrix*, as illustrated in the following:

$$A = \begin{bmatrix} 9 & 0 & 2j \\ 0 & 3 & -1+j \\ 0 & 0 & 5 \end{bmatrix}$$

is upper triangular, and

$$B = \begin{bmatrix} 9 & 0 & 0 \\ 0 & 3+2j & 0 \\ 2-j & 4 & 5 \end{bmatrix}$$

is lower triangular.

The *identity* matrix, represented as I, is a square matrix having ones along its main diagonal and zeros elsewhere. For example, the 3×3 identity matrix is

$$I = \begin{bmatrix} 1 & 0 & 0 \\ 0 & 1 & 0 \\ 0 & 0 & 1 \end{bmatrix}.$$

The *complex conjugate* of a matrix (vector) is obtained by taking the complex conjugate of each of its elements. For instance, the complex conjugate of the above matrix B is

$$B^* = \begin{bmatrix} 9 & 0 & 0 \\ 0 & 3 - 2j & 0 \\ 2 + j & 4 & 5 \end{bmatrix}.$$

The *derivative of a matrix* is given by the derivatives of each of its components. For instance, if

$$A = \begin{bmatrix} 1 & 3t^3 + t & \cos(j2\pi t) \\ j & \sin(2t) & t \\ 2t + 3 & 2 - j & -t^2 \end{bmatrix},$$

then

$$\frac{dA}{dt} = \begin{bmatrix} 0 & 9t^2 + 1 & -(j2\pi)\sin(j2\pi t) \\ 0 & 2\cos(2t) & 1 \\ 2 & 0 & -2t \end{bmatrix}.$$

Given an $N \times M$ matrix A and an $M \times N$ matrix B, the *product* between A and B, indicated as $C = AB$, is defined as

$$c_{i,j} = \sum_{k=1}^{M} a_{i,k} b_{k,j}. \tag{2.11}$$

Given an $N \times N$ matrix A, its *quadratic form* is defined by the polynomial

$$Q(x_1, x_2, \ldots, x_N) = \vec{x}^T A \vec{x}$$

where $\vec{x} = (x_1, x_2, \ldots, x_N)$. For instance, if $A = \begin{bmatrix} 2 & -1 \\ 1 & 3 \end{bmatrix}$, then

$$Q(x_1, x_2) = \begin{bmatrix} x_1 & x_2 \end{bmatrix} \begin{bmatrix} 2 & -1 \\ 1 & 3 \end{bmatrix} \begin{bmatrix} x_1 \\ x_2 \end{bmatrix} = \begin{bmatrix} x_1 & x_2 \end{bmatrix} \begin{bmatrix} 2x_1 - x_2 \\ x_1 + 3x_2 \end{bmatrix} = 2x_1^2 + 3x_2^2.$$

If $Q(x_1, x_2, \ldots, x_N) = \vec{x}^T A \vec{x} > 0$, for any $\vec{x}$ except for $\vec{x} = \vec{0}$, then matrix A is said to be *positive definite*. In case $Q(x_1, x_2, \ldots, x_N) = \vec{x}^T A \vec{x} \geqslant 0$, then A is called *positive semidefinite*.

Some Important Properties

Complex matrices obey the following properties, where A, B, and C are any complex matrices and a and b are scalar values:

$$A + B = B + A \qquad \text{(addition is } commutative\text{)}$$
$$A + (B + C) = (A + B) + C \quad \text{(addition is } associative\text{)}$$
$$a(A + B) = aA + aB \qquad \text{(distributive law)}$$
$$(a + b)A = aA + bA$$

$$\text{usually } AB \neq BA \qquad \text{(multiplication is } not\ commutative\text{)}$$
$$A(BC) = (AB)C \qquad \text{(multiplication is } associative\text{)}$$
$$a(AB) = (aA)B = A(aB)$$
$$A(B + C) = AB + AC \qquad (distributive\ law)$$

$$(A + B)^T = A^T + B^T$$
$$(AB)^T = B^T A^T$$
$$(aA)^* = a^* A^*$$
$$(A^*)^T = \left(A^T\right)^*$$
$$(A + B)^* = A^* + B^*$$
$$(AB)^* = A^* B^*$$

If $A^T = A$	then A is said to be *symmetric*
If $A = (A^*)^T$	then A is said to be *Hermitian*
If $A^T = -A$	then A is said to be *skew symmetric*
If $AA = A$	then A is said to be *idempotent*
If $A^* A = AA^*$	then A is said to be *normal*
If $A^T = A^{-1}$	then A is said to be *orthogonal*
If $(A^*)^T = A^{-1}$	then A is said to be *unitary*

Determinants

The determinant is a transformation that associates a scalar value to every square matrix. The *determinant of a 2 × 2 matrix* can be calculated as follows:

$$\det\left\{\begin{bmatrix} a_{1,1} & a_{1,2} \\ a_{2,1} & a_{2,2} \end{bmatrix}\right\} = \begin{vmatrix} a_{1,1} & a_{1,2} \\ a_{2,1} & a_{2,2} \end{vmatrix} = a_{1,1}a_{2,2} - a_{1,2}a_{2,1}.$$

The *determinant of a 3 × 3 matrix* is

$$\det\left\{\begin{bmatrix} a_{1,1} & a_{1,2} & a_{1,3} \\ a_{2,1} & a_{2,2} & a_{2,3} \\ a_{3,1} & a_{3,2} & a_{3,3} \end{bmatrix}\right\} = \begin{vmatrix} a_{1,1} & a_{1,2} & a_{1,3} \\ a_{2,1} & a_{2,2} & a_{2,3} \\ a_{3,1} & a_{3,2} & a_{3,3} \end{vmatrix}$$

$$= a_{1,1}a_{2,2}a_{3,3} + a_{2,1}a_{3,2}a_{1,3} + a_{3,1}a_{1,2}a_{2,3} - a_{1,3}a_{2,2}a_{3,1} - a_{2,3}a_{3,2}a_{1,1} - a_{3,3}a_{1,2}a_{2,1}.$$

Some properties of the determinant of an $N \times N$ matrix A are listed in the following:

① $\det(AB) = \det(A)\det(B)$;

② if A is an upper or lower triangular matrix, then $\det(A) = a_{1,1}a_{2,2}\ldots a_{N,N}$;

③ $\det\left(A^{\mathrm{T}}\right) = \det(A)$;

④ $\det(aA) = a^{N}\det(A)$;

⑤ the vectors defined by each row of A are linearly dependent if and only $\det(A) = 0$;

⑥ if all the elements in one of the rows (or columns) of A are zero, then $\det(A) = 0$;

⑦ if A is orthogonal (unitary), then $|\det(A)| = 1$.

A matrix whose determinant is zero, which is called *singular*, is not invertible. If A is a square matrix defining a linear transformation, its determinant $\det(A)$ measures the magnification of the area (in the case of $\mathbb{R}^2$) or volume (in the case of $\mathbb{R}^3$) implemented by the transformation. Consider the rectangle with area $area_1$ in Figure 2.36 (a). When the vectors representing each of the vertices of this rectangle are transformed by a linear transformation T represented by the square matrix A, each of the respectively obtained vectors define the vertices of the polyhedra shown in (b), having area $area_2$. The determinant of A indicates how much the area has been scaled, i.e., the $\det(A) = \frac{area_2}{area_1}$. In the case of linear transformations on $\mathbb{R}^3$, we have $\det(A) = \frac{volume_2}{volume_1}$.

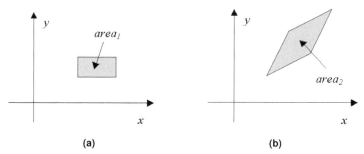

Figure 2.36: *The transformation T, defined by the square matrix A, taking a rectangle in* (a) *into a polygon in* (b). *The determinant of A indicates the area magnification.*

A matrix whose rows are formed by linearly dependent vectors necessarily has null determinant and is singular. In case the determinant is non-null, the vectors corresponding to the rows are linearly independent. For instance, if $\vec{v}_1 = (1, 2, -1)$, $\vec{v}_2 = (1, 0, 0)$, and $\vec{v}_3 = (2, 2, -1)$, then

$$\begin{vmatrix} 1 & 2 & -1 \\ 1 & 0 & 0 \\ 2 & 2 & -1 \end{vmatrix} = 0 + 0 - 2 - 0 - 0 + 2 = 0,$$

indicating that the three vectors are linearly dependent.

Orthogonality and Rotations

This section considers vectors in $\mathbb{R}^2$ and $\mathbb{R}^3$. A linear transformation having an orthogonal transformation matrix A is called an *orthogonal transformation*. In this case, we necessarily have $|\det(A)| = 1$, which means that the area (or volume) is conserved by the transformation. It is important to observe that orthogonal transformations are particularly important since they also preserve the norm of the transformed vectors. In addition, in case $\det(A) = 1$ the transformation does not change the structure of the coordinate system (i.e., the relative senses of the coordinate axes are preserved) and we have a *special orthogonal transformation*. For instance, the following transformation matrix:

$$A = \begin{bmatrix} 1 & 0 & 0 \\ 0 & \cos(\theta) & -\sin(\theta) \\ 0 & \sin(\theta) & \cos(\theta) \end{bmatrix}$$

has determinant

$$\det(A) = \begin{vmatrix} 1 & 0 & 0 \\ 0 & \cos(\theta) & -\sin(\theta) \\ 0 & \sin(\theta) & \cos(\theta) \end{vmatrix} = \cos^2(\theta) + \sin^2(\theta) = 1$$

and consequently does not change the structure of the coordinate system, as illustrated in Figure 2.37.

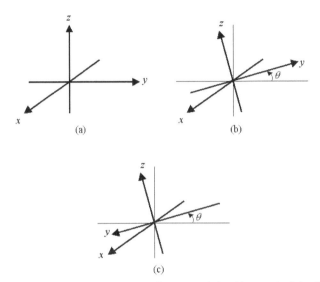

(a)

(b)

(c)

Figure 2.37: *The orthogonal transformation defined by matrix A implements a rotation of the canonical coordinate system* (a) *in the y − z plane by and angle of θ, with respect to the y-axis. The fact that* $\det(A) = 1$ *indicates that the senses of all axes are preserved. This is not verified for the transformation defined by matrix B* (c).

Therefore, the transformation implemented by A, i.e., $T : \vec{v} \in \mathbb{R}^3 \to \vec{q} \in \mathbb{R}^3 \mid \vec{q} = A\vec{v}$, is a *special orthogonal transformation*. However, the linear transformation defined by matrix:

$$B = \begin{bmatrix} 1 & 0 & 0 \\ 0 & -\cos(\theta) & \sin(\theta) \\ 0 & \sin(\theta) & \cos(\theta) \end{bmatrix}, \tag{2.12}$$

which implies $\det(B) = -1$, is not a special orthogonal transformation, because the sense of the y-axis is reversed, as shown in Figure 2.37 (c). Rotation transformations can be bijectively associated to *special orthogonal transformations*, i.e., to every rotation corresponds a special orthogonal transformation, and vice versa.

Vector Product

An important concept is that of the *vector* (or *cross*) *product*. Given two vectors $\vec{v} = (v_1, v_2, v_3)$ and $\vec{r} = (r_1, r_2, r_3)$ in $\mathbb{R}^3$, their vector product $\vec{v} \wedge \vec{r}$, which is itself a vector, can be calculated in terms of the following determinant:

$$\vec{v} \wedge \vec{r} = \begin{vmatrix} \hat{\imath} & \hat{\jmath} & \hat{k} \\ v_1 & v_2 & v_3 \\ r_1 & r_2 & r_3 \end{vmatrix} = (v_2 r_3 - v_3 r_2)\,\hat{\imath} + (v_3 r_1 - v_1 r_3)\,\hat{\jmath} + (v_1 r_2 - v_2 r_1)\,\hat{k}$$

where $\hat{\imath} = (1, 0, 0)$, $\hat{\jmath} = (0, 1, 0)$, and $\hat{k} = (0, 0, 1)$ are the elements of the canonical basis of $\mathbb{R}^3$. For instance, if $\vec{v} = (1, 1, 1)$ and $\vec{r} = (1, 2, 0)$, then

$$\vec{v} \wedge \vec{r} = \begin{vmatrix} \vec{\imath} & \hat{\jmath} & \hat{k} \\ 1 & 1 & 1 \\ 1 & 2 & 0 \end{vmatrix} = (0 - 2)\,\hat{\imath} + (1 - 0)\,\hat{\jmath} + (2 - 1)\,\hat{k} = -2\hat{\imath} + \hat{\jmath} + \hat{k}$$

It can be easily verified that the vector product presents the following properties:

① $\langle \vec{v} \wedge \vec{r}, \vec{v} \rangle = \langle \vec{v} \wedge \vec{r}, \vec{r} \rangle = 0$;

② $\vec{r} \wedge \vec{v} = -\vec{v} \wedge \vec{r}$;

③ $\vec{v} \wedge (a\vec{v}) = 0, a \in \mathbb{R}$;

④ $\hat{\imath} \wedge \hat{\jmath} = \hat{k}$;

⑤ $\hat{\jmath} \wedge \hat{k} = \hat{\imath}$;

⑥ $\hat{k} \wedge \hat{\imath} = \hat{\jmath}$.

The vector $\vec{q} = \vec{v} \wedge \vec{r}$ is therefore normal to the plane spanned (see Section 2.2.2) by vectors $\vec{v}$ and $\vec{r}$. The sense of the vector $\vec{q}$ can be obtained by the *corkscrew rule*: turn a hypothetic corkscrew from the tip of $\vec{v}$ towards the tip of $\vec{r}$, as illustrated in Figure 2.38, and the advance of the corkscrew will correspond to the sense of $\vec{q} = \vec{v} \wedge \vec{r}$.

Rank and Echelon Form of Matrices

Any $N \times M$ matrix A can be placed in *echelon form*, which is useful for solving linear systems and calculating the determinant and rank of a matrix. For a square matrix A, this process can be understood as aimed at obtaining an upper triangular

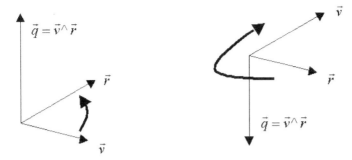

Figure 2.38: *Examples of vector products.*

version of A. It should be observed that a matrix is not typically equal to its echelon form. A practical procedure for placing a matrix into this form consists of applying the operations below (called *elementary operations*) until all the elements below $a_{i,i}$ are zero:

① multiply any row of the matrix by a nonzero constant a;

② add any two rows;

③ interchange any two rows.

This process is illustrated in the accompanying box. It is clear that there are many (actually infinite) valid echelon forms for the same matrix. An interesting property of the echelon form is that the *rank of a matrix A*, indicated as rank (A), and hence of the linear transformation it represents (see Section 2.2.3), corresponds to the number of nonzero rows (or columns) in any of the echelon forms of A. In the same example, the rank of the matrix A is 2, which implies that the respectively implemented linear transformation will take vectors from $\mathbb{R}^3$ into vectors in $\mathbb{R}^2$. An algorithm for calculating echelon forms of a matrix can be found, for instance, in [Kreyszig, 1993].

Example: *Echelon Form*

Place the matrix $A = \begin{bmatrix} 1 & 3 & 2 \\ 0 & 2 & -1 \\ 1 & 7 & 0 \end{bmatrix}$ in echelon form.

Solution:

① Multiply the first row by -1 and add to the third row:

$$
\begin{bmatrix} 1 & 3 & 2 \\ 0 & 2 & -1 \\ 1 & 7 & 0 \end{bmatrix} \rightarrow \begin{bmatrix} 1 & 3 & 2 \\ 0 & 2 & -1 \\ 0 & 4 & -2 \end{bmatrix}.
$$

② Multiply the second row by -2 and add to the third row:

$$
\begin{bmatrix} 1 & 3 & 2 \\ 0 & 2 & -1 \\ 0 & 4 & -2 \end{bmatrix} \rightarrow \begin{bmatrix} 1 & 3 & 2 \\ 0 & 2 & -1 \\ 0 & 0 & 0 \end{bmatrix}.
$$

③ As all elements below $a_{i,i}$ are zero, the matrix is in echelon form, and its rank is equal to 2.

Hadamard Product

Given two matrices A and B with the same dimension $N \times M$, the *Hadamard* (or *Schur*) *product* between these two matrices, represented as $A \otimes B$, is an $N \times M$ matrix C such that $c_{i,j} = a_{i,j} b_{i,j}$. For instance, if

$$
A = \begin{bmatrix} 3 & -j & 1 \\ -6 & 3 & 2 \\ 0 & 2 & 1+j \end{bmatrix} \quad \text{and} \quad B = \begin{bmatrix} 3 & 10 & 2 \\ -2 & -2 & 1 \\ 1 & 5 & 2 \end{bmatrix},
$$

then $A \otimes B = \begin{bmatrix} (3)(3) & (-j)(10) & (1)(2) \\ (-6)(-2) & (3)(-2) & (2)(1) \\ (0)(1) & (2)(5) & (1+j)(2) \end{bmatrix} = \begin{bmatrix} 9 & -10j & 2 \\ 12 & -6 & 2 \\ 0 & 10 & 2+2j \end{bmatrix}$.

The properties of the Hadamard product include the following:

① $A \otimes B = B \otimes A$

② $(A \otimes B)^* = A^* \otimes B^*$;

③ $(A \otimes B)^T = A^T \otimes B^T$;

④ $A \otimes (B \otimes C) = (A \otimes B) \otimes C$;

⑤ $A \otimes (B + C) = (A \otimes B) + (A \otimes C)$.

Inverse and Pseudoinverse of a Matrix

The *inverse* of an $N \times N$ matrix A is the $N \times N$ matrix A^{-1} such that $A^{-1}A = AA^{-1} = I$. Three properties of the inverse matrix are listed in the following:

① $(AB)^{-1} = B^{-1}A^{-1}$;

② $\det\left(A^{-1}\right) = \frac{1}{\det(A)}$;

③ $\left(A^{\mathrm{T}}\right)^{-1} = \left(A^{-1}\right)^{\mathrm{T}}$.

A matrix A of dimension $N \times N$ is *invertible* if and only if its determinant $\det(A)$ is nonzero, i.e., the matrix is nonsingular.

Now, consider the matrix equation

$$A\vec{x} = \vec{b}$$

where $\vec{x} = (x_1, x_2, \ldots, x_N)$ is a vector of unknown values, and $\vec{b} = (b_1, b_2, \ldots, b_N)$ is a constant vector. It can be easily verified that this matrix equation represents the following familiar system of linear equations:

$$\begin{cases} a_{1,1}x_1 + a_{1,2}x_2 + \cdots + a_{1,N}x_N = b_1 \\ a_{2,1}x_1 + a_{2,2}x_2 + \cdots + a_{2,N}x_N = b_2 \\ \qquad\qquad\qquad \vdots \\ a_{N,1}x_1 + a_{N,2}x_2 + \cdots + a_{N,N}x_N = b_N \end{cases}$$

In case a unique solution to such a system exists, it can be obtained by

$$\vec{x} = A^{-1}\vec{b}. \tag{2.13}$$

Methods for determining the inverse of a matrix can be found, for instance, in [Jennings, 1977] and readily available in many mathematical applicatives such as MATLAB® and Mathematica®.

Consider now an $M \times N$ matrix C, with $M > N$, and the matrix equation:

$$C\vec{x} = \vec{b}, \tag{2.14}$$

where $\vec{x} = (x_1, x_2, \ldots, x_N)$ is a vector of unknown values (i.e., variables), and $\vec{b} = (b_1, b_2, \ldots, b_M)$ is a constant vector. Observe that this situation corresponds to a system having more equations than variables, which is called *overdetermined*. By multiplying (at the left side) both sides of the above equation by $\left(C^{\mathrm{T}}\right)^*$, we obtain

$$\left(C^{\mathrm{T}}\right)^* C\vec{x} = \left(C^{\mathrm{T}}\right)^* \vec{b}. \tag{2.15}$$

Since $\left(C^{T}\right)^{*} C$ is a square matrix with dimension $N \times N$, it is now possible to consider its inverse, which leads to the following solution for (2.14):

$$\vec{x} = \left(\left(C^{T}\right)^{*} C\right)^{-1} \left(C^{T}\right)^{*} \vec{b}. \tag{2.16}$$

It can be verified [Bronson, 1988; Jennings, 1977] that the above equation provides the best solution, in the minimum mean square root sense, of the overdetermined system in (2.14). Since the matrix $\left(\left(C^{T}\right)^{*} C\right)^{-1} \left(C^{T}\right)^{*}$ plays the role of the inverse matrix A^{-1} in (2.13), it is called the *pseudoinverse* (or *generalized inverse,* or *Moore-Penrose inverse*) of C. It should be observed that the pseudoinverse can be obtained for more general matrix (in fact, any matrix) by using a slightly more sophisticated procedure (see, for instance, [Bronson, 1988]).

Example: *Fitting Functions by Using the Pseudoinverse Matrix*

Find the cubic polynomial best fitting, in the least square sense, the following data:

x	y
−2.2483	−26.3373
1.4309	17.0237
3.4775	78.9667
0.1553	17.4468
−3.7234	−67.3074
0.2376	9.4529
−3.9384	−61.5023
−3.4653	−38.0609

Solution:

A generic cubic polynomial is defined by the equation

$$y = a_0 + a_1 x + a_2 x^2 + a_3 x^3$$

Therefore, we have to find the coefficients a_i. We have

$$
C = \begin{bmatrix}
1 & x_1 & x_1^2 & x_1^3 \\
1 & x_2 & x_2^2 & x_2^3 \\
1 & x_3 & x_3^2 & x_3^3 \\
1 & x_4 & x_4^2 & x_4^3 \\
1 & x_5 & x_5^2 & x_5^3 \\
1 & x_6 & x_6^2 & x_6^3 \\
1 & x_7 & x_7^2 & x_7^3 \\
1 & x_8 & x_8^2 & x_8^3
\end{bmatrix}
=
\begin{bmatrix}
1 & -2.2483 & 5.0550 & -11.3652 \\
1 & 1.4309 & 2.0475 & 2.9298 \\
1 & 3.4775 & 12.0933 & 42.0550 \\
1 & 0.1553 & 0.0241 & 0.0037 \\
1 & -3.7234 & 13.8639 & -51.6211 \\
1 & 0.2376 & 0.0565 & 0.0134 \\
1 & -3.9384 & 15.5111 & -61.0892 \\
1 & -3.4653 & 12.0080 & -41.6110
\end{bmatrix},
$$

$$
\vec{b} = \begin{bmatrix}
y_1 \\
y_2 \\
y_3 \\
y_4 \\
y_5 \\
y_6 \\
y_7 \\
y_8
\end{bmatrix}
=
\begin{bmatrix}
-26.3373 \\
17.0273 \\
78.9667 \\
17.4668 \\
-67.3074 \\
9.4529 \\
-61.5023 \\
-38.0609
\end{bmatrix},
$$

and

$$
\vec{x} = \begin{bmatrix}
a_0 \\
a_1 \\
a_2 \\
a_3
\end{bmatrix}.
$$

Applying (2.16) and observing that all data are real values

$$
\vec{x} = \left(C^{\mathrm{T}}C\right)^{-1}\left(C^{\mathrm{T}}\vec{b}\right) \Rightarrow
$$

$$
\Rightarrow \vec{x} =
\begin{bmatrix}
0.0008 & -0.0008 & 0.0061 & -0.0121 \\
-0.0008 & 0.0061 & -0.0121 & 0.0753 \\
0.0061 & -0.0121 & 0.0753 & -0.1706 \\
-0.0121 & 0.0753 & -0.1706 & 1.0034
\end{bmatrix}^{-1}
\begin{bmatrix}
-70 \\
988 \\
-1487 \\
12486
\end{bmatrix}
=
\begin{bmatrix}
6.9749 \\
9.5220 \\
0.6183 \\
0.7187
\end{bmatrix}.
$$

Consequently, the cubic polynomial best fitting the above data is

$$
y = 6.9749 + 9.522x + 0.6183x^2 + 0.7187x^3
$$

Figure 2.39 illustrates the original points (squares) and the obtained cubic polynomial.

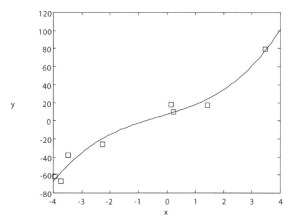

Figure 2.39: *The original points (squares) and the obtained cubic polynomial.*

The procedure illustrated in the above example can be generalized to any polynomial or function. As a matter of fact, in the particular case of straight line fitting, we have the general line equation

$$y = a_0 + a_1 x$$

and, therefore

$$C = \begin{bmatrix} 1 & x_1 \\ 1 & x_2 \\ \vdots & \vdots \\ 1 & x_N \end{bmatrix} ; \quad \vec{b} = \begin{bmatrix} y_1 \\ y_2 \\ \vdots \\ y_N \end{bmatrix} ; \quad \text{and} \quad \vec{x} = \begin{bmatrix} a_0 \\ a_1 \end{bmatrix} .$$

Applying (2.15)

$$\left(C^{\mathrm{T}} C \right) \vec{x} = C^{\mathrm{T}} \vec{b} \Rightarrow \left(\begin{bmatrix} 1 & 1 & \cdots & 1 \\ x_1 & x_2 & \cdots & x_N \end{bmatrix} \begin{bmatrix} 1 & x_1 \\ 1 & x_2 \\ \vdots & \vdots \\ 1 & x_N \end{bmatrix} \right) \begin{bmatrix} a_0 \\ a_1 \end{bmatrix} =$$

$$= \left(\left[\begin{array}{cccc} 1 & 1 & \cdots & 1 \\ x_1 & x_2 & \cdots & x_N \end{array} \right] \left[\begin{array}{c} y_1 \\ y_2 \\ \vdots \\ y_N \end{array} \right] \right) \Rightarrow \left[\begin{array}{cc} N & \sum_{i=1}^{N} x_i \\ \sum_{i=1}^{N} x_i & \sum_{i=1}^{N} x_i^2 \end{array} \right] \left[\begin{array}{c} a_0 \\ a_1 \end{array} \right] = \left[\begin{array}{c} \sum_{i=1}^{N} y_i \\ \sum_{i=1}^{N} x_i y_i \end{array} \right]$$

which implies

$$a_0 = \frac{\sum_{i=1}^{N} y_i \sum_{i=1}^{N} x_i^2 - \sum_{i=1}^{N} x_i \sum_{i=1}^{N} x_i y_i}{N \sum_{i=1}^{N} x_i^2 - \left(\sum_{i=1}^{N} x_i \right)^2}$$

and

$$a_1 = \frac{N \sum_{i=1}^{N} x_i y_i - \sum_{i=1}^{N} x_i \sum_{i=1}^{N} y_i}{N \sum_{i=1}^{N} x_i^2 - \left(\sum_{i=1}^{N} x_i \right)^2}$$

This pair of equations corresponds to the broadly known and useful *linear least squares approximation* (or *fitting*).

Eigenvalues and Eigenvectors

Given a square matrix with dimension $N \times N$ and the respectively defined linear transformation $A\vec{x}$, we can write the following equation:

$$A\vec{x} = \lambda\vec{x}, \tag{2.17}$$

where λ is a complex scalar value. In this equation, $\vec{x}$ corresponds to one or more vectors whose orientation is not affected by the linear transformation defined by A. In this case $\vec{x}$ is called an *eigenvector* of A. Observe that $a\vec{x}$, where a is any complex value, is also an eigenvector since:

$$A(a\vec{x}) = \lambda(a\vec{x}) \Leftrightarrow aA\vec{x} = a\lambda\vec{x} \Leftrightarrow A\vec{x} = \lambda\vec{x}$$

Let us now rewrite (2.17) as

$$A\vec{x} = \lambda\vec{x} \Leftrightarrow A\vec{x} - \lambda\vec{x} = \vec{0} \Leftrightarrow (A - \lambda I)\vec{x} = \vec{0}$$

This is immediately recognized to be a homogeneous system of linear equations, which is known to have non-trivial solution (i.e., $\vec{x} \neq \vec{0}$) if and only $\det(A - \lambda I) = 0$. This equation, which is a polynomial on λ, is known as the *characteristic equation* of A. The solutions of this equation are the so-called *eigenvalues* of A. In practice, one usually starts by calculating the eigenvalues and then finding the respective eigenvectors, by solving linear systems. Consider the example in the accompanying box.

Example: *Eigenvalues and Eigenvectors*

Calculate the eigenvalues and eigenvectors of $A = \begin{bmatrix} 1 & 3 \\ 3 & 1 \end{bmatrix}$.

Solution:

$$(A - \lambda I)\vec{x} = \vec{0} \Rightarrow \begin{bmatrix} 1 - \lambda & 3 \\ 3 & 1 - \lambda \end{bmatrix} \vec{x} = \vec{0}$$

$$\det(A - \lambda I) = 0 \Rightarrow \det\left\{ \begin{bmatrix} 1 - \lambda & 3 \\ 3 & 1 - \lambda \end{bmatrix} \right\} = (1 - \lambda)^2 - 9 = 0$$

whose solutions, namely $\lambda_1 = -2$ and $\lambda_2 = 4$ are the sought eigenvalues of A.
The eigenvectors are obtained respectively to each eigenvalue:
For $\lambda_1 = -2$:

$$A\vec{x}_1 = \lambda_1 \vec{x}_1 \Rightarrow \begin{bmatrix} 1 & 3 \\ 3 & 1 \end{bmatrix}\begin{bmatrix} x \\ y \end{bmatrix} = -2\begin{bmatrix} x \\ y \end{bmatrix}$$

$$\Rightarrow \begin{cases} x + 3y = -2x \\ 3x + y = -2y \end{cases} \Rightarrow x = -y \Rightarrow \vec{x}_1 = \begin{bmatrix} a \\ -a \end{bmatrix}$$

which is the eigenvector of A with respect to $\lambda_1 = -2$, where a is any nonzero complex value.
For $\lambda_2 = 4$

$$A\vec{x}_2 = \lambda_2 \vec{x}_2 \Rightarrow \begin{bmatrix} 1 & 3 \\ 3 & 1 \end{bmatrix}\begin{bmatrix} x \\ y \end{bmatrix} = 4\begin{bmatrix} x \\ y \end{bmatrix} \Rightarrow \begin{cases} x + 3y = 4x \\ 3x + y = 4y \end{cases} \Rightarrow x = y \Rightarrow \vec{x}_2 = \begin{bmatrix} b \\ b \end{bmatrix}$$

which is the eigenvector of A with respect to $\lambda_2 = 4$, where b is any nonzero complex value.

The process of calculating eigenvalues and eigenvectors rarely is as simple as in the above example. For instance, it may happen that one of the eigenvalues repeats itself (multiple root of the characteristic equation), which will demand special treatment. The reader interested in additional theoretical and practical related aspects is referred to [Press et al., 1989] and to [Kreyszig, 1993]. In practice, the eigenvalues and eigenvectors are typically obtained by using specialized libraries, such as those found in mathematical applicatives.

Some particularly important related properties of eigenvalues and eigenvectors are listed in the following

① A and A^T have the same eigenvalues;

② $\det(A) = \lambda_1\lambda_2 \ldots \lambda_N$ (considering eigenvalue multiplicity);

③ A is singular if and only it has a null eigenvalue;

④ eigenvectors corresponding to distinct eigenvalues are linearly independent;

⑤ all the eigenvalues of a Hermitian (symmetric) matrix are real;

⑥ all the eigenvalues of a unitary matrix have unit magnitude;

⑦ an $N \times N$ Hermitian (symmetric) or unitary (orthogonal) matrix has N distinct eigenvalues and the respective eigenvectors form a unitary (orthonormal) set.

Tricks with Matrices

We have already seen that matrices allow us to implement a series of operations in a compact and elegant fashion. This section presents some additional "tricks" with matrices which are particularly suitable for matrix-oriented environments, such as MATLAB®, Scilab® and R®.

Parallel Linear Transformations   It is possible to implement linear transforms of a series of vectors "in parallel", i.e., in the sense that all the transforms are processed through a single matrix product, instead of several independent products. Suppose you want to apply the linear transformation represented by matrix A separately over M vectors $\vec{v}_i$, i.e., $\vec{q}_i = A\vec{v}_i$. This can be performed by using an auxiliary matrix B, having as columns each of the vectors $\vec{v}_i$. The sought respective transformations $\vec{q}_i$ can be obtained by making $C = AB$, where each of the columns of C will correspond to the respective vectors $\vec{q}_i$. Consider the following example:

Let $A = \begin{bmatrix} 1 & 2 \\ -1 & -3 \end{bmatrix}$, $\vec{v}_1 = \begin{bmatrix} 0 \\ 2 \end{bmatrix}$, $\vec{v}_2 = \begin{bmatrix} 3 \\ -1 \end{bmatrix}$, $\vec{v}_3 = \begin{bmatrix} 1 \\ 2 \end{bmatrix}$, and $\vec{v}_4 = \begin{bmatrix} -4 \\ 5 \end{bmatrix}$.

The sought transformations can be obtained at once as

$$C = AB = \begin{bmatrix} 1 & 2 \\ -1 & -3 \end{bmatrix}\begin{bmatrix} 0 & 3 & 1 & -4 \\ 2 & -1 & 2 & 5 \end{bmatrix} = \begin{bmatrix} 4 & 1 & 5 & 6 \\ -6 & 0 & -7 & -11 \end{bmatrix},$$

implying

$$\vec{q}_1 = \begin{bmatrix} 4 \\ -6 \end{bmatrix}, \quad \vec{q}_2 = \begin{bmatrix} 1 \\ 0 \end{bmatrix}, \quad \vec{q}_3 = \begin{bmatrix} 5 \\ -7 \end{bmatrix}, \quad \text{and} \quad \vec{q}_4 = \begin{bmatrix} 6 \\ -11 \end{bmatrix}$$

Building Matrices by Columns   Consider now that you want to obtain an $N \times M$ matrix A having all elements zero, except for the element $a_{i,j}$, which should be equal to a real or complex value r. Define an $N \times 1$ vector $\vec{v}$ as initially filled with zeros, except for $v_{i,1} = r$, and another $M \times 1$ vector $\vec{q}$ also filled with zeros, except for $q_{j,1} = 1$. The desired matrix A can now be conveniently obtained as $A = \vec{v}\vec{q}^{\mathrm{T}}$. For instance, for $N = 3$, $M = 4$, and $a_{2,3} = 4$, we have $\vec{v} = (0, 4, 0)$, $\vec{q} = (0, 0, 1, 0)$, in such a way that

$$ A = \vec{v}\vec{q}^{\mathrm{T}} = \begin{bmatrix} 0 \\ 4 \\ 0 \end{bmatrix} \begin{bmatrix} 0 & 0 & 1 & 0 \end{bmatrix} = \begin{bmatrix} 0 & 0 & 0 & 0 \\ 0 & 0 & 4 & 0 \\ 0 & 0 & 0 & 0 \end{bmatrix} $$

The same strategy can be used to copy vector $\vec{v}$ at a specific column j of A:

$$ A = \vec{v}\vec{q}^{\mathrm{T}} = \begin{bmatrix} 1 \\ 4 \\ -2 \end{bmatrix} \begin{bmatrix} 0 & 0 & 1 & 0 \end{bmatrix} = \begin{bmatrix} 0 & 0 & 1 & 0 \\ 0 & 0 & 4 & 0 \\ 0 & 0 & -2 & 0 \end{bmatrix} $$

In case we want multiple copies of $\vec{v}$ placed at several columns $j_1, j_2, \ldots, j_k$ of A, weighted by respective weights $c_1, c_2, \ldots, c_k$, all we have to do is to place these weights at the respective columns of $\vec{q}$, as illustrated in the following:

$$ A = \vec{v}\vec{q}^{\mathrm{T}} = \begin{bmatrix} 1 \\ 4 \\ -2 \end{bmatrix} \begin{bmatrix} -0.5 & 0 & 1 & 2 \end{bmatrix} = \begin{bmatrix} -0.5 & 0 & 1 & 2 \\ -2 & 0 & 4 & 8 \\ 1 & 0 & -2 & -4 \end{bmatrix} $$

Observe that this "trick" allows us to compact a matrix having NM elements in terms of a product of two vectors having a total of $N + M$ elements. The above strategy can be easily extended to build matrices by rows, instead of columns, by using matrix transposition.

Parallel Eigenvalue-Eigenvector Equation   Consider that we want to copy N vectors $\vec{v}_j$ of dimension $N \times 1$, weighted by respective weights $\lambda_1, \lambda_2, \ldots, \lambda_N$ into subsequent columns of an $N \times N$ matrix B. This can be done by defining the matrices Λ and V as follows:

$$ \Lambda = \begin{bmatrix} \lambda_1 & 0 & \cdots & 0 \\ 0 & \lambda_2 & & 0 \\ \vdots & & \ddots & \vdots \\ 0 & 0 & \cdots & \lambda_N \end{bmatrix} ; V = \begin{bmatrix} \uparrow & \uparrow & \cdots & \uparrow \\ \vec{v}_1 & \vec{v}_2 & \cdots & \vec{v}_N \\ \downarrow & \downarrow & \cdots & \downarrow \end{bmatrix} $$

and making $B = V\Lambda$:

$$B = V\Lambda = \begin{bmatrix} \uparrow & \uparrow & \cdots & \uparrow \\ \vec{v}_1 & \vec{v}_2 & \cdots & \vec{v}_N \\ \downarrow & \downarrow & \cdots & \downarrow \end{bmatrix} \begin{bmatrix} \lambda_1 & 0 & \cdots & 0 \\ 0 & \lambda_2 & & 0 \\ \vdots & & \ddots & \vdots \\ 0 & 0 & \cdots & \lambda_N \end{bmatrix} =$$

$$= \begin{bmatrix} \uparrow & \uparrow & \cdots & \uparrow \\ (\lambda_1 \vec{v}_1) & (\lambda_2 \vec{v}_2) & \cdots & (\lambda_N \vec{v}_N) \\ \downarrow & \downarrow & \cdots & \downarrow \end{bmatrix}$$

If $\vec{v}_j$ are the eigenvectors of an $N \times N$ matrix A, and the weights are the respective eigenvalues, this methodology, combined with the above "parallel" linear transformation technique, allows us to write

$$AV = B = V\Lambda, \tag{2.18}$$

which is the "parallel" version of (2.17). An illustration of this methodology can be found in the box titled *Parallel Eigenvectors*, given in the following:

Example: *Parallel Eigenvectors*

Represent in "parallel" fashion the eigenvalue/eigenvector equations in the Box above.

Solution:

By making $\Lambda = \begin{bmatrix} -2 & 0 \\ 0 & 4 \end{bmatrix}$ and $V = \begin{bmatrix} a & b \\ -a & b \end{bmatrix}$ we immediately have

$$AV = V\Lambda \Leftrightarrow \begin{bmatrix} 1 & 3 \\ 3 & 1 \end{bmatrix} \begin{bmatrix} a & b \\ -a & b \end{bmatrix} = \begin{bmatrix} a & b \\ -a & b \end{bmatrix} \begin{bmatrix} -2 & 0 \\ 0 & 4 \end{bmatrix}$$

Designing Matrices to Have Specific Eigenvectors   The situation where we wanted to identify the eigenvalues and eigenvectors of a specific square matrix A has been previously discussed. Here we present how to build a matrix A having a specific $N \times 1$ eigenvector $\vec{v}$ or a set of $k \leqslant N$ orthogonal eigenvectors $\vec{v}_i$ with dimension $N \times 1$. In the former case, the sought matrix is $A = \vec{v}\vec{v}^T$, since:

$$A = \vec{v}\vec{v}^T \Rightarrow A\vec{v} = \left(\vec{v}\vec{v}^T\right)\vec{v} = \vec{v}\left(\vec{v}^T\vec{v}\right) = r\vec{v}, r = \vec{v}^T\vec{v} \in \mathbb{R}$$

Observe that the matrix product $\vec{v}\vec{v}^{\mathrm{T}}$ can be understood in terms of the above concept of building a matrix by columns. This product implies that vector $\vec{v}$ is copied into a subsequent column j of A weighted by each of its respective coordinates v_j. This implies that matrix A columns are all equal except for a multiplicative constant, and therefore A necessarily has rank 1.

In the latter case, i.e., we want $k \leqslant N$ orthogonal eigenvectors $\vec{v}_i$; $i = 1, 2, \ldots, k$; the matrix A is also easily obtained as

$$A = \vec{v}_1 \vec{v}_1^{\mathrm{T}} + \vec{v}_2 \vec{v}_2^{\mathrm{T}} + \cdots + \vec{v}_k \vec{v}_k^{\mathrm{T}} \Rightarrow A\vec{v}_i = \left(\vec{v}_1 \vec{v}_1^{\mathrm{T}} + \vec{v}_2 \vec{v}_2^{\mathrm{T}} + \cdots + \vec{v}_k \vec{v}_k^{\mathrm{T}} \right) \vec{v}_i =$$
$$= \vec{v}_i \left(\vec{v}_i^{\mathrm{T}} \vec{v}_i \right) = r\vec{v}_i, r = \vec{v}_i^{\mathrm{T}} \vec{v}_i \in \mathbb{R}$$

It can be verified that the so-obtained matrix A has rank k.

To probe further: *Functions, Matrices and Linear Algebra*

A good and relatively comprehensive introduction to many of the covered issues, including propositional logic, functions, linear algebra, matrices, calculus and complex numbers can be found in [James, 1996]. A more advanced reference on mathematical concepts include the outstanding textbook by [Kreyszig, 1993], which covers linear algebra, calculus, complex numbers, and much more. Other good general references are [Bell, 1990] and [Fong et al., 1997]. Interesting references covering complementary aspects related to mathematical physics, including variational calculus, are provided by [Boas, 1996] and [Dettman, 1988]. For those interested in probing further into function vector spaces (i.e., functional analysis), the books [Halmos, 1958; Kreyszig, 1993; Michel & Herget, 1981; Oden, 1979] provide excellent reading, also including good reviews of basic concepts. An interesting approach to complex number and analysis, based on visualization of the involved concepts, is provided by [Needham, 1997]. Good references on vectors, matrices and linear algebra are provided in [Hoffman & Kunze, 1971; Lawson, 1996]. More in-depth treatment of matrices, often including numerical related aspects, can be found in [Barnett, 1992; Bronson, 1988, 1991; Golub & Loan, 1989; Jennings, 1977].

2.3 Differential Geometry

Differential geometry is arguably one of the most beautiful areas of mathematics, especially because it is located at the interface between the immensely important areas of linear algebra, differential calculus and geometry. Therefore, it allows several important concepts from calculus and linear algebra to be expressed in geometric, i.e., visual terms. Typical concepts studied in differential geometry include curves, surfaces, tangent and normal fields, curvature, torsion and so on. While older and more traditional developments in differential geometry, which has a memorable history going back to Descartes and even the ancient Greeks, are characterized by a more local nature, in the sense that only the neighborhood of a curve or surface is taken into account, more recent approaches have targeted global treatments [do Carmo, 1976]. An example of a global problem is to infer general properties of a closed contour, such as its perimeter, in terms of local properties (e.g., derivatives).

Given its geometric and visual nature, differential geometry constitutes a particularly relevant bedrock for image and shape analysis. For instance, several of the important properties of shapes, such as salient points, smoothness, area and perimeter, to name a few, can be nicely represented and analyzed in terms of differential geometry. The present section concentrates on parametric curves in two-dimensional spaces, which represents only a small yet important sub-area of differential geometry. As a matter of fact, the comprehensive treatment of surfaces in image and shape analysis is still relatively incipient, due especially to the implied high computational cost.

2.3.1 2D Parametric Curves

A parametric curve can be understood in terms of the evolution of a point (or a very small particle) as it moves along a 2D space. Mathematically, the position of the particle on the plane can be expressed in terms of the position vector $\vec{p}(t) = (x(t), y(t))$, where t is a real value called the *parameter* of the curve. Let us illustrate such a concept in terms of the specific case:

$$\vec{p}(t) = (x(t), y(t)) = (\cos(t), \sin(t)).\qquad(2.19)$$

Figure 2.40 shows the trajectory of the particle as specified by $\vec{p}(t)$, considering $0 \leqslant t < 2\pi$. Observe that, for this interval, the curve starts at $t = 0$, defining a reference along its trajectory. It can be immediately verified that the trajectory is

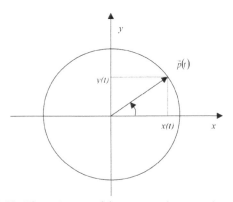

Figure 2.40: *The trajectory of the parametric curve given by* (2.19).

the unitary radius circle, and that the parameter t corresponds to the angles with respect to the x-axis.

Other interesting types of parametric curves having an angle as parameter are those in the epitrochoid and hypotrochoid families, whose basic equations and specific cases are given in Table 2.3, where $m = a + b$ and $-\pi \leqslant t \leqslant \pi$.

	Epitrochoids	Hypotrochoids
Equations	$\begin{cases} x = m\cos(t) - h\cos\left(\frac{m}{b}t\right) \\ y = m\sin(t) - h\sin\left(\frac{m}{b}t\right) \end{cases}$	$\begin{cases} x = m\cos(t) + h\cos\left(\frac{m}{b}t\right) \\ y = m\sin(t) - h\sin\left(\frac{m}{b}t\right) \end{cases}$
Special cases	$a = 0 \rightarrow$ circle $a = b \rightarrow$ limaçon $h = b \rightarrow$ epicycloid	$a = 2b \rightarrow$ ellipse $h = b \rightarrow$ hypocycloid

Table 2.3: *Parametric equations of epitrochoids and hypotrochoids.*

A series of different parametric curves are obtained for combinations of the involved parameters. In general, the curves are completely contained in the circle of radius $|r| \leqslant m + h$ and, in case $\frac{m}{b}$ is an integer number, the curve presents $\frac{m}{b-1}$ loops. Figures 2.41 and 2.42 illustrate a series of epitrochoids and hypotrochoids with respect to several combinations of the m and h parameters.

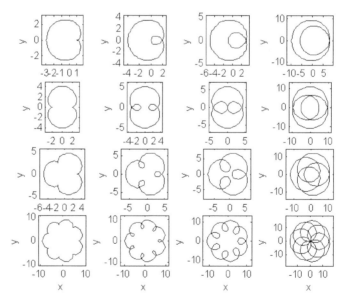

Figure 2.41: *Epitrochoids defined for several combinations of the parameters m and h.*

In addition to angles, other types of parameters are also possible for parametric curves, including time and arc length. It is thus clear that a parametric curve can

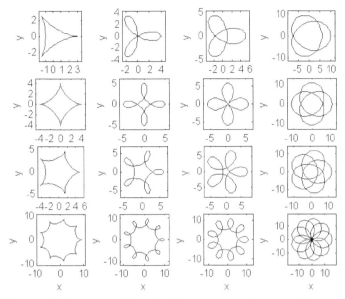

Figure 2.42: *Hypotrochoids defined for several combinations of the parameters m and h.*

be generally understood as a mapping from the real line into the 2D (or even ND) space, i.e., $r : \mathbb{R} \to \mathbb{R}^2$. Other examples of parametric curves include

- $\vec{r}_1(t) = (at + c, bt + d)$, for $t \in \mathbb{R}$, which defines a straight line in $\mathbb{R}^2$ having slope $\frac{b}{a}$ and passing through the point (c, d);

- $\vec{r}_2(t) = (a\cos(t), b\sin(t))$, for $t \in [0, 2\pi)$, which defines a generic ellipse in $\mathbb{R}^2$;

- $\vec{r}_3(t) = (a\cos(t), b\sin(t), t)$, for $t \in \mathbb{R}$, which defines a helix in $\mathbb{R}^3$.

By being a particularly useful curve, the above straight line parametric equation $\vec{r}_1(t) = (at + c, bt + d)$ is graphically shown in Figure 2.43. Observe that there are

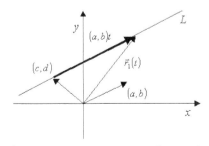

Figure 2.43: *The parametric representation of a generic straight line L.*

infinite possibilities for choosing the point (c, d), which can be any point on the line. The choice of this point determines the zero reference for t along the straight line.

An important observation is that the "drawing" (or trajectory) defined as the particle moves, i.e., the image of the mapping implemented by the curve, is not the same as the parametric curve itself. The obtained figure is usually called the *trace* of the parametric curve. Indeed, the same trace can be produced by many distinct parametric curves. For instance, the straight line trace obtained from the parametric curve $\vec{r}_1$ in the above example can also be produced by $\vec{r}_4(t) = (-at + c, -bt + d)$ or $\vec{r}_5(t) = (3at + c, 3bt + d)$. The parametric curve $\vec{r}_4$ produces the same trace as $\vec{r}_1$, but with reverse sense. On the other hand, the trace produced by $\vec{r}_5$ has the same sense as that obtained for $\vec{r}_1$, but its speed is three times larger. Similar situations can be verified for the above curves $\vec{r}_2$ and $\vec{r}_3$. Such facts indicate that the difference between curves with the same trace is implied by the way in which the curve evolves along the trace, including its speed.

2.3.2 Arc Length, Speed and Tangent Fields

Given a parametric curve, the *arc length* from an initial point t_0 to a generic point t of the curve can be calculated as

$$s(t) = \int_{t_0}^{t} ds = \int_{t_0}^{t} \sqrt{\dot{x}^2 + \dot{y}^2}\, dt = \int_{t_0}^{t} \|\dot{r}(t)\|\, dt$$

which is graphically illustrated in Figure 2.44.

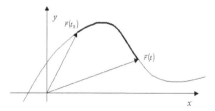

Figure 2.44: *The arc length between the points $\vec{r}(t_0)$ and $\vec{r}(t)$.*

For instance, the arc length $s_1(t)$ from $t_0 = 0$ up to the point identified by t along the curve r_1 is given by

$$s_1(t) = \int_{0}^{t} \sqrt{a^2 + b^2}\, dt = \sqrt{a^2 + b^2}\, t$$

Consider now the situation depicted in Figure 2.45, indicating the particle position at a given parameter value t, represented as $\vec{p}(t) = (x(t), y(t))$, and the position at $t + dt$, i.e., $\vec{p}(t + dt)$.

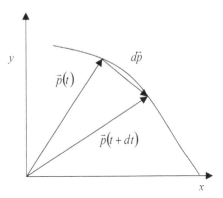

Figure 2.45: *Two nearby points along the trajectory of a generic paramet-*
ric curve.

By relating these two positions as $\vec{p}(t + dt) = \vec{p}(t) + d\vec{p}$, and assuming that the curve is differentiable, the *speed* of the particle at the parameter value t is defined as

$$\dot{\vec{p}}(t) = \frac{d\vec{p}}{dt} = \lim_{dt \to 0} \frac{\vec{p}(t + dt) - \vec{p}(t)}{dt} = \left[\begin{array}{c} \lim_{dt \to 0} \frac{x(t+dt)-x(t)}{dt} \\ \lim_{dt \to 0} \frac{y(t+dt)-y(t)}{dt} \end{array} \right] = \left[\begin{array}{c} \dot{x}(t) \\ \dot{y}(t) \end{array} \right]$$

For instance, the speed of the parametric curve given by (2.19) is

$$\dot{\vec{p}}(t) = (\dot{x}(t), \dot{y}(t)) = (-\sin(t), \cos(t))$$

In case the speed of a parametric curve is never zero (i.e., the particle never stops), the curve is said to be a *regular curve*. In case the curve never crosses itself, the curve is said to be a *Jordan curve*.

One of the most important properties of the speed of a regular parametric curve is the fact that it is a vector tangent to the curve at each of its points. This can be easily verified by observing in Figure 2.45 that, as $\vec{p}(t + dt)$ approaches $\vec{p}(t)$, the difference vector $d\vec{p}$ tends to the tangent to the curve at the point t. The set of all possible tangent vectors to a curve is called a *tangent field*. Provided the curve is regular, it is possible to normalize its speed in order to obtain unit magnitude for all the tangent vectors along the curve:

$$\vec{\alpha}(t) = \frac{\dot{\vec{p}}(t)}{\left\| \dot{\vec{p}}(t) \right\|} \Rightarrow \left\| \vec{\alpha}(t) \right\| = 1. \tag{2.20}$$

It is now clear why non-regular curves are excluded, since they would imply zero denominator for some t. In the related literature, unitary tangent fields are sometimes referred to simply as tangent fields. Figure 2.46 illustrates the unitary tangent field for the curve in Figure 2.45.

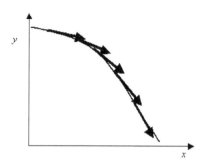

Figure 2.46: *Unit tangent field for the curve in Figure* 2.45. *Only a few of
the infinite unit tangent vectors are shown. All vectors have
unit magnitude.*

As shown in the previous section, the same trace can be defined by many (actually an infinite number of) different parametric curves. A particularly important
and useful parameterization is that characterized by unitary magnitude speed. Such
a curve is said to be *parameterized by arc length*. It is easy to verify that, in this
case:

$$s(t) = \int_0^t \|\dot{r}(t)\| \, dt = \int_0^t 1 \, dt = t$$

i.e., the parameter t corresponds to the arc length. Consequently, a curve is in arc
length parameterization if and only its tangent field has unitary magnitude. The
process of transforming the analytical equation of a curve into arc length parameterization is called *arc length reparameterization*, which consists in expressing the
parameter t as a function of the arc length, and substituting t in the original curve
for this function (for more details see [Davies & Samuels, 1996; do Carmo, 1976;
Guggenheimer, 1977]). Unfortunately, only a few curves can be analytically repa-
rameterized. It is also important to note that the procedure for obtaining unitary
tangent fields based on Equation 2.20 does not immediately supply the reparame-
terized curve, but only the normalized unit field. In practice, the reparameterization
of curves usually involves some interpolating scheme.

2.3.3 | Normal Fields and Curvature

It is known from classical mechanics [Goldstein, 1980; Marion & Thornton, 1995]
that the second derivative of the parameterized curve $\vec{p}(t)$ defining the evolution
of a small particle gives its respective *acceleration* (see Figure 2.47). This acceleration, as the speed, is also a vector quantity typically decomposed into two
components, one parallel to the speed $\dot{\vec{p}}(t)$, called the *tangent acceleration* and
represented as $\vec{a}_T(t)$, and the other a component normal to the speed, the so-called
radial acceleration $\vec{a}_R(t)$. Thus, $\ddot{\vec{p}}(t) = \vec{a}_T(t) + \vec{a}_R(t)$.

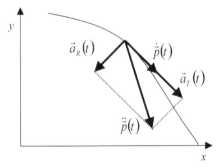

Figure 2.47: *The speed, acceleration, and their components at the point indicated by $\vec{p}(t)$.*

The tangent acceleration expresses changes of the *speed magnitude*, and the normal acceleration expresses changes of *speed orientation*. Observe that these components are orthogonal vectors.

Now, let $\vec{u}(t) = (x(t), y(t))$ be a curve parameterized by arc length, in such a way that its first derivative $\dot{\vec{u}}(t) = (\dot{x}(t), \dot{y}(t))$ defines a unit tangent field. Since the speed magnitude is always unitary, the tangent acceleration is zero, and $\ddot{\vec{u}}(t) = \vec{a}_R(t)$ expresses the orientation change in the unitary tangent field. Consequently, for a curve parameterized by arc length, the second derivative of the curve is always *normal* to the curve (i.e., normal to the tangent of the curve), defining a *normal field* along the points of the curve. The respective unit normal field at the values of t for which $\left\| \ddot{\vec{p}}(t) \right\| \neq 0$ can be obtained as

$$\vec{n}(t) = \frac{\ddot{\vec{p}}(t)}{\left\| \ddot{\vec{p}}(t) \right\|}$$

Figure 2.48 illustrates the unit normal field for the curve in Figure 2.45. For an

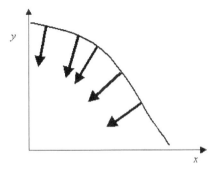

Figure 2.48: *Unitary normal field for the curve in Figure 2.45. Only a few of the infinite unit normal vectors are shown. All vectors have unit magnitude.*

arc length parameterized curve $\vec{p}(t)$, the magnitude of its second derivative, i.e., $k(t) = \left\| \ddot{\vec{p}}(t) \right\| = \| a_R(t) \|$, is said to be the *curvature* of the curve at t. Observe that this expression always produces a non-negative curvature value. In the general case, i.e., the curve is not necessarily arc length parameterized, the curvature of a plane curve can be calculated as:

$$k(t) = \frac{\dot{x}(t)\,\ddot{y}(t) - \dot{y}(t)\,\ddot{x}(t)}{\left(\dot{x}^2(t) + \dot{y}^2(t)\right)^{\frac{3}{2}}}. \tag{2.21}$$

Unlike in the previous equation, the signal of the curvature calculated by Equation 2.21 can be positive or negative, indicating the respective local concavity (see below). The curvature is an extremely important concept because it nicely expresses the local "geometric nature" of a curve. For instance, if zero curvature is observed along a portion of a curve, this portion will correspond to a straight line segment. On the other hand, a constant curvature value indicates a circle or an arc of circle. Generally, the curvature value is proportional to the local variation of the curve. More precisely, as defined above, it corresponds to the radial acceleration magnitude of the arc length parameterized version of the curve, therefore indicating how fast the tangent vector changes its orientation. Another interesting feature exhibited by curvature is the fact that it is *invariant* to rotations, translations and reflections of the original curve (observe that it is not invariant to scaling). Moreover, the curvature is *information preserving* in the sense that it allows the original curve to be recovered, up to a rigid body transformation (i.e., combinations of translations, rotations and reflections that do not alter the size of the shape—see Section 4.9.3). Thus we have that, if $k(t)$ is a differentiable function expressing the curvature of a curve from t_0 to t, its reconstruction can be obtained as

$$\vec{p}(t) = \left(\int_{t_0}^{t} \cos(\alpha(r))\,dr + c_1, \int_{t_0}^{t} \sin(\alpha(r))\,dr + c_2 \right),$$

where $\alpha(t) = \int_{t_0}^{t} k(r)\,dr + c_3$ and (c_1, c_2) and c_3 represent the translation vector and the rotation angle, respectively.

Although it is clear from the above curvature definition $k(t) = \left\| \ddot{\vec{p}}(t) \right\| = \| a_R(t) \|$ that its values are non-negative, it is often interesting to consider an alternative definition allowing negative curvature values. This is done by considering the standard coordinate system $(\hat{i}, \hat{j}, \hat{k})$ of $\mathbb{R}^3$ (i.e., $\hat{k} = \hat{i} \wedge \hat{j}$). The signed curvature $k_s(t)$, which can be calculated by Equation 2.21, can be defined as

$$k_s(t) = \mathrm{sgn}\left[\left\langle \dot{\vec{p}}(t) \wedge \vec{n}(t), \hat{i} \wedge \hat{j} \right\rangle \right] k(t)$$

This means that positive curvature will be obtained whenever the sense of the vector $\dot{\vec{p}}(t) \wedge \vec{n}(t)$ agrees with that of the unit vector $\hat{i} \wedge \hat{j}$. Negative curvature is obtained otherwise.

An immediate advantage allowed by the signed curvature is that its sign provides indication about the *concavity* at each of the curve points. It should however be taken into account that the sign of $k_s(t)$ depends on the sense of the curve, and will change with the sense in which the curve is followed and with the sign of t. Figure 2.49 illustrates the change of curvature sign considering two senses along a closed curve, and the respective concavity criteria. A point where $k_s(t) = 0$ and $\dot{k}_s(t) \neq 0$ is said to be an *ordinary inflection point*. Such a point corresponds to a change of concavity along the curve.

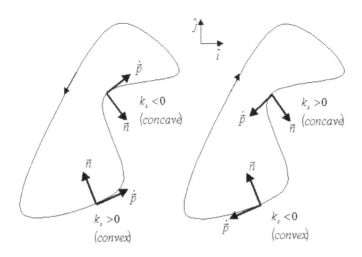

Figure 2.49: *The sign of the signed curvature k_s changes as the sense of the curve is inverted. The concavity criterion also depends on the adopted curve sense.*

The curvature can also be geometrically understood in terms of osculating circles and radius of curvature. Consider Figure 2.50 and assume that the curvature is never zero. The circle having radius $r(t) = \frac{1}{k}(t)$, called *radius of curvature*, and centered at $\frac{\vec{u}(t)+\vec{n}(t)}{k(t)}$, where $\vec{u}(t)$ is an arc length parameterized curve and $\vec{n}(t)$ is the unit normal field to $\vec{u}(t)$, is called the *osculating circle* at t.

It is interesting to observe that the term "osculating" comes from the Latin "osculari," meaning to kiss. In differential geometry, such a concept is related to the number of contact points between curves (see, for instance, [Giblin & Bruce, 1992]).

The trace defined by the center of the osculating circle along the curve, assuming $k(t) \neq 0$, is called the *evolute* of the curve. Consequently, the evolute $\vec{e}(t)$ of a curve $\vec{u}(t)$ parameterized by arc length, with unit normal vector $\vec{n}(t)$, can be

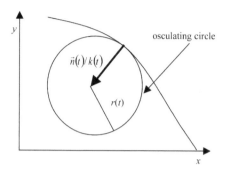

Figure 2.50: *Osculating circle and radius of curvature at a point t.*

obtained as the trace of the following parametric curve:

$$\vec{e}(t) = \vec{u}(t) + \frac{1}{k(t)}\vec{n}(t) = \vec{u}(t) + r(t)\,\vec{n}(t)$$

The curve defined by $\vec{g}(t) = \vec{u}(t) + a\vec{n}(t)$, where a is a real constant, represents one of the infinite *parallels* to the original curve.

To probe further: *Differential Geometry*

A classical and comprehensive textbook on differential geometry is [do Carmo, 1976]. Additional references include [Davies & Samuels, 1996; Graustein, 1966; Guggenheimer, 1977; Kreyszig, 1991]. Although focusing on singularities theory, the book [Giblin & Bruce, 1992] also provides a good introduction to differential geometry. Similarly, the Physics books [Faber, 1983; Göckeler & Schücker, 1987] include a nice and particularly accessible introduction to this area.

2.4 | Multivariate Calculus

In this section we review, in an introductory fashion, the main concepts from multivariate calculus that are particularly relevant to image and shape processing and analysis.

2.4.1 | Multivariate Functions

A multivariate function is a function involving more than a single free variable. For instance $z = g(x, y) = x^2 + 2y^2$ is a multivariate (in this case, bivariate) function, which is graphically illustrated in Figure 2.51. It should be observed that the dimension refers to the function domain and not to its graphical aspect. For instance,

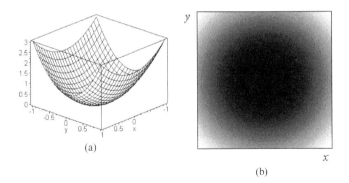

(a)

(b)

Figure 2.51: *Graphical representation of the function $z = g(x, y) = x^2 + 2y^2$ as a surface in the 3D space* (a) *and as a gray-level image* (b).

the surface defined by the above bivariate function shown in Figure 2.51 (a) ex-tends through the 3D space, but its domain is $\mathbb{R}^2$. Figure 2.51 (b) depicts the same function as a gray-level image, where the function values are represented by gray levels, the lower values, as usually, being represented by darker levels. As a matter of fact, images can be understood as bivariate functions. In this book, we shall be particularly interested in 1D and 2D functions, used to represent and transform planar shapes.

The concepts of scalar and vector fields are particularly important in image and shape analysis. Basically, a *real scalar field* is any function f that associates a real value to each point in a domain $D \subset \mathbb{R}^N$, i.e., $f : D \subset \mathbb{R}^N \to \mathbb{R}$. A *real vector field* is any function g that associates a real vector in $\mathbb{R}^M$ to each point in a domain $D \subset \mathbb{R}^N$, i.e., $g : D \subset \mathbb{R}^N \to \mathbb{R}^M$. For instance, the function $g(x, y) = x^2 + 2y^2$ is a real scalar field of the type $g : \mathbb{R}^2 \to \mathbb{R}$, and $\vec{f}(x, y) = 2x\hat{\imath} + 4y\hat{\jmath} = (2x, 4y)$ is a real vector field $f : \mathbb{R}^2 \to \mathbb{R}^2$. For simplicity's sake, we will refer to real scalar (and vector) fields simply as scalar (and vector) fields, reserving the extended names to indicate other situations, such as complex scalar fields. Observe that a vector field can be thought of as a vector having as elements several scalar fields. For instance, in the above example we have $\vec{f}(x, y) = \left(f_x(x, y), f_y(x, y) \right)$, where $f_x(x, y) = 2x$ and $f_y(x, y) = 4y$. Vector fields can also be thought of as a generalization of the concept of linear transformations, introduced in Section 2.2.3. More specifically, the linear transformation defined by the matrix A, as follows,

$$A = \begin{bmatrix} -1 & 2 \\ 0 & -1 \end{bmatrix}, \quad \text{i.e.,} \quad \begin{bmatrix} \tilde{x} \\ \tilde{y} \end{bmatrix} = \begin{bmatrix} -1 & 2 \\ 0 & 1 \end{bmatrix} \begin{bmatrix} x \\ y \end{bmatrix}$$

can be represented in terms of the vector field

$$\vec{f}(x, y) = (\tilde{x}(x, y), \tilde{y}(x, y)) = (-x + 2y, y).$$

However, vector fields can also be used to represent nonlinear transformations such as $\vec{f}(x, y) = (x^2, xy)$. Some especially important 2D scalar fields are presented in Table 2.4.

A particularly simple and important real bivariate scalar field $z = g(x, y)$ is that defining a generic *plane P*. Its implicit form is given by

$$P : c_1 x + c_2 y + z = c_3,$$

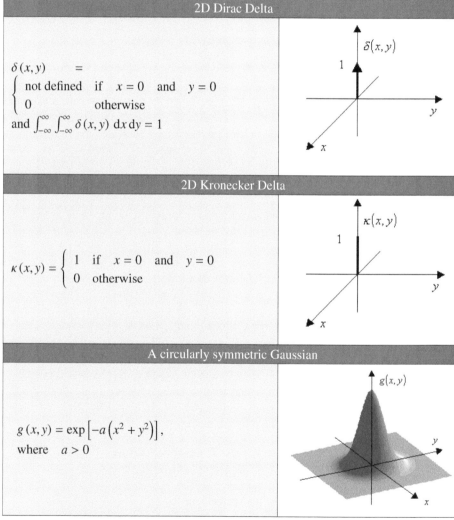

2D Dirac Delta

$\delta(x, y) \quad =$
$\begin{cases} \text{not defined} & \text{if} \quad x = 0 \quad \text{and} \quad y = 0 \\ 0 & \text{otherwise} \end{cases}$
and $\int_{-\infty}^{\infty} \int_{-\infty}^{\infty} \delta(x, y) \, dx \, dy = 1$

2D Kronecker Delta

$\kappa(x, y) = \begin{cases} 1 & \text{if} \quad x = 0 \quad \text{and} \quad y = 0 \\ 0 & \text{otherwise} \end{cases}$

A circularly symmetric Gaussian

$g(x, y) = \exp\left[-a\left(x^2 + y^2\right)\right],$
where $\quad a > 0$

Table 2.4: *Some particularly important bivariate real scalar fields.*

where c_1, c_2, and c_3 are real constants. By *implicit* it is meant that the function is not in the form $z = g(x, y)$, such as in $z = g(x, y) = xy^2$, but as an equation of the type $h(x, y, z) = c$, where c is a constant. It can be verified that the plane defined by the above implicit function has as normal vector $\vec{n} = (c_1, c_2, 1)$, which defines a whole family of planes that are parallel to P, including P itself. The equation of the plane P_0, which is parallel to these planes and contains the origin of the coordinate system, can be obtained by imposing that any vector $\vec{v} = (x, y, z)$ contained in this plane is orthogonal to the normal vector $\vec{n} = (c_1, c_2, 1)$, i.e., $\langle \vec{v}, \vec{n} \rangle = c_1 x + c_2 y + z = 0$, and hence $z = -xc_1 - yc_2$. Now, the whole family of planes parallel to P_0 can be obtained by adding the constant c_3 to the latter equation, yielding $z = -xc_1 - yc_2 + c_3$ and consequently we get $z + xc_1 + yc_2 = c_3$, which corresponds to the above implicit equation of the plane. It is sometimes possible to rewrite an implicit function into explicit form. For instance, the above implicit function can be rewritten as $P : z = -c_1 x - c_2 y + c_3$.

A real vector field $\vec{g} : \mathbb{R}^N \rightarrow \mathbb{R}^M \mid \vec{q} = \vec{g}(\vec{v})$ is said to be *continuous at a point* $\vec{q}_0 \in \mathbb{R}^M$ if for each open ball B_ε with radius ε centered at $\vec{q}_0$ (i.e., the vectors $\vec{q} \in \mathbb{R}^M$ such as $\|\vec{q} - \vec{q}_0\| < \varepsilon$), it is always possible to find an open ball B_δ with radius δ centered at $\vec{v}_0 \in \mathbb{R}^N$ (i.e., the vectors $\vec{v} \in \mathbb{R}^N$ such as $\|\vec{v} - \vec{v}_0\| < \delta$), such as the mapping of this ball by the vector field $\vec{g}$, i.e., $\vec{g}(B_\delta)$, falls completely inside B_ε. A vector field that is continuous at all the points of its domain is simply said to be *continuous*. The continuity of a vector field can be understood as a particular case of the above definition in the case $M = 1$.

Given a bivariate function $z = g(x, y)$, we can think about this function in terms of unidimensional functions by taking slices of $g(x, y)$ along planes perpendicular to the (x, y) plane. Observe that any of such planes is completely specified by the straight line L defined by the intersection of this perpendicular plane with the plane (x, y). It is particularly useful to define such lines in a parametric fashion (see Section 2.3.1), which can be done by imposing that these lines are parallel to a vector $\vec{v} = (a, b)$ and contain a specific point, identified by the vector $\vec{p}_0 = (c, d)$. Therefore the general form of these straight lines is $L : \vec{p} = \vec{v}t + \vec{p}_0 = (at + c, bt + d)$.

Since this line will be used to follow the line along the slice, defining a function of a single variable, unit speed (see Section 2.3.2), and therefore arc length parameterization, is required. This can be easily achieved by imposing that $a^2 + b^2 = 1$. Now, the values of the function g along the slice can easily be obtained by substituting the x and y coordinates of the positions defining the line L into the function $z = g(x, y)$ to yield $z = g(at + c, bt + d)$, which is a function on the single variable t. The box entitled *Slicing a Circularly Symmetric Bivariate Gaussian* provides an example about scalar fields slicing.

Example: *Slicing a Circularly Symmetric Bivariate Gaussian*

Consider the Gaussian scalar field $F(x, y) = \exp\left[\frac{-(x^2 + y^2)}{4} \right]$. Obtain the univariate function defined by slicing this field along the plane that is orthogonal to the

(x, y) plane and contains the line L, which is parallel to the vector $\vec{v} = (1, 0.5)$ and passes onto the point $\vec{b} = (0, -2)$, which defines the origin along the slice.

Solution:

First, we obtain the equation of the line L. In order to have arc length parameterization, we impose $\|\vec{v}\| = 1 \Rightarrow \tilde{v} = \frac{\vec{v}}{\|\vec{v}\|} = \left(\frac{1}{\sqrt{1.25}}, \frac{0.5}{\sqrt{1.25}} \right)$. Now, the sought line equation can be expressed as

$$\vec{p}(s) = (x(s), y(s)) = \vec{b} + \tilde{v}s = (0, -2) + \left(\frac{1}{\sqrt{1.25}}, \frac{0.5}{\sqrt{1.25}} \right) s \Leftrightarrow \begin{cases} x(s) = \frac{s}{\sqrt{1.25}} \\ y(s) = -2 + \frac{0.5s}{\sqrt{1.25}} \end{cases}$$

Substituting these coordinates into the scalar field we obtain the following:

$$F(x(s), y(s)) = f(s) =$$
$$= \exp \left\{ - \left[\left(\frac{s}{\sqrt{1.25}} \right)^2 + \left(-2 + \frac{0.5s}{\sqrt{1.25}} \right)^2 \right] / 4 \right\} = \exp \left(-1 + \frac{s}{2\sqrt{1.25}} - 0.25s^2 \right)$$

The above scalar field, indicating the considered slice, as well as the obtained univariate function $f(s)$, are shown below.

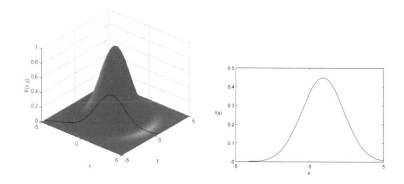

2.4.2 Directional, Partial and Total Derivatives

Given a 2D function $z = g(x, y)$, its *directional derivative* along the straight line L (in arc length parameterization) is given by

$$\left.\frac{dg}{dt}\right|_L = \frac{d}{dt} g(t)$$

For the Gaussian in the previous example, i.e., $F(x, y) = \exp\left[\frac{-(x^2+y^2)}{4}\right]$, we have

$$f(t) = \exp\left(-1 + \frac{t}{2\sqrt{1.25}} - 0.25t^2\right)$$

and hence

$$\frac{df(t)}{dt} = \left(\frac{1}{2\sqrt{1.25}} - 0.5t\right)\exp\left(-1 + \frac{t}{2\sqrt{1.25}} - 0.25t^2\right)$$

The directional derivatives obtained at the special situations where the line L corresponds to the x- and y-axis are defined to be the *partial derivatives* of the function $g(x, y)$:

$$\left.\frac{dg}{ds}\right|_{L:\begin{array}{c} t = x \\ 0 \end{array}} = \frac{\partial g}{\partial x},$$

the partial derivative of g with respect to x and

$$\left.\frac{dg}{dy}\right|_{L:\begin{array}{c} 0 \\ t = y \end{array}} = \frac{\partial g}{\partial y},$$

the partial derivative of g with respect to y.

Given a real vector field

$$\vec{g} : \mathbb{R}^N \to \mathbb{R}^M \mid \vec{g}(\vec{v}) = \vec{g}[g_1(\vec{v}), g_2(\vec{v}), \dots, g_M(\vec{v})],$$

its *Jacobian matrix* is an $N \times M$ matrix $J(\vec{v})$ having as elements $\left[\frac{\partial g_i(\vec{v})}{\partial x_j}\right]$. In case $N = M$, $\det(J)$ is said to be the *Jacobian* of the vector field. The box entitled *Jacobian Matrix and Jacobian of a Vector Field* provides a complete example of these concepts. The Jacobian matrix defines a linear transformation known as *total derivative* of $\vec{g} : \mathbb{R}^N \to \mathbb{R}^M$, having the general form $\vec{\Delta}_{\vec{q}} = J\vec{\Delta}_{\vec{v}}$. The vector $\vec{\Delta}_{\vec{q}} \in \mathbb{R}^M$, corresponding to the output of the transformation to the input vector $\vec{\Delta}_{\vec{v}} \in \mathbb{R}^N$, indicates the variation $\vec{\Delta}_{\vec{q}}$ of the vector field as response to the small variation $\vec{\Delta}_{\vec{v}}$ in its argument (actually, the equality is only verified for infinitesimally small displacements, more generally we only have $\vec{\Delta}_{\vec{q}} \cong J\vec{\Delta}_{\vec{v}}$). In addition, observe that $g\left(\vec{v} + \vec{\Delta}_{\vec{v}}\right) \cong J\vec{\Delta}_{\vec{v}} + g\left(\vec{v}\right)$, which is a natural extension of the univariate Taylor series (i.e., $g\left(t + \Delta t\right) \cong \dot{g}\left(t\right)\Delta t + g\left(t\right)$). This important concept is illustrated in Figure 2.52. This result can be immediately extended to scalar fields, in which case $\vec{\Delta}_{\vec{q}} = J\vec{\Delta}_{\vec{v}} = \left\langle \vec{\nabla}g, \vec{\Delta}_{\vec{v}} \right\rangle$ (see the next section).

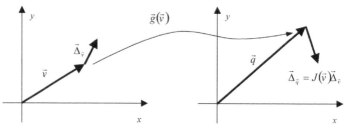

Figure 2.52: *The vector field variation $\vec{\Delta}_{\vec{q}}$ obtained as a consequence of a small perturbation $\vec{\Delta}_{\vec{v}}$ of its argument is given by the linear transformation $\vec{\Delta}_{\vec{q}} = J\vec{\Delta}_{\vec{v}}$.*

Example: *Jacobian Matrix and Jacobian of a Vector Field*

① Calculate the Jacobian matrix J and the Jacobian of the vector field $\vec{g} : \mathbb{R}^2 \to \mathbb{R}^2 \mid \vec{g}\left(\vec{v}\right) = \vec{g}\left(2xy, x^3y\right)$.

② Knowing that $\vec{q} = \vec{g}\left(\vec{v} = (2, 1)\right) = (4, 8)$, estimate $\tilde{\vec{q}} = \vec{g}\left[\vec{v} = (2.05, 0.998)\right]$ without direct substitution into the vector field equation, and by taking into account $J\left[\vec{v} = (2, 1)\right]$.

Solution:

① By definition we have

$$J\left[\vec{v} = (x, y)\right] = \begin{bmatrix} \frac{\partial g_1(x,y)}{\partial x} & \frac{\partial g_1(x,y)}{\partial y} \\ \frac{\partial g_2(x,y)}{\partial x} & \frac{\partial g_2(x,y)}{\partial y} \end{bmatrix} = \begin{bmatrix} 2y & 2x \\ 3x^2y & x^3 \end{bmatrix}$$

and, consequently: $\det\left(J\right) = 2x^3y - 6x^3y = -4x^3y$

② Now, the new vector $\tilde{q} = \vec{g}[\vec{v} = (2.05, 0.998) = \vec{v} + \vec{\Delta}_{\vec{v}}]$ can be estimated as $\tilde{q} \cong \vec{q} + \vec{\Delta}_{\vec{q}}$, where

$$\vec{\Delta}_{\vec{q}} = J(\vec{v})\,\vec{\Delta}_{\vec{v}} = \begin{bmatrix} 2 & 4 \\ 12 & 8 \end{bmatrix}\begin{bmatrix} 0.05 \\ -0.002 \end{bmatrix} = \begin{bmatrix} 0.092 \\ 0.584 \end{bmatrix}.$$

Hence

$$\tilde{q} \cong \vec{q} + \vec{\Delta}_{\vec{q}} = \begin{bmatrix} 4 \\ 8 \end{bmatrix} + \begin{bmatrix} 0.092 \\ 0.584 \end{bmatrix} = \begin{bmatrix} 4.092 \\ 8.584 \end{bmatrix}.$$

Since we know from the definition of the vector field that $\tilde{q} = (4.0918, 8.5979)$, it is clear that the above estimation is reasonably accurate. Of course, the accuracy increases for smaller displacements.

2.4.3 Differential Operators

The partial derivatives can also be used to define some extremely important differential *operators*, such as the gradient, the Laplacian, and the divergent, which are presented in the following.

The *gradient* of a differentiable scalar field $z = g(x, y)$ is the vector field given by the following

$$\vec{\nabla}g(x, y) = \frac{\partial g}{\partial x}\hat{\imath} + \frac{\partial g}{\partial y}\hat{\jmath} = \frac{\partial g}{\partial x}\begin{bmatrix} 1 \\ 0 \end{bmatrix} + \frac{\partial g}{\partial y}\begin{bmatrix} 0 \\ 1 \end{bmatrix} = \left[g_x(x, y), g_y(x, y) \right].$$

A particularly important property of this operator is that it always points towards the direction of maximum variation of the function, at each specific point (x, y). For instance, consider $g(x, y) = x^2 + 2y^2$ (see Figure 2.51). The gradient of this function is $\vec{\nabla}g(x, y) = 2x\hat{\imath} + 4y\hat{\jmath}$ which, at any point (x, y), points towards the steepest direction along the surface defined by the function, i.e., away from the minimum value of the function at $(0, 0)$. It is often interesting to decompose the gradient in terms of its magnitude,

$$\left\| \vec{\nabla}g(x, y) \right\| = \sqrt{g_x^2(x, y) + g_y^2(x, y)}$$

and angle with the x axis (clockwise sense)

$$\phi\left[\vec{\nabla}g(x, y) \right] = \arctan\left[\frac{g_y(x, y)}{g_x(x, y)} \right].$$

It can be verified that at extrema values (i.e., local maxima or minima) of the function we necessarily have $\left\| \vec{\nabla}g(x, y) \right\| = 0$, but not the other way around (e.g., along a plateau).

The *Laplacian* of a bivariate function $z = g(x, y)$ is the scalar field given by

$$\nabla^2 g(x, y) = \frac{\partial^2 g}{\partial x^2} + \frac{\partial^2 g}{\partial y^2}.$$

For example, the Laplacian of the function $g(x, y) = x^2 + 2y^2$ is $\nabla^2 g(x, y) = 2x + 4y$, which is immediately verified to be a plane.

Finally, the *divergent* of a vector field $\vec{g}(x, y) = \left[g_x(x, y), g_y(x, y) \right]$ is the scalar field defined as

$$\vec{\nabla}.\vec{g}(x, y) = \frac{\partial g_x(x, y)}{\partial x} + \frac{\partial g_y(x, y)}{\partial y}.$$

For instance, the divergent of the vector function $\vec{h}(x, y) = 2x\hat{\imath} + 4y\hat{\jmath}$ is the constant function $\vec{\nabla}.\vec{h}(x, y) = 6$.

To probe further: *Multivariate Calculus*

Excellent classical textbooks on multivariate calculus include [Apostol, 1969; Edwards & Penney, 1998; Leithold, 1990; Williamson & Trotter, 1996]. The comprehensive textbook [Kreyszig, 1993] also covers several related aspects, while [Schey, 1997] provides a nice introduction to the main differential operators.

2.5 | Convolution and Correlation

The convolution and correlation are both operations involving two functions, let us say $g(t)$ and $h(t)$, and producing as result a third function. Informally speaking, these two operations provide a means for "combining" or "mixing" the two functions as to allow important properties, such as the convolution and correlation theorems to be presented in Section 2.7.3. In addition, convolution provides the basis for several filters, and correlation provides a means for comparing two functions. These operations are presented in the following, first with respect to continuous domains, then to discrete domains.

2.5.1 Continuous Convolution and Correlation

Let $g(t)$ and $h(t)$ be two real or complex functions. The *convolution* between these functions is the univariate function resulting from the operation defined as

$$
\begin{aligned}
q(\tau) &= g(\tau) * h(\tau) \\
&= (g * h)(\tau) \\
&= \int_{-\infty}^{\infty} g(t)\, h(\tau - t)\, dt.
\end{aligned}
\tag{2.22}
$$

The *correlation* between two real or complex functions $g(t)$ and $h(t)$ is the function defined as

$$
\begin{aligned}
q(\tau) &= g(\tau) \circ h(\tau) \\
&= (g \circ h)(\tau) \\
&= \int_{-\infty}^{\infty} g^*(t)\, h(\tau + t)\, dt.
\end{aligned}
\tag{2.23}
$$

As is clear from the above equations, the correlation and convolution operations are similar, except that in the latter the first function is conjugated and the signal of the free variable t in the argument of $h(t)$ is inverted. As a consequence, while the convolution can be verified to be commutative, i.e.,

$$
(g * h)(\tau) = \int_{-\infty}^{\infty} g(t) h(\tau - t)\, dt \overset{a=\tau-t}{=} \int_{-\infty}^{\infty} g(\tau - a) h(a)\, da = (h * g)(\tau)
$$

we have that the correlation is not, i.e.,

$$
(g \circ h)(\tau) = \int_{-\infty}^{\infty} g^*(t) h(\tau + t)\, dt \overset{a=\tau+t}{=} \int_{-\infty}^{\infty} g^*(a - \tau) h(a)\, da \neq (h \circ g)(\tau)
$$

However, in case both $g(t)$ and $h(t)$ are real, we have

$$
(g \circ h)(\tau) = \int_{-\infty}^{\infty} g(t) h(\tau + t)\, dt \overset{a=\tau+t}{=} \int_{-\infty}^{\infty} g(a - \tau) h(a)\, da = (h \circ g)(-\tau)
$$

In other words, although the correlation of two real functions is not commutative, we still have $(g \circ h)(\tau) = (h \circ g)(-\tau)$. In case both $g(t)$ and $h(t)$ are real and even, then $(g \circ h)(\tau) = (h \circ g)(\tau)$. For real functions, the convolution and correlation are related as

$$g(\tau) * h(-\tau) = \int_{-\infty}^{\infty} g(t) h(t - \tau) \, dt \stackrel{a = t - \tau}{=} \int_{-\infty}^{\infty} g(a + \tau) h(a) \, da = h(\tau) \circ g(\tau)$$

If, in addition, $h(t)$ is even, we have

$$g(\tau) * h(\tau) = \int_{-\infty}^{\infty} g(t) h(t - \tau) \, dt = h(\tau) \circ g(\tau) = (g \circ h)(-\tau)$$

An interesting property is that the convolution of any function $g(t)$ with the Dirac delta reproduces the function $g(t)$, i.e.,

$$(g * \delta)(\tau) = \int_{-\infty}^{\infty} g(t) \delta(\tau - t) \, dt = \int_{-\infty}^{\infty} g(\tau) \delta(\tau - t) \, dt = g(\tau) \int_{-\infty}^{\infty} \delta(\tau - t) \, dt = g(\tau)$$

An effective way to achieve a sound conceptual understanding of the convolution and correlation operations is through graphical developments, which is done in the following with respect to the convolution. Let $g(t)$ and $h(t)$ be given by Equations 2.24 and 2.25, as illustrated in Figure 2.53.

$$g(t) = \begin{cases} 1.5 & \text{if } -1 < t \leqslant 0, \\ 0 & \text{otherwise,} \end{cases} \tag{2.24}$$

and

$$h(t) = \begin{cases} 2 & \text{if } 0 < t \leqslant 2, \\ 0 & \text{otherwise.} \end{cases} \tag{2.25}$$

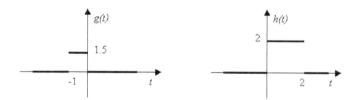

Figure 2.53: *Two functions $g(t)$ and $h(t)$ to be convolved.*

The first step required in order to obtain the convolution between these two functions consists in determining $h(-t)$, which is achieved, as discussed in Section 2.1.3, by reflecting $h(t)$ with respect to the y-axis, as illustrated in Figure 2.54 (a).

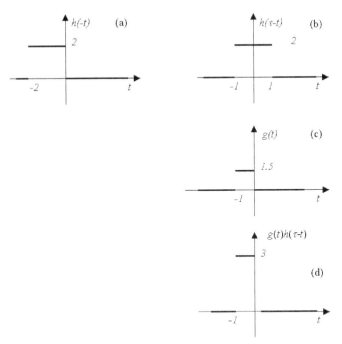

Figure 2.54: *Illustration of the basic operations involved in the convolution of the functions g(t) and h(t). See text for explanation.*

Let us now restrict our attention to a specific value of the variable τ, for example, $\tau = 1$. When this value is added to the argument of $h(-t)$, the function shifts to the right, as illustrated in Figure 2.54 (b). According to Equation 2.22, the function $h(\tau - t)$ is now multiplied by $g(t)$, shown again in Figure 2.54 (c), thus yielding the function $g(t)h(\tau - t)$ in Figure 2.54 (d).

The convolution at $\tau = 1$ is finally given by the integral of $g(t)h(\tau - t)$, which corresponds to the area below the function in Figure 2.54 (d):

$$(g * h)(\tau) = \int_{-\infty}^{\infty} g(t)\,h(\tau - t)\,\mathrm{d}t = 3$$

Thus we have obtained the convolution value for $\tau = 1$, as shown in Figure 2.55.

By repeating the above procedure for all (and infinite) possible values of τ, we get the complete convolution shown in Figure 2.56.

The correlation can be understood in a similar manner, except for the fact that the second function is not reflected and, for complex functions, by the conjugation of the first function. Figure 2.57 shows the correlation of the above real functions, i.e., $g(t) \circ h(t)$.

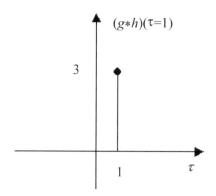

Figure 2.55: *The convolution* $(g * h)(\tau)$ *for* $\tau = 1$.

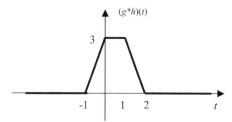

Figure 2.56: *The complete convolution* $(g * h)(t)$.

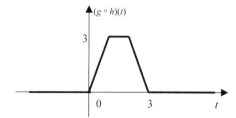

Figure 2.57: *The correlation* $(g \circ h)(t)$.

Let us now consider that both $g(t)$ and $h(t)$ have finite extension along the domain, i.e., $g(t), h(t) = 0$ for $t < r$ and $t > s$. Recall from Section 2.2.4 that the inner product between two functions $g(t)$ and $h(t)$ with respect to the interval $[a, b]$ is given by

$$\langle g, h \rangle = \int_a^b g^*(t)\, h(t)\, dt$$

Observe that this equation is similar to the correlation equation, except that the latter includes the parameter τ in the argument of the second function, which allows the second function to be shifted along the x-axis with respect to the first function. As a matter of fact, for each fixed value of τ, the correlation equation becomes an inner product between the first function and the respectively shifted version of the second. This property allows us to interpret the correlation as a measure of the 'similarity' between the two functions with respect to a series of relative shifts between these functions. Figure 2.58 illustrates this fact. The correlation $f(t) = (g \circ h)(t)$

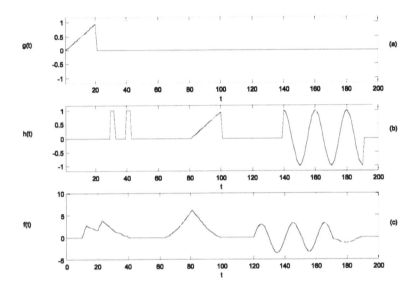

Figure 2.58: *The correlation* (c) *between two time limited functions* $g(t)$ (a) *and* $h(t)$ (b) *provides an indication about the similarity between the several pieces along the functions. Observe that in* (c) *the correlation* $f(t) = (g \circ h)(t)$ *is shown only for* $t \geqslant 0$, *the other half being omitted for simplicity's sake.*

slides the "template" function $g(t)$ along the function $h(t)$, calculating the inner product for each of these situations, in such a way that each correlation intensity provides an indication of the 'similarity' between the functions. The maximum intensity is verified for $t = 80$, which corresponds to the reference position of the sought pattern in the function $h(t)$. In other words, the correlation allows us to seek for the position where the two functions are most similar (in the sense of larger inner product).

However, it should be borne in mind that the inner products implemented by the correlation only make sense when the functions have their amplitudes properly *normalized*, or at least nearly so. For instance, if the first peak in Figure 2.58(b), centered at $t = 31$, were high enough, the correlation function would peak near $t = 10$, causing a false alarm. Among the many possible normalization schemes, it would be possible to apply the affine transformation described in Section 2.1.4

in order to map both functions to be correlated into the [0, 1] interval, or to use the statistic normal transformation described in Section 2.6.2.

The convolution and correlation can be straightforwardly extended to 2D functions $g(x, y)$ and $h(x, y)$ as presented in the following:

2D Convolution:

$$(g * h)(\alpha, \beta) = \int_{-\infty}^{\infty} \int_{-\infty}^{\infty} g(x, y) h(\alpha - x, \beta - y) \, dx \, dy$$

2D Correlation:

$$(g \circ h)(\alpha, \beta) = \int_{-\infty}^{\infty} \int_{-\infty}^{\infty} g^*(x, y) h(x + \alpha, y + \beta) \, dx \, dy$$

See Chapter 3 for more detail on the application of these operations to images.

2.5.2 Discrete Convolution and Correlation

Let $g(i)$ and $h(i)$ be discrete domain (i.e., i is an integer value) functions which are zero outside $a \leqslant i \leqslant b$ and $c \leqslant i \leqslant d$, respectively, where a, b, c and d are integer values, as illustrated in Figure 2.53. Our objective in this section is to develop numerical procedures for calculating the convolution and correlation between these two functions. The discrete correlation is developed graphically in the following, in order to illustrate the basic numeric approach.

Let us start by observing that the functions g and h have lengths $N = b-a+1$ and $M = d-c+1$, respectively. These functions are illustrated in Figure 2.59 (a) and (b), respectively. Next, the function $h(i)$ is padded with $N - 1$ zeros at both its right and left sides, i.e., $h(i) = 0$ for $cc = c-N+1 \leqslant i \leqslant c-1$ and $d+1 \leqslant i \leqslant D+N-1 = dd$, yielding the new extended function $\tilde{h}(i)$, shown in Figure 2.59 (c).

Observe that, since $c - d < b - a$ is always verified (in fact, by construction $c < d$, implying that $c - d$ is always negative; and $a < b$, implying that $b - a$ is always positive), we ensure that $c-b < d-a$. Thus, the correlation can be organized as $\left(g \circ \tilde{h}\right)(k)$ for $c - b \leqslant k \leqslant d - a$, as illustrated in Figure 2.60. The correlation is obtained by shifting the padded function $\tilde{h}(i)$ according to the above values of k, which is then multiplied by the function $g(i)$ for $a \leqslant i \leqslant b$. Finally, the results are added.

Observe that, consequently, the discrete correlation $\left(g \circ \tilde{h}\right)(k)$ has length L given as

$$L = (b - c)-(a - d)+1 = b-a+d-c+1 = (b - a + 1)+(d - c + 1)-1 = M+N-1$$

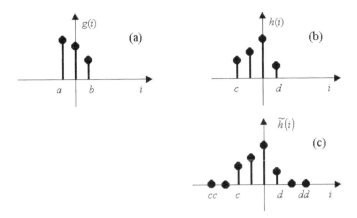

Figure 2.59: *The two discrete functions g(i) (a) and h(i) (b) to be corre-lated. The padded function $\tilde{h}(i)$ (c).*

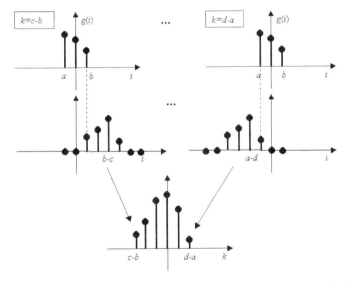

Figure 2.60: *Graphical development of the discrete correlation $(g \circ h)(k)$. Since $c - d < b - a$ we have $c - b \leqslant k \leqslant d - a$. The correlation is obtained by shifting the padded function $\tilde{h}(i)$ according to the above values of k, multiplying it by the function g(i), and then adding the results.*

The *discrete correlation* $(g \circ h)(k)$ can now be expressed as

$$(g \circ h)(k) = \sum_{i=a}^{b} g^*(i)\, \tilde{h}(k+i) \quad \text{for} \quad c - b \leqslant k \leqslant d - a. \tag{2.26}$$

By using a similar development, the *discrete convolution* $(g * h)(k)$ can be verified to be

$$(g * h)(k) = \sum_{i=a}^{b} g(i) \, \tilde{h}(k - i) \quad \text{for} \quad c + a \leqslant k \leqslant d + b. \tag{2.27}$$

In case the adopted programming language does not allow negative indexes, the strategy presented in the box in Section 2.2.1 can be used. For discrete correlation, we have the following algorithm:

Algorithm: *Discrete Correlation*

1. **for** $k \leftarrow c - b$ **to** $d - a$
2.   **do**
3. $aux \leftarrow 0$;
4.   **for** $i = a$ **to** b
5.   **do**
6. $aux \leftarrow \text{CONJ}(g(i - a + 1)) * h(k + i - cc + 1) + aux$;
7. $corr(1, k - c + b + 1) \leftarrow aux$;

---

where *conj* stands for the complex conjugate. For discrete convolution, we have the following algorithm:

Algorithm: *Discrete Convolution*

1. **for** $k \leftarrow c + a$ **to** $d + b$
2.   **do**
3. $aux \leftarrow 0$;
4.   **for** $i \leftarrow a$ **to** b
5.   **do**
6. $aux \leftarrow g(i - a + 1) * h(k - i - cc + 1) + aux$;
7. $conv(1, k - c - a + 1) \leftarrow aux$;

Observe that the above pseudo-codes, which have been prepared in order to favor intelligibility and ensure strictly positive indexing, can be further optimized.

In case the discrete functions have subsequent elements separated by Δ instead of 1, as is the situation assumed in the above developments, the discrete correlation and convolution equations should be rewritten as equations (2.28) and (2.29), respectively:

$$(g \circ h)(k) = \Delta \sum_{i=a}^{b} g^*(i) \, \tilde{h}(k + i) \quad \text{for } c - b \leqslant k \leqslant d - a, \tag{2.28}$$

$$(g * h)(k) = \Delta \sum_{i=a}^{b} g(i)\,\tilde{h}(k-i) \quad \text{for } c + a \leqslant k \leqslant d + b. \tag{2.29}$$

2.5.3 Nonlinear Correlation as a Coincidence Operator

While the correlation is frequently used as a means to compare (or match) two signals (see Figure 2.58), it presents the serious shortcoming that its result is largely affected by the signal amplitude. For instance, consider that we want to compare the two discrete signals g and h below:

$g = 0\,0\,1\,1\,1\,1\,0\,0\,0\,0\,0\,0\,4\,4\,4\,4\,0\,0\,0\,0\,0\,0\,0\,0\,0\,0$

$h = 0\,0\,0\,0\,0\,0\,4\,4\,4\,4\,0\,0\,0\,0\,0\,0\,0\,0\,0\,0\,1\,1\,1\,1\,0\,0$

Figure 2.61 presents the above two sequences, shown in (a) and (b), respectively, and the result of the standard correlation between these two signals (c). Recall from Section 2.5.1 that, in principle, each peak produced by the correlation indicates a possible coincidence (or match) between portions of the signals. However, three peaks are observed in (c). Indeed, a false intermediate peak has been obtained because of the interference between the two groups of "1s" and "4s." Moreover, the two external peaks indeed corresponding to the matches between portions of the signals present different amplitudes although they refer to partial matches of the same length (i.e., 4 values).

Much improved (actually exact) coincidence detection can be obtained by using the methodology first described in [Felsenstein et al., 1982], which involves the decomposition of the discrete signals into binary signals, yielding a nonlinear correlation. First, each signal is decomposed into a series of binary subsignals s_V, one for each of the M possible nonzero values of the elements in the original signal (observe that in the above signals $M = 2$). "1s" in one such subsignal corresponding to the value V indicate that the respective positions contained the value V, as illustrated below for the above discrete signal $g(t)$:

$$
\begin{aligned}
g &= 0011110000004444000000000000\\
g_1 &= 001111000000000000000000000000\\
g_4 &= 000000000000111100000000000
\end{aligned}
$$

Once both signals have been decomposed, each pair of subsignals is correlated by using the standard linear correlation, yielding $s_V(t) = g_V(t) \circ h_V(t)$. The final coincidence $u(t)$ is obtained simply by adding all such correlations, i.e., $u(t) = \sum_{V=1}^{M} s_V(t)$. Figure 2.61 (d) presents the coincidence between the two above signals $g(t)$ and $h(t)$ obtained by using the above described methodology. It is clear that now exactly two peaks, each with maximum value of 4, have been obtained as a precise identification of the position and extension of the matches between the two discrete signals. In addition to being fully precise, the coincidence operator can be performed in $O(N \log(N))$ by using the correlation theorem in Section 2.7.3 and the fast Fourier transform to calculate the involved correlations. Further improvements to this technique have been described in [Cheever et al., 1991].

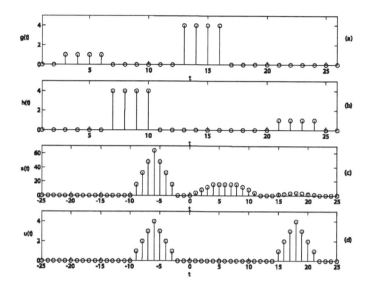

Figure 2.61: *The discrete signals g(t)* (a) *and h(t)* (b). *While the result*
of the standard correlation between these two signals (c)
provides peaks with different heights for each group of "1s"
and "4s," as well as false alarms between the two peaks, the
coincidence operator allows a fully precise result (d).

To probe further: *Correlation and Convolution*

The correlation and convolution operations are usually covered in signal and image
processing textbooks, such as [Brigham, 1988; Gonzalez & Woods, 1993; Morrison, 1994; Oppenheim & Schafer, 1975, 1989].

2.6 Probability and Statistics

Probability and statistics play a key role in image and shape analysis because of the
variability of shapes and images as well as the diverse types of noise and artifacts,
which can often be statistically modeled. Concepts and techniques from these two
important areas are also used to treat situations of which we do not have complete
knowledge. For instance, a face recognition system can be trained to recognize a
set of specific faces, but it is not generally informed about all other possible faces.
The current section starts by presenting the key concepts of events and probabilities

and follows by introducing random variables and vectors, distributions and density functions, basic concepts in estimation, random processes, cross- and autocorrelation, and the Karhunen-Loève transform.

2.6.1 Events and Probability

Probability deals with observations and measurements made during *experiments* involving outcomes that cannot be fully predicted, i.e., that are *random*. The set S containing all possible outcomes in a given experiment is called its *sample space* (or *ensemble*), and an *event* obtained in an experiment can be any subset of S. Table 2.5 presents examples of these concepts.

Experiment	E_1: Measuring your weight in the morning	E_2: Throwing a dice
Sample space	$S = \{$a real number between 30 Kg and 200 Kg$\}$	$S = \{1, 2, 3, 4, 5, 6\}$
Some possible events	$A_1 = \{71\,\text{Kg}\}$ $A_2 = \{\text{weight} \geqslant 70\,\text{Kg}\}$ $A_3 = \{65\,\text{Kg} < \text{weight} < 69\,\text{Kg}\}$	$B_1 = \{1\}$ $B_2 = \{2, 3, 4\}$ $B_3 = S$
Outcomes	One of the infinite elements of S	$\{1\}\ \{2\}\ \{3\}\ \{4\}\ \{5\}\ \{6\}$

Table 2.5: *Elementary concepts in probability.*

Random experiments can be classified into two major classes according to the nature of their sample space: *continuous* (e.g., E_1 in Table 2.5), and *discrete* (e.g., E_2 in Table 2.5). In the case of *discrete* experiments involving outcomes which are equally likely, the *probability* of a specific event A is defined as

$$P(A) = \frac{\text{number of elements in } A}{\text{number of elements in } S}.$$

Thus, the probabilities for the events of the second experiment in Table 2.5 are

$$P(B_1) = \frac{1}{6},$$
$$P(B_2) = \frac{3}{6} = \frac{1}{2}, \quad \text{and}$$
$$P(B_3) = \frac{6}{6} = 1.$$

More generally, the probability of an outcome/event is defined as the limit of the relative frequencies of respective observations when the number of observations tends to infinity. A well-defined probabilistic model should satisfy the following axioms:

Axiom 1: $P(S) = 1$;

Axiom 2: $0 \leqslant P(A) \leqslant 1$;

Axiom 3: For two *mutually exclusive events A and B* (i.e., $A \cap B = \emptyset$):

$$P(A \cup B) = P(A) + P(B).$$

In many practical situations, it is interesting to specify the probability of a specific event B given that another event A has already occurred. For instance, we may be interested in the following situations:

① The outcome of throwing a coin and observing head (event B) given that the last throw produced a tail (event A);

② In an urn containing 10 balls, numbered from 1 to 10, to extract an even ball (event A) and then verify that this number is 3 or 4 (event B).

Figure 2.62 graphically illustrates the important concept of conditional probability.

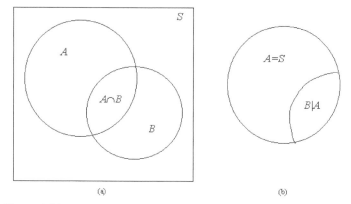

(a)   (b)

Figure 2.62: *Venn diagram illustrating the events A, B, and their intersection (a). The fact that A has occurred redefines the sample space (b).*

The initial universe set, shown in (a), corresponds to the sample space S, of which both A and B are subsets, and thus events. Given that A has already taken place, the observation of B, an event represented as $B \mid A$ (B given A), is referred to as the event B conditioned to A. It should be observed that, although $B \mid A$ corresponds to the intersection between A and B, we do not necessarily have $P(B \mid A) = P(B \cap A)$ because the fact that A has already happened defines a new universe set for the event $B \mid A$, as illustrated in (b). The probability associated to the conditional event $B \mid A$ is therefore calculated as

$$P(B \mid A) = \frac{P(A \cap B)}{P(A)}$$

A further illustration of the use of the Venn diagram to represent conditional probabilities, with respect to the above example ②, is presented in Figure 2.63.

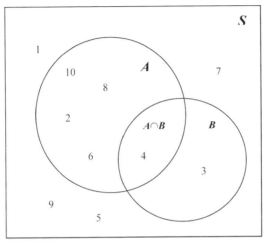

Figure 2.63: *Venn diagram illustration of the situation in the above example (b), i.e., extracting balls from an urn. In case the extracted ball is known to be even (represented as the event A), the probability of the ball number being 4 is* $P(B \mid A) = \frac{P(A \cap B)}{P(A)} = \frac{1}{5}$.

The following relationship between conditional probabilities is known as *Bayes's rule*:

$$P(A \mid B) = \frac{P(B \mid A)\, P(A)}{P(B)}. \tag{2.30}$$

In case $P(A \cap B) = P(A) \cdot P(B)$, thus implying $P(B) = P(B \mid A)$, the events B and A are said to be *independent* (not to be confused with mutually exclusive). For instance, the two events in the above example ①, i.e., coin throwing, are independent in the sense that the fact of observing the outcome of a coin throw does not affect the next observations.

2.6.2   Random Variables and Probability Distributions

The concept of random variable is essential in probability because it allows algebraic treatment of the involved concepts and models, such as density and distribution functions, which characterize the probabilistic behavior of the respective random variables.

Random Variables

One of the most important concepts in probability is that of a *random variable*, which is used to represent the value resulting as outcome of a specific experiment.

For instance, in the first example of Table 2.5, the random variable is the morning weight, which can assume different values on different days. Henceforth, we shall adopt the convention of representing the *name* of a random variable in *upper case* and its *observed values* in *lower case*. For instance, in the case of the above example, the morning weight random variable is W, and its observation on February 1, 2008 can be expressed as w.

Random variables can be classified as continuous or discrete depending on the type of measurement. For instance, your morning weight is a continuous random value, but the number of coins in the purse of a New Yorker is a discrete random variable. Although continuous random variables are restricted to the theoretical side of probability and statistics, their study is very important for defining practical concepts and techniques. That is why we first consider probabilistic modeling of continuous random variables, of which discrete random variables will be considered a particular case (by using the Dirac delta function—see Section 2.1.4).

Density Functions

The key concept involved in random variables is that of a *density function*, which is any function $p(x)$, where x stands for observations of a random variable X, satisfying the following criteria:

$$p(x) \geqslant 0 \quad \text{for any real } x$$

and

$$\int_{-\infty}^{\infty} p(x) \, dx = 1.$$

Given a density probability function $p(x)$, the respective *probability distribution* $P(x)$ can be defined as

$$P(x) = P(X \leqslant x) = \int_{-\infty}^{x} p(s) \, ds, \qquad (2.31)$$

hence

$$p(x) = \frac{dP}{dx} = P'(x).$$

The importance of the density function associated with a random variable X resides in the fact that it tells us everything that is possible to know about the behavior of that random variable. Particularly, we know from equation (2.31) that

$$P(a < X \leqslant b) = P(b) - P(a) = \int_{a}^{b} p(s) \, ds. \qquad (2.32)$$

In other words, the probability of observing a value of a random variable within the interval $[a, b)$ corresponds to the area under the density function along that interval. Figure 2.64 illustrates this concept with respect to a Gaussian density function $g(x)$.

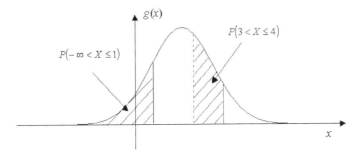

Figure 2.64: *A Gaussian density function characterizing the random variable X. The probability of finding an observation comprised in an interval is equal to the area below the function along that interval.*

In addition, it follows from the basic postulates of probability that

$$P(X \leqslant 1) \quad \text{or} \quad P(3 < X \leqslant 4) = P(X \leqslant 1) + P(3 < X \leqslant 4).$$

An important implication of (2.32) is that

$$P(X = a) = \int_{a}^{a} p(s)\, ds = 0,$$

which indicates that the probability of observing an exact real value is null.

Statistical Moments

The *moments* and *central moments* of the random variable X, characterized by the density function $p(x)$, are defined in terms of the expectance of X, i.e., $E[X] = \int_{-\infty}^{\infty} xp(x)\, dx$, as indicated in the following

k-th moment:

$$M_k[X] = E\left[X^k\right] = \int_{-\infty}^{\infty} x^k p(x)\, dx;$$

k-th central moment:

$$\widetilde{M}_k[X] = E\left[(X - E[X])^k\right] = \int_{-\infty}^{\infty} (x - E[X])^k\, p(x)\, dx.$$

Such moments are particularly important because they provide global information about the behavior of the random variable X. For instance, the *first moment*, which is commonly known as the *average* (or *expectance*) of X, is calculated as

$$E[X] = \mu_X = \int_{-\infty}^{\infty} xp(x)\,dx.$$

The *second central moment* of X, known as the *variance* of X, is calculated as

$$\widetilde{M}_2[X] = \text{var}(X) = \sigma X^2 = E\left[(X - \mu_X)^2\right] = \int_{-\infty}^{\infty} (x - \mu_X)^2\, p(x)\,dx.$$

It can be verified that

$$\text{var}(X) = E\left[X^2\right] - (E[X])^2.$$

The *standard deviation* σ_X of the random variable X is given by the positive square root of the respective variance, i.e., $\sigma_X = +\sqrt{\sigma_X^2} = \sqrt{\text{var}(X)}$. Both the variance and the standard deviation are particularly important because they provide an indication about the variation (or dispersion) of X around the average value. In addition, together with the average, the standard deviation completely characterizes a Gaussian density, which is one of the most important functions in probability.

Some Density Functions

Table 2.6 presents three particularly important density probability functions, namely the *uniform*, *exponential* and *Gaussian* distributions and their respective means and variances, as well as their graphical illustrations.

Recall from Section 2.1.4 that the standard deviation of the Gaussian corresponds to the displacement from the average where the maximum of the Gaussian falls by the factor $1/\sqrt{e}$. It can be shown that one standard deviation around the average comprises 68.27% of the area of the Gaussian (i.e., one standard deviation to the left and one to the right), and the substantial percentage of 99.73% of the area of the Gaussian is verified for three standard deviations around the average.

Random Variables Transformations, Conditional Density Functions and Discrete Random Variables

Given a random variable X, it is possible to transform it in order to obtain a new random variable Y. A particularly useful *random variable transformation*, called *normal transformation* or *standardization*, is obtained by applying the following equation:

$$Y = \frac{X - E[X]}{\sigma_X}.$$

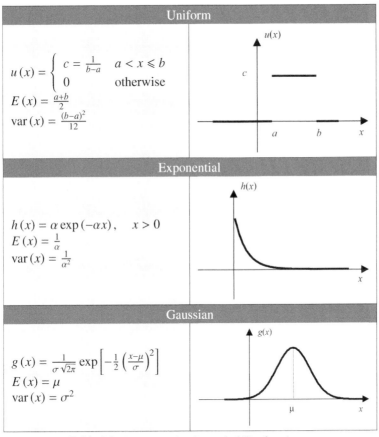

Table 2.6: *Important density probability functions.*

It can be verified that the new random variable Y has zero mean and unit standard deviation.

Going back to conditional probabilities, it should be observed that it is also possible to define conditional density functions, such as $p(X \mid A)$, meaning that the density function of X is restricted by the fact that the event A has occurred. For instance, one might be interested in the height of an individual (the random variable X) given that this person is a female (the event A). The Bayes' rule can be immediately extended to conditional density functions:

$$p(x \mid A) = \frac{P(A \mid x) \, p(x)}{P(A)}. \tag{2.33}$$

Now that we have acquainted ourselves with the main concepts in continuous random variables, it is time to proceed to *discrete random variables*. Let us return to the dice throwing example presented in Section 2.6.1, where the observed value is represented as the discrete random variable X. We have already verified that

$$P(x = 1) = P(x = 2) = P(x = 3) = P(x = 4) = P(x = 5) = P(x = 6) = \frac{1}{6}.$$

Therefore, by using the Dirac delta function to represent the probabilities at the discrete values, the respective density probability function modeling this experiment can be obtained as

$$p(x) = \frac{1}{6} \left[\delta(x - 1) + \delta(x - 2) + \delta(x - 3) + \delta(x - 4) + \delta(x - 5) + \delta(x - 6) \right].$$

For instance, the probability of having the event $A = \{2, 3, 4\}$ can now be immediately calculated as

$$P(A) = \int_{1.5}^{4.5} p(x) \, dt =$$
$$= \int_{1.5}^{4.5} \frac{1}{6} \left[\delta(x - 1) + \delta(x - 2) + \delta(x - 3) + \delta(x - 4) + \delta(x - 5) + \delta(x - 6) \right] dt =$$
$$= \frac{1}{6} \int_{1.5}^{4.5} \left[\delta(x - 2) + \delta(x - 3) + \delta(x - 4) \right] dt = \frac{1}{6}3 = \frac{1}{2}.$$

2.6.3 | Random Vectors and Joint Distributions

We have thus far been restricted to scalar random variables. In order to apply the important and useful concepts of probability and statistics to image and shape processing and analysis, it is essential to get acquainted with some basic principles in multivariate random variables.

A *random vector* is nothing else but a vector of random variables. For instance, the temperature (T), pressure (P), and humidity (H) at noon in the North Pole on a certain day can be represented by the following random vector:

$$p = \begin{bmatrix} T \\ P \\ H \end{bmatrix} = (T, P, H).$$

As in the one-dimensional case, the behavior of a random vector is completely specified in terms of the associated density function, which now becomes a *multivariate function*. Let us henceforth assume a generic random vector including N random variables, i.e.,

$$\vec{X} = \begin{bmatrix} X_1 \\ X_2 \\ \vdots \\ X_N \end{bmatrix} = (X_1, X_2, \ldots, X_N).$$

The *joint density function* characterizing the behavior of such a random vector has the general form

$$p(\vec{x}) = p(x_1, x_2, \ldots, x_N),$$

which must satisfy the following conditions:

$$p(\vec{x}) \geq 0$$

and

$$\int_{-\infty}^{\infty} \int_{-\infty}^{\infty} \cdots \int_{-\infty}^{\infty} p(x_1, x_2, \ldots, x_N) \, dx_1 \, dx_2 \ldots dx_N = 1.$$

Given a density function $g(\vec{x}) = g(x_1, x_2, \ldots, x_N)$, it is possible to define *marginal density functions* for any of its random variables x_i, integrating along all the other variables, i.e.,

$$p(x_i) = \int_{-\infty}^{\infty} \int_{-\infty}^{\infty} \cdots \int_{-\infty}^{\infty} g(x_1, x_2, \ldots, x_i, \ldots, x_N) \, dx_1 \, dx_2 \ldots dx_{i-1} \, dx_{i+1} \ldots dx_N.$$

An example of joint density function is the multivariate Gaussian, given by the following

$$p(\vec{x}) = \frac{1}{(2\pi)^{\frac{N}{2}} \sqrt{\det(K)}} \exp\left[-\frac{1}{2}\left(\vec{x} - \vec{\mu}_{\vec{x}}\right)^{\mathrm{T}} K^{-1} \left(\vec{x} - \vec{\mu}_{\vec{x}}\right)\right].$$

This density function is completely specified by the mean vector $\vec{\mu}_{\vec{x}}$ and the covariance matrix K (see below).

The moments and central moments of an $N \times 1$ random vector $\vec{X}$, modelled by the joint density function $p(\vec{x})$, are defined as

$(n_1, n_2, \ldots, n_N)$-**th moment:**

$$M_{(n_1, n_2, \ldots, n_N)}\left[\vec{X}\right] = E\left[X_1^{n_1} X_2^{n_2} \ldots X_N^{n_N}\right] =$$
$$= \int_{-\infty}^{\infty} \int_{-\infty}^{\infty} \cdots \int_{-\infty}^{\infty} x_1^{n_1} x_2^{n_2} \ldots x_N^{n_N} p(\vec{x}) \, dx_1 \, dx_2 \ldots dx_N,$$

$(n_1, n_2, \ldots, n_N)$-**th central moment:**

$$M_{(n_1, n_2, \ldots, n_N)}\left[\vec{X}\right] = E\{(X_1 - E[X_1])^{n_1} (X_2 - E[X_2])^{n_2} \ldots (X_N - E[X_N])^{n_N}\} =$$
$$= \int_{-\infty}^{\infty} \int_{-\infty}^{\infty} \cdots \int_{-\infty}^{\infty} (x_1 - E[X_1])^{n_1} \ldots (x_N - E[X_N])^{n_N} p(\vec{x}) \, dx_1 \, dx_2 \ldots dx_N,$$

where $n_1, n_2, \ldots, n_N \in \{0, 1, 2, \ldots\}$. As with scalar random variables, such moments provide *global* information about the behavior of the random vector $\vec{X}$. For instance, the *mean vector* for $\vec{X}$ is the vector defined by the averages of each of its components, i.e.,

$$\vec{\mu}_{\vec{X}} = E\left[\vec{X}\right] = \begin{bmatrix} \mu_{X_1} \\ \mu_{X_2} \\ \vdots \\ \mu_{X_N} \end{bmatrix}.$$

Two particularly useful special cases of these moments are the so-called *correlation* and *covariance* between two scalar random variables X_i and X_j, which are respectively defined as

$$\mathrm{corr}\left(X_i, X_j\right) = E\left[X_i X_j\right] = \int_{-\infty}^{\infty} \int_{-\infty}^{\infty} \cdots \int_{-\infty}^{\infty} x_i x_j p(\vec{x}) \, dx_1 \, dx_2 \ldots dx_N$$

and

$$\mathrm{cov}\left(X_i, X_j\right) = E\left[(X_i - E[X_i])\left(X_j - E\left[X_j\right]\right)\right] =$$
$$= \int_{-\infty}^{\infty} \int_{-\infty}^{\infty} \cdots \int_{-\infty}^{\infty} (x_i - E[X_i])\left(x_j - E\left[X_j\right]\right) p(\vec{x}) \, dx_1 \, dx_2 \ldots dx_N.$$

For bivariate density functions $p(x, y)$, these equations simplify to

$$\mathrm{corr}(X, Y) = E[XY] = \int_{-\infty}^{\infty} \int_{-\infty}^{\infty} xy p(x, y) \, dx \, dy$$

and

$$\mathrm{cov}(X, Y) = E[(X - E[X])(Y - E[Y])] = \int_{-\infty}^{\infty} \int_{-\infty}^{\infty} (x - E[X])(y - E[Y]) p(x, y) \, dx \, dy.$$

The covariance between two random variables is a particularly relevant concept in probability. Conceptually, the covariance between X and Y can be understood as the tendency for these two random variables to *"vary together."* For instance, it is natural to expect that the weight and height of people tend to vary together in the sense that the higher a person is, the heavier that person *tends* to be. In such a case, the two random variables, height and weight, are said to be *positively correlated*. But two random variables can also be *negatively correlated*. For instance, your life expectancy is negatively correlated to the time you spend climbing high mountains.

In case $\text{cov}(X, Y) = 0$, we say that the random variables X and Y are *uncorrelated*. It is important to note that the fact that two random variables X and Y are uncorrelated does not imply that they are independent (see Section 2.6.1), but independence implies uncorrelation. The covariance can be alternatively expressed as

$$\text{cov}\left(X_i, X_j\right) = E\left[X_i X_j\right] - E\left[X_i\right] E\left[X_j\right].$$

In addition, it is interesting to observe that

$$\sigma_{X_i}^2 = \text{var}\left(X_i\right) = \text{cov}\left(X_i, X_i\right)$$

and, consequently, the standard deviation of the random variable X_i can be alternatively expressed as

$$\sigma_{X_i} = +\sqrt{\text{cov}\left(X_i, X_i\right)}.$$

Since the covariance between two random variables is not dimensionless, it becomes interesting to define the *correlation coefficient*, which provides a dimensionless and relative measure of the correlation between the two variables. The correlation coefficient $\text{CorrCoef}\left(X_i, X_j\right)$ is defined as

$$\text{CorrCoef}\left(X_i, X_j\right) = E\left[\left(\frac{X_i - \mu_{X_i}}{\sigma_{X_i}}\right)\left(\frac{X_j - \mu_{X_j}}{\sigma_{X_j}}\right)\right] = \frac{\text{cov}\left(X_i, X_j\right)}{\sigma_{X_i}\sigma_{X_j}}.$$

An important property of the correlation coefficient is that $\left|\text{CorrCoef}\left(X_i, X_j\right)\right| \leqslant 1$.

It should be observed that when the means of all the involved random variables are zero, the correlation between two variables becomes equal to the respective covariance. Similarly, the covariance becomes equal to the correlation coefficient when the standard deviations of all involved random variables have unit value. When all means are zero and all standard deviations are one, the correlations are equal to the covariances, which in turn are equal to the correlation coefficients. In other words, the covariance and correlation coefficient between random variables can be understood in terms of correlations between versions of those random variables transformed in order to have zero means and unit standard deviations (i.e., a normal transformation), respectively. These important and practical properties are summarized in the following:

X_i	$\widetilde{X}_i = \left(X_i - \vec{\mu}_{\vec{X}} \right)$	$\hat{X}_i = \left(\frac{X_i - \vec{\mu}_{\vec{X}}}{\sigma_{X_i}} \right)$
$\mathrm{corr}\left(X_i, X_j \right)$	$\mathrm{corr}\left(\widetilde{X}_i, \widetilde{X}_j \right) = \mathrm{cov}\left(X_i, X_j \right)$	$\mathrm{corr}\left(\hat{X}_i, \hat{X}_j \right) =$ $= \mathrm{cov}\left(\hat{X}_i, \hat{X}_j \right) =$ $= \mathrm{CorrCoef}\left(X_i, X_j \right)$

The pairwise correlations between all the components of a random vector $\vec{X}$ can be conveniently represented in terms of the *correlation matrix* of $\vec{X}$:

$$R = \left[\mathrm{corr}\left(X_i, X_j \right) \right].$$

The *covariance matrix* K of $\vec{X}$ and the *correlation coefficient matrix* C are obtained in an analogue fashion:

$$K = \left[\mathrm{cov}\left(X_i, X_j \right) \right] \tag{2.34}$$

and

$$C = \left[\mathrm{CorrCoef}\left(X_i, X_j \right) \right]. \tag{2.35}$$

It is interesting to observe that these three matrices are Hermitian (symmetric, in the case of real random variables). In addition, the elements along the main diagonal of the covariance matrix K can be immediately verified to correspond to var(X_i). The correlation and covariance matrices can be alternatively calculated as

$$R = E\left(\vec{X}\vec{X}^{\mathrm{T}} \right) \quad \text{and} \quad K = E\left[\left(\vec{X} - \vec{\mu}_{\vec{X}} \right)\left(\vec{X} - \vec{\mu}_{\vec{X}} \right)^{\mathrm{T}} \right].$$

Hence

$$K = E\left(\vec{X}\vec{X}^{\mathrm{T}} \right) - \vec{\mu}_{\vec{X}}\vec{\mu}_{\vec{X}}^{\mathrm{T}} = R - \vec{\mu}_{\vec{X}}\vec{\mu}_{\vec{X}}^{\mathrm{T}}. \tag{2.36}$$

2.6.4 Estimation

As shown in the last two sections, the knowledge of the density function associated with a random variable, or the joint density function associated with a random vector, is essential and enough for the complete characterization of the statistical behavior of the random phenomenon under study. However, while we have seen how the density functions can be used to calculate probabilities and moments, little has been said so far about how such functions can be obtained or estimated in practice. Basically, such functions are defined by the intrinsic properties of each random phenomenon. For instance, the positions $\vec{p} = (x, y)$ of electrons being projected by an electron beam over a screen can be analytically verified to follow a Gaussian joint density function. In such cases, we know the type of density function, but may have to infer the value of the involved parameters (e.g., the mean vector and covariance matrix of a multivariate Gaussian), a situation usually referred to as *parametric estimation*. However, since it is not always possible to analytically infer the type of density function, we often have to infer the type of density function that better models the investigated situation, which is called *non-parametric estimation*. The current section presents some of the most important basic concepts related to estimation.

Nonparametric Estimation and Parzen Windows

Let X be a random variable taking any value in the interval $[a, b]$. One of the most natural means to gain insight about the behavior of X consists in calculating the *relative frequencies* of the observations of X. In order to do so, we divide the interval $[a, b]$ into M subintervals of equal length $\Delta = \frac{b-a}{M}$, and associate a bin to each such subinterval. Thus, the i-th bin, $i = 1, 2, \ldots, M$, will be associated to the interval $[a + (i-1)\Delta, a + i\Delta)$. The value of a specific bin, which is initially zero, is incremented every time an observation falls inside the respective interval. After a large number N of observations have been processed, a *histogram* is obtained describing the distribution of the observed values. The histogram of relative frequencies f_R of the values in each interval can then be calculated by dividing the number of observations stored into each bin by M. It is convenient to define the relative frequency as a function of the middle value of each interval, i.e.,

$$f_R\left(x_i = a + i\Delta - \frac{\Delta}{2}\right) = \frac{\text{count in bin } i}{M}.$$

These concepts are exemplified in the box entitled *Relative Frequency Histograms*.

Example: *Relative Frequency Histograms*

In order to investigate the distribution of sizes of carps in a lake, several specimens were captured and measured, yielding the following table:

Carp #	Size (cm)	Carp #	Size (cm)	Carp #	Size (cm)	Carp #	Size (cm)
1	15.82	6	18.98	11	2.76	16	6.80
2	13.13	7	6.51	12	16.22	17	8.19
3	10.37	8	10.29	13	6.49	18	9.32
4	11.75	9	11.32	14	3.64	19	3.25
5	18.48	10	14.35	15	12.88	20	14.92

Obtain and graphically illustrate the histogram of relative frequencies of the random variable X = size of the carp.

Solution:

First, we observe that the values range from $a = 2.76$ to $b = 18.98$. Considering $M = 10$ bins, we have

$$\Delta = \frac{b - a}{M} = \frac{18.98 - 2.76}{10} = 1.62.$$

Next, we complete the following table:

Bin #	Subinterval	Number of Observations	Relative Frequency
1	[2.77, 4.38)	3	0.15
2	[4.38, 6.00)	0	0
3	[6.00, 7.62)	3	0.15
4	[7.62, 9.25)	1	0.05
5	[9.25, 10.87)	3	0.15
6	[10.87, 12.49)	2	0.1
7	[12.49, 14.11)	2	0.1
8	[14.11, 15.73)	2	0.1
9	[15.73, 17.35)	2	0.1
10	[17.35, 18.98]	2	0.1

The obtained histogram of relative frequencies is presented in Figure 2.65.

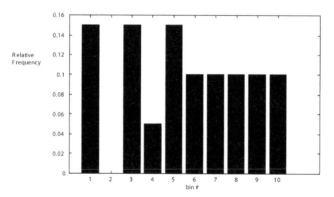

Figure 2.65: *The relative frequency histogram.*

It is interesting to observe that the relative frequency histogram can vary substantially with the number of observations. For instance, if $N = 200$ carps had been observed, the histogram of relative frequencies shown in Figure 2.66 could have been obtained.

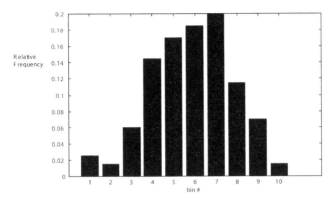

Figure 2.66: *The observation of a larger number of samples could lead to a different relative frequency histogram.*

The important property of the relative frequencies is that, when the size of the intervals (i.e., Δ) tends to zero, and a very large number of observations is available, the frequency distribution tends to the density function governing the respective random variable. Therefore, in addition to providing an insightful conceptual basis for properly understanding the meaning of density functions, the relative frequency distribution also provides a quantized (in the sense that the function is defined with respect to intervals and not points) approximation of the respective density functions, as illustrated in Figure 2.67.

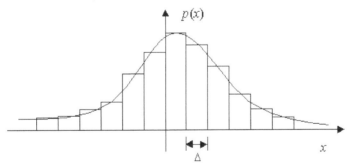

Figure 2.67: *The relative frequency histogram of observations of a random variable X provides a discrete approximation to the density function p(x) characterizing that variable.*

The above histogram approach can be nicely extended to provide a continuous estimation of the respective density functions through the technique known as *Parzen estimation*. This simple and elegant technique consists in approximating the sought continuous density function $p(x)$ in terms of a sum of windowing functions, such as a rectangular or Gaussian with unit area (i.e., the normal density), centered at each observation. The average of such estimated density functions is proven to converge to $p(x)$ under not so restrictive conditions. More information about the

Parzen approach can be found in [Duda & Hart, 1973; Duda et al., 2000; Therrien, 1992].

Parametric Estimation

Unlike the situation considered in the previous section, in parametric estimation one knows the mathematical nature of the density function (e.g., if it is Gaussian), and must only determine the respective parameters (the mean and standard deviation in case of a univariate Gaussian). While the theory of parameter estimation has a long tradition in statistics and has been thoroughly formalized and validated (e.g., [Duda & Hart, 1973; Duda et al., 2000; Therrien, 1992]), here we limit ourselves to presenting the equations for, given a set of observations of one or more random variables, estimating the mean, (2.37); correlation, (2.38); covariance, (2.39); and correlation coefficient (2.40). Recall that the variance is a special case of the covariance. The accompanying box illustrates these concepts with respect to a real situation. If the involved random variables or vectors are known to follow Gaussian density functions, it is preferable to use $N - 1$ instead of N in equations (2.38), (2.39) and (2.40), in order to avoid biased estimation.

Parameter	Estimate	
Mean	$$\tilde{\mu}_X = \frac{1}{N} \sum_{i=1}^{N} x$$	(2.37)
Correlation	$$\text{corr}(X, Y) = \frac{1}{N} \sum_{i=1}^{N} xy$$	(2.38)
Covariance	$$\text{cov}(X, Y) = \frac{1}{N} \sum_{i=1}^{N} (x - \mu_X)(y - \mu_Y)$$	(2.39)
Correlation Coefficient	$$\text{CorrCoef}(X, Y) = \frac{1}{N} \sum_{i=1}^{N} \frac{(x - \mu_X)}{\sigma_x} \frac{(y - \mu_Y)}{\sigma_Y}$$	(2.40)

Example: *Parameter Estimation*

While working on a excavation, an archaeologist decided to investigate the many coins that were being found. In order to do so, he measured the diameters and weights of 15 coins, having obtained the following data:

Coin #	Diameter (cm)	Weight (g)
1	1.6729	10.8251
2	1.5188	9.0747
3	1.3296	6.9096
4	1.6747	10.2153
5	1.4124	6.8538
6	1.4141	8.3434
7	1.2795	6.0283
8	1.5792	8.7062
9	1.3070	6.2545
10	1.5337	9.5365
11	1.1069	5.1879
12	1.3511	7.3627
13	1.3895	7.9017
14	1.3361	7.4823
15	1.7218	10.8841

Calculate the correlation, covariance and correlation coefficient matrices for such measures and discuss the achieved results.

Solution:

Let us represent the total of observed coins as $N = 15$. We clearly have the following two random variables: (a) diameter (X_1); and (b) weight (X_2). Let us represent these random variables in terms of the random vector $\vec{X} = (X_1, X_2)$. The mean vector therefore is

$$\vec{\mu}_{\vec{X}} = (\tilde{\mu}_{X_1}, \tilde{\mu}_{X_2}) = (1.4418, 8.1044).$$

Instead of obtaining the correlation matrix R by considering (2.38) for each random vector observation, it is more practical to understand the above table as a 15×2 matrix S. Then, we have

$$R = \frac{1}{N} S^T S = \begin{bmatrix} 2.1063 & 11.9557 \\ 11.9557 & 68.5617 \end{bmatrix}.$$

Now, by using (2.36), we have:

$$K = R - \vec{\mu}_{\vec{X}}\left(\vec{\mu}_{\vec{X}}\right)^{\mathrm{T}} = \begin{bmatrix} 2.1063 & 11.9557 \\ 11.9557 & 68.5617 \end{bmatrix} - \begin{bmatrix} 2.0789 & 11.6851 \\ 11.6851 & 65.6814 \end{bmatrix} =$$

$$= \begin{bmatrix} 0.0275 & 0.2706 \\ 0.2706 & 2.8803 \end{bmatrix}.$$

The correlation coefficient matrix can now be obtained by applying (2.35):

$$C = \begin{bmatrix} \dfrac{0.0275}{\sqrt{(0.0275)(0.0275)}} & \dfrac{0.2706}{\sqrt{(0.0275)(2.8803)}} \\ \dfrac{0.2706}{\sqrt{(0.0275)(2.8803)}} & \dfrac{2.8803}{\sqrt{(2.8803)(2.8803)}} \end{bmatrix} = \begin{bmatrix} 1 & 0.9616 \\ 0.9616 & 1 \end{bmatrix}.$$

In order to better understand such results, it is interesting to examine the graphical aspect of the data. Figure 2.68 illustrates the observed measures and includes the mean vector, corresponding to the point defined by the intersection of the two arrows.

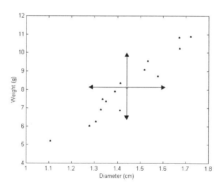

Figure 2.68: *The graphical representation of the observed random values, the mean and the standard deviations (represented by the arrows).*

Since the mean vector is not null, the correlation matrix is affected by the respective means of the diameter and weight. In such a situation, it is more useful to consider the variance and correlation coefficient matrices, which do not depend on the mean vector. The elements $K_{1,1}$ and $K_{2,2}$ are immediately verified to correspond to the diameter and weight variances, respectively, in such a way that their positive square roots indicate the respective standard deviations $\sigma_{X_1} = 0.17$ and $\sigma_{X_2} = 1.70$, which can be related to the dispersion of the observations along the x- and y-axis in the above picture (indicated in the above graph by the "arrowed" intervals centered at the mean vector). The elements $K_{1,2} = K_{2,1} = 0.2706$ correspond to the covariance between the diameter and weight values. Since the diameter and weight values vary along reasonably different ranges (while the diameters vary from 1.1069 to 1.7218, the weights vary from 5.1879 to 10.8841), it becomes difficult to interpret the covariance matrix. The correlation coefficient matrix, however, provides a normalized version of the covariance matrix, corresponding to normal transformations of the diameter and weight random variables (see Section 2.6.2), which is particularly useful for interpreting the covariance between the two variables. In this example, the correlation coefficient between the diameter and weight is observed to be $C_{1,2} = C_{2,1} = 0.9616$, which indicates an almost linear dependency between the two variables. Such linear relationship is graphically expressed by the fact that the cloud of points in Figure 2.68 tends to present an elongated and straight shape.

In addition to illustrating some important basic concepts in parametric estimation, the above example also provides insight about possible misleading interpretations of statistical data. Indeed, a more careful analysis of the situation addressed by the archaeologist, namely the measure of diameters and weights of coins, indicate that these two measures can in fact be related by the following equation:

$$\text{weight} = 2\pi \left(\frac{\text{diameter}}{2} \right)^2 (\text{width})(\text{volumetric density}).$$

In other words, except for fluctuations implied by possibly varying volumetric densities (e.g., in case the coins are made of different materials or the material quality varies), allied to the fact that coins are not exact cylinders and exhibit defects caused at production or by wearing out, it is expected that the weight would be a quadratic function of the diameter, and not linear as hinted by the above correlation coefficient (i.e., $C_{1,2} = C_{2,1} = 0.9616$). This fact is illustrated in Figure 2.69, which indicates a possible situation that would have been observed in case more coins (especially with diameter exceeding the previous range) were considered. Observe that the conclusion about a possible linear relationship between the random variables was a consequence of limiting our observations to a small range of diameter and volume values.

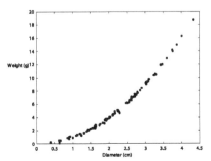

Figure 2.69: *A nonlinear relationship between the weight and diameter could have been obtained in case more observations were considered in the previous example.*

Although the above characterized misinterpretation does not properly constitute a mistake provided the considered data remain within the considered ranges (i.e., a parabola does approximate a straight line along a small portion of its domain), it indicates that every care should be taken while extrapolating and drawing general conclusions based on statistical observations and measures along limited intervals.

2.6.5 | Stochastic Processes and Autocorrelation

A sequence of random variables occurring along a domain (typically time or space) is called a *stochastic process*. For instance, the number of coins a New Yorker carries on each successive day can be understood as a stochastic process having time as domain. In case the number of coins on each day i is represented in terms of the random variable X_i, the stochastic process can be represented as $X = \ldots, X_{i-1}, X_i, X_{i+1}, \ldots$, and the joint density fully characterizing this process is $g(\ldots, X_{i-1}, X_i, X_{i+1}, \ldots)$. The distribution of gray-level values along images is another example, but now the respective bivariate domain corresponds to the positions (x, y) in space. The set of all possible observations is called the *ensemble* of the stochastic process, and a specific observed sequence is called an *observation path*. A stochastic process is said to be *stationary* in case its statistical description (be it in terms of the joint density or moments) does not vary with displacements of the sequence around the domain origin (i.e., when calculated along different domain positions). More informally, the statistical properties of a stationary process do not vary from one position to another along the domain. A stationary process whose average and correlations are the same along the whole domain is said to be *wide-sense stationary*. In case the estimation of a specific moment of a stationary process calculated along the observation paths is equal to the estimation along the ensemble, the process is said to be *ergodic* with respect to that moment.

Let X be a stochastic process assuming real values. The correlations between any of its two elements, hence $\text{corr}(X_i, X_j)$, is called the *autocorrelation* of X. Note that this concept can be immediately extended, considering that each of the random variables is taken from two different stochastic processes, to yield the cross-correlation between those processes. As illustrated in the current section, the autocorrelation concept is particularly important for signal and image analysis, since it indicates how much each point along the observation path is correlated (and therefore redundant) to its neighbors. In case X is wide-sense stationary, we have that the autocorrelation is a function only of the lag $n = j - i$, i.e.,

$$\text{corr}(n) = \text{corr}(X_i, X_{i+n}) \quad \text{for} \quad n = \ldots, -2, -1, 0, 1, 2, \ldots$$

Observe that the lag represents the distance between pairs of elements along the observation path. If in addition X is ergodic in correlation, we can estimate the above correlations by observing any of its specific observation paths. Although we have thus far concentrated on stochastic process defined along infinite extensions of its domain (e.g., from minus infinity to infinity), practical situations imply these observations are limited along time. In such cases, the sequences must be long enough to provide a reasonable representation of the process. In such a situation, i.e., for a stochastic process X with finite duration, henceforth represented as $X(i)$, $a \leqslant i \leqslant b$ (a and b are integer values), it can be easily verified that the respective autocorrelation function can be calculated in terms of the discrete correlation operation presented in Section 2.5.2.

Figure 2.70 illustrates two images (a) and (c) and their respective circular autocorrelations (b) and (d).

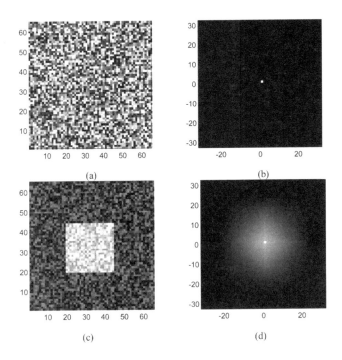

(a)

(b)

(c)

(d)

Figure 2.70: *Two images* (a) *and* (c) *and their respective circular autocorrelations* (b) *and* (d). *See text for discussion.*

Recall that the autocorrelation domain corresponds to all the possible vector lags (i.e., relative positions) between any two pixels in the image. In addition, circular autocorrelation considers the signal as being periodic, i.e., a toroid, in the sense that its extremities are merged. The fact that the gray level of each of the pixels in image (a) differs substantially from most of the other pixels in the image is properly expressed by the sharp peak at lag $(0, 0)$ in (b). In other words, the pixels' gray levels are only similar when considering the pixels in a one-by-one fashion. On the other hand, the pixels in image (c) tend to have higher and more similar gray levels within the middle square, which is suitably captured by the wider cloud of points in respective autocorrelation in (d). The above example also illustrates the property that the autocorrelation is always maximum at zero lag (i.e., the origin of the autocorrelation domain).

It is important to observe that *correlation*, *redundancy* and *smoothness* are related concepts. For instance, a stochastic sequence characterized by intense correlations for several lags will tend to be redundant and smooth. A sinusoidal wave, for instance, is highly correlated and smooth. This can be conceptually perceived by the fact that the nearest neighbors at each point along the sinusoidal sequence tend to vary together (i.e., increase or decrease together), except at pairs of points placed immediately at each side of a zero crossing, where negative correlation is verified. As a matter of fact, most images of natural and human-made objects and scenes are highly correlated and redundant, which explains the importance of the object contours for image analysis. On the other hand, noisy signals are highly uncorrelated. A particularly interesting and fast way to calculate the correlation and autocorrelation by using the Fourier transform is described in Section 2.7.3.

2.6.6   The Karhunen–Loève Transform

Let $\vec{X}$ be a random vector with covariance matrix K. Since K is Hermitian, its eigenvalues are distinct and can be represented as real values, and its eigenvectors are orthonormal. Let the eigenvectors $\vec{v}_i$ ($i = 1, 2, \ldots, N$) of K be represented in terms of the following matrix:

$$\Omega = \begin{bmatrix} \leftarrow \vec{v}_1 \rightarrow \\ \leftarrow \vec{v}_2 \rightarrow \\ \vdots \\ \leftarrow \vec{v}_N \rightarrow \end{bmatrix}.$$

Consider now the linear transformation of the random vector $\vec{X}$ implemented by this matrix, which is known as the *Karhunen–Loève transform*, i.e.

$$\hat{\vec{X}} = \Omega \vec{X}. \tag{2.41}$$

We know from Section 2.2.5 that, by being orthogonal, this matrix implements a coordinate transformation that rotates the old axes towards a new orientation in the N-dimensional space where the random vector $\vec{X}$ is defined. The important property of this rotation is that it is done in such a way as to maximize the dispersions (e.g., the variances or standard deviations) along each of the N new axes. By representing the N eigenvalues of K as λ_i, and assuming that they are sorted in decreasing order, the covariance matrix of the new random vector $\hat{\vec{X}}$ can be verified to be given by

$$\hat{K} = \begin{bmatrix} \lambda_1 & 0 & 0 & 0 \\ 0 & \lambda_2 & 0 & 0 \\ \vdots & \vdots & \ddots & \vdots \\ 0 & 0 & 0 & \lambda_N \end{bmatrix}.$$

Since $\hat{K}$ is a diagonal matrix, it is clear that the Karhunen–Loève transform acts in the sense of fully uncorrelating the scalar random variables x_i in $\vec{X}$, i.e., all the correlations involving distinct random variables are null.

As we saw in Section 2.2.5, if λ_i is an eigenvalue, so is $-\lambda_i$. This fact implies that the Karhunen–Loève transform is not unique. Indeed, although there are 2^N possible transforms in an N-dimensional space, all these transforms will imply the same $\hat{K}$. Consider the example in the accompanying box.

Example: *Karhunen-Loève Transform*

Obtain the Karhunen-Loève transformation matrix for the data presented in the Example of Section 2.6.4.

Solution:

We start with the already calculated covariance matrix

$$K = \begin{bmatrix} 0.0275 & 0.2706 \\ 0.2706 & 2.8803 \end{bmatrix}.$$

The eigenvectors (normalized to have unit norm) are

$$\Omega_1 = \begin{bmatrix} 0.9956 & -0.0936 \\ 0.0936 & 0.9956 \end{bmatrix}.$$

The orientations o_1 and o_2 defined respectively by the eigenvectors $\vec{v}_1$ and $\vec{v}_2$ are shown by the dashed lines in Figure 2.71, starting from the mean $\vec{\mu}_{\vec{X}}$.

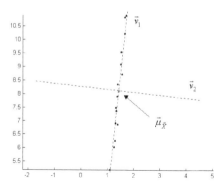

Figure 2.71: *The two main orientations of the observed data, given by the eigenvectors of the respective covariance matrix.*

The other three possible matrices are

$$\Omega_2 = \begin{bmatrix} 0.9956 & 0.0936 \\ 0.0936 & -0.9956 \end{bmatrix},$$

$$\Omega_3 = \begin{bmatrix} -0.9956 & -0.0936 \\ -0.0936 & 0.9956 \end{bmatrix},$$

$$\text{and} \quad \Omega_4 = \begin{bmatrix} -0.9956 & 0.0936 \\ -0.0936 & -0.9956 \end{bmatrix}.$$

The new covariance matrix, obtained from the transformed data, for any of such cases is

$$\hat{K} = \begin{bmatrix} 2.9057 & 0 \\ 0 & 0.0021 \end{bmatrix}$$

To probe further: *Probability and Statistics*

Several good introductions to probability and statistics are available in [Davenport-Jr, 1970; Kreyszig, 1993; Larson & Shubert, 1979]. More advanced aspects, including stochastic processes, multivariate statistics, and the Karhunen-Loève transform, can be found in [Johnson & Wichern, 1998; Shanmugan & Breipohl, 1988; Therrien, 1992].

2.7 Fourier Analysis

The Fourier series and transform, the objects of the present section, represent some of the most important and interesting approaches in applied mathematics, allowing particularly useful applications in signal analysis, image processing and analysis, computer vision, differential statistics, dynamic systems, vibration analysis, psychology, acoustics and telecommunications, to name a few. Part of the importance of the Fourier approach arises from the fact that it allows a representation of a broad class of functions in terms of a linear combination of sine, cosine or complex exponential basic functions. Moreover, unlike most alternative representations of functions in terms of an orthogonal kernel, the Fourier approach exhibits an inherent and special compatibility with the signals typically found in nature, especially regarding their oscillatory and highly correlated features. As a matter of fact, the Fourier transform allows the compaction of signals by performing a decorrelation process similar and almost as effective as the statistically optimal Karhunen-Loève transform. Besides, the Fourier transform has frequently been related to the human perceptual system. For instance, our inner ears can be understood as performing spectral analysis in a way that is directly related to the Fourier series. In addition, several models of human visual perception have considered the Fourier transform as an essential component underlying processing and analysis. One of the main topics covered in this book, namely the numerical approaches to multi-resolution curvature estimation (Chapter 7), is founded on the useful but less frequently used properties of the Fourier transform, known as the derivative property. In addition, much insight about other important transforms, such as the cosine and wavelet transform, can be gained by treating them in terms of the Fourier transform.

Because of its special relevance to image and shape processing and analysis, the Fourier approach is treated to a considerable depth in this section. Although we have addressed most of the information needed for the proper understanding of the concepts and applications developed in this book, the issues related to the Fourier transform are particularly broad and can by no means be covered in an exhaustive manner here. Fortunately, there are several excellent textbooks, to which the reader is referred, providing clear and comprehensive treatment of more sophisticated related topics (see the *To Probe Further* box at the end of this section).

After some brief historical remarks, this section introduces the Fourier series, the continuous Fourier transform, its properties, frequency filtering concepts, and the discrete Fourier transform. It is observed that the sound understanding of the continuous Fourier transform, especially its properties, is *essential* for the proper application of the discrete Fourier transform to practical problems, which is discussed in some detail in the final sections of this chapter.

2.7.1 Brief Historical Remarks

The history of Fourier analysis (see, for instance [Davis & Hersh, 1999; Gullberg, 1997]) can be traced back to pioneering approaches by L. D'Alembert (1717–

1783), L. Euler (1707–1783) and D. Bernoulli (1700–1782) to the solution of the wave equation, such as that governing a vibrating string, which is a differential equation of the type

$$\frac{\partial^2 \Psi(x,t)\phi}{dx^2} = \alpha \frac{\partial^2 \Psi(x,t)\phi}{dt^2},$$

where x is position, t time, and α a constant. For the first time, Bernoulli's approach represented the initial position of the string in terms of an infinite sum of sine functions with varying frequencies—which represents the main underlying concept in the Fourier series. However, it was left to Euler to find a convenient formula for calculating the now-called *Fourier coefficients*. The interesting study of representing functions as a series of sines and cosines was resumed much later by Joseph Fourier (1768–1830), while dealing with the heat equation, a differential equation of the type

$$\frac{\partial^2 \Psi(x,t)\phi}{dx^2} = \alpha \frac{\partial \Psi(x,t)\phi}{dt},$$

as reported in his "Théorie Analytique de la Chaleur." The main contribution of Fourier was to prove, although in an inconsistent fashion, that functions defined by pieces could also be represented in terms of infinite series of sines and cosines, the now famous *Fourier series* representation.

2.7.2 The Fourier Series

The Fourier series (or expansion) of a periodic function $g(t)$, with period $2L$, whenever it exists, is given by

$$g(t) = a_0 + \sum_{n=1}^{\infty} \left[a_n \cos\left(\frac{n\pi t}{L}\right) + b_n \sin\left(\frac{n\pi t}{L}\right) \right],$$

where

$$a_0 = \frac{1}{2L} \int_{-L}^{L} g(t)\, dt$$

$$a_n = \frac{1}{L} \int_{-L}^{L} g(t) \cos\left(\frac{n\pi t}{L}\right) dt \qquad (2.42)$$

$$b_n = \frac{1}{L} \int_{-L}^{L} g(t) \sin\left(\frac{n\pi t}{L}\right) dt,$$

where $n = 1, 2, \ldots$

These values are known as *Fourier coefficients*, and the involved sine and cosine functions are known as *kernel functions*. The frequency f of the sine and cosine functions in the above equations can be immediately obtained by equating the argument of those functions, i.e., $\frac{n\pi t}{L}$, with $2\pi ft$, since the frequency of the functions $\cos(2\pi ft)$ or $\sin(2\pi ft)$ is, by definition, f. Therefore

$$\frac{n\pi t}{L} = 2\pi ft \Leftrightarrow f = \frac{n}{2L}.$$

Figure 2.72 illustrates the function $\cos(2\pi ft)$ for $f = 1, 2, 3, 4, 5$ and 6.

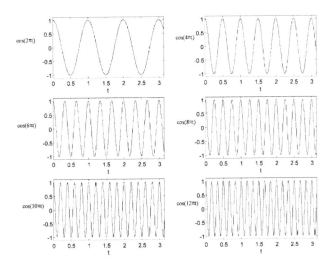

Figure 2.72: *Several instances of the function $\cos(2\pi ft)$ respectively defined by $f = 1, 2, 3, 4, 5$ and 6.*

Observe that the kernel function for a specific value of n has period $T = \frac{1}{f} = \frac{2L}{n}$, implying that *any* of the kernel functions are periodic of period $2L$. Since both the function g_i and the kernel functions have the same period $2L$, the products between these functions in equation (2.42) are also periodic of period $2L$ (the product of two periodical functions is a periodic function with the same period), and the integrals in equation (2.42) can be calculated not only between $-L$ and L, but along any interval with extent $2L$ along the time domain, as illustrated below:

$$a_n = \frac{1}{L} \int_{-L}^{L} g(t) \cos\left(\frac{n\pi t}{L}\right) dt = \frac{1}{L} \int_{0}^{2L} g(t) \cos\left(\frac{n\pi t}{L}\right) dt = \frac{1}{L} \int_{-L+2}^{L+2} g(t) \cos\left(\frac{n\pi t}{L}\right) dt.$$

It is clear from the above developments that the Fourier series of a function $g(t)$ represents a means for expressing this function as a linear combination of sine and cosine functions of distinct frequencies. It should be observed that, as implied by equation (2.42), the term a_0 corresponds to the average (or "direct current"—DC) value of the original function along the period $2L$. Although in principle limited to periodic functions, it should be observed that any function defined over a finite domain can be expanded as a periodic function, as illustrated in the box entitled *Fourier Series*. The reader is referred to specialized literature (e.g., [Tolstov, 1976]) for the convergence conditions of the Fourier series.

An informal way to understand the Fourier series is as a "cookie recipe." If the function is understood as the "cookie," the Fourier coefficients can be understood as the amount of each ingredient (i.e., the amplitude of each sine and cosine functions of several frequencies) that have to be added in order to produce the cookie (i.e., the function). Table 2.7 summarizes this analogy.

Cookie Recipe	Fourier Series
The cookie.	The function.
Ingredients (i.e., flour, sugar, chocolate, etc.).	The kernel functions (i.e., sines and cosines with several frequencies).
The amount of each ingredient.	The amplitude of each kernel function, specified by the respective Fourier coefficients.
The cookie is obtained by adding together the specific amounts of ingredients.	The function is obtained as the addition of the kernel functions weighted by the Fourier coefficients (i.e., a linear combination).

Table 2.7: *The analogy between a cookie recipe and the Fourier series.*

Consider the following example:

Example: *Fourier Series*

Calculate the Fourier series of the rectangular function $g(t)$ given by

$$g(t) = \begin{cases} 1 & \text{if } -a \leqslant t < a, \\ 0 & \text{otherwise.} \end{cases} \qquad (2.43)$$

Solution:

Since this function is not periodic, the first step consists in transforming it into a periodic function $h(t)$. A suitable period is $2L = 4a$, i.e., $L = 2a$, which yields $h(t) = g(t)$ for $-2a \leqslant t < 2a$ and $h(t) = h(t + 4ak)$, $k = \ldots, -2, -1, 0, 1, 2, \ldots$ Figure 2.73 illustrates this situation with respect to $a = 1$.

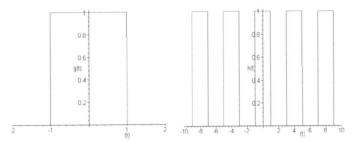

Figure 2.73: *The non-periodic original function g(t) and its periodic version h(t), for a = 1.*

Now the Fourier coefficients can be calculated as

$$a_0 = \frac{1}{2L} \int_{-L}^{L} h(t)\, dt = \frac{1}{4a} \int_{-2a}^{2a} h(t)\, dt = \frac{1}{4a} \int_{-a}^{a} 1\, dt = \frac{1}{2},$$

$$a_n = \frac{1}{L} \int_{-L}^{L} h(t) \cos\left(\frac{n\pi t}{L}\right) dt = \frac{1}{2a} \int_{-2a}^{2a} h(t) \cos\left(\frac{n\pi t}{2a}\right) dt = \frac{1}{2a} \int_{-a}^{a} \cos\left(\frac{n\pi t}{2a}\right) dt =$$

$$= \frac{1}{2a} \left[\frac{2a}{n\pi} \sin\left(\frac{n\pi t}{2a}\right) \right]_{-a}^{a} \overset{a=1}{=} \frac{1}{n\pi} \left[\sin\left(\frac{n\pi}{2}\right) - \sin\left(-\frac{n\pi}{2}\right) \right] =$$

$$= \frac{1}{n\pi} \left[\sin\left(\frac{n\pi}{2}\right) + \sin\left(\frac{n\pi}{2}\right) \right] = \frac{2}{n\pi} \sin\left(\frac{n\pi}{2}\right) = \text{sinc}\left(\frac{n}{2}\right),$$

$$b_n = \frac{1}{L} \int_{-L}^{L} h(t) \sin\left(\frac{n\pi t}{L}\right) dt = \frac{1}{2a} \int_{-2a}^{2a} h(t) \sin\left(\frac{n\pi t}{2a}\right) dt = \frac{1}{2a} \int_{-a}^{a} \sin\left(\frac{n\pi t}{2a}\right) dt =$$

$$= -\frac{1}{2a} \left[\frac{2a}{n\pi} \cos\left(\frac{n\pi t}{2a}\right) \right]_{-a}^{a} \overset{a=1}{=} -\frac{1}{n\pi} \left[\cos\left(\frac{n\pi}{2}\right) - \cos\left(-\frac{n\pi}{2}\right) \right] =$$

$$= -\frac{1}{n\pi} \left[\cos\left(\frac{n\pi}{2}\right) - \cos\left(\frac{n\pi}{2}\right) \right] = 0,$$

where $\text{sinc}(x) = \sin(\pi x)/(\pi x)$. It is interesting to observe that a_n is zero for $n = 2, 4, 6, \ldots$, and that the obtained coefficients are independent of the parameter a which, however, reappears in the function reconstruction

$$h(t) = \frac{1}{2} + \sum_{n=1}^{\infty} \left[\text{sinc}\left(\frac{n}{2}\right) \cos\left(\frac{n\pi t}{2a}\right) \right] \qquad (2.44)$$

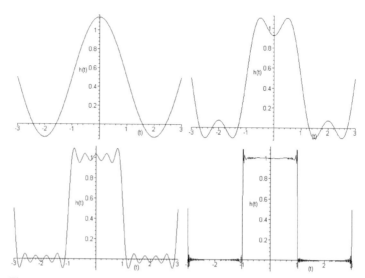

Figure 2.74: *Reconstruction of the rectangular function by including an increasing number of components.*

It is important to bear in mind that the above coefficients and series representation is specific to the periodical function $h(t)$, and not to $g(t)$, which can nevertheless be extracted from any of the periods of $h(t)$. At continuous points, the Fourier series tends to the function $h(t)$ as more and more terms are included—as a matter of fact, the exact convergence is only typically obtained for an infinite number of terms. At each discontinuity point P of $g(t)$, the Fourier series tends to the mean value of the respective left and right limit values, but also implies an oscillation of the series expansion around P, which is known as *Gibbs phenomenon* (see, for instance, [Kreyszig, 1993]). Figure 2.74 presents four series approximations to the above rectangular function considering increasing numbers of terms and $a = 1$.

The Fourier series can be represented in a more compact fashion by using the complex exponential as a kernel function. Given a function $g(t)$, the respective complex Fourier coefficients are given by equation (2.46), which allows the original function to be recovered by using equation (2.45).

$$g(t) = \sum_{n=-\infty}^{\infty} \left[c_n \exp\left\{ \frac{jn\pi t}{L} \right\} \right] \tag{2.45}$$

$$c_n = \frac{1}{2L} \int_{-L}^{L} g(t) \exp\left\{ -\frac{jn\pi t}{L} \right\} dt \tag{2.46}$$

where $n = \ldots, -2, -1, 0, 1, 2, \ldots$

As observed for the real Fourier series, the integration in equation (2.46) can be performed over any full period $2L$ of $g(t)$. For instance:

$$c_n = \frac{1}{2L} \int_0^{2L} g(t) \exp\left\{-\frac{jn\pi t}{L}\right\} dt \qquad (2.47)$$

Consider the following example:

Example: *Complex Fourier Series*

Calculate the complex Fourier series of the function $g(t)$ in equation (2.43).

Solution:

First, the function $g(t)$ has to be made periodic, for instance by imposing period $2L = 4a$, as before, yielding $h(t)$. The complex coefficients can be calculated by using equation (2.46):

$$c_n = \frac{1}{2L} \int_{-L}^{L} h(t) \exp\left(-\frac{jn\pi t}{L}\right) dt = \frac{1}{4a} \int_{-2a}^{2a} h(t) \exp\left(-\frac{jn\pi t}{2a}\right) dt =$$

$$= \frac{1}{4a} \int_{-a}^{a} \exp\left(-\frac{jn\pi t}{2a}\right) dt = \frac{1}{4a}\left[-\frac{2a}{jn\pi} \exp\left(-\frac{jn\pi t}{2a}\right)\right]_{-a}^{a} \overset{a=1}{=}$$

$$= -\frac{1}{j2n\pi}\left[\exp\left(-\frac{jn\pi}{2}\right) - \exp\left(\frac{jn\pi}{2}\right)\right] =$$

$$= -\frac{1}{j2n\pi}\left[\cos\left(\frac{n\pi}{2}\right) - j\sin\left(\frac{n\pi}{2}\right) - \cos\left(\frac{n\pi}{2}\right) - j\sin\left(\frac{n\pi}{2}\right)\right] =$$

$$= \frac{1}{n\pi}\sin\left(\frac{n\pi}{2}\right) = \frac{1}{2}\,\text{sinc}\left(\frac{n}{2}\right)$$

It should be observed that these coefficients are zero for $n = \pm 2, \pm 4, \ldots$, and that $c_0 = 0.5$.

The function $h(t)$ can then be represented as

$$h(t) = \frac{1}{2} \sum_{n=-\infty}^{\infty}\left[\text{sinc}\left(\frac{n}{2}\right)\exp\left\{\frac{jn\pi t}{2a}\right\}\right]$$

which is equivalent to that in equation (2.44).

2.7.3 | The Continuous One-Dimensional Fourier Transform

Let $g(t)$ be a complex and not necessarily periodic function. In case its *Fourier transform* exists, it is given by

$$G(f) = \Im \{g(t)\} = \int_{-\infty}^{\infty} g(t) \exp\{-j2\pi ft\} \, dt, \qquad (2.48)$$

where the variables t and f are usually called *time* and *frequency*, respectively. The *inverse Fourier transform* of $G(f)$, which returns $g(t)$, is given by

$$g(t) = \Im^{-1} \{G(f)\} = \int_{-\infty}^{\infty} G(f) \exp\{j2\pi ft\} \, df. \qquad (2.49)$$

The original function and its inverse are usually represented as the *Fourier pair*:

$$g(t) \leftrightarrow G(f).$$

Observe that both $g(t)$ and $G(f)$ are, in general, complex. The function $G(f)$ is usually expressed in one of the two following representations (i) *real* and *imaginary* parts, i.e., $Re\{G(f)\}$ and $Im\{G(f)\}$, and (ii) *magnitude* (or *modulus*) and *phase*, given by

$$|G(f)| = \sqrt{[Re\{G(f)\}]^2 + [Im\{G(f)\}]^2}$$

and

$$\Phi\{G(f)\} = \arctan\left\{\frac{Im\{G(f)\}}{Re\{G(f)\}}\right\}.$$

In addition, observe that if $g(t) \leftrightarrow G(f)$, then $g(-t) \leftrightarrow G(-f)$ and, in case $g(t)$ is real we also have $g(-t) \leftrightarrow G^*(f)$.

Since the existence of the Fourier transform is verified in practice for most functions, this topic is not covered here and the reader is referred to the literature (e.g., [Brigham, 1988]) for theoretical conditions for its existence. It is important to observe that there are several alternative definitions for the Fourier transform and its inverse [Brigham, 1988] to be found in the literature, all of which should be compatible with the Laplace transform and the energy conservation principle, i.e., Parseval's theorem.

There is an interesting analogy between the Fourier transform and the Fourier series in the sense that equation (2.48) can be understood as producing the continuous Fourier coefficients $G(f)$, which can then be used to represent the function through the inverse Fourier transform in equation (2.49). This similarity becomes evident when equations (2.45) and (2.46) are compared to equations (2.48) and (2.49), respectively, except for the change of the sum symbol in equation (2.45) into integral in equation (2.49). In other words, the Fourier transform and its inverse can be thought of as playing the role of analysis and synthesis of the original signal, respectively. The difference between the Fourier series and transform is that in the latter the "Fourier coefficients" $G(f)$ merge to form the continuous Fourier transform of the original function instead of a series of discrete values as is the case with the Fourier series. Indeed, it can be shown (e.g., [Brigham, 1988]) that the Fourier transform can be understood as a limit situation of the Fourier series where the spacing between the Fourier coefficients tends to zero. It is precisely this fact that allows the Fourier transform to be defined for non-periodical functions (i.e., functions with infinite period).

Being directly related to the Fourier series, the Fourier transform can also be understood in terms of the "cookie recipe" analogy introduced in Section 2.7.2. That is to say, the continuous Fourier coefficients $G(f)$ provide the amount of each ingredient (again the complex exponential kernel functions) that must be linearly combined, through equation (2.49) in order to prepare the "cookie," i.e., the original signal. The *spectral composition* of the signal is usually represented in terms of the *power spectrum* $P_g\{f\}$ of the original signal $g(t)$, which is defined as:

$$P_g|f| = |G(f)|^2 = G(f)^*G(f).$$

An important property of the power spectrum is that it does not change as the original function is shifted along its domain, which is explored by the so-called Fourier descriptors for shape analysis (see Section 6.5).

Consider the following example:

Example: *Fourier Transform I: Periodic Functions*

Calculate the Fourier transform and power spectrum of the function

$$g(t) = \exp\{-t\}, \quad 0 \leqslant t < \infty.$$

First, we apply equation (2.48):

$$\Im\{g(t)\} = \int_0^\infty \exp\{-t\} \exp\{-j2\pi ft\} \, dt = \int_0^\infty \exp\{-t(j2\pi f + 1)\} \, dt =$$

$$= -\frac{1}{1 + j2\pi f} \left[\exp\{-t(j2\pi f + 1)\}\right]\Big|_0^\infty =$$

$$= -\left(\frac{1}{1 + j2\pi f}\right)[0 - 1] = \frac{1}{1 + j2\pi f}\left(\frac{1 - j2\pi f}{1 - j2\pi f}\right) = \frac{1 - j2\pi f}{1 + (2\pi f)^2} = G(f).$$

Thus, the Fourier transform $G(f)$ of $g(t)$ is a complex function with the following real and imaginary parts, shown in Figure 2.75:

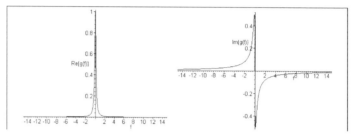

Figure 2.75: *The real and imaginary parts of $G(f)$.*

$$\mathrm{Re}\,\{G(f)\} = \frac{1}{1 + (2\pi f)^2} \quad \text{and} \quad \mathrm{Im}\,\{G(f)\} = \frac{-2\pi f}{1 + (2\pi f)^2}.$$

Alternatively, in the magnitude and phase representation, we have

$$|G(f)| = \sqrt{\frac{1 + 4\pi^2 f^2}{\left[1 + (2\pi f)^2\right]^2}} \quad \text{and} \quad \Phi\,\{G(f)\} = \arctan\,\{-2\pi f\}.$$

The power spectrum can be calculated as

$$P_g\{f\} = G(f)^* G(f) = \frac{1 - j2\pi f}{1 + (2\pi f)^2}\,\frac{1 + j2\pi f}{1 + (2\pi f)^2} = \frac{1 + (2\pi f)^2}{\left(1 + (2\pi f)^2\right)^2} = |G(f)|^2$$

and is illustrated in Figure 2.76.

Although the Fourier transform of a complex function is usually (as in the above example) a complex function, it can also be a purely real (or imaginary) function. On the other hand, observe that the power spectrum is always a real function of the frequency. Consider the following example

Example: *Fourier Transform II: Aperiodic Functions*

Calculate the Fourier transform of the function

$$g(t) = \begin{cases} 1 & \text{if } -a \leqslant t < a \\ 0 & \text{otherwise.} \end{cases}$$

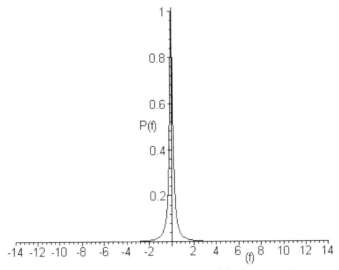

Figure 2.76: *The power spectrum of the function g(t).*

Applying equation (2.48)

$$\Im\{g(t)\} = \int_{-\infty}^{\infty} g(t)\,\exp\{-j2\pi ft\}\,dt = \int_{-a}^{a} 1.\exp\{-j2\pi ft\}\,dt = \left[\frac{-1}{j2\pi f}\exp\{-j2\pi ft\}\right]\Bigg|_{-a}^{a} =$$

$$= \left(\frac{-1}{j2\pi f}\right)[\exp\{-j2\pi af\} - \exp\{j2\pi af\}] =$$

$$= \left(\frac{-1}{j2\pi f}\right)[\cos(2\pi af) - j\sin(2\pi af) - \cos(2\pi af) - j\sin(2\pi af)] =$$

$$= \frac{\sin(2\pi af)}{\pi f} = 2a\frac{\sin(2\pi af)}{2\pi af} = 2a\,\mathrm{sinc}\,(2af),$$

which is the purely real function shown in Figure 2.77, together with its respective power spectrum, considering $a = 1$.

The Dirac delta function is particularly useful in Fourier analysis. Its transform can easily be calculated as:

$$\Im\{\delta(t)\} = \int_{-\infty}^{\infty} \delta(t)\,\exp\{-j2\pi ft\}\,dt = \int_{-\infty}^{\infty} \delta(t)\,\exp\{-j2\pi f.0\}\,dt = \int_{-\infty}^{\infty} \delta(t)\,dt = 1.$$

Since the respective inverse can be verified to exist, we have:

$$\delta(t) \leftrightarrow 1.$$

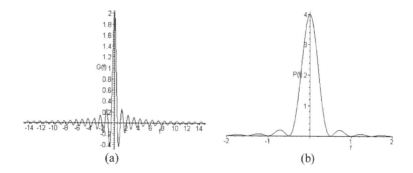

Figure 2.77: *The purely real Fourier transform of the function g(t)* **(a)** *and its respective power spectrum* **(b)** *for a* = 1. *Observe the different scales for the x-axis.*

Another especially relevant function in Fourier analysis is the Gaussian, which defines the following Fourier pair:

$$g_\sigma(t) = \frac{1}{\sigma\sqrt{2\pi}}\exp\left\{-\frac{1}{2}\left(\frac{t}{\sigma}\right)^2\right\} \quad\leftrightarrow\quad G(f) = \exp\left\{-\frac{1}{2}\left(\frac{t}{\sigma_f}\right)^2\right\}, \qquad (2.50)$$

where $\sigma_f = \frac{1}{2\pi\sigma}$. Therefore, the Fourier transform of a normalized Gaussian function is a non-normalized Gaussian function.

The Fourier transform exhibits a series of extremely useful and practical properties in signal processing and analysis, which are also essential for properly applying the discrete Fourier transform. The most important of such properties are presented and exemplified in the following sections.

Symmetry

Let

$$g(t) \leftrightarrow G(f).$$

Then

$$G(t) \leftrightarrow g(-f).$$

This property provides an interesting possibility for obtaining new transforms directly from previously known transforms, as illustrated in the following example. We have already seen that $g(t) = \delta(t) \leftrightarrow G(f) = 1$. By using the above property, we have that

$$G(t) = 1 \quad\leftrightarrow\quad g(-f) = \delta(-f) = \delta(f),$$

implying the new Fourier pair $1 \leftrightarrow \delta(f)$.

Time Shifting

Let

$$g(t) \leftrightarrow G(f).$$

Then

$$g(t - t_0) \leftrightarrow G(f) \exp\{-j2\pi f t_0\}.$$

Thus, the effect of shifting a function in the time domain implies that the respective Fourier transform is modulated by the complex exponential with frequency equal to the time shift value. Observe that:

$$\left| G(f) \exp\{-j2\pi f t_0\} \right|^2 = |G(f)|^2 \left| \exp\{-j2\pi f t_0\} \right|^2 = |G(f)|^2,$$

i.e., the power spectrum is not modified by time shiftings of the original function.

Example 1: Given $\delta(t) \leftrightarrow 1$, calculate the Fourier transform of $\delta(t - t_0)$. This is immediately provided by the above time shifting property as $\delta(t - t_0) \leftrightarrow \exp\{-j2\pi f t_0\}$.

Example 2: Given $g(t) = \exp\{-|t|\} \leftrightarrow \frac{2}{1+(2\pi f)^2} = G(f)$, calculate the Fourier transform of $h(t) = g(t - 2) = \exp\{-|t - 2|\}$. Again, by applying the time shifting property we obtain

$$\exp\{-|t - 2|\} \quad \leftrightarrow \quad \frac{2 \exp\{-4\pi j f\}}{1 + (2\pi f)^2},$$

that is graphically shown in Figure 2.78.

Time Scaling

Let

$$g(t) \leftrightarrow G(f).$$

Then

$$g(at) \leftrightarrow \frac{1}{|a|} G\left(\frac{f}{a}\right).$$

Therefore, a compression (extension) of the function along the time domain implies an extension (compression) of the respective Fourier transform. This property is particularly interesting from the point-of-view of multiscale analysis (especially wavelets), since it relates the Fourier transform of scaled versions of a signal.

Example: Given $g(t) = \exp\{-|t|\} \leftrightarrow \frac{2}{1+(2\pi f)^2} = G(f)$, calculate the Fourier transform of $h(t) = g(3t) = \exp\{-|3t|\}$. By applying the above property:

$$\exp\{-|3t|\} \leftrightarrow \frac{2}{3} \frac{1}{1 + \left(\frac{2}{3}\pi f\right)^2},$$

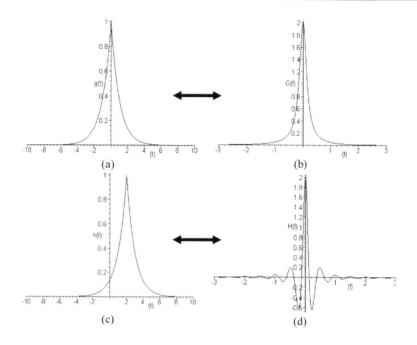

Figure 2.78: *The function $g(t) = \exp\{-|t|\}$ **(a)** and its respective Fourier transform **(b)**. A time shifted version of the function $g(t) = \exp\{-|t|\}$ **(c)** and the real part of its respective Fourier transform **(d)**.*

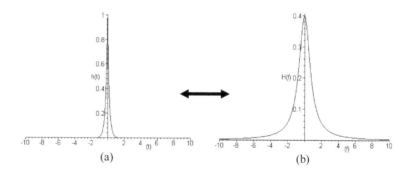

Figure 2.79: *The scaled version of the function $g(t) = \exp\{-|t|\}$ **(a)** and its respective Fourier transform **(b)**. Compare with Figure 2.78. The compression in the time domain implies an expansion in the frequency domain.*

which is illustrated in Figure 2.79. It is clear from the above example that the effect of compressing a function in the time domain implies that its respective Fourier transform expands in the frequency domain, and viceversa.

Frequency Shifting

Let

$$g(t) \leftrightarrow G(f).$$

Then

$$g(t) \exp\{j2\pi f_0 t\} \leftrightarrow G(f - f_0).$$

This property can be understood similarly to the time shifting property presented above.

Frequency Scaling

Let

$$g(t) \leftrightarrow G(f).$$

Then

$$\frac{1}{|a|} g\left(\frac{t}{a}\right) \leftrightarrow G(af).$$

This property can be understood similarly to the time scaling property presented above.

Linearity

The Fourier transform is linear, i.e.,

$$\Im\{ag(t) + bh(t)\} = a\Im\{g(t)\} + b\Im\{h(t)\}.$$

This important property of the Fourier transform is very useful in practice. It can also be used to calculate new Fourier transforms. For instance, it allows the Fourier transform of the sine and cosine transform to be easily calculated as follows.

Since

$$\sin(2\pi f_0 t) = \frac{\exp\{j2\pi f_0 t\} - \exp\{-j2\pi f_0 t\}}{2j},$$

we have

$$\Im\{\sin(2\pi f_0 t)\} = \Im\left\{\frac{\exp\{j2\pi f_0 t\} - \exp\{-j2\pi f_0 t\}}{2j}\right\} =$$

$$= \frac{1}{2j}\Im\{\exp\{j2\pi f_0 t\}\} - \frac{1}{2j}\Im\{\exp\{-j2\pi f_0 t\}\}$$

$$= \frac{1}{2j}(\delta(f - f_0) - \delta(f + f_0)).$$

Since the inverse can be verified to exist, we have the Fourier pair

$$\sin(2\pi f_0 t) \quad \leftrightarrow \quad \frac{j}{2}\delta(f + f_0) - \frac{j}{2}\delta(f - f_0).$$

In a similar fashion

$$\Im\{\cos(2\pi f_0 t)\} = \Im\left\{\frac{\exp\{j2\pi f_0 t\} + \exp\{-j2\pi f_0 t\}}{2}\right\} =$$

$$= \frac{1}{2}\Im\{\exp\{j2\pi f_0 t\}\} + \frac{1}{2}\Im\{\exp\{-j2\pi f_0 t\}\}$$

$$= \left(\frac{1}{2}\delta(f - f_0) + \frac{1}{2}\delta(f + f_0)\right)$$

and, therefore $\cos(2\pi f_0 t) \leftrightarrow \frac{1}{2}\delta(f + f_0) + \frac{1}{2}\delta(f - f_0)$.

The Convolution Theorem

This important property of the Fourier transform is expressed as follows.

Let

$$g(t) \leftrightarrow G(f) \quad \text{and} \quad h(t) \leftrightarrow H(f).$$

Then

$$(g * h)(t) \leftrightarrow G(f)H(f)$$

and

$$g(t)h(t) \leftrightarrow (G * H)(f),$$

where $g(t)$ and $h(t)$ are generic complex functions. See Sections 2.7.4 and 7.2 for applications of this theorem.

The Correlation Theorem

Let $g(t)$ and $h(t)$ be real functions defining the Fourier pairs $g(t) \leftrightarrow G(f)$ and $h(t) \leftrightarrow H(f)$. Then $(g \circ h)(t) \leftrightarrow G^*(f) H(f)$.

The Derivative Property

Let the generic Fourier pair $g(t) \leftrightarrow G(f)$ and a be any non-negative real value. Then

$$\frac{d^a g(t)}{dt^a} \leftrightarrow D_a(f) G(f), \tag{2.51}$$

where $D_a(f) = (j2\pi f)^a$. This interesting property, which is used extensively in the present book (see Section 7.2), allows not only the calculation of many derivatives in terms of the respective Fourier transforms, but also the definition of fractionary derivatives such as $\frac{d^{0.5} g(t)}{dt^{0.5}}$ and $\frac{d^{\pi} g(t)}{dt^{\pi}}$. This property can also be used to calculate integrals, which is done by using $a < 0$.

Example: Given $g(t) = \cos(2\pi f_0 t) \leftrightarrow \frac{1}{2}(\delta(f + f_0) + \delta(f - f_0)) = G(f)$, calculate the first derivative of $g(t)$. By applying the above property:

$$
\begin{aligned}
g'(t) &\leftrightarrow \frac{(j2\pi f)}{2} (\delta(f + f_0) + \delta(f - f_0)) \\
&= \frac{(j2\pi)}{2} (f\delta(f + f_0) + f\delta(f - f_0)) \\
&= \frac{(j2\pi)}{2} (f_0\delta(f + f_0) - f_0\delta(f - f_0)) \\
&= (j2\pi f_0) \frac{j}{2} (-\delta(f + f_0) + \delta(f - f_0)) \\
&\leftrightarrow -2\pi f_0 \sin(2\pi f_0 t).
\end{aligned}
$$

Parseval's Theorem

Let

$$g(t) \leftrightarrow G(f).$$

Then

$$\int_{-\infty}^{\infty} |g(t)|^2 \, dt = \int_{-\infty}^{\infty} |G(f)|^2 \, df.$$

This property indicates that the Fourier transform preserves the "energy" of the function. As a matter of fact, this is an ultimate consequence that the norm of a signal is preserved by an orthonormal transformation.

Parity-Related Properties

The Fourier transform of a function is determined by the parity and nature of the function $g(t)$ to be transformed (i.e., real, imaginary, or complex). Some of the most useful of such properties are summarized in Table 2.8.

$g(t)$	$G(f)$
Real and even	Real and even
Real and odd	Imaginary and odd
Imaginary and even	Imaginary and even
Imaginary and odd	Real and odd

Table 2.8: *Some of the parity properties of the Fourier transform.*

Discrete and Periodical Functions

First, consider the sampling function $\Psi_{\Delta t}(t)$ defined as the sum of equally spaced (by Δt) Dirac deltas, i.e., $\Psi_{\Delta t}(t) = \sum_{n=-\infty}^{\infty} \delta(t - n\Delta t)$, which defines the Fourier pair

$$\Psi_{\Delta t}(t) = \sum_{n=-\infty}^{\infty} \delta(t - n\Delta t) \leftrightarrow \frac{1}{\Delta t}\Psi_{1/\Delta t}(f) = \frac{1}{\Delta t}\sum_{n=-\infty}^{\infty} \delta\left(f - n\frac{1}{\Delta t}\right),$$

which is illustrated in Figure 2.80.

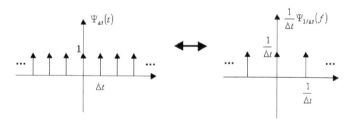

Figure 2.80: *The sampling function **(a)** and its respective Fourier transform **(b)**, which is also a sampling function, but with different features.*

Now, let $g(t)$ be a non-periodic function entirely contained in a finite interval $r \leqslant t < s$ along its domain, such as

$$g(t) = \begin{cases} 1 & \text{if } -a \leqslant t \leqslant a, \\ 0 & \text{otherwise,} \end{cases}$$

where $r < -a$ and $a < s$. Let also $G(f)$ be the respective Fourier transform of $g(t)$, in this case $G(f) = 2a \, sinc\,(2af)$. A periodic version $h(t)$ of the function $g(t)$, with period $2L > 2a$ can be obtained by convolving $g(t)$ with the sampling function $\Psi_{2L}(t) = 2L \sum_{n=-\infty}^{\infty} \delta\,(t - 2nL)$, i.e., $h(t) = g(t) * [2L\Psi_{2L}(t)]$. The coefficient $2L$ adopted for the sampling function avoids the otherwise implied scaling of the Fourier transform by a factor of $\frac{1}{2L}$. The functions $g(t)$ and $h(t)$ are illustrated in Figure 2.81 (a) and (c), respectively, considering $a = 1$ and $L = 2$. Since $g(t)$ is

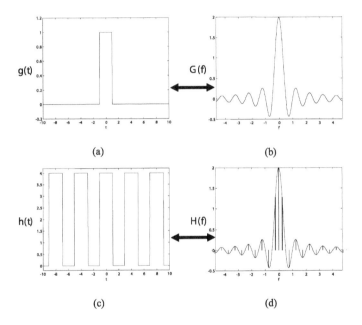

(a) (b) (c) (d)

Figure 2.81: *The function $g(t)$ (a) and its Fourier transform $G(f)$ (b). The periodical version $h(t) = g(t) * \Psi_{2L}(t)$ of $g(t)$, for $a = 1$ and $L = 2$ (c), and its respective Fourier transform $H(f)$ (d).*

contained in a limited interval along its domain and $2L > 2a$, this process corresponds to copying $g(t)$ at the values of t that are multiples of the period $2L$. By the convolution theorem, the Fourier transform of $h(t)$ is given by the product between $G(f)$ and the Fourier transform of the sampling function, i.e.,

$$H(f) = G(f)\,[\Psi_{1/2L}(f)] = \sum_{i=-\infty}^{\infty} G\left(f - i\frac{1}{2L}\right)\delta\left(f - i\frac{1}{2L}\right).$$

The periodical function $h(t)$ and its respective Fourier transform $H(f)$ are shown in Figures 2.81 (b) and (d), respectively, considering $a = 1$ and $L = 2$.

A completely similar effect is observed by sampling the function $g(t)$, implying the respective Fourier transform to be periodical. The above results are summarized below:

$g(t)$	$G(f)$
Periodical	Discrete
Discrete	Periodical

It should be observed that the Fourier transform can also be applied to periodical functions, producing as a result a necessarily *discrete* Fourier transform, i.e., a collection of Dirac deltas along the frequency domain. As a matter of fact, the Fourier transform of a periodical function $h(t)$ can be verified to produce a Fourier transform that is identical to the Fourier series of $h(t)$ [Brigham, 1988]. In other words, *for periodical functions the Fourier series becomes equal to the Fourier transform*, and the resulting transform or series is always *quantized* (or *discrete*) in the frequency space. In this sense, the Fourier series can be thought of as a particular case of the Fourier transform when the input function is periodical.

2.7.4 Frequency Filtering

One of the many important practical applications of the Fourier transform is as a means for implementing filters. To *frequency filter* a specific function or signal is henceforth understood as modifying its Fourier coefficients in a specific fashion. In this section, we consider the following three main types of filters: low-pass, high-pass and band-pass. The implementation of such filters in the frequency domain, however, is common to all these types and is achieved by multiplying the Fourier transform of the analyzed signal with a *filtering function*, and taking as result the inverse Fourier transform. That is to say, if $h(t)$ is the function to be filtered, with respective continuous Fourier transform $H(f)$, and $V(f)$ is the filtering function, the filtered version of $h(t)$, henceforth represented as $q(t)$, can be obtained as

$$q(t) = \Im^{-1}\{H(f)V(f)\}. \tag{2.52}$$

By considering the convolution theorem, such a filtering process can be verified to correspond to convolving, in the time domain, the function $h(t)$ with the inverse Fourier transform $v(t)$ of the filtering function $V(F)$. It is observed that there are, at least in principle, no restrictions to the type of filtering function (e.g., continuous, strictly positive, differentiable, etc.). Let us now consider each of the three types of filters individually.

As we understand from its name, a *low-pass filter* acts by attenuating the magnitude of the high frequency components in the signal, while the low frequency components are allowed to pass. Therefore, the respective filter function is expected to decrease for high values of frequency magnitude. It is important to note that such effect is relative, i.e., what matters is to attenuate the high frequency components *relative* to the low frequency components, even if all components are attenuated or magnified as a consequence. Figure 2.82 presents two possible low-pass filtering functions.

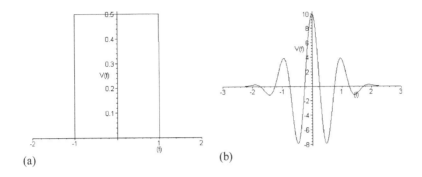

Figure 2.82: *Two possible low-pass filtering functions.*

Observe that the low-pass filter in Figure 2.82 (a) attenuates all frequencies, but the attenuation is smaller for the lower frequencies. Low-pass filtering tends to produce functions that are *smoother* and *more intensely correlated* than the original function $h(t)$ (see Section 2.6.5).

A typical low-pass filtering function is the zero-mean Gaussian (see Section 2.1.4). It is interesting to relate the Gaussian filtering function to its inverse Fourier transform, since this allows us to understand the filtering effect in terms of the standard deviation σ of the Gaussian respectively defined in the time domain (the higher this value, the more intense the low-pass filtering effect). Recall from Section 2.7.3 and equation (2.50) that the Gaussian in the frequency domain has as parameter $\sigma_f = 1/(2\pi\sigma)$. The henceforth adopted Gaussian filtering function $V(f)$ and its respective inverse Fourier transform (which is a Gaussian in the strict sense), are given in terms of the following Fourier transform pair:

$$g_\sigma(t) = \frac{1}{\sigma\sqrt{2\pi}} \exp\left\{-\frac{1}{2}\left(\frac{t}{\sigma}\right)^2\right\} \leftrightarrow V(f) = \exp\left\{-2\left(\pi\sigma f\right)^2\right\}.$$

Observe that the above Gaussian filter function $V(f)$ always varies between 0 and 1. Figure 2.83 illustrates the process of Gaussian low-pass filtering.

The Fourier transform $H(f)$ (b) of the function $h(t)$ to be filtered (a) is multiplied by the filtering function $V(f)$ (c), which in this case is the Gaussian $V(f) = \exp\left\{-2\left(\pi\sigma f\right)^2\right\}$ with $\sigma = 0.1$, and the filtered function (d) is obtained by taking the inverse Fourier transform of $H(f)\,V(f)$. The effect of this filtering process over the original function, a cosine function corrupted by additive uniform noise, is clear in the sense that the higher frequency components of $h(t)$, i.e., the sharp oscillations along the cosine function, have been substantially attenuated, although at the expense of a substantial change in the amplitude of $h(t)$. An additional discussion about Gaussian filtering, in the context of contour processing, is presented in Section 7.2.3.

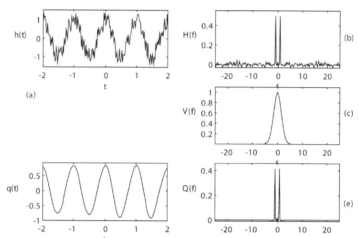

Figure 2.83: *The function h(t) to be low-pass filtered* **(a)**, *its respective Fourier transform* **(b)**, *the filtering function (in Fourier domain)* **(c)**, *the filtered function q(t)* **(d)**, *and its respective Fourier transform* **(e)**.

The second class of filters, known as *high-pass filters*, act conversely to the low-pass filters, i.e., by attenuating the magnitude of the low frequency components of the signal, while the higher frequency components are allowed to pass. Such an attenuation should again be understood in *relative* terms. An example of high-pass filter is the *complemented Gaussian V(f)*, defined as

$$g_\sigma(t) = \delta(t) - \frac{1}{\sigma\sqrt{2\pi}} \exp\left\{-\frac{1}{2}\left(\frac{t}{\sigma}\right)^2\right\} \leftrightarrow V(f) = 1 - \exp\left\{-2\left(\pi\sigma f\right)^2\right\}.$$

It is interesting to observe that the complemented Gaussian filter function always varies between 0 and 1. This function is illustrated in Figure 2.84 for $\sigma = 0.25$.

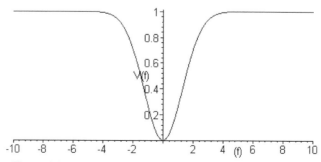

Figure 2.84: *The complemented Gaussian function for* $\sigma = 0.25$.

As illustrated in Figure 2.85, a high-pass filter tends to accentuate the most abrupt variations in the function being filtered, i.e., the regions where the derivative magnitude is high (in image processing and analysis, such abrupt variations are related to the image contrast). In other words, high-pass filtering reduces the

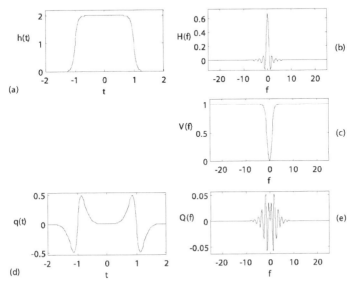

Figure 2.85: *The function h(t) to be high-pass filtered (a), its respective Fourier transform (b), the filtering function (c), the filtered function (d), and its respective Fourier transform (e).*

correlation and redundancy degree in the original signal.

A particularly interesting type of high-pass filter in the context of the present book is related to the derivative property of the Fourier transform (see Section 2.7.3). We have already seen that, in order to obtain the first derivative of a function $h(t)$, all that we need to do is to multiply its Fourier transform by the purely imaginary function $D_1(f) = (j2\pi\sigma f)$ and take the inverse transform. As shown in Figure 2.86, this filter function presents the general shape of a high-pass filter, attenuating the low-frequency components relative to the higher frequency components. Figure 2.87 illustrates the use of this function in order to differentiate a function $h(t)$. Since the differentiation can substantially enhance high frequency noise, such an operation is usually performed by using as filter function the product of the function $D_1(f) = (j2\pi\sigma f)$ by a Gaussian function.

The filters under the category known as *band-pass* act by relatively accentuating the frequency components along a specific portion of the frequency domain. Therefore, a low-pass filter can be understood as a particular case of a band-pass filter centered at zero frequency. Gaussian functions with non-zero means in the frequency domain provide a good example of band-pass filter functions. Figure 2.88 illustrates the filtering of the function $h(t) = \cos(2\pi f_1 t) + \cos(2\pi f_2 t)$, $f_2 = 4f_1$,

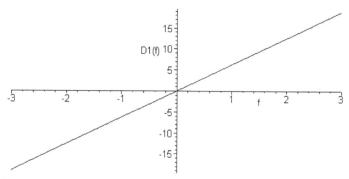

Figure 2.86: *The first derivative filter function, which is a purely imaginary function.*

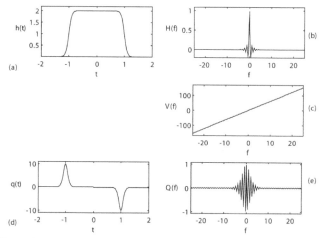

Figure 2.87: *The function h(t) to be high-pass filtered by the first derivative filter (a), its respective Fourier transform (b), the filtering function (c), and the filtered function (d) and its respective Fourier transform (e). Observe that V(f) is a pure imaginary function.*

by using as two band-pass Gaussian filtering functions centered respectively at $-f_2$ and f_2, as shown in Figure 2.88 (c). Since the filter removes almost completely the lower frequency component (i.e., $\cos(2\pi f_1 t)$), the resulting filtered function consists almost exclusively of $\cos(2\pi f_2 t)$. It should be observed that, in this specific example, a similar effect could have been obtained by using a zero-mean complemented Gaussian narrow enough to attenuate the low frequencies.

Although filters are usually applied with a specific objective, such as smoothing a function, situations arise where some undesirable filtering has already taken place

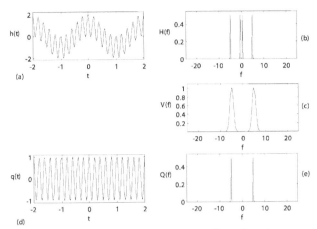

Figure 2.88: *The function h(t) to be band-pass filtered (a), its respective Fourier transform (b), the filter function (c), and the filtered function (d) and its respective Fourier transform (e).*

and we want to recover the original function. Such a problem, which is relatively common in image processing and analysis, is called *deconvolution* (this follows from the fact that the filtering process can be alternatively understood as a convolution in the time space). If the original function $h(t)$ was filtered by a function $V(f)$, yielding $q(t)$, we may attempt to recover the original function by dividing the Fourier transform $Q(f)$ of the filtered function by the filter function $V(f)$ and taking the inverse Fourier transform as the result. Thus, the sought recovered function would be obtained as $h(t) = \Im^{-1}\{Q(f)/V(f)\}$. However, this process is not possible whenever $V(f)$ assumes zero value. In practice, the situation is complicated by the presence of noise in the signal and numeric calculation. Consequently, effective deconvolution involves more sophisticated procedures such as Wiener filtering (see, for instance, [Castleman, 1996]).

2.7.5 The Discrete One-Dimensional Fourier Transform

In order to be numerically processed by digital computers, and to be compatible with the discrete signals produced by digital measuring systems, the Fourier transform has to be reformulated into a suitable discrete version, henceforth called *discrete Fourier transform—DFT*.

First, the function g_i to be Fourier transformed is assumed to be a uniformly sampled (spaced by Δt) series of measures along time, which can be modelled in terms of multiplication of the original, continuous function $\tilde{g}(t)$ with the sampling function $\Psi_{\Delta t}(t) = \sum_{i=-\infty}^{\infty} \delta(t - i\Delta t)$. Second, by being the result of some measuring process (such as the recording of a sound signal) the function g_i is assumed to have *finite duration* along the time domain, let us say from time $a = i_a \Delta t$ to $b = i_b \Delta t$.

The function g_i is henceforth represented as:

$$g_i = \tilde{g}\,(i\Delta t).$$

Observe that the discrete function g_i can be conveniently represented in terms of the vector $\vec{g} = (g_{i_a}, g_{i_a+1}, \ldots, g_{i-1}, g_i, g_{i+1}, \ldots, g_{i_b}-1, g_{i_b})$. Figure 2.89 illustrates the generic appearance (i.e., sampled) of the function g_i.

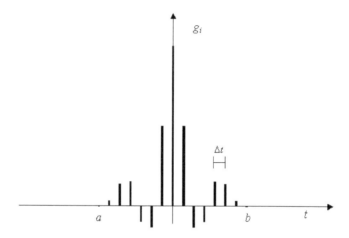

Figure 2.89: *Example of a sampled function g_i to be Fourier transformed.*

As seen in the previous section, the fact that the function g_i is discrete implies that *the DFT output $H(f)$ is always periodical of period* $\frac{1}{\Delta t}$. This important property is illustrated in Figure 2.90, which also takes into account the fact that the resulting Fourier transform is discrete (see below).

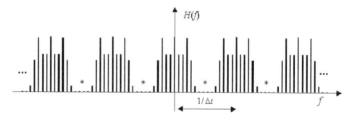

Figure 2.90: *The DFT output function $H(f)$ is always discrete and periodic of period $1/\Delta t$.*

Figure 2.90 also allows us to immediately derive an important result in signal processing, known as the *sampling theorem*, which relates the maximum frequency of a sampled signal g_i that can be represented in terms of the adopted sampling step Δt. Since the function $H(f)$ is periodic of period $\frac{1}{\Delta t}$, and the period centered at zero frequency extends from $-f_{\max}$ to $f_{\max}$, we have that $f_{\max} = \frac{1}{2\Delta t}$, which is known as the *Nyquist rate*. Therefore, any higher frequency contained in the original continuous signal $\tilde{g}(t)$ will not be properly represented in the Fourier transform of the respective sampled signal g_i. Indeed, such too high frequencies will imply an overlapping of each of the borders of the basic period, identified by asterisks in Figure 2.90, a phenomenon known as *aliasing*. The best strategy to reduce this unwanted effect is to use a smaller value for Δt.

To be represented in a digital computer (e.g., as a vector), the Fourier transform output function $H(f)$ has to be sampled in the frequency space (the samples are assumed to be uniformly spaced by Δf), i.e., multiplied by the sampling function $\Psi_{\Delta f}(t) = \sum_{i=-\infty}^{\infty} \delta(t - i\Delta f)$, implying the periodical extension $h(t)$ (see the previous section) of g_i. Consequently, *the DFT input $h(t)$ is always a periodical extension of g_i with period* $\frac{1}{\Delta f}$. This fact is illustrated in Figure 2.91.

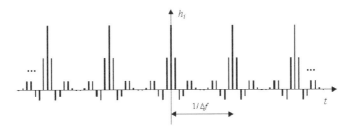

Figure 2.91: *The DFT input function h_i is always discrete and periodic of period* $1/\Delta f$.

As seen in the previous section, the fact that the input function is periodical implies that the DFT represents a numerical approximation of the *Fourier series* of $h(t)$. These important facts are summarized below:

Discrete input function g_i implies the DFT *output* $H(f)$ to be *periodical* of period $\frac{1}{\Delta t}$.
Discrete DFT *output* $G(f)$ implies the *input* function g_i to become a *periodical* function $h(t)$ with period $\frac{1}{\Delta f}$.
The DFT *input* (and also the *output*) function can be represented as a *vector*.
The *DFT* corresponds to a numerical approximation of the *Fourier series*.

Since each period of $h(t)$ has length $\frac{1}{\Delta f} = b - a = 2L$ and the samples are equally spaced by Δt, and by assuming that the period $2L$ is an integer multiple of Δt, the number N of sampling points representing the input function g_i along one period is given by:

$$N = \frac{1/\Delta f}{\Delta t} = \frac{1}{\Delta t \Delta f}.$$

Observe that we have $N = \frac{1}{\Delta t \Delta f}$ instead of $N = \frac{1}{\Delta t \Delta f + 1}$ because we want to avoid repetition at the extremity of the period, i.e., the function is sampled along the interval $[a, b)$. The number M of sampling points in any period of the output function $H(f)$ is similarly given by

$$M = \frac{1/\Delta t}{\Delta f} = \frac{1}{\Delta t \Delta f}.$$

By considering $N = M$, i.e., the number of sampling points representing the input and output DFT functions are the same (which implies vectors of equal sizes in the DFT), we have

$$N = M = \frac{1}{\Delta t \, \Delta f}. \tag{2.53}$$

Since the input function is always periodical, the DFT can be numerically approximated in terms of the Fourier series, which can be calculated by considering any full period of the input function $h(t)$. In order to be numerically processed, the Fourier series given by equation (2.47) can be rewritten as follows. First, the integral is replaced by the sum symbol and the continuous functions are replaced by the above sampled input and output functions. In addition, this sum is multiplied by Δt because of the numerical integration and the relationship $n = 2Lf$ is taken into account, yielding

$$G_k = G\left(k \Delta f\right)$$

$$= c_{n=2Lf} = \frac{1}{2L} \int_{0}^{2L} h(t) \, \exp\left\{-\frac{j\pi n t}{L}\right\}$$

$$= \Delta t \frac{1}{2L} \sum_{i=0}^{N-1} h\left(i \Delta t\right) \exp\left\{-\frac{j\pi \left(2Lf\right)\left(i \Delta t\right)}{L}\right\}$$

$$= \Delta t \frac{1}{2L} \sum_{i=0}^{N-1} h\left(i \Delta t\right) \exp\left\{-j2\pi \left(k \Delta f\right)\left(i \Delta t\right)\right\}.$$

Observe that we have considered the time interval $[0, 2L)$, in order to avoid redundancies. By considering the input function as having period $2L = \frac{1}{\Delta f}$, we obtain

$$H_k = H\left(k \Delta f\right) = \Delta t \, \Delta f \sum_{i=0}^{N-1} h\left(i \Delta t\right) \exp\left\{-j2\pi \left(k \Delta f\right)\left(i \Delta t\right)\right\}.$$

Now, from equation (2.53) we have $\Delta t \, \Delta f = \frac{1}{N}$, which implies that

$$H_k = H\left(k \Delta f\right) = \frac{1}{N} \sum_{i=0}^{N-1} h\left(i \Delta t\right) \exp\left\{-\frac{j2\pi i k}{N}\right\}. \tag{2.54}$$

This equation, which is commonly known as the *discrete Fourier transform equation*, allows us to numerically estimate the Fourier series of the periodical function $h(t)$. It is easily verified that the computational execution of equation (2.54) for each specific value of k demands N basic steps, being therefore an algorithm of complexity order $O(N)$. Since the complete Fourier series involves N calculations of this equation (i.e., $k = 0, 1, \ldots, N-1$), the overall number of basic operations in the DFT algorithm is of $O(N^2)$.

2.7.6 Matrix Formulation of the DFT

Equation (2.54) can be compactly represented in matrix form, which is developed in the following. By defining the abbreviations:

$$w_{k,i} = \exp\left\{-\frac{j2\pi ik}{N}\right\}, \quad h_i = h(i\,\Delta t), \quad \text{and} \quad H_k = H(k\,\Delta f),$$

equation (2.54) can be rewritten as:

$$H_k = \frac{1}{N}\sum_{i=0}^{N-1} w_{k,i}\,h_i. \tag{2.55}$$

Before proceeding with the derivation of the matrix form of the DFT, it is interesting to have a closer look at the discretized kernel function $w_{k,i} = \exp\left\{-\frac{j2\pi ik}{N}\right\}$ (see Figure 2.92). Let us introduce $w_{ki} = w_{k,i}$ and observe that $w_{k,i} = w_{i,k}$; for instance $w_4 = w_{1,4} = w_{4,1} = w_{2,2}$. From Section 2.1, it is easy to see that the complex exponential kernel function $w_{i,k}$ in the above equation can be understood as the sequence of complex points uniformly distributed along the unit circle in the Argand plane. For instance, for $N = 8$, $i = 1$ and $k = 0, 1, \ldots, N-1$, we have the result shown in Figure 2.92.

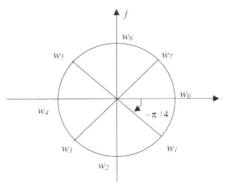

Figure 2.92: *The sampled Fourier kernel function* $w_{k,i} = \exp\left\{-\frac{j2\pi ik}{N}\right\}$ *for* $N = 8$ *and* $i = 1$.

It is clear from such a construction and graphical representation that $w_{ki} = w_{k,i}$ is periodic of period N, i.e., $w_{ki} = w_{ki+pN}$, $p = \ldots, -2, -1, 0, 1, 2, \ldots$ It is also interesting to observe that larger values of k imply larger angular spacings between the sampled kernel values and larger frequencies for the sampled complex exponential function. The reader can easily verify that for $k > \frac{N}{2}$ the sampled complex values $w_{k,i}$ rotate in the reverse sense (i.e., counterclockwise). For instance, for $k = 7$ we have $w_{7,1} = w_7$; $w_{7,2} = w_{14} = w_6$; $w_{7,3} = w_{21} = w_5$; and so on. Ultimately, this is the reason why high speed wheels in western movies seem to rotate backwards (see also the sampling theorem in Section 2.7.5).

But now it is time to return to the matrix representation of the DFT. It follows directly from the definition of the product between matrices (equation (2.11)) that equation (2.55) can be effectively represented as the product between the $N \times N$ matrix $W_N = [w_{i,k}]$ and the $N \times 1$ vector $\vec{h} = [h_i]$, $i = 0, 1, \ldots, N - 1$, resulting the $N \times 1$ vector $\vec{H} = [H_i]$, i.e.,

$$\vec{H} = \frac{1}{N} W_N \, \vec{h}. \tag{2.56}$$

Hence we have obtained an elegant representation of the DFT in terms of a simple matrix multiplication. In addition to providing such a compact representation, this formulation makes it clear that the DFT is indeed a *linear transformation*. Since such transformations are completely characterized by the nature of the transformation matrix, it is important to have a closer look at the matrix W_N and its properties. To begin with, we observe that this matrix is symmetric, but not Hermitian (see Section 2.2.5). Next, it is easy to verify that

$$W_N \left(W_N^* \right)^T = NI,$$

which means that the matrix W_N is *almost* unitary (Section 2.2.5). As a consequence, we can write

$$\vec{H} = \frac{1}{N} W_N \vec{h} \Rightarrow \left(W_N^* \right)^T \vec{H} = \frac{1}{N} \left(W_N^* \right)^T W_N \vec{h} \Rightarrow \left(W_N^* \right)^T \vec{H} = \frac{1}{N} (NI) \vec{h} \Rightarrow \vec{h} = \left(W_N^* \right)^T \vec{H}.$$

But, as the matrix W_N is symmetric, so is its conjugate, and we can write

$$\vec{h} = \left(W_N^* \right)^T \vec{H} \Rightarrow \vec{h} = W_N^* \vec{H}.$$

This equation, which provides a means for recovering the input vector $\vec{h}$ from the DFT output $\vec{H}$, corresponds to the *inverse discrete Fourier transform*, henceforth abbreviated as *IDFT*, which clearly is also a linear transformation.

Consider the following example:

Example: *DFT*

Calculate the DFT of the sampled ($\Delta t = 1$) and periodic function $h(t)$ defined as

$$h(t) = \delta(t - 1) \quad \text{and} \quad h(t) = h(t + 4).$$

As always, it is interesting to start by visualizing the involved function, which is shown in Figure 2.93.

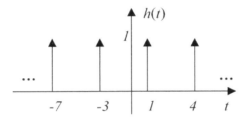

Figure 2.93: *Sampled function to be Fourier transformed.*

It is clear that $h(t)$ is periodic of period 4. We need to consider only the period defined from 0 to $N - 1$, yielding $\vec{h} = (0, 1, 0, 0)^{\mathrm{T}}$. Now, the DFT can be calculated by using equation (2.55) as

$$\vec{H} = \frac{1}{N}W_N\vec{h} = \frac{1}{N}\begin{bmatrix} w_{0,0} & w_{0,1} & w_{0,2} & w_{0,3} \\ w_{1,0} & w_{1,1} & w_{1,2} & w_{1,3} \\ w_{2,0} & w_{2,1} & w_{2,2} & w_{2,3} \\ w_{3,0} & w_{3,1} & w_{3,2} & w_{3,3} \end{bmatrix}\begin{bmatrix} 0 \\ 1 \\ 0 \\ 0 \end{bmatrix}$$

$$= \frac{1}{N}\begin{bmatrix} w_0 & w_0 & w_0 & w_0 \\ w_0 & w_1 & w_2 & w_3 \\ w_0 & w_2 & w_4 & w_6 \\ w_0 & w_3 & w_6 & w_9 \end{bmatrix}\begin{bmatrix} 0 \\ 1 \\ 0 \\ 0 \end{bmatrix}.$$

The above matrix can be straightforwardly obtained by considering the graphical representation of the sampled complex exponential function for $N = 4$ shown in Figure 2.94.

Therefore

$$\vec{H} = \frac{1}{4}\begin{bmatrix} 1 & 1 & 1 & 1 \\ 1 & -j & -1 & j \\ 1 & -1 & 1 & -1 \\ 1 & j & -1 & -j \end{bmatrix}\begin{bmatrix} 0 \\ 1 \\ 0 \\ 0 \end{bmatrix} = \frac{1}{4}\begin{bmatrix} 1 \\ -j \\ -1 \\ j \end{bmatrix}$$

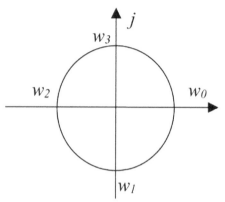

Figure 2.94: *The graphical representation of the* 4 × 4 *Fourier matrix elements.*

and the original signal can be immediately recovered as

$$\vec{h} = W_N^* \vec{H} = \frac{1}{4} \begin{bmatrix} 1 & 1 & 1 & 1 \\ 1 & j & -1 & -j \\ 1 & -1 & 1 & -1 \\ 1 & -j & -1 & j \end{bmatrix} \begin{bmatrix} 1 \\ -j \\ -1 \\ j \end{bmatrix} = \begin{bmatrix} 0 \\ 1 \\ 0 \\ 0 \end{bmatrix}$$

2.7.7 Applying the DFT

In the previous developments we assumed that the function $\tilde{g}(t)$ had already been obtained from g_i by the use of a sampling and time-limiting processes. It is now time to have a closer look at how such sampled and time-limited signals can be extracted in practical situations. This problem is particularly relevant to fully understanding several situations in image processing and analysis, such as in the process of acquiring digital images by using a camera or a scanner. The basic steps involved in obtaining g_i from $\tilde{g}(t)$ are illustrated in Figure 2.95.

The above diagram also makes it clear that the acquisition process presents as parameters the total number N of observations and the time interval Δt between successive observations, as well as the initial and final times, a and b, respectively, with $b - a = 2L$. Let us clarify this process by considering a practical example. Suppose we want to analyze the sound of a flute which is being played continuously at a specific pitch and intensity. We can use a microphone connected to an A/D converter interfaced to a digital computer in order to obtain the sampled and time-limited signal g_i. We start recording the signal at time a and stop at time b. Once the original continuous signal has been time limited and sampled, it is ready

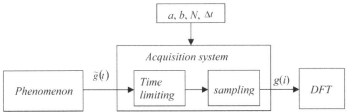

Figure 2.95: *A continuous signal $\tilde{g}(t)$ has to be time-limited and previously sampled, yielding the approximated signal g_i, before being processed by the DFT. The signal g_i has N sample points, spaced by Δt, extending from time a to b, with respective extension $b - a = 2L$.*

to be processed by the DFT. By choosing different values for these parameters, we can obtain different representations of the continuous signals $\tilde{g}(t)$ and, as discussed next, to have drastic effects over the quality of the obtained results. For instance, the choice of an improper value of Δt (e.g., too large) may not allow the proper representation of the involved high frequencies. In addition, observe that by considering equation (2.53), the frequency interval (and resolution) is automatically determined for each specific value of N and Δt as $\Delta f = \frac{1}{N\Delta t} = \frac{1}{b-a} = \frac{1}{2L}$.

Let us consider the acquisition process more carefully in terms of the hypothetical function $\tilde{g}(t) = \cos(2\pi t) + \cos\left(4\sqrt{2}\pi t\right)$, which is not completely unlike the signal that would be produced by a wood flute[1]. This signal clearly involves two frequencies, i.e., $f_1 = 1\,\text{Hz}$ and $f_2 = 2\sqrt{2}\,\text{Hz}$. Figure 2.96 shows this continuous signal together with its respective Fourier transform, given by

$$\widetilde{G}(f) = 0.5\left[\delta\left(f + f_1\right) + \delta\left(f - f_1\right) + \delta\left(f + f_2\right) + \delta\left(f - f_2\right)\right].$$

The fact that the observed signal has to be limited along time can be modelled by multiplying $\tilde{g}(t)$ by a windowing function $\phi(t)$ such as the rectangular function (see Section 2.1.4):

$$\phi(t) = r(t) = \begin{cases} 1 & \text{if } -a \leqslant t < a, \\ 0 & \text{otherwise,} \end{cases}$$

whose Fourier transform was calculated in Section 2.7.3. Observe that although we have adopted $a = b$ for simplicity's sake, the more generic situation involving $a \neq b$ can be easily developed by the reader. Figure 2.97 presents the signal $\tilde{g}(t)$ after being windowed by the above function, considering $a = 2$ seconds, as well

---

[1]The Fourier representation of sounds such as those produced by a flute is known to involve frequencies that are multiples of the smallest frequency (the so-called *fundamental*). However, this law of multiples can become distorted as a consequence of non-linearities involved in sound production. In addition, the Fourier coefficients respective to the multiple frequencies, unlike in our hypothetical example, tend to decrease with the frequency.

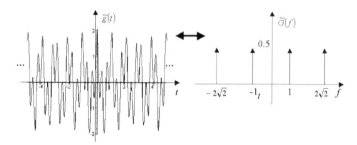

Figure 2.96: *A hypothetic continuous signal $\tilde{g}(t)$ produced by a wood flute being played continuously at a specific pitch and intensity **(a)** and its respective Fourier transform $\widetilde{G}(f)$ **(b)**.*

as its respective Fourier transform, which is obtained by convolving the Fourier transforms of $\tilde{g}(t)$ and $\phi(t)$, i.e., $\widetilde{G}(f) * \Phi(f)$.

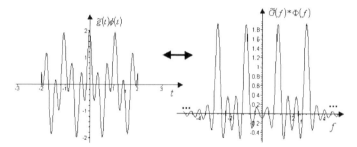

Figure 2.97: *The windowed version of $\tilde{g}(t)$ and its respective Fourier transform given by $\widetilde{G}(f) * \Phi(f)$.*

It is clear from the oscillating nature of the Fourier transform of the rectangular function, i.e., $\widetilde{G}(f)$, that the windowing of the signal $\tilde{g}(t)$ implies a *ripple* effect onto its respective Fourier transform. This unwanted effect can be minimized by using larger values of a or by using a smoother windowing function such as the Gaussian. As a matter of fact, observe that when a tends to infinity, its Fourier transforms tend to the Dirac delta function, and no ripple effect is implied.

Now that we have obtained a time-limited version $\tilde{g}(t)\,\phi(t)$ of the possibly infinite original signal $\tilde{g}(t)$, it has to be uniformly sampled before it can be represented as a vector suitable to be used in the DFT equation (2.55). As discussed in Section 2.7.5, such a sampling can be obtained by multiplying the function $\tilde{g}(t)\,\phi(t)$ by the sampling function $\Psi_{\Delta t}(t) = \sum_{i=-\infty}^{\infty} \delta(t - i\,\Delta t)$, i.e., the acquired signal finally can be represented as $g(t) = \tilde{g}(t)\,\phi(t)\,\Psi_{\Delta t}(t)$.

Figure 2.98 illustrates the sampled version of $\tilde{g}(t)\,\phi(t)$, assuming $\Delta t = 0.05$ second, as well as its respective Fourier transform $G(t) = \widetilde{G}(t) * \Phi(f) * \left(\frac{1}{\Delta t}\right)\Psi_{1/\Delta t}(f)$. Since the convolution is commutative, the order of the sampling and time-limiting operations become irrelevant.

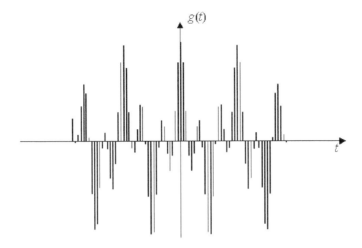

Figure 2.98: *The sampled function* $g(t) = \tilde{g}(t)\,\phi(t)\,\Psi_{\Delta t}(t)$, *assuming* $\Delta t = 0.05$ *second, and its respective Fourier transform* $G(f)$.

It is clear that the number N of samples not cancelled by the windowing function can be calculated as

$$N = \text{floor}\left\{\frac{2a}{\Delta t}\right\}.$$

In the above case, since $\Delta t = 0.05$ and $a = 2$, we have $N = \frac{2a}{\Delta t} = \frac{4}{0.05} = 80$ sampling points. We also have from equation (2.53) that $\Delta f = \frac{1}{N\Delta t} = 0.25$, and from the sampling theorem that $f_{\max} = \frac{1}{2\Delta t} = 10$ Hz. As discussed in Section 2.7.5, the sampling process implies the Fourier transform of g_i to be periodical with period $\frac{1}{\Delta t} = 20$ Hz. It can be verified that the Fourier transform $G(f)$ provides a good approximation for the original Fourier transform $\widetilde{G}(f)$, except for the ripple effect caused by the windowing operation.

The DFT of g_i can now be determined by applying equation (2.54), i.e.,

$$G(k\,\Delta f) = \frac{1}{N}\sum_{i=0}^{N-1} g(i\,\Delta t)\,\exp\left\{-\frac{j2\pi ik}{N}\right\}.$$

A complete period (starting at zero frequency) of the respective DFT output $H(k\,\Delta f)$ is shown in Figure 2.99.

A more careful analysis of the Dirac delta approximations obtained in Figure 2.99 indicates that the lower frequency peaks (marked with asterisks), with respect to $f_1 = 1$ Hz, have been better represented (there is less ripple around it

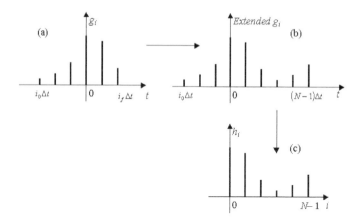

Figure 2.99: *The DFT output function $H(k \Delta f)$. The pair of lower frequency peaks for $f_1 = 1$ Hz are indicated by asterisks.*

and the amplitude is exactly as expected, i.e., 0.5) than the deltas for $f_2 = 2\sqrt{2}$ Hz.
This is because f_1 is an integer multiple of $\Delta f = \frac{1}{N\Delta t} = 0.25$ Hz, while f_2 is not.
Indeed, a complete cancellation of the ripple effect is observed in such a multiple
situation, because the zero crossings of $\tilde{G}(f)$ can be verified to coincide with the
sampling points. However, this welcomed effect cannot usually be guaranteed in
practical situations, and the unwanted rippling effect has to be somehow alleviated,
for instance by using a smoother window function. Figure 2.100 illustrates this possibility considering as windowing function $\phi(t) = \exp\left\{-t^2\right\} r(t)$, i.e., the product of
a Gaussian with the rectangular function (a truncated Gaussian function).

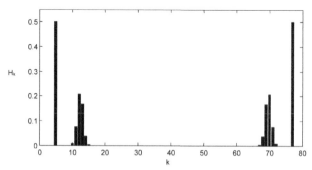

Figure 2.100: *DFT obtained by windowing $\tilde{g}(t)$ with the truncated Gaussian function $\phi(t) = \exp\left\{-t^2\right\} r(t)$.*

The ripple attenuation is readily verified, though at the expense of a decrease
in the amplitude of the coefficients related to $f_2 = 2\sqrt{2}$ Hz. The reader is referred
to the literature (e.g., [Ingle & Proakis, 1997; Kamen & Heck, 1997; Papoulis,

1984]) for a more detailed discussion about several windowing functions and their respective properties.

Another important practical aspect related to the DFT application concerns the fact that in most DFT algorithms (and especially many FFT algorithms—see Section 2.7.8) the function g_i is assumed to initiate at time zero, i.e., the origin of the time axis, and extend up to $N - 1$, so that g_i can be represented as the vector $\vec{h} \mid h_i = g(i\,\Delta t)$, $i = 0, 1, \ldots, N - 1$. If this is indeed the case, the DFT execution is immediate, being only needed to use the number of samples N. However, special attention is required when dealing with functions extending into the negative portion of the time domain, such as that considered in the above examples. It should be borne in mind that the time zero while acquiring the signal is a relative reference that is important and must be taken into account, otherwise the time shifting effect described in Section 2.7.3 will imply a modulation of the respective Fourier transform.

Let us now consider the situation where the signal extends into the negative portion of its domain in more detail. Basically, what is needed is to move the left portion of g_i, i.e., that in the negative portion of the time axis (excluding the zero), to the right-hand side of the function (see Figure 2.101). If $g_i = g(i\,\Delta t)$ starts at

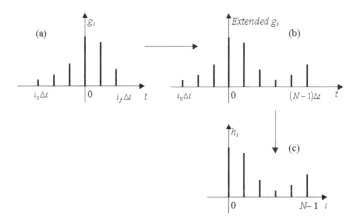

Figure 2.101: *Most DFT (and FFT) algorithms require the input function h(t) to be defined from 0 to N − 1. In case the function g_i, from which h_i is obtained, extends into the negative time domain (a), it needs to be conveniently reorganized (b) and (c). The vector $\vec{h}$ to be used as input to the DFT can then be obtained by taking the sampled points in the non-negative region of the time axis.*

i_a and terminates at i_b, i.e., $i = i_a, \ldots, -1, 0, 1, \ldots, i_b$ (observe that $i_a < 0$), the above mentioned translation operation can be implemented by using the following algorithm:

1. $N \leftarrow i_b - i_a + 1$;
2. **for** $k \leftarrow i_a$ **to** -1
3.   **do**
4. $g_{(k+N)\Delta t} = g_{k\Delta t}$;

Once this extended version of g_i is obtained, all that remains to be done is to copy it into the DFT input vector $\vec{h} \mid h_i = g(i\Delta t)$, $i = 0, 1, \ldots, N - 1$, which, as discussed in this section, will be considered as being periodical of period N. This process is illustrated in Figure 2.101.

Figure 2.102 presents the DFT input function $h(i)$ (a) and its respective Fourier transform (b) as typically produced by DFT algorithms.

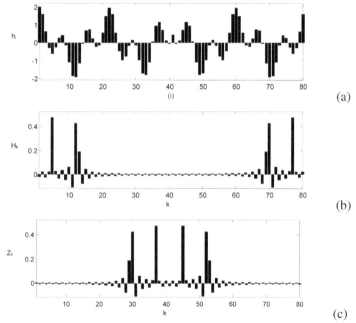

(a)

(b)

(c)

Figure 2.102: *The input (a) and output (b) functions for a typical DFT algorithm. The obtained output can be represented with the zero frequency at its middle (c).*

It is clear that the leftmost point in the DFT output is at zero frequency. In case we want to have the zero frequency in the middle of function, a transposition process similar to the above described has to be applied, yielding the reorganized signal Z_k. A possible strategy for doing so is presented in terms of the following algorithm:

1. $n \leftarrow$ FLOOR$((N-1)/2)$;
2. **for** $k \leftarrow 0$ **to** $N - n - 1$
3.   **do**
4. $Z_{k+n} \leftarrow H_k$;
5. **for** $k \leftarrow N - n$ **to** $N - 1$
6.   **do**
7. $Z_{k+n-N} \leftarrow H_k$;

Let us conclude this section by characterizing two alternative practical situations typically met while applying the DFT, which are presented in the following:

Alternative 1: Imposed frequency sampling interval.	
1	Estimate the maximum frequency f_{max} to be found in the continuous signal $\tilde{g}(t)$. In case this is not possible, assume the highest possible value for f_{max}.
2	Calculate $\Delta t = \frac{1}{2 f_{max}}$.
3	Define a suitable frequency sampling interval Δf. The frequency resolution is then $\frac{\Delta f}{2}$.
4	Determine the number of samples (i.e., the dimension of the DFT input vector) as $N = \frac{1}{\Delta t \Delta f}$.
Alternative 2: Imposed number of samples.	
1	Estimate the maximum frequency f_{max} to be found in the continuous signal $\tilde{g}(t)$. In case this is not possible, assume the highest possible value for f_{max}.
2	Calculate $\Delta t = \frac{1}{2 f_{max}}$.
3	Define a suitable value for N. The higher this value, the slower the DFT.
4	Determine the frequency sampling interval as $\Delta f = \frac{1}{N \Delta t}$.

The choice between these two alternatives will depend on practical constraints. First, it should be observed that the DFT parameters Δt, Δf and N are linked by equation (2.53), i.e., $N = \frac{1}{\Delta t \Delta f}$. Thus, having chosen two of these parameters, the third is automatically defined. Ideally, we would wish to have both Δf and Δt as small as possible, in order to allow maximum frequency resolution and the largest maximum representable frequency. However, the problem is that both these facts imply higher values for N, and thus the DFT becomes slower. The first alternative, by defining N as a consequence of the choice of Δf, is therefore suitable for situations where there is not too much concern about the execution time. The second alternative should be used otherwise.

Before proceeding to the next section, it is important to note that, although in this section all the Fourier transforms were purely real (because the considered time signals were all real and even), this is by no means the general case. In other words, the DFT results generally involve both real and imaginary parts. In addition, although we were limited to purely real input functions, it is also possible to use complex functions. The above presented developments, however, are immediately applicable to such situations.

2.7.8 The Fast Fourier Transform

The *fast Fourier transform*, hence *FFT*, should be understood simply as a numeric method for computing the DFT in a fast manner, its result being identical except for round-off noise (i.e., the noise produced by the limited precision in representing numbers and performing operations in digital computers). The effectiveness of this technique—indeed of this class of techniques, since there are many FFT algorithms—cannot be overlooked, making a real difference in practice, especially when the number of samples representing the signals is relatively large. Since the time savings allowed by the FFT is considerable, the matrix calculation of the DFT in equation (2.56) is rarely used in practice.

The main advantage of the FFT methods arises from the fact that they remove many of the redundancies in the DFT matrix calculation (see [Brigham, 1988], for instance). As a matter of fact, the number of basic operations (i.e., additions and multiplications) involved in the FFT algorithm is of order $O(N \log N)$, while (as seen in Section 2.7.5), the standard DFT implies $O(N^2)$.

Several FFT algorithms, including the classical Cooley and Tukey's approach, require the value of N to be an integer power of two, such as 32, 64, 128 and so on. In such cases, the method is said to be of *radix* 2. It is also possible to have alternative radixes, such as 4, 5, 6 and so on [Brigham, 1988]. In the general case, N must be an integer power of the radix. In practice, this requirement of having $N = (\text{radix})^k, k = 0, 1, 2, \ldots$ can be easily met by using the smallest value of k such as the vector size is smaller than 2^k and filling up the unused positions in the DFT input vector $\vec{h}$ with zeros. However, observe that this procedure may cause a discontinuity and, consequently, introduce oscillations in the recovered signal because of the Gibbs effect. A method to alleviate this problem consists in filling up the first half of the unused portion of the vector with the same value as the last in the original function, and the second half as the first value in the original function (recall that the DFT implies that the function represented by this vector is periodical).

The simplicity and small number of numerical operations implied by the FFT, involving mostly complex products, has motivated the whole family of new devices known as *digital signal processors*, namely circuits or integrated circuits capable of processing the FFT very quickly by using dedicated hardware components (such as multipliers). Such a tendency has allowed the FFT to be processed at very high speeds allowing real-time applications for most situations.

In spite of the FFT importance and usefulness, we do not present an algorithm for its calculation in the present book for the following three reasons: (a) the proper explanation of FFT algorithms is relatively extensive; (b) there are excellent books covering the FFT (for instance, see [Brigham, 1988]); and (c) nowadays, it is unlikely that the reader will have to implement the FFT, since it is broadly available in mathematical environments (e.g., MATLAB®, Scilab®, Maple® and so on) and also as ready-to-use libraries and objects to be used in programming languages and environments.

2.7.9 Discrete Convolution Performed in the Frequency Domain

One of the most frequent applications of the discrete Fourier transform is as a means of performing fast convolution or correlation between two sampled functions or signals. Such operations are typically applied to compare two signals (correlation) or to filter a signal (see Section 2.7.4). While Section 2.5.2 has presented equations and algorithms for performing convolution and correlation between two discrete signals, the calculation of such operations in the time domain can be too slow, especially when the signals involve many samples. The computational advantage of performing convolution in the frequency domain by using the FFT is still more substantial for higher dimensional signals, such as images.

We have already seen in Section 2.7.7 how to perform the DFT over discrete signals. In addition to the procedures presented, the calculation of the convolution (or correlation) by using the FFT (or DFT) demands particular attention to the fact that the DFT input signal is necessarily periodical. Therefore, the convolution in the frequency domain implies the input signal to be treated as if it were a closed loop or ring. As an example, consider that we want to low-pass filter a discrete signal h_i by using a Gaussian as the filter function. As we have already seen, this corresponds to convolving the signal with the inverse Fourier transform of the Gaussian, which is also a Gaussian whose discrete and time-limited version is henceforth represented as g_i. The convolution procedure involves multiplying the signal h_i by time shifted versions of the function g_i—recall that since the Gaussian is even we have $g_{-i} = g_i$. Because in the DFT the signals are always periodical, when such time-displaced Gaussian functions are near the period extremities, they tend to wrap over to the other side of the signal, which is known as *circular convolution*. In such a way, the signal extremities interact with one another. While this is sometimes exactly what is required from the convolution (such as in the case of closed contours in Section 5.2.1), the circular convolution implemented by the DFT will not generally produce the same results as those obtained in the time domain by using the algorithms described in Section 2.5.2. Fortunately, the noncircular convolution can be easily calculated by the DFT simply by padding the functions with N zeros between each subsequent period of the signals being convolved (for additional information, see [Castleman, 1996]). The same approach can be applied to obtain noncircular correlation.

An interesting and relevant situation where the DFT is used to calculate the (generally noncircular) correlation occurs in statistics. We have seen in Section 2.6.5 that the cross-correlation and autocorrelation of stochastic signals can provide important information about the respective signals (or processes). Indeed, observe that the cross- and autocorrelation can be performed in the Fourier domain by using the correlation theorem (Section 2.7.3). As a matter of fact, the autocorrelation of a signal can be obtained as the inverse Fourier transform of its power spectrum. This possibility is illustrated for the case of images in Section 2.6.5. Additional information about the use of the Fourier transform in statistics can be obtained in [Papoulis, 1962; Therrien, 1992].

To probe further: *Fourier Analysis*

The importance of Fourier analysis as a theoretical and applied tool has been fully substantiated by an ever growing number of related references. Nice coverages of Fourier series, including many examples, can be found in [Tolstov, 1976] and [Kreyszig, 1993]. Interesting introductions to the Fourier transform, including applications, are provided by [James, 1995] and [Sneddon, 1995], and [Spiegel, 1995] includes several exercises (many of which are solved) on Fourier analysis. Good references on Fourier analysis from the perspective of signal processing include [Lynn, 1984; Oppenheim & Schafer, 1975; Papoulis, 1984] and [Burrus et al., 1994] provides many computer-based exercises on signal processing, including discrete Fourier transforms and fast Fourier transforms. Additional hands-on approaches to digital signal processing include [Ingle & Proakis, 1997; Kamen & Heck, 1997]. The classical reference on the fast Fourier transform is [Brigham, 1988]. A good reference on Fourier analysis from the perspective of linear systems can be found in [Frederick & Carlson, 1971], and [Körner, 1996] includes a series of more advanced topics on Fourier analysis. References on the application of Fourier analysis to imaging and computer vision include [Castleman, 1996; Gonzalez & Woods, 1993; Schalkoff, 1989].

2.8 Graphs and Complex Networks

Graphs[1] are discrete data structures involving nodes (or vertices) and links (or edges), so that emphasis is placed on the *connectivity*. Despite their intrinsic simplicity, graphs are particularly general in the sense that most other discrete data structures can be derived from them. For instance, lists, queues, trees, lattices, among many other structures, are but particular instances of graphs. As such, graphs are the natural choice for representing and modeling most (or even all) systems which involve components and relationships between these components. Figure 2.103 shows a simple graph composed of 8 nodes and 10 edges.

Examples of representations of real-world systems in terms of graphs include

---

[1]Though graphs and networks are understood as different structure in graph theory, with networks being normally associated to flows, for historic reasons these two terms have been used indistinctively in the area of complex networks. In this book graphs and networks are treated as synonyms.

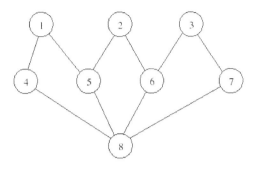

Figure 2.103: *A graph containing 8 nodes and 10 edges.*

the Internet, WWW, maps of towns or roads, flight routes, power distribution, artistic and scientific collaboration, disease spreading, social relations, protein–protein interaction, neuronal networks, amongst many other possibilities. For instance, in the case of scientific collaborations, each node may represent a scientist, while the edges correspond to collaborations. Actually, it is hard to think of an interesting system that cannot be represented as a graph. Even continuous structures such as a bridge or the wing of an airplane can be partitioned into components connected in terms of their physical adjacency (the approach used in finite difference and finite element approaches to simulation).

Image analysis and vision are no exception to the generality and power of graph representation and modeling. Actually, there are two main ways in which graphs have been applied to these areas. First, the visual information can be decomposed into several components and represented as a graph. For instance, a digital image with $N_x \times N_y$ pixels together with a neighborhood system (e.g., 4-neighborhood) defines a respective graph where each pixel is represented as a node and each of the neighborhoods defines a respective edge. Higher-level instances of the visual information — e.g., involving borders, regions or objects — can also be naturally transformed into graphs. At this point it is important to observe that the concept of graph as nodes and edges can be enhanced in several ways, for instance by allowing different types of edges (representing different relationships), different types of nodes, incorporating information inside the nodes (e.g., the R, G, and B values of each pixel), and so on. Therefore, versatile and powerful structures such as the semantic networks used to represent relationships in scenes [Niemann et al., 1990] can also be understood as graphs. In shape analysis, graphs can be used to represent a single shape (e.g., piecewise-linear representations, where the nodes are the straight parts and the edges stand for the adjacency between these parts, (e.g., piecewise-linear representations, where the nodes are the straight parts and the edges stand for the adjacency between these parts, see [Niemann et al., 1990]) or multiple shapes (each shape is a node, while the edges express the relationships between the individual shapes). Indeed, networks can be used to model visual information at several time and spatial scales, from pixels to objects, and even relationships between such scales. By making explicit the important components in the scene as well as the relevant relationships between such components, graphs are a primary choice for representing, analysing and modeling visual information and shapes.

The second way in which graphs can be used in image analysis and vision is as a resource for pattern recognition. While the feature vectors representing visual entities (e.g., properties of pixels, regions or objects) are naturally mapped into multidimensional feature spaces as clouds of isolated points, it is also possible to transform such clouds into graphs so that graph-based methods can be used for the separation and identification of the relevant clusters. A possible way to do this is to connect every pair of points (each point corresponds to a mapped feature vector) whose respective feature vectors are at a distance smaller than or equal to a given threshold. Weighted networks can also be considered, where the weights are associated to the distance or similarity between feature vectors. Therefore, regions with high density of points, natural candidates for clusters, will imply in densely connected graphs. The use of graphs in pattern recognition received growing attention along the last decade (e.g., [Kandel et al., 2007; Marchette, 2004]).

More recently, the area of *complex networks* established itself as one of the most dynamic and important possibilities for technological and scientific advances. Its origins can be traced back to the systematic theoretical study of uniformly random graphs by Flory [Flory, 1941] and Erdős-Rényi [Erdos & Renyi, 1960; Erdős & Rényi, 1961]. Such graphs are characterized by fixed probability of connection between any possible pair of nodes. Consequently, several nodes result in the same degree (i.e., the number of connections of a given node) so that the connectivity of the graph can be reasonably characterized in terms of its average node degree. As such, these random graphs are almost *regular*[1]. Observe that these graphs have a stochastic nature, rather than the deterministic (predictable) connectivity of discrete structures such as lattices and meshes. Unfortunately, there are very few real-world or theoretical systems underlain by such an indiscriminate type of stochastic connectivity. As a consequence, the comprehensive studies on uniformly random graphs remained mostly in the realm of theory.

During the 1960s, Milgram investigated networks of social acquaintances and found an average separation of 6 contacts between people in the United States [Milgram, 1967]. Such networks have been called *small world*, being characterized also by high clustering between nodes (see Section 2.8.1). The small-world property is intrinsically related to the concept of a *path* between two nodes of a graph. Two

---

[1]Regular graphs have all nodes with the same degree.

nodes are said to be adjacent if they share an edge. Analogously, two edges are adjacent whenever they share a node. A *walk* is a sequence of adjacent edges and nodes. Walks which never repeat an edge or node are called *paths*. The *length* of a walk or path is often defined as being equal to its number of edges. The *shortest path* between two nodes is the path connecting the two nodes which involves the smallest number of edges. Observe that more than one shortest path can be eventually found possible between any two nodes. The *shortest path length* between two nodes is the length of the respective shortest path. Informally speaking, a small-world network is such that we have a relatively small shortest path length between most pairs of nodes. More strictly speaking, a small-world network requires that the average shortest path scales logarithmically with the number of nodes [Newman, 2003]. Random networks, for instance, are also small-world networks as their shortest path length scales in logarithmic fashion with the number of nodes in the network.

The area of complex networks acquired its impetus from the late 1990s with the identification of power-law distributions of node degree on the Internet [Faloutsos et al., 1999] and WWW [Barabási & Albert, 1999]. More specifically, it was shown [Faloutsos et al., 1999] that the distribution of the node degrees on the Internet followed a power law, i.e.,

$$n(k) \propto k^{-\alpha} \qquad (2.57)$$

Consequently, a log-log representation of $n(k)$, i.e., number of nodes with each specific degree k, in terms of k yields a straight line. Because no special distinguishing features can be found along a straight line, no inherent changes can be noticed in the shape of the respective distribution (a straight line) as one moves along the node degree axis, which motivates the denomination of such distributions as being *scale free*. Consequently, networks underlain by power law degree distributions have been called *scale free networks*.

One of the important implications of having a network with node degrees obeying a power law is that the chances of getting nodes with many connections become substantially larger than in an uniformly random counterpart (i.e., a uniformly random network with the same number of nodes and save average degree). Such nodes with particularly high degree have been called *hubs* [Albert & Barabási, 2002; Costa et al., 2007b; Dorogovtsev & Mendes, 2002; Newman, 2003]. Hubs are all important for both the structure and dynamics of complex networks. For instance, because a hub connects to many other nodes, it provides an immediate bypass of length two between any of those nodes, therefore substantially decreasing the average shortest path length in the network. At the same time, a hub is critical for infection spreading because it interacts with a large number of connected nodes. The special relevance of scale-free networks became evident after several important systems such as protein–protein interactions, flight routes, scientific collaborations, among many others, were found to have power law distribution of node degrees (e.g., [Costa et al., 2007a]).

2.8.1 Basic Concepts

In this section, we present the main concepts and measurements which have been traditionally used in complex networks. We start by describing how graphs and complex networks can be represented in terms of matrices and follow by presenting some of the most traditionally used measurements of complex networks topology.

Complex Networks Representation

Complex networks can be classified into several categories according to their properties. For instance, networks with directed or undirected edges are respectively called *directed* and *undirected networks*. *Weighted networks* are characterized by having weights associated to each of their edges; otherwise they are called *unweighted*. A directed, unweighted complex network (or graph) is composed by N nodes and E directed edges. Such a network can be fully represented in terms of its *adjacency matrix* K by making $K(j,i) = 1$ whenever an edge is found from node i to node j; otherwise $K(j,i) = 0$. The seemingly counter-intuitive way to represent the indices in the adjacency matrix is adopted here in order to allow for compatibility with operations performed with this matrix. An undirected network can also be represented in terms of its respective adjacency matrix, which is now necessarily symmetric. An undirected edge can be thought of as corresponding to two directed edges with different directions, implying that the number of undirected edges is twice as large as the number of directed edges. In the case of an undirected weighted network, it can be completely represented in terms of its *weight matrix* W, such that each directed edge extending from node i to node j, with respective weight a, implies $W(j,i) = a$.

The adjacency matrix of the undirected, unweighted graph in Figure 2.103 can be easily found to be given as

$$
K = \begin{pmatrix}
0 & 0 & 0 & 1 & 1 & 0 & 0 & 0 \\
0 & 0 & 0 & 0 & 1 & 1 & 0 & 0 \\
0 & 0 & 0 & 0 & 0 & 1 & 1 & 0 \\
1 & 0 & 0 & 0 & 0 & 0 & 0 & 1 \\
1 & 1 & 0 & 0 & 0 & 0 & 0 & 1 \\
0 & 1 & 1 & 0 & 0 & 0 & 0 & 1 \\
0 & 0 & 1 & 0 & 0 & 0 & 0 & 1 \\
0 & 0 & 0 & 1 & 1 & 1 & 1 & 0
\end{pmatrix}
\tag{2.58}
$$

Observe that this matrix is symmetric as a consequence of the undirected nature of the network. It is also possible to fully represent a network by listing the pairs of nodes constituting the respective edges. For instance, the network in Figure 2.103 can be represented by the following list of edges:

$$L = \{(1,4);(1,5);(2,5);(2,6);(3,6);(3,7);(4,1);(4,8);$$
$$(5,1);(5,2);(5,8);(6,2);(6,3);(6,8);(7,3);(7,8)\} \tag{2.59}$$

In the case of undirected networks, it is enough to represent only one of each pair of directed edges in the list, i.e., it would be enough to include only the directed edge $(1,4)$ in the above list instead of storing both $(1,4)$ and $(4,1)$.

Measurements of Complex Networks Topology

In order to better understand a network, its topology can be quantified in terms of several respective measurements. The most traditionally used measurement is the *degree* of a node $k(i)$, which corresponds to the number of edges attached to a given node i. For example, for the network in Figure 2.103 we have $k(1) = 2$; $k(5) = 3$ and $k(8) = 4$. Nodes with a particularly high degree are called *hubs*. In the case of directed networks, two degrees become necessary to characterize the connectivity of a given node i, the *indegree* and the *outdegree*, corresponding to the number of edges which arrive and depart from that node, respectively. These two measurements are henceforth abbreviated as $k_{in}(i)$ and $k_{out}(i)$. While the degree of a node can be understood as a *local* measurement of the network under analysis, it is also possible to use the average or standard deviation (actually any statistical moment) of the degrees considering all nodes in order to obtain a *global* characterization of the connectivity of the network. Another important characterization of complex networks is provided by the histogram of the degrees of the nodes of a complex network (e.g., scale-free networks yield straight lines in log-log representations of their node degree distributions).

Another important measurement of the local connectivity in complex networks is the *clustering coefficient* of a node i. Let the *immediate neighbors* of node i be those nodes which are directly attached to i, i.e., through paths of length 1 (a single edge). For instance, the immediate neighbors of node 5 in the network in Figure 2.103 are the nodes 1, 2 and 8. The number of immediate neighbors of a node i is henceforth indicated as $n_{im}(i)$. Also, let $e_{im}(i)$ be the number of interconnections between the immediate neighbors of node i. The clustering coefficient of node i can now be defined as the number of connections between the immediate neighbors of i and the maximum total number of connections between those nodes, i.e.,

$$cc(i) = \frac{2e_{im}(i)}{n_{im}(i)(n_{im}(i) - 1)}. \tag{2.60}$$

As such, the clustering coefficient of a node tells us how much interconnected its neighbors are. Observe that $0 \le cc(i) \le 1$, with $cc(i) = 0$ indicating no interconnection between the immediate neighbors of i and $cc(i) = 1$ expressing full interconnectivity between those nodes.

2.8.2 Some Representative Network Models

The characterization of real-world networks soon revealed that these structures tend to have different connectivity properties as expressed by their respective measurements. Small-world networks, for instance, are characterized by small average shortest path lengths and high average clustering coefficient. Many other types of networks have been so far identified. In an attempt to understand and reproduce these different types of network connectivity, researchers have developed several theoretical models of complex networks which can be computationally simulated in order to produce networks with specific connectivity properties. The more traditional of these models are reviewed as follows.

Uniformly Random Networks: Erdös-Rényi

This is one of the oldest network models, considered by Flory [Flory, 1941] and by Erdős and Rényi [Erdos & Renyi, 1960; Erdős & Rényi, 1961]. In this model, henceforth called random networks or Erdős-Rényi (ER), one starts with N disconnected nodes and then joins pairs of nodes with constant probability γ. ER networks are well described by their average degree, which implies that this type of network is almost regular[1]. As a consequence of the uniform connectivity of ER networks, they do not reflect the more heterogeneous and structured organization of real-world networks. As a matter of fact, the term *complex* in *complex networks* stands precisely for the structured and intricate connectivity of several real-world structures, contrasting with the regularity of the ER model. Because of their inherently uniform connectivity pattern, random networks are often considered as a reference while characterizing other types of networks.

Small-World Networks

The concept of small-world networks was formalized by S. Milgram in the late 1960s [Travers & Milgram, 1969]. In this experiment, persons from specific American cities were randomly selected and queried about specific contact people. In case the person knew the contact at the personal level, he/she should send a letter to them, otherwise he/she should send a letter to somebody who they believe could know the contacts and inform the researchers about this. Out of the relatively small fraction of letters that reached their final destinations, the average length of contacts was of only 5.5 letters from origin to correct final destination. This was understood as the average social separation between people in the United States. Subsequent investigations not only confirmed, but even decreased the social separation between American people. For obvious reasons, this property has been called *small-world property*. Interestingly, small world networks also tend to be characterized by a relatively high average clustering coefficient. More recently [Newman, 2003], the small world property was formalized in the sense that the average shortest path length should scale in logarithmic fashion with the size of the network.

---

[1] A *regular* network is a network in which all nodes have the same degree.

One of the most famous theoretical models capable of generating networks re-producing the small world property was proposed by Watts and Strogatz [Watts & Strogatz, 1998]. In this model, one starts with a regular network (e.g., nodes distributed along a circle, with each node being connected to the p previous and subsequent nodes (in clock and anti-clockwise fashion) and then rewires a small percentage (e.g., 1%) of the edges. As a consequence of such a rewiring, the aver-age shortest path length of the original network is drastically reduced.

Geographical Networks

Geographical networks are characterized by the fact that their nodes have well-defined coordinates in an embedding space (any dimension), so that distances (e.g., Euclidean) can be calculated between any pair of nodes. Often, these distances are used to defined the weights of the respective edges. For instance, highways networks where each town or destination is represented as a node and the routes between such sites are represented as edges are geographical networks because each node has a well-defined position along the surface of earth. Frequently in geo-graphical networks, the probability of connection between two nodes is affected by their respective proximity and/or adjacency. For instance, it is more likely that two nearby towns are connected through a highway than two distant places. It is also possible to define the connectivity between geographical nodes while considering their adjacency as revealed by some partitioning scheme such as the Voronoi tes-sellation [Costa et al., 2007b], i.e., every adjacent pair of nodes are interconnected.

Scale-Free Networks: Barabási-Albert

The identification of the important facts that the Internet and WWW have degree distributions which follow power laws motivated the search for theoretical models of network growth which could yield structures with similar properties. One of the most important such models is that called *rich get richer* [Barabási & Albert, 1999; Simon, 1955], which is founded on preferential attachment of new nodes. In this model, one starts with $m0$ nodes and then successively incorporate new nodes with m links each. Every time such a new node is included, each of its m connections are attached to the existing nodes in the network with probability proportional to their respective degrees. Therefore, nodes which are more intensely connected tend to receive more connections. This type of preferential attachment implies power law distribution of node degree, with average degree equal to $2m$.

Knitted Networks: The Path-Regular Model

Introduced recently [Costa, 2007a,b], the family of *knitted networks* is character-ized by being composed of paths, i.e., sequences (or chains) of sequentially con-nected nodes with degree two. So, while several other models build networks edge by edge, knitted networks have paths as basic constituting elements. As such, this type of networks is inherently suitable for modeling several real-world systems such as transportation routes, where connection is typically planned and established in terms of paths going through a pre-defined set of places. In computer vision and shape analysis, knitted networks are potentially related to gaze movement, required for integrating salient points along time and space [Costa, 2006b]. In other words, the sequence of gaze fixations intrinsically corresponds to a path and therefore can be modeled by knitted networks. The use of complex network for modeling gaze movement is further discussed in Section 9.8.3.

2.8.3 Community Finding

The heterogeneity in complex networks connectivity can also be characterized in terms of their *modules* or *communities*, i.e., groups of nodes which are more intensely interconnected one another than with the remainder of the network. Networks with communities are said to be *modular*. Completely uniform networks such as lattices have no communities. However, a great deal of real-world networks exhibit modular structure. For instance, author collaboration networks tend to present communities which are intrinsically related to the respective areas and/or spatial distribution of the authors and institutions [Costa et al., 2008]. Therefore, the identification of the communities in a given complex network can provide valuable information about the nature and organization of the respectively represented systems.

Interestingly, the identification of communities is closely related to clusterization in pattern recognition. However, while the relationships between the objects in the latter situation have to be inferred such as in terms of separation distances, the relationships between the nodes in the communities to be detected are explicitly given in terms of the respective edges. Figure 2.104 illustrates this intrinsic relationship between pattern recognition (clustering) and community detection. It is thus not surprising that pattern recognition methods have been proposed for the identification of communities in complex networks (e.g., [Zhou, 2003]). However, the mainstream of methods for that finality are founded on the concept of *modularity* [Girvan & Newman, 2002].

Additional resources: *Complex networks online resources*

Useful complex networks resources are available at `http://cyvision.if. sc.usp.br/~bant/hierarchical/`.

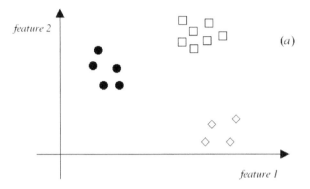

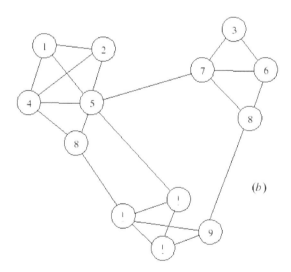

Figure 2.104: *Pattern recognition (clustering) corresponds to finding clus-ters of objects (a) while taking into account inferred rela-tionships, such as the distances between objects in the mea-surement space. The detection of communities in a complex network (b) also involves the identification of clusters (the communities or modules), but with the difference that the relationships between objects are explicitly given in terms of the respective edges.*

The greatest masterpieces were once only pigments on a palette.

Henry S. Haskins

3

Shape Acquisition and Processing

Chapter Overview

THIS CHAPTER PRESENTS an overview of the main concepts and techniques for acquiring and processing images of shapes. Starting with image formation and representation by computers, the present chapter reviews several algorithms for image enhancement, noise filtering, edge detection and image segmentation, which can be applied to produce suitable representations of the shapes to be further analyzed.

This chapter introduces some of the basic tools that are commonly used in image processing and shape analysis. Digital image processing is an important and well developed research field, and, as will soon become clear, a typical approach to shape analysis involves a series of operations to be carried out on the original (raw) image in order to reduce both unwanted distortions and secondary information from the image, while emphasizing the most relevant data. In a typical shape analysis application the researcher is normally interested in isolating an object (e.g., a cell) from other less important objects in the image before it can be properly analyzed. Despite the introductory nature of the present chapter, it is intended to provide a sound approach to this field, allowing the reader to implement the discussed techniques and to smoothly proceed to more in-depth developments presented in the suggested bibliography, which include the standard textbooks [Castleman, 1996; Gonzalez & Woods, 1993; Jain, 1989; Pratt, 1991; Schalkoff, 1989]. It is important to observe that image processing techniques can be classified into two large groups, i.e., *specific* and *generic*. Specific techniques are characterized by the careful identification of a suitable mathematical model of the particular phenomenon of interest while designing image processing algorithms. For instance, the specific type of noise deteriorating a set of images can be identified (e.g., white of speckle noise) and used in order to obtain optimal noise filters, which is usually done by using variational calculus[1]. While such an approach is generally more powerful and effective, its implementation demands a substantial mathematical basis, and its application to generic problems is naturally limited. On the contrary, generic approaches to image processing are aimed at obtaining algorithms that can be applied to a broader scope of problems. Since several of the simplest and most popular image processing methods, which are often capable of reasonable performance, belong to the generic class, the present chapter is limited to this type of approach.

3.1 Image Representation

3.1.1 Image Formation and Gray-Level Images

The fact that computers represent information in a discrete way implies that images must be translated into this discrete form before undergoing any subsequent computational processing. Figure 3.1 illustrates the discretization of the square shape image shown in (a). In order to represent such an image, the continuous image must first be made discrete, which can be done by projecting it onto a grid of photodetector cells, as illustrated in Figure 3.1 (b). Each photodetector cell calculates the average light energy projected onto it, therefore spatially discretizing the continuous image. The respective spatial discretization of the image in Figure 3.1 (b) is shown in (c).

In order to represent the discretized image in a computer, it is also necessary to decide how to represent the continuous averaged light intensities of the discrete image elements, which accounts for the differences between the gray levels in each cell of Figure 3.1 (c). This can be done by assigning a number indicating the light intensity to each of these points, e.g., by assigning 255 to white and 0 to black, in such a way that intermediate values represent a discrete sequence of gray levels. After such additional discretization, the *digital image* version of Figure 3.1 (a) into an array of discrete values, presented in Figure 3.1 (d), is obtained. Each element of the digital image is called a *pixel*, which stands for "picture element." If each pixel is allowed to assume a set of different values, say integer values between 0 and 255, then we have a monochrome or gray level image, where each pixel assumes a *gray level* value. It is worth emphasizing that digital images are therefore obtained from continuous images through *two* discretization processes, one for the spatial points and another for the colors (or gray levels) assigned to each pixel. These two processes are called *sampling* and *quantization*, respectively. From the above discussion, a gray level digital image can be defined as follows:

Definition (Gray-Level Digital Images). *Gray-level digital images are represented as 2D discrete matrices g, of dimension $P \times Q$, where each pixel assumes a gray*

---

[1] In variational calculus, one is interested in analytically or numerically finding a function that minimizes (or maximizes) a specific merit figure.

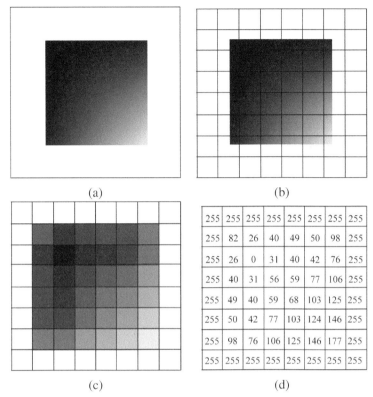

Figure 3.1: *(a) Image of a graded square; (b) the original image* (a) *projected onto a discrete photoreceptor grid; (c) the average light intensities produced by the photoreceptors; and (d) digital image representation as an integer array.*

level integer value n, $0 \leqslant n < N$. An image where n can only take 0 and 1 as values (i.e., $N = 2$) is said to be a BINARY IMAGE.

Therefore, considering the example of Figure 3.1 (d), we have $P = 8$, $Q = 8$, $N = 256$, and the digital image is represented as

$$g = \begin{bmatrix} 255 & 255 & 255 & 255 & 255 & 255 & 255 & 255 \\ 255 & 82 & 26 & 40 & 49 & 50 & 98 & 255 \\ 255 & 26 & 0 & 31 & 40 & 42 & 76 & 255 \\ 255 & 40 & 31 & 56 & 59 & 77 & 106 & 255 \\ 255 & 49 & 40 & 59 & 68 & 103 & 125 & 255 \\ 255 & 50 & 42 & 77 & 103 & 124 & 146 & 255 \\ 255 & 98 & 76 & 106 & 125 & 146 & 177 & 255 \\ 255 & 255 & 255 & 255 & 255 & 255 & 255 & 255 \end{bmatrix}$$

Other important types of digital images are obtained by considering different attributes (e.g., ultra-violet, infra-red, distance, etc.) in order to represent the scene, but they basically follow the same underlying idea of gray level 2D digital images. Some of the most important alternative types of images are discussed in the following sections. Meanwhile, let us concentrate on the problem of how a scene from the real world can be transformed into a digital image. Although this process ultimately depends on the image acquisition device (i.e., if done by a scanner, camera, etc.), a simplified model of image formation in an ordinary camera can be considered in order to give some insight into this operation. In this case, light reflects on real world objects and is projected through an aperture fitted with a lens onto a surface inside the camera. The lens normally affects the image formation by blurring it (Gaussian filtering is typically assumed as a model for this effect, see Section 3.2.4). The formation of a spatially sampled image can be modelled by assuming that the surface onto which the patterns of light are projected inside the light-proof chamber of the camera contains an array of photodetectors, such as in a CCD (charge coupled device). Each photodetector, in turn, samples a value corresponding to a respective pixel in Figure 3.1 (c), and outputs a response proportional to the average intensity of light falling onto its area. The quantization of this response yields the digital image.

To probe further: *Image Acquisition*

Given the broad variety of image acquisition devices, this chapter has overlooked some important image-related acquisition issues. Chapter 8 of Schalkoff's book [Schalkoff, 1989] provides some information on this issue, while specific aspects of graphics hardware can be found in computer graphics books, such as in Chapter 4 of Foley et al. [1994]. The book of Travis [1991] may also be of interest to the reader. Castleman's book [Castleman, 1996] presents some discussion on image acquisition devices (Chapter 2) and image display (Chapter 3). Important characteristics that should be considered for product specification are

☞ frame rates (i.e., allowed frames per second);

☞ resolution (related to the number of pixels);

☞ available software and drivers;

☞ output data format;

☞ camera control features.

However, the intense advances in hardware technology constantly make new imaging devices available, and the reader interested in solving a specific problem should be well informed about the current technologies related to his/her particular field. The reader should take great care while specifying the necessary image acquisition devices and lighting accessories, since the proper acquisition of the images is essential for any further image processing and analysis. Nowadays, the WWW provides an outstanding source of images. Although mostly disorganized, the WWW is an extremely rich image database, possibly the largest one, to such an extent that research on visual querying of the Web has already become established as a promising research area [Chen et al., 1996; Delogne, 1996; Hearst, 1997; Lynch, 1997; Scientific American, 1997; Stix, 1997]. Some good starting points for image acquisition products on the WWW include

☞ `http://www.vision1.com/`;

☞ `http://www.ime.usp.br/~cesar/revision/produc.htm`;

☞ `http://www.cs.ubc.ca/spider/lowe/vision.html`;

☞ `http://www.eeng.dcu.ie/~whelanp/resources/resources.html`.

3.1.2 Case Study: Image Sampling

Figure 3.2 illustrates the impact of spatial sampling in image acquisition, showing the same scene sampled with 512×326 (a), 256×163 (b), 128×82 (c) and 64×41 (d) pixels. Observe that the size of the image does not change, which means that the size of each sampling cell (i.e., pixel) increases for each subsequent image, implying that the resolution decreases. Note also that, as the sampling rate decreases, the small details in the image are progressively removed, while larger structures are emphasized. For instance, the cow's and the owl's eyes, as well as the pattern on the body of the penguin statuette, disappear altogether. On the other hand, the presence of the three objects can still be perceived in the images with lower resolution. This analysis bears a strong relationship to the so-called multiscale approach, a concept that is discussed further throughout this book.

An important concept of multiscale image analysis that emerges from the previous discussion is that of an *image pyramid*, where the image is successively downsampled by the disposal of some samples (i.e., pixels), e.g., by discarding the even samples while keeping the odd ones or vice versa. A sequence of images obtained in this way is shown in Figures 3.3 (a)-(d), which have the same number of pixels as their respective counterparts shown in Figures 3.2 (a)-(d), except that now the printed pixel size is held constant for all images.

As a matter of fact, a filtering step is generally involved in the generation of image pyramids. The pyramid represented by Figures 3.3 (a)-(d) starts with the image presenting small-scale structures in Figure 3.3 (a), and proceeds by filtering out such details as the analyzing scale increases along Figures 3.3 (b)-(d). As the smaller scale structures are filtered out, larger scale elements (like the bodies of the three objects) are emphasized.

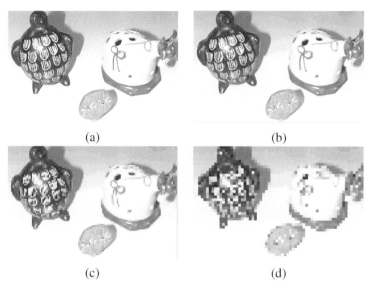

(a)   (b)

(c)   (d)

Figure 3.2: *(a)-(d) Images sampled respectively with 512 × 326, 256 ×
163, 128 × 82 and 64 × 41.*

To probe further: *Image Sampling*

The main ideas discussed in the last section can be formalized in terms of the theory of continuous signals sampling [Brigham, 1988; Oppenheim & Schafer, 1989]. Additional material about sampling and quantization can be found in the majority of image processing books, as well as in the literature related to signal processing. A particularly interesting approach to this problem is based on the convolution between the input image and the function known as *point spread function*, which allows the more complete modeling of the image sampling process, including the blurring implied by the image acquisition optical system and the fact that each pixel gray level is the result of integrating the light intensity along the photodetector sensible region. Further information on these topics can be found, for instance, in Chassery & Montanvert [1991].

3.1.3   Binary Images

Binary images correspond to the simplest and yet one of the most useful image types. In general, a binary image is a representation of scenes with only two possible gray values for each pixel, typically 0 and 1. Often, binary images can be understood intuitively as including only two types of elements: the object(s), which define the *foreground*, and the *background*. Once only two types of elements are present, they can be represented by an array containing only two gray levels, i.e.,

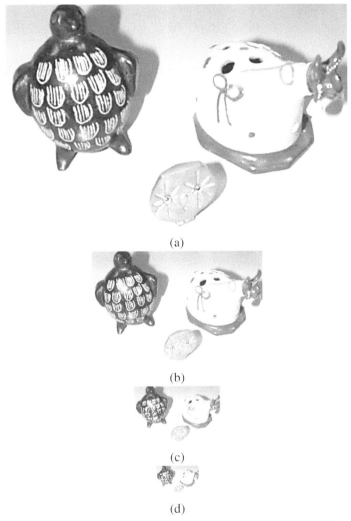

(a)

(b)

(c)

(d)

Figure 3.3: *Image pyramid. Each image has the same number of pixels as the images in Figure 3.2.*

black and white. Whether the objects are represented by the white or by the black pixels is just a matter of convention, which must be defined and clearly indicated *a priori*. In this book it is henceforth assumed, in binary images, that the foreground is represented by 1, while the background is represented by 0. Since higher pixel values are usually understood to represent higher luminosity, this convention implies the objects will be represented in white, and the background pixels in black. However, because typical images usually have a higher percentage of background pixels than those representing objects, it is interesting to change the above convention in order to have the objects (represented by value 1) in black and the foreground (represented by value 0) in white, which contributes to a more esthetic presentation of the figures and a more rational use of ink. It should also be observed that the convention of assigning 0 to the background pixels and 1 to the object pixels has a logical intuitive interpretation, in the sense that the value 0 ("false" or "off") indicates lack of objects (i.e., the background), while the value 1 ("true" or "on") indicates presence of an object at that pixel position. Of course, these are just conventions, and all image processing algorithms can be adapted to cope with different schemes. Table 3.1 summarizes the conventions for gray level and binary image representations that will be adopted throughout this book.

Gray level images	Binary images
0 (dark)	0 = white = background
255 (bright)	1 = black = foreground

Table 3.1: *Pixel value conventions adopted in this book.*

Binary images naturally arise in image processing problems in a number of different situations. First, there are some kinds of images that are originally binary, and may have already been acquired in this way. Examples include line drawings and characters on a uniform background, which can be acquired with the aid of a scanner (most scanner software allows the user to specify the acquisition of a binary image, instead of a gray level or a color image) or with an optical pen, for instance. As far as biomedical applications are concerned, standard procedures, such as drawing neurons with the aid of a *camera lucida* attached to a microscope, also yield binary images. Alternatively, binary images often result from color or gray level images after applying specific image processing techniques such as image segmentation. A simple example illustrating how to obtain binary images as a result of segmentation is presented in Figure 3.4. Figure 3.4 (a) presents a gray level image. Two possible binary representations of this gray level scene are presented in Figures 3.4 (b) and (c). The former has been manually obtained by a human operator who has painted the black silhouettes over the original image using image processing software. This picture can be considered a nearly *ideal* binary representation[1] of Figure 3.4 (a), as it correctly captures the object outlines and is compatible with the way humans perceive such features (observe that, as a consequence of being produced by a human, this figure is somewhat biased and subjective, i.e., someone else could have produced a slightly different version of it). On the other hand, Figure 3.4 (c) is the result of *automated image segmentation*. It is important to note that the automatic segmentation algorithm has generated a resulting image (Figure 3.4 (c)) that does not precisely correspond to the expected objects' silhouettes (Figure 3.4 (b)), which illustrates a potential difficulty often encountered in shape analysis. This example illustrates how difficult automatic image segmentation can be, an issue to be discussed further in Section 3.3. It is interesting to note that, because of the *a priori* human operator knowledge, the human-based

---

[1]The concept of "ideal" depends on human perception, which varies from one person to another.

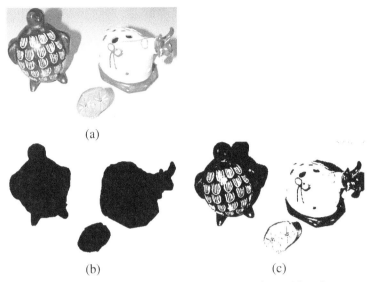

Figure 3.4: *(a) Gray-level image, (b) binary image obtained by a human, and (c) binary image obtained by an automatic algorithm.*

segmentation has been able to successfully eliminate the accentuated shading effects from Figure 3.4 (a), a task which has completely failed in the segmentation by the automatic thresholding algorithm.

Binary images are especially important in the context of this book, mainly because shapes are herein understood as connected sets of points. Consequently, the pixels of an image can either belong to a shape, being marked as "true" ("1"), or not, indicated as "false" ("0"). It is worth noting that objects to be processed using 2D shape analysis techniques can frequently be represented as a binary image. Such a binary image provides a silhouette-like representation of the object, or of its constituent parts (compare Figures 3.4 (a) and (b)) . For instance, we can easily apply efficient algorithms to count how many objects are present in a binary image such as that in Figure 3.4 (b), and to obtain their respective area, perimeter and center of mass. Clearly, gray level shape analysis can also be explored, but 2D shape analysis from binary images remains a particularly popular approach in many practical situations.

3.1.4 Shape Sampling

As discussed in more depth in Chapter 4, shapes can be understood as connected sets of points and classified into two main categories: *thin* and *thick*. Informally speaking, thin shapes are composed of contours or curves, while thick shapes involve filled portions (such as a region). A particularly important problem in shape analysis consists in, given a continuous shape (such as those found in nature or mathematics), obtaining a spatially sampled representation of it. This problem, which is henceforth called *shape sampling*, can be more properly understood by considering first thin shapes involving only contours or curves, and then proceeding to the problem of sampling thick shapes. Consider the situation where a continuous curve lies over a square grid associated with a digital image (refer to Figure 3.5). In order to represent the curve in the discrete image, it is necessary to determine which pixels should be "turned on" and "off," in such a way that the curve is approximated by a sequence of discrete pixels, yielding a binary image. The two main schemes for performing this task, namely the square-box and the grid-intersect quantization, are explained below.

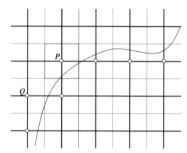

Figure 3.5: *The square-box quantization scheme (SBQ).*

The first quantization approach is the *square-box quantization* (SBQ), which can be easily understood from Figure 3.5. Consider that the continuous curve is to be spatially sampled over the square grid (solid lines). Each pixel is associated with one of the grid intersections between a horizontal and a vertical line. The SBQ considers a square box with the same side size as the grid spacing and centered on each pixel, as illustrated by the shaded box around pixel P in Figure 3.5. Let S_P denote the square box associated with pixel P. The SBQ scheme implies P to be "turned on" whenever the curve passes through the respectively associated S_P. In Figure 3.5, the selected pixels that represent the discrete version of the superposed curve are indicated by circles. The respectively obtained binary image is

$$Im = \begin{bmatrix} 0 & 0 & 0 & 0 & 0 \\ 0 & 1 & 1 & 1 & 1 \\ 1 & 1 & 0 & 0 & 0 \\ 1 & 0 & 0 & 0 & 0 \end{bmatrix}$$

The second quantization scheme for sampling curves is known as *grid-intersect quantization* (GIQ), which is illustrated in Figure 3.6.

In this approach whenever a portion of the continuous curve to be sampled intersects the grid, the nearest pixel to that intersection is marked. In other words, a pixel such as P in Figure 3.6 should be selected if and only if the continuous curve intersects the grid half segments that are nearest to it (indicated by the arrows in

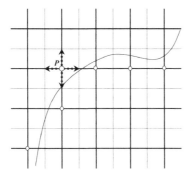

Figure 3.6: *The grid-intersect quantization scheme (GIQ).*

Figure 3.6). It should be observed that different spatial quantizations of the same continuous curve can be obtained by using the SBQ or GIQ, as is actually the case in Figure 3.5 and Figure 3.6, which considered the same original continuous curve.

It is important to note that both the SBQ and GIQ are defined as curve quantization schemes, not applicable to thick shapes. A third shape quantization scheme, which can be adopted to obtain the internal contours of thick shapes, is the so-called *object boundary quantization* (OBQ). Such region-based shapes can be quantized by taking into account their boundary points. Figure 3.7 (a) shows a thick shape to be quantized. Observe that whenever the shape boundary intercepts the grid,

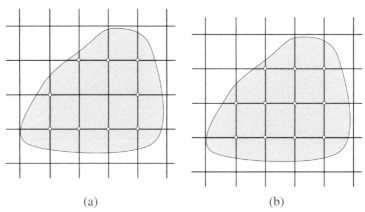

(a)   (b)

Figure 3.7: *The object boundary quantization scheme (OBQ) of a thick shape* **(a)** *and the whole quantized shape after region-filling* **(b)**.

there is an internal and an external pixel associated with that intersection point. In the OBQ scheme, the internal pixel is taken (see Figure 3.7 (a)) for every possible intersection. In case the whole shape, and not only its contours, is to be quantized, all internal pixels are added to the binary image (Figure 3.7 (b)). This can be accomplished by filling inside the quantized closed contour, which is known as region-filling (see Section 3.4.2). A similar approach, called *background boundary quantization* (BBQ) is defined analogously, but takes instead the external pixels associated with the boundary interceptions.

Other possibilities for sampling thick continuous shapes is to first obtain their contours (see Chapter 4), apply the GIQ or SBQ, and then fill the internal contours in the same way as above discussed for the OBQ case. The reader is referred to Freeman [1974]; Groen & Verbeek [1978] for additional detail about shape quantization schemes.

3.1.5 Some Useful Concepts from Discrete Geometry

A key concept for the application of many shape analysis algorithms is that of *neighborhood of a pixel*, which is especially important because it underlies several concepts in discrete shape analysis, including connectivity, paths between two points, digital contours and so on. Two alternative definitions of neighborhood are commonly considered in image analysis: 4- and 8-neighborhood. These definitions are exemplified in Figures 3.8 (a) and (b), respectively. The four 4-neighbors (denoted as V_0, V_1, V_2 and V_3) of pixel P are shown in Figure 3.8 (a), and the eight 8-neighbors (denoted as V_0 through V_7) of pixel P are presented in Figure 3.8 (b). As can be readily seen, the 4-neighbors of a pixel P are the pixels immediately to the right (V_0), above (V_1), to the left (V_2) and below (V_3) the reference pixel P, while the 8-neighbors include the 4-neighbors and the pixels on the immediate diagonals. If pixel P has coordinates (r, s) (i.e., row r and column s), then the 4-neighbors present the following coordinates:

$$V_0 = (r, s + 1), V_1 = (r - 1, s), V_2 = (r, s - 1), \quad \text{and} \quad V_3 = (r + 1, s)$$

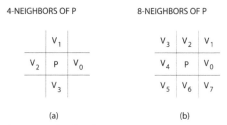

(a)   (b)

Figure 3.8: *The 4-neighborhood* **(a)** *and the 8-neighborhood* **(b)** *of a pixel* P.

The 8-neighbors are

$$V_0 = (r, s + 1), V_1 = (r - 1, s + 1), V_2 = (r - 1, s), V_3 = (r - 1, s - 1), V_4 = (r, s - 1),$$

$$V_5 = (r + 1, s - 1), V_6 = (r + 1, s), \quad \text{and} \quad V_7 = (r + 1, s + 1)$$

Observe that the above coordinates are associated with rows and columns (matrix-like), and not with the traditional (x, y) Cartesian coordinates.

A *connected path* between two pixels A and B is a sequence of N pixels p_1, $p_2, \ldots, p_N$ where each pair of consecutive pixels p_i, p_{i+1} is such that p_i is a neighbor of p_{i+1}, with $p_1 = A$ and $p_N = B$. Observe that this definition is relative to the type of considered neighborhood scheme, such as the above discussed 4- or 8-neighborhoods. A *connected component* is a set of pixels such that there is a connected path between any pair of pixels in that set. Commonly, binary images contain connected components corresponding to objects or object parts. In order to distinguish such distinct connected components in a binary image, a *labeling algorithm* can be used to assign the same label to all pixels of each object. Observe that a different label is used for each object. For instance, consider the binary image represented by the array in Figure 3.9 (a).

Figure 3.9 (b) shows a connected path between A and B, assuming 4-neighborhood. This binary image is composed of 4 connected components, as indicated in Figure 3.9 (c). A labeling algorithm that labels each connected component with a different integer number will produce a result such as that shown in Figure 3.9 (d). In this image, each connected component has a different integer label (ranging from 1 to 4), as well as the background (labeled 0). On the other hand, if 8-neighborhood is considered, then the image is composed of two connected components, as shown in Figure 3.9 (e). In this case, the labeled image will be as shown in Figure 3.9 (f). It should be borne in mind that several labelings can be obtained for the same image by exchanging the label values, i.e., the labeling order is arbitrary (see Figure 9.9(d) for an example).

3.1.6 Color Digital Images

Color images represent an especially important type of image since the human visual system is much more sensitive to color variations than to gray level variations. Let us recall that the standard model for image formation involves measuring the light intensity reflected by the objects in a scene and projected onto a surface constituted by photodetectors. Visible light is composed of electromagnetic waves of different wavelengths. Physiological and psychophysical results indicate that the analysis of the wavelength content of incident light onto our photoreceptors (three types of cones) gives rise to the perception of color. Our color perception is based on the integration of the response of three types of cones tuned to three different light wavelengths (therefore, when you look at your "red" t-shirt, it does not mean that you are perceiving "pure" red, but an integration of the response of your photoreceptors responding to many different light frequencies). The perception of color by the human visual system is a very important task since it usually helps us to define the shapes in a perceived scene.

Color images are represented by adopting a color model involving three or more values for each pixel. Therefore, at least three 2D matrices are required to represent a color image. Some of the most popular color spaces are the RGB ("Red, Green and Blue"), CMY ("Cyan, Magenta and Yellow"), and HSI ("Hue, Saturation and Intensity"). In a system such as the RGB, images can be represented in terms of

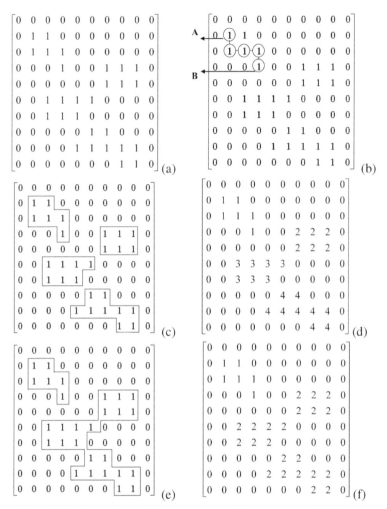

Figure 3.9: *Binary image **(a)**, a connected path between A and B **(b)**, con-*
*nected components adopting 4-neighborhood **(c)**, respec-*
*tive labeled image **(d)**, connected components adopting 8-*
*neighborhood **(e)**, respective labeled image **(f)**.*

3D signals $g(x, y, i)$ where i corresponds to each component of the color model, or
by three 2D signals, e.g., $r(x, y)$ (red component), $g(x, y)$ (green component), and
$b(x, y)$ (blue component). These concepts are illustrated in Figures 3.10 and 3.11.

Figure 3.10 illustrates how to represent a color *micrograph*, i.e., digital image
obtained by a camera attached to a microscope. The zoomed detail indicates that
the numerical representation of each pixel is actually composed of three values, one
for each RGB component. On the other hand, the gray level version of that image,

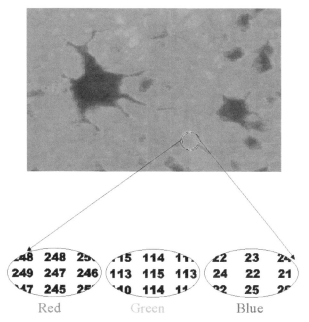

Figure 3.10: *Color image representation scheme.*

that is shown in Figure 3.11, only needs one value per pixel, as already discussed in Section 3.1.1.

The consideration of color can be essential for several different vision problems. For instance, Yang and Waibel have defined an approach to face-tracking in video sequences based on color [Yang & Waibel, 1996]. In their approach, a simple model of the color distribution of human faces is devised, and the image is searched for sets of pixels belonging to such a color distribution, which can therefore be considered as face candidates. As already commented, color can also be used for image segmentation, providing an alternative to more traditional approaches based only on monochromatic images. Other applications include image enhancement and restoration (see Chapter 21 of Castleman [1996]) and object classification. For instance, mature fruits can sometimes be distinguished from immature ones by color. In addition, color often provides an essential resource for segmenting and analyzing biological cells and tissues in micrographs.

To probe further: *Color Images*

Color images have attracted much attention in the literature, and important related research fields include color image segmentation and integration with other image properties, such as texture. These features are of interest, for instance, for visual information retrieval in multimedia databases [Mojsilovic et al., 2000]. The earlier

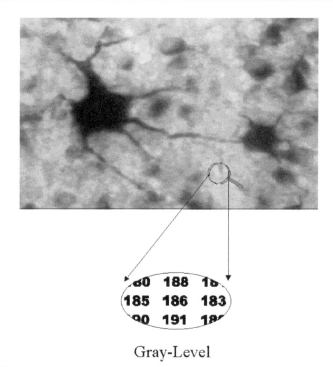

Gray-Level

Figure 3.11: *Gray-level image representation of the image in Figure* 3.10.

experiments by B. Julesz may also be of interest to the reader [Julesz, 1971]. Additional topics related to color images, such as the advantages and disadvantages of each color model, the concepts of chromaticity, pseudo-colors [Gonzalez & Woods, 1993], and color processing by the human visual system can be found in the above referenced image processing books as well as in Foley et al. [1994]; Hearn & Baker [1986]; Hubel [1995]; Klinker [1993]; Levine [1985]; Wandell [1995]; Wyszecki & Stiles [1982].

3.1.7 Video Sequences

Vision and image processing applications can be divided into two main classes, those involving *static images* and those involving *dynamic images*[1], i.e., video sequences. The latter concept can be easily understood by considering animated cartoons, which are actually sequences of static images that, when displayed continuously and sequentially, give rise to the perception of motion, known as the animation effect. Video sequences are therefore composed of a sequence of images acquired at different and subsequent time instants, as illustrated in Figure 3.12,

---

[1]The term *spatio-temporal images* is also sometimes used in this context.

Figure 3.12: *Example of a video sequence.*

which presents six frames in a video sequence. These images are represented by a 3D signal $g(x, y, t)$, corresponding to a scalar field, where the parameters x and y represent the ordinary spatial coordinates and t denotes time [Schalkoff, 1989]. For each time instant t_0, a *frame* $g(x, y, t_0)$, which is actually a simple 2D digital gray level image, is obtained. Video sequences are central to the active vision approach [Aloimonos, 1993], being particularly important in many applications such as robotic vision, surveillance, material sciences (crystal formation) and biology (growing of cells, organs and tissue). Potential applications of shape analysis are defined by each of such situations. Furthermore, recent technological developments (e.g., cheaper digital cameras, framegrabbers and powerful storage devices) have contributed to popularize video sequences, from their acquisition (inexpensive cameras) and storage to processing and display. For instance, digital videos have become popular in computer networks like the WWW.

Because of the extra dimension (time) available in video sequences, this type of image presents intrinsic peculiarities. First, they normally involve a much larger amount of data, demanding more powerful hardware and algorithms. Nevertheless, nice features are also allowed. To begin with, such images are generally characterized by presenting a higher degree of redundancy, which can be reduced and/or explored. For instance, in a video sequence of a moving object over a static background, most background pixels remain nearly constant throughout the video sequence, thus computational efforts can be concentrated on the moving object. Figure 3.13 illustrates this fact by showing, in the left column, 4 frames of a video sequence with a moving person in front of a static background. The right column of Figure 3.13 shows the images obtained by subtracting consecutive frames from the left column, e.g., Figure 3.13 (e) corresponds to the difference between Figures 3.13 (a) and (b). As can be easily noted, the main differences between the consecutive frames are due to the head movements. Background subtraction may also be used to detect moving objects (please refer to Lara & Hirata Jr [2006] for additional information).

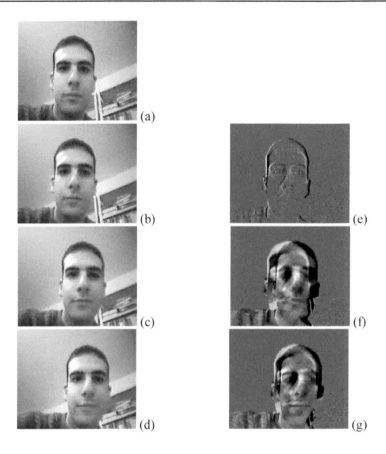

Figure 3.13: *Video sequence* **(a)** *through* **(d)**, *and respective difference images* **(e)** *through* **(g)**.

Another important feature of video sequences is that they provide information that cannot be conveyed by static images, mainly with respect to 3D structure and time-related features. An example of the latter is the analysis of shape growing or shape movement, especially in biological entities. On the other hand, as far as 3D structures are concerned, recall that image formation normally involves the projection of a 3D scene onto a 2D plane, which implies some information will be lost. There are many approaches to the problem of recovering 3D information from 2D images, and one of the most important is based on the analysis of 3D structure from the moving patterns present in video sequences. This is called the *shape-from-motion* problem [Leyton, 1992].

There are some important tasks involved in the analysis of video sequences. Let us recall the example of a moving object in front of a static background (Figure 3.13). The first important problem to be tackled is the detection of the moving objects, i.e., to determine which pixels belong to the objects that are moving along the image. This problem is called *detection*. Once an object has been detected in a frame, the next important problem is to *track* this object through the video sequence, a task known as *tracking*. In addition, an important feature of biological visual systems can be applied in order to analyze video sequences: *selective attention*. This can be easily understood as a strategy of paying different attention (which is equivalent to dedicating larger/smaller amount of computation) to different regions of the image at different times. A straightforward example of this process is text reading, where the focus of attention is shifted from a word to the next as reading proceeds. Of course, selective attention can be applied to static images, but such a strategy is particularly important in video sequences because of the dynamic nature and larger amount of data normally involved.

3.1.8 Multispectral Images

As explained in Section 3.1.1, image formation can be understood and modelled as the process of light reflection from objects in a scene and subsequent projection onto a plane surface containing photodetectors that respond proportionally to the intensity of the incident light. It is important to note that different photodetectors respond differently to distinct wavelengths of the light spectrum. For instance, the photoreceptors (cones and rods) in our retina respond to light wavelengths from 370 to 730 nm, i.e., from blue to red. On the other hand, we are not able to perceive light of wavelengths below (e.g., ultraviolet) or above (e.g., infrared) this range. In this context, the RGB color model can be thought of as a collection of three images formed with respect to different bands of the light spectrum. The generalization of this idea, i.e., images formed on arbitrary different bands of the light spectrum, leads to the idea of multispectral images, which are collections of images, each one for a specific band. This is a common type of data in some applications, especially remote sensing. These images can be represented by 3D signals $g(x, y, i)$ where i indexes each band of the considered light spectrum, just as with video sequences, except that the index does not refer to time, but to a specific spectral band, and that all images are acquired at the same time.

3.1.9 Voxels

Many image processing applications, such as in medicine, agriculture and biology, often involve volumetric data processing, which can be understood as a generalization of the concept of digital images. In such applications, a set of images corresponding to slices of an object under analysis (such as in computerized tomography) is obtained. The volumetric representation can be formed by aligning and stacking up these slices. Each slice may be represented as a digital image and the corresponding pixels in the stacked volume are called *voxels* (volume pixels). These images can be represented by 3D signals $g(x, y, i)$, where i indexes each slice of the volume data, just as with video sequences, except that the index does not refer to time but to a third spatial dimension.

To probe further: *Image Types*

Further material about the above discussed several image types can be found mainly in scientific papers and conference proceedings, since they are not so thoroughly explored by most classical textbooks. The reader can find more information about spatio-temporal images, motion analysis and active vision in Aloimonos [1993]; Hildreth [1983]; Schalkoff [1989]. Some information about video processing can be found in del Bimbo [1999]. Some equally important image types include *range* [Trucco & Verri, 1998], *foveal* and *log-polar images* [Araújo & Dias, 1996], which can largely be addressed by using adapted versions of the techniques presented and discussed herein.

3.2 Image Processing and Filtering

The term *image processing* is generally applied to methods that receive an image as input and produce a modified image as output. For example, the algorithm can input an image with poor contrast and will produce as a result an enhanced image. In addition, image processing can be used to remove noise from an image, a process classified as *image filtering*. In fact, image processing corresponds to one of the initial operations performed in a vision system, in which important features are emphasized, and redundant (or unimportant) information and noise are reduced or even eliminated. Such processes transform the input image in order to produce an improved image, hence performing image processing. A simple example is to add a constant value to each pixel in order to produce a brighter image (see Figures 3.14 (a) and (b)).

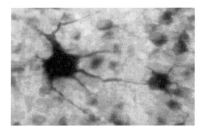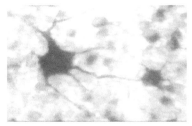

Figure 3.14: *Simple image processing: adding a positive constant to the gray levels brightens the image.*

SIMPLE TYPICAL PIPELINE FOR SHAPE ANALYSIS

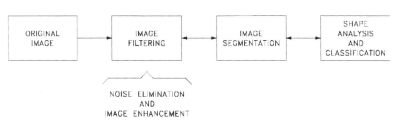

Figure 3.15: *The main initial steps in shape analysis and classification. Observe that image processing techniques include image filtering and segmentation.*

The initial basic steps in shape analysis, with emphasis on image processing, are shown in Figure 3.15. It is important to note that the steps represented in this pipeline are not necessarily sequential and one-way, but they are presented in this form for didactical purposes. In fact, the processes within each box, and even inter-boxes, may be carried out in parallel or involve feedback. For instance, specific knowledge about the problem in hand can be explored. Suppose an image segmentation process produces objects with holes, and such holes are smaller than a given minimum size. In such a case, image filtering tools can be applied in order to close the small noisy holes (for instance, by applying the morphological closing operation, see Section 3.5). In addition, it is possible to place the image acquisition under control of the higher level modules of Figure 3.15 (e.g., the shape analysis module), in order to allow adaptive image acquisition. An example of such a possibility involves the control of the camera responsible for acquiring the images in order to follow a moving object while it is being recognized, which characterizes a simple active vision system [Aloimonos, 1993].

Considering all that has been said, image processing can be viewed as an intermediate step for obtaining new meaningful image representations, which can be discussed from the following three perspectives:

Noise filtering: If the image is corrupted by one or more types of noise, then it should be filtered by appropriate techniques. The application of special filters can generate less noisy images.

Reduction of redundancy and unimportant information: The image aspects that are not important for the specific shape analysis problem in hand should not be taken into account and eliminated. For instance, suppose that a shape classification system has to discriminate between squares and circles presenting similar textures. In this case, the texture should be eliminated from the objects before applying the shape classification algorithm.

Emphasize important information: It is often important to accentuate image aspects that can be especially relevant for each particular problem. For example, edge enhancement operations can be applied to the input image in order to emphasize the shapes in an image, especially if it has low contrast.

This section explains some traditional and modern techniques for image filtering aimed at noise reduction and image enhancement using both linear and nonlinear methods. The concepts presented in this section are also central to several shape analysis tools to be discussed in subsequent chapters of this book. However, much work has been done on the subject of image filtering, and the reader is referred to classical textbooks in order to have a broader perspective about the main approaches to this area [Castleman, 1996; Gonzalez & Wintz, 1987; Gonzalez & Woods, 1993; Jain, 1989; Pitas & Venetsanopoulos, 1990; Schalkoff, 1989].

Note: *Image Processing Software and Tools*

There is a large variety of image processing software available to the interested reader, who should take into account the following while selecting a specific product:

☞ Does the software allow new functionalities to be added to the original program, e.g., new filtering methods? Programs like MATLAB®, Mathematica®, Scilab® and R® allow programming scripts that are interpreted by the software, thus allowing a flexible programming environment for image processing applications. Also, additional functions can be added to programs like GIMP® and Photoshop® as plugins or toolboxes. It is observed that many imaging programs simply do not allow new procedures to be included.

☞ Are scripts, toolboxes or plugins already available, e.g., through the Internet? Can GUI-based interfaces be created? Which file formats are supported? Standard, widely used image file formats include pbm, pgm, ppm, jpg, gif®, tif, bmp®, pcx, pcd, psd, wmf, …

☞ How does the software communicate with other programs running under the same operational system? Are cut and paste and clipboard operations available?

Some of the representative modern systems suitable for image processing software currently available include xv®, GIMP®, MATLAB®, Scilab®, Lview®, PaintShopPro®, Adobe PhotoShop®, Khoros®, PBMPlus®, SDC Morphology Toolbox® for MATLAB®. Information about each of these software programs can be obtained from the URLs listed at http://www.ime.usp.br/~cesar/shape_crc/chap3.html.

3.2.1   Histograms and Pixel Manipulation

Among the simplest and most useful techniques for treating digital images are those based on *histograms*. A histogram is a mapping that assigns a numerical value $h(n)$ to each gray level n of the image. These values correspond to the number of times each specific gray level n occurs in the image. Let g be a digital image of size PQ, where each pixel assumes a gray level $0 \leqslant n < N$. Therefore, the histogram h of g can be calculated by the following algorithm:

Algorithm: *Histogram Generation*

HISTOGRAM(g, N)

1.   **for** $i \leftarrow 0$ to $N - 1$
2.   **do**
3. $h(i) \leftarrow 0$;
4.   **for** each pixel $g(p, q)$
5.   **do**
6. $h(g(p, q)) \leftarrow h(g(p, q)) + 1$;

Figure 3.16 illustrates the concept of the image histogram. Figure 3.16 (b) shows the relative frequency gray level histogram (see below) and Figure 3.16 (c) illustrates the population histogram (or simply histogram) of the image of Figure 3.16 (a). Observe that the relative frequency histogram can be obtained from the population histogram by dividing each of its values by the total number of pixels in the image. Figure 3.16 (d) shows the logarithm of the histogram of Figure 3.16 (c). It is observed that an arbitrary constant, 1, has been added to all the values of the original histogram so that logarithms of 0 are avoided. This alternative visualization is useful in situations where the histogram values vary so much that small details become imperceptible. Compare the plots of the histograms in Figures 3.16 (c) and (d). While in the plot of Figure 3.16 (c) only two peaks or *modes* can be easily identified, the plot in Figure 3.16 (d) shows a third mode between those two. A simple analysis suggests that mode A is associated to the darker pixels that constitute the floppy disk, mode B is associated to the small brighter round region at the floppy disk center and, mode C corresponds to the background pixels. This logarithm technique is general and can be applied to many other situations, such as image displaying.

Figure 3.17 presents another example, where the histogram of the image in Figure 3.17 (a) is shown in Figure 3.17 (b), and its logarithmic version is presented in Figure 3.17 (c). Note that, in this example, the modes are not so clearly separated. An analysis of the image Figure 3.17 (a) reveals that it is relatively easy to separate the upper-left darker object, the head and the basis of the right object. On

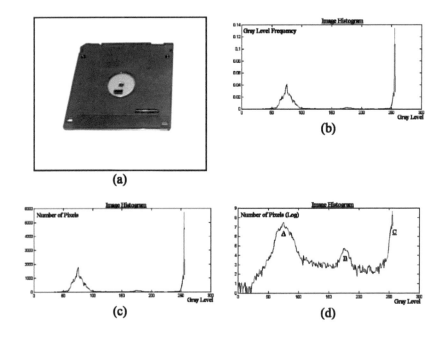

Figure 3.16: *A digital image (a); its respective relative frequency gray level histogram (b) and gray level population histogram (c). A logarithmic version of* (c) *is presented in* (**d**).

the other hand, many of the pixels that constitute the body of this right object, as well as the lower object, have gray levels rather similar to those in the background, which accounts for the merging of the respective peaks. Because of its importance to image segmentation by thresholding, this problem is further discussed in Section 3.4.1.

In some situations, it is important to create the histogram representing the *relative frequency* of each gray level instead of the distribution of number of pixels per gray level. In fact, the gray level frequency is obtained by dividing the number of pixels per gray level by the size of the image (i.e., its total number of pixels, given by PQ). It is observed that this operation does not change the shape of the histogram, but only the scale of the ordinate axis. An algorithm for obtaining the relative frequency histogram of an image is presented next. Figure 3.16 (b) shows the relative frequency histogram obtained for the image of Figure 3.16 (a), which illustrates the above discussion.

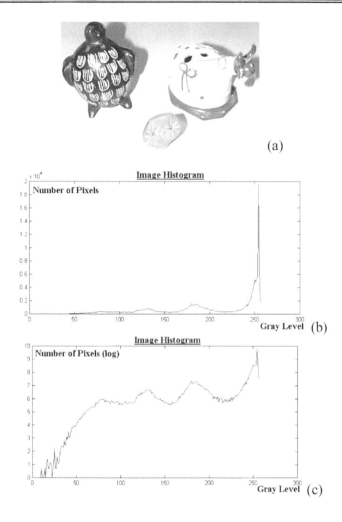

Figure 3.17: *A digital image (a) and corresponding histogram (b). The logarithmic visualization of* (b) *is shown in* (c).

Algorithm: *Relative Frequency Histogram Generation*

HISTOGRAM2(g, N)

1.   **for** i = 0 to N - 1
2.   **do**
3. $h(i) = 0$;
4.   **for** each pixel $g(p, q)$
5.   **do**
6. $h(g(p, q)) = h(g(p, q)) + 1$;
7.   **for** i = 0 to N - 1
8.   **do**
9. $h(i) = h(i)/(P * Q)$;

Note: *Histogram-Based Measures*

It is important to note that measures extracted from the image histogram cannot properly characterize the shapes in the image because the histogram does not depend on the geometrical arrangement of the image pixels. Nevertheless, histogram measures (or features) can become important for shape classification when the objects differ with respect to their brightness (see the example of leaves classification in Chapter 1). Furthermore, histogram features extracted from blurred images can provide additional resources for shape characterization, as explained in Chapter 6.

There are many useful image processing operations that can be carried out based on histograms, one of the most useful being *histogram equalization*. The basic idea underlying this operation, under a practical point of view, is the production of an image that takes full advantage of the allowed gray levels. Often, digital image gray levels are concentrated around some part of the histogram, a consequence of the fact that the image pixels do not assume all the possible gray levels, implying lower contrast. A sensible solution is to spread the gray level distribution in the histograms as much as possible (ideally, an equalized histogram should present an uniform distribution; see Gonzalez & Woods [1993]). In the case of digital images, histogram equalization aims at obtaining an output image with uniform distribution of gray levels.

Let g be a digital image, where each pixel assumes a gray level $0 \leqslant n < N$. In order to perform the equalization of g, it is necessary to find a mapping $f : n \rightarrow f(n)$, where $f(n)$ is the new gray level. In other words, once $f(n)$ has been obtained, each pixel with gray level n in the input image will be replaced by a gray level $f(n)$ in the output image. This operation is called *histogram transformation*. In the case of histogram equalization, the mapping $f(n)$ is incrementally constructed based on the relative frequency histogram $h(n)$ of the original image, i.e.,

$$f(n) = (N - 1) \sum_{k=0}^{n} h(k)$$

The following algorithm summarizes a possible implementation (adapted from Myler & Weeks [1993]) of histogram equalization.

Algorithm: *Histogram Equalization*

HISTOGRAM2(g, N)

1.   **for** $n \leftarrow 0$ **to** $N - 1$
2.   **do**
3. $f(n) \leftarrow$ ROUND$(sum(h, n) * (N - 1))$;
4.   **for** each pixel $g(p, q)$
5.   **do**
6. $i(p, q) \leftarrow f(g(p, q))$;

In the above algorithm, *sum(h, n)* is a function that adds the first $n + 1$ elements of the histogram h (i.e., from $h(0)$ to $h(n)$), and *round(x)* is a function that rounds its argument (see Section 2.1.4). Histogram equalization is useful in many practical problems. A very common application is for increasing the contrast of images for the finality of human inspection, which can lead to impressive results (see, for instance, Gonzalez & Woods [1993]). Histogram equalization can also be applied to normalize illumination variations in image understanding problems. For instance, local equalization is used with this purpose in the face detection approach described in Rowley et al. [1998]. Histogram equalization is but one of the operations that can be applied to obtain new images based on histogram modification or specification, and the reader is referred to Chapter 5 of Myler & Weeks [1993] for some examples of other similar techniques, as well as to Chapter 6 of Castleman [1996], which introduces different point operations. Finally, Gonzalez & Woods [1993] provides an additional good reference on this subject.

Note: *Algebraic Operations on Digital Images—Noise Reduction through Image Averaging*

Other useful operations that may be carried out on digital images are called *algebraic operations*, which typically involve algebraic operations between two images. For instance, two important algebraic operations between two images f and g, assumed to have the same sizes, are the addition and the difference between them, i.e.,

$$h = f + g$$
$$h = f - g$$

An important practical application of algebraic operations to digital images is noise reduction by *image averaging*. Frequently, digital images obtained by acquisition devices (e.g., cameras) are noisy and can be modelled as

$$g = f + n$$

where g is the acquired noisy image, f is the original (noise free) scene and n is the noise. The above equation assumes that the noise has been added to the original image in a pixelwise fashion. By assuming that the original scene does not change along time (i.e., a static image), a series of noisy images g_i can be acquired from the same scene and the following sum can be defined:

$$g_i = f + n_i, i = 0, 1, \ldots, N - 1$$

If some assumptions about the noise are met (the noise n_i at different acquisitions is not correlated, the noise has null mean and is not correlated with the original scene f), then averaging leads to a less noisy image:

$$f \cong \bar{g} = \frac{1}{N} \sum_{i=0}^{N-1} g_i(p, q)$$

However simple this technique can be, it generally leads to good results. It fails, for instance, if the scene changes between successive acquisitions, which would require a previous application of an image registration process. Further information about image averaging, including an explanation of why $\bar{g}$ converges to f can be found in Section 4.2.4 of Gonzalez & Woods [1993]. Algebraic operations between images are discussed in more detail in Chapter 7 of Castleman [1996].

3.2.2 Local or Neighborhood Processing

Local processing is often required for the analysis of visual scenes, which is natural since the characteristics of an image to be processed can significantly vary from one region to another. In fact, it is frequently important to separate the information between different portions of the image, so that the information in a region does not significantly affect the processing of another region. In this context, the output image is obtained by processing the input image by regions.

Unlike the methods discussed in the last section (techniques such as histograms are said to be *global*), where the value of each output pixel depends only on the value of the respective input pixel, there are many useful image operations where the output pixels depend (mainly) on a local neighborhood of the respective input pixels. Indeed, there are a multitude of approaches based on this underlying idea, usually involving the use of sliding windows and ranging from convolution to mathematical morphology. Some of the more relevant of such approaches are presented and discussed in the following sections.

3.2.3 Average Filtering

The first neighborhood-based technique to be discussed is the *average filtering*, a particular case of the linear filters. One of the simplest local operations over an image is to replace each of its pixels by the average value of gray levels in its neighborhood. Figure 3.18 presents a region of a digital image centered at one of the pixels to be processed (marked within a darker square). The corresponding

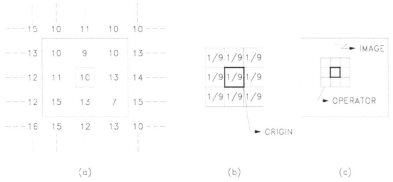

(a)   (b)   (c)

Figure 3.18: *The average filtering basic structure.*

output pixel is calculated by adding the gray levels of its neighbors (delimited by the outer square in Figure 3.18) and itself, and dividing the result by 9, i.e.,

$$(10 + 9 + 10 + 11 + 10 + 13 + 15 + 13 + 7)/9 \cong 11$$

This type of filtering reduces the relative difference between the gray level value of neighboring pixels, decreasing the amount of noise as well as the contrast in the image, which characterizes a *smoothing* operation. This idea is commonly generalized through the definition of a sliding window such as that shown in Figure 3.18 (b). It involves a set of *weights*, and an *origin* is defined and used as positioning reference. In the present case, the window is 3×3, all its weights are equal to $\frac{1}{9}$, and the origin is defined at the central pixel. The implemented operation corresponds simply to adding the products of each weight with the corresponding pixels of the image. In order to process the entire image, the window is shifted by positioning its origin onto each of the image pixels, generating the corresponding output pixels (see Figure 3.18 (c)).

Clearly, the size of the window can be changed, implying neighborhoods of different sizes to be taken into account. For instance, a neighborhood of size 5×5 would work in an analogous way, except that the weights should be $\frac{1}{25}$. Such a larger neighborhood allows more distant pixels to also influence the output value. In other words, the *analyzing spatial scale* becomes larger. The practical effect is that the output image becomes smoother. It is observed that choosing a suitable analyzing scale is not a problem that has a unique or simple solution.

It is important to note that processing the pixels on the border of the image presents a special problem because when the origin of the operator is placed over the border pixels, some positions of the window fall outside the image. In fact, not only the outermost pixels are affected, but those near (but not on) the border also present this problem (the number of outer pixel layers affected by this problem depends on the size of the operator). The following three typical solutions can be considered in such situations:

① The outer layers are simply ignored, i.e., the filter operation is carried out only for the inner pixels. As a consequence, the resulting image is typically smaller than the original, although its size can be kept by filling up the borders with zeros.

② The image is augmented with outer layers in order to complement the necessary number of neighboring pixels required by the filtering process. The value of these extra pixels have to be set arbitrarily (e.g., to zero), which can adversely affect the results.

③ The image is assumed to be periodic (namely a thorus), so that if a position of the operator falls off the image, the corresponding operation is carried out over the pixel at the other side of the image. This type of structure is naturally implemented by linear filtering techniques based on the Fourier transform [Morrison, 1994].

The choice of one of the above solutions depends on each specific problem. Algorithms can be easily altered in order to implement any of the above discussed solutions. In this sense, average filtering is a special case of *linear filtering*, which can be modelled by 2D convolutions:

$$f(p, q) = \frac{1}{MN} \sum_m \sum_n h(m, n) g(p - m, q - n)$$

where $M \times N$ is the size of the input image.

In fact, there are many image processing tasks carried out in an analogous way, varying only as far as the filtering function is concerned. The definition of different sets of weights can lead to completely different results. For instance, special operators can be defined in order to differentiate the image or to analyze its local frequency content. Furthermore, the generation of such operators generally relies on the assumption of a mathematical model that endows the linear filtering with a special and useful set of properties.

3.2.4 Gaussian Smoothing

It is interesting to observe that the above-discussed average filter corresponds to a low-pass filtering with a square-box weight function, whose Fourier transform is the 2D sinc, as discussed in Chapter 2. In the process of averaging with the square box, only the points of the considered neighborhood are taken into account, all of them being treated with the same weight in the averaging process. In this sense, given the point about which the average is to be calculated, the influence of the other image points is defined in a somewhat "hard" way, i.e., they affect or do not affect the image. In many situations it would be interesting to consider graded weights, assigning larger weights to pixels that are closer to the central reference point in order to emphasize their special importance to the average. All that is required to achieve this goal is to substitute the square-box function with a more suitable one. In fact, the most popular function used in such cases is the circularly symmetric 2D Gaussian, which is defined as

$$g(p,q) = \exp\left(-\frac{\left(p^2 + q^2\right)}{2a^2}\right)$$

A typical plot of such a type of Gaussian is shown in Figure 3.19.

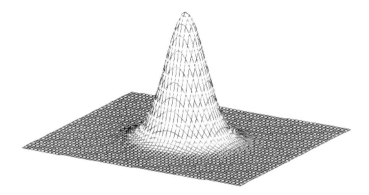

Figure 3.19: *A circularly symmetric 2D Gaussian.*

There are several reasons for choosing the Gaussian kernel for image smoothing and filtering. The first important fact to be considered regarding Gaussian filtering is the role played by the Gaussian spread parameter, especially its relation to the amount of blurring of the image and the concept of *scale*. The Gaussian spread parameter a of the Gaussian equation controls the width of the Gaussian shape, in the sense that the larger the value of a, the wider the Gaussian will be. Therefore, the standard deviation parameter is directly related to the spatial *scale* of the Gaussian. In fact, the Gaussian spread parameter a will be henceforth referred to as the *scale parameter*.

An additional important fact is that, as the scale parameter increases, the Gaussian becomes wider in the space domain, while the Fourier domain counterpart (also a Gaussian) becomes narrower. Let us now consider what happens to a signal or an image when it is filtered through convolution with Gaussians of different scales. Recall that the convolution of a signal (image) in the time domain is equivalent to multiplication in the frequency domain (from the convolution theorem). This means that, as the Gaussian scale increases, the higher spatial frequency content of the signal being filtered is progressively eliminated. This analysis explains the low-pass filtering mechanism implemented by the Gaussian filtering.

As far as computational aspects are concerned, Gaussians are separable functions (i.e., can be expressed as the product of a function of x and a function of y), in such a way that 2D Gaussian filtering can be implemented by convolving the image rows with a 1D filter, followed by a similar procedure applied to the image columns. An example of an approximation filter is

1	4	6	4	1

The book by Jain et al. [1995] presents an extensive discussion of 2D Gaussian filtering implementation.

At this point, it is also important to discuss the effects of Gaussian filtering over the image structures. Figure 3.20 (a) shows an image that has been filtered with three different Gaussian kernels of increasing scales, the respective results being shown in Figures 3.20 (b), (c), and (d). One of the most important effects

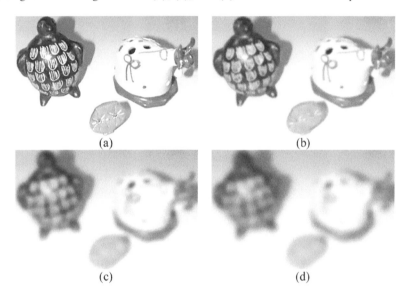

(a)   (b)

(c)   (d)

Figure 3.20: *(a) Digital image; (b) through (d) corresponding Gaussian filtering with increasing scales.*

of multiscale image filtering is that small-scale structures are eliminated as the filtering scale increases, which has the additional effect of emphasizing the largest scale structures. For instance, refer to the small details in the head of the upper-right object of Figure 3.20 (a) (e.g., the eyes). The smallest details have already been attenuated, or even eliminated, in Figure 3.20 (b), which is the result of small-scale Gaussian filtering (e.g., the black dots within the eyes). These small-scale structures have nearly vanished for the medium-scale Gaussian filtering presented in Figure 3.20 (c) (e.g., the eyes can no longer be seen). Also, only the largest-scale structures can be clearly identified in Figure 3.20 (d). Note that even medium-sized structures can merge as a consequence of the filtering process, as is the case for the upper-right and lower objects.

The above example also illustrates what is frequently a drawback of Gaussian filtering. It is important to recall that one of the main goals of image filtering is typically noise elimination. Clearly, if noise is present in an image as small structures (e.g., small spots or local perturbations on the pixels' gray levels), these tend to be attenuated by Gaussian filtering, mainly when the filtering scale increases. However, as the Gaussian filter cannot distinguish between noise and real image detail, both are usually attenuated by the filtering process, which is often strongly undesirable. The classic example of this problem is the image edges: as discussed in Section 3.3, the image edges represent a particularly important resource for image understanding. One of the most important problems of Gaussian filtering is that it generally displaces the object edges in an accentuated manner. Chapter 7 discusses some related issues in the context of scale-space theory and multiscale filtering.

3.2.5 Fourier-Based Filtering

Whenever it exists, the 2D continuous Fourier transform of a signal $g(x, y)$ is defined as

$$G(u, v) = \Im\{g(x, y)\} = \int\limits_{-\infty}^{\infty} \int\limits_{-\infty}^{\infty} g(x, y) \exp\{-j2\pi(ux + vy)\}\,dxdy$$

The inverse Fourier transform of $G(u, v)$ is defined as

$$g(x, y) = \Im^{-1}\{G(u, v)\} = \int\limits_{-\infty}^{\infty} \int\limits_{-\infty}^{\infty} G(u, v) \exp\{j2\pi(ux + vy)\}\,dudv$$

As in the one-dimensional case, we define the *Fourier pair* as follows:

$$g(x, y) \leftrightarrow G(u, v)$$

As an example, the 2D Dirac delta function is defined as

$$\delta(x, y) = \begin{cases} \text{not defined} & \text{if } x = 0 \text{ and } y = 0 \\ 0 & \text{otherwise} \end{cases}$$

$$\int\limits_{-\infty}^{\infty} \int\limits_{-\infty}^{\infty} \delta(x, y)\,dxdy = 1$$

Property	Description
Separability (DFT)	The discrete Fourier transform can be computed in terms of 1D Fourier transforms of the image rows followed by 1D transforms of the columns (or vice versa).
Spatial Translation (Shifting)	$g(x - x_0, y - y_0) \quad \leftrightarrow$ $\exp\left[-j2\pi(ux_0 + vy_0)\right] G(u, v)$
Frequency Translation (Shifting)	$\exp\left[j2\pi(xu_0 + yv_0)\right] g(x, y) \quad \leftrightarrow$ $G(u - u_0, v - v_0)$
Conjugate Symmetry	If $g(x, y)$ is real, then $G(u, v) = G^*(-u, -v)$
Rotation by θ	$g(x\cos\theta + y\sin\theta, -x\sin\theta + y\cos\theta) \leftrightarrow$ $\leftrightarrow G(u\cos\theta + v\sin\theta, -u\sin\theta + v\cos\theta)$
Linearity—Sum	$g_1(x, y) + g_2(x, y) \leftrightarrow G_1(u, v) + G_2(u, v)$
Linearity—Multiplication by Scalars	$ag(x, y) \leftrightarrow aG(u, v)$
Scaling	$g(ax, by) \leftrightarrow \frac{1}{|ab|} G\left(\frac{u}{a}, \frac{v}{b}\right)$
Average Value	The image average value is directly proportional to $G(0, 0)$ (the so-called DC component)
Convolution Theorem	$g(x, y) * h(x, y) \leftrightarrow G(u, v)H(u, v)$ and $g(x, y) h(x, y) \leftrightarrow G(u, v) * H(u, v)$
Correlation Theorem	$g(x, y) \circ h(x, y) \leftrightarrow G^*(u, v)H(u, v)$ and $g^*(x, y) h(x, y) \leftrightarrow G(u, v) \circ H(u, v)$
Differentiation	$\left(\frac{\partial}{\partial x}\right)^m \left(\frac{\partial}{\partial y}\right)^n g(x, y) \quad \leftrightarrow$ $(j2\pi u)^m (j2\pi v)^n G(u, v)$

Table 3.2: *2D Fourier transform properties assuming $g(x, y) \leftrightarrow G(u, v)$.*

and its Fourier transform is calculated as

$$\Im\{\delta(x, y)\} = \int_{-\infty}^{\infty}\int_{-\infty}^{\infty} \delta(x, y) \exp\{-j2\pi(ux + vy)\}\,dxdy = \int_{-\infty}^{\infty}\int_{-\infty}^{\infty} \delta(x, y) \exp\{0\}\,dxdy =$$
$$= \int_{-\infty}^{\infty}\int_{-\infty}^{\infty} \delta(x, y)\,dxdy = 1$$

Since the respective inverse can be verified to exist, we have $\delta(x, y) \leftrightarrow 1$.

The 2D Fourier transform exhibits a series of useful and practical properties in image processing and analysis, most of which are analogous to the 1D Fourier properties explained in Chapter 2. Table 3.2 summarizes some of the most useful 2D Fourier properties (see also Castleman [1996]; Gonzalez & Woods [1993]).

As far as image and shape processing is concerned, at least the following three good reasons justify the special importance of the Fourier transform:

① Understanding linear filtering: Fourier transform is the core of linear signal processing and linear signal expansion, providing a solid mathematical background for the understanding of several important imaging procedures, such as numerical differentiation, edge detection, noise reduction and multiscale processing, to name a few. Fourier theory provides not only an important conceptual frame of reference for the initiated reader, but also solid conditions to the development of new algorithms and theories.

② Designing image filters in the frequency domain: The previous sections have discussed a series of linear filtering techniques based on convolution (or correlation) between the input image and a filtering window. From the convolution theorem, this approach is equivalent to multiplying the corresponding Fourier transforms (of the image and the mask). Situations often arise where it is easier to design the filter directly in the frequency domain, instead of the spatial domain.

③ Efficient implementation of filtering algorithms: The direct implementation of 2D linear filters can be carried out efficiently in the image domain if the filter window is small. Nevertheless, for larger windows, linear filtering in the spatial domain is particularly inefficient. The Fourier transform is an important tool to circumvent this problem because, by applying the convolution theorem, linear filtering can be carried out by using FFT algorithms, which have complexity order $O(n \log n)$.

④ The Fourier transform of typical (correlated) images concentrates most of its energy along the low-frequency coefficients.

The 2D discrete Fourier transform (DFT) of a discrete signal $g_{p,q}$ is defined as follows:

$$G_{r,s} = \Im\left\{g_{p,q}\right\} = \frac{1}{MN} \sum_{p=0}^{M-1} \sum_{q=0}^{N-1} g_{p,q} \exp\left\{-j2\pi\left(\frac{pr}{M} + \frac{qs}{N}\right)\right\}$$

The inverse Fourier transform of $G(u, v)$, which returns $g(x, y)$, is given by

$$g_{p,q} = \Im^{-1}\left\{G_{r,s}\right\} = \sum_{r=0}^{M-1} \sum_{s=0}^{N-1} G_{r,s} \exp\left\{j2\pi\left(\frac{pr}{M} + \frac{qs}{N}\right)\right\}$$

Analogously to the 1D case, we define the *Fourier transform pair*: $g_{p,q} \leftrightarrow G_{r,s}$.

The 2D Fourier transform can be implemented with complexity order of $N^2 \log N$ by using 1D FFT algorithms applied first along the image rows, and then along the columns (or vice versa). Frequency filtering of an image f by a filter g is carried out by multiplying F by G, followed by the application of the inverse Fourier transform, i.e.,

Algorithm: *Frequency Filtering*

1. Choose $G(r, s)$;
2. Calculate the Fourier transform $F(r, s)$;
3. $H(r, s) \leftarrow F(r, s)G(r, s)$;
4. Calculate the inverse Fourier Transform $h(p, q)$;

It is worth noting that G can be calculated as the Fourier transform of a spatial filter g, as well as by direct design in the frequency domain, as above mentioned. In the above algorithm h, is the resulting filtered image.

An illustrative example of frequency filtering is presented in the following. Consider the image of Figure 3.21 (a), whose 2D Fourier transform modulus is shown in Figure 3.21 (b).

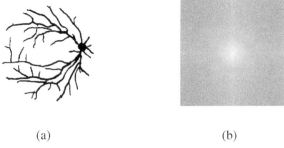

(a)   (b)

Figure 3.21: *Digital image **(a)**; and its corresponding modulus of its respective 2D Fourier transform **(b)***

As in the 1D case, the 2D Fourier representation has been properly shifted so that the origin is placed in the center of the image, for visualization's sake, as shown in Figure 3.21 (b). The first considered frequency filter is presented below (which is analogous to the 1D box filter function):

$$G_{r,s} = \begin{cases} 1, & \text{if} \quad \left(r^2 + s^2 \right) \leqslant T \\ 0, & \text{if} \quad \left(r^2 + s^2 \right) > T \end{cases} \tag{3.1}$$

This filter is shown as an image in Figure 3.22 (a), also placed in the center of the image. The modulus (observe that the filtering processes are actually applied to the complex coefficients) of $H = FG$ is shown in Figure 3.22 (b), while the filtered image h is presented in Figure 3.22 (c). As can be seen, the resulting filtered image has been smoothed, but oscillations have been introduced in the image, as shown more clearly in Figure 3.22 (d). This figure corresponds to the equalized version of the image in Figure 3.22 (c). Such oscillations are due to the fact that filtering

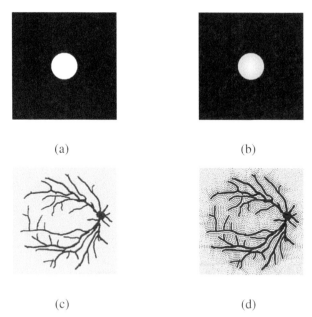

(a)   (b)

(c)   (d)

Figure 3.22: *Frequency domain 2D filter defined by equation* (3.1) *(a);*
Fourier transform of the image in Figure 3.21 (b) *filtered*
by this filter function (b); the resulting filtered image (c).
The histogram equalized version of (c) *is shown in* (d). *It*
is important to observe that while (b) *shows only the filtered*
Fourier modulus for visualization's sake, the filtering is actu-
ally carried out on the complex coefficients.

with the above function in equation (3.1) is equivalent to convolving the image
with an oscillating function (the inverse Fourier transform of the filtering function
of equation (3.1)). This undesirable effect can be circumvented by applying a 2D
Gaussian filter instead of the above function. The Fourier domain 2D Gaussian
filtering function shown in Figure 3.23 (a), is

$$G_{r,s} = \exp\left(-\frac{\left(r^2 + s^2\right)}{2\sigma^2}\right)$$

The filtered Fourier coefficients are shown in Figure 3.23 (b). The inverse
Fourier transform of the Gaussian filtered image is shown in Figure 3.23 (c), while
its equalized version is illustrated in Figure 3.23 (d). As we can see in both images, the input image has been smoothed out without being contaminated by the
oscillation effect.

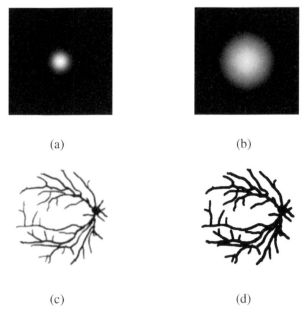

(a)   (b)

(c)   (d)

Figure 3.23: *Frequency domain 2D Gaussian filter **(a)**; Fourier transform of Figure 3.21 (b) filtered by this Gaussian function **(b)**; and the resulting filtered image **(c)**. The histogram equalized version of* (c) *is shown in **(d)**.*

3.2.6 Median and Other Nonlinear Filters

Another classic technique for image filtering is known as *median filtering*. This technique can be understood similarly to the previously discussed average filtering, though the mathematical implications are different because of its nonlinear nature. Recall that the linear filtering approach is based on placing the origin of an operator at each pixel of the image and carrying out a weighted sum between the mask weights and the respective pixels under the mask. As far as median filtering is concerned, the window operator does not have any weight. Instead, the pixels under the operator are sorted[1], and the middle value (i.e., the median) is selected to substitute the reference pixel. Figure 3.24 illustrates this process. In this example, as in the sorted sequence of Figure 3.24, the median of the pixels under the operator is 10, which substitutes the origin pixel with original value 15.

Median filtering is a nonlinear operation commonly used to eliminate noisy pixels that differ significantly from other pixels in its neighborhood, such as salt-and-pepper (i.e., isolated points with different gray levels scattered over uniform regions). Natural variations of this procedure may be defined by picking the max-

---

[1]Information about efficient sorting algorithms can be found in Cormen et al. [1990]; Langsam et al. [1996].

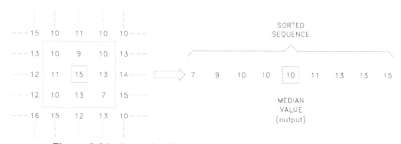

Figure 3.24: *Example of the basic operation in median filtering.*

imum value of the pixels under the operator, or the minimum value, or the most frequent value, and so on. A reference covering such methods is Pitas & Venetsanopoulos [1990].

3.3 Image Segmentation: Edge Detection

The previous sections about image processing have presented a series of useful tools that can be used to perform image filtering, to be applied, for instance, to noise reduction. A subsequent important step in most image analysis approaches is the partitioning of the image in terms of its components, or patches that define the different foreground objects and the background of a scene. As an example, refer to Figure 3.25 (a), which presents a gray level image containing several leaves. In order to perform shape analysis over the image objects (i.e., the leaves), it is necessary to *segment* them from the background, leading to a binary image (see Figure 3.25 (b)). This operation, commonly referred to as *image segmentation*,

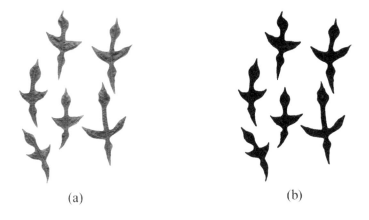

(a)   (b)

Figure 3.25: *Digital image (a) and respective segmentation (b).*

allows the detection and separation of the set of pixels that constitute the object(s) under analysis.

Frequently, an important visual element considered in image segmentation is the contrast between the image objects and the background, i.e., the difference between their respective gray levels. One of the most popular approaches for detecting such high contrast regions is *edge detection*, i.e., the identification of the sets of points (or curves or contours) corresponding to the more pronounced gray level variations. Figure 3.26 (a) illustrates this concept with respect to an image containing two neural cells. The obtained edges are shown in Figure 3.26 (b). Several

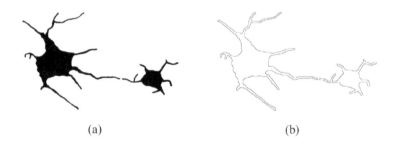

(a)   (b)

Figure 3.26: *Digital image (a) and respective edge detection (b).*

algorithms for detecting such boundaries between image structures are presented in the following sections. As far as gray level images are concerned, the edge detection procedures are based either on first- (gradient) or second-order (Laplacian) derivatives. In the gradient case, the edges are detected by thresholding the magnitude of the gradient of the image, since edge points correspond to high gradient values. On the other hand, edges are detected as *zero-crossings* of the Laplacian of the image. The edge map, obtained by one of these approaches (first-order differentiation maxima or second-order differentiation zero-crossings) is usually noisy in practice, with many broken edge lines, frequently requiring edge tracking and linking in order to obtain closed boundaries for the image objects (see Ballard & Brown [1982]; Marshall [1989] for additional discussion on such issues).

3.3.1 Edge Detection in Binary Images

The simplest edge detection situation is defined for binary images. In this case, only two gray levels are present in the image, e.g., 1 for object pixels and 0 for background pixels, and the edges are actually the object boundaries. It should be borne in mind that, as explained in Section 3.1.3, this is just a convention. There are different algorithms that can be used to detect binary edges, the simplest one being based on the fact that a boundary pixel is a pixel that belongs to the object (i.e., it is valued 1) and has at least one background pixel as a neighbor. In case an 8-neighborhood is adopted, then 4-neighborhood connected edges are obtained.

Otherwise, if 4-neighborhood is adopted instead, then the resulting edges are 8-connected. Figure 3.27 illustrates this concept. The *interior pixels* belong to the

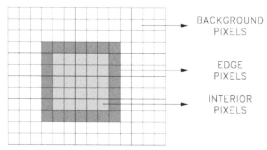

Figure 3.27: *The shape boundary concept.*

object and are surrounded by pixels also belonging to the object. Considering that each pixel $p(x, y)$ has four neighbors, i.e., above $p(x, y + 1)$, below $p(x, y - 1)$, to the right $p(x + 1, y)$ and to the left $p(x - 1, y)$, the following algorithm takes as input a binary image p and produces another binary image h with boundary points valued 1.

Algorithm: *Edge Detection in Binary Images*

---

1.   **for** each pixel $p(x, y)$
2.   **do**
3. $h(x, y) = 0$
4.   **if** $p(x, y) = 1$
5.   **then**
6.   **if** $(p(x, y + 1) = 0$ **OR** $p(x, y - 1) = 0$
                      OR $p(x + 1, y) = 0$ **OR** $p(x - 1, y) = 0)$
7.   **then**
8. $h(x, y) = 1$

The above algorithm can be easily modified in order to deal with alternative neighborhoods.

3.3.2 Gray-Level Edge Detection

As previously commented, edge detection in binary images is a straightforward issue. But how can edges be detected in gray level images? A binary image is used to motivate this more general and difficult situation. Figure 3.28 shows an image that is half black and half white (it could be, for instance, a zoomed detail of an edge region of the binary objects in Figure 3.26 (a)). The matrix representation of this digital image is also shown in Figure 3.28, where the black region corresponds to the gray level 0 and the white region to the gray level 255.

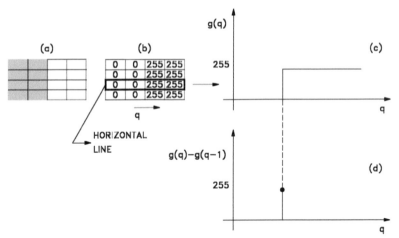

Figure 3.28: *The concept of edge detection through image differentiation.*

In order to better understand the process of edge detection, let $g(q)$ be the gray levels along a horizontal row of the image in Figure 3.28 (b), which is also shown in Figure 3.28 (c). As can be readily seen, the edge region corresponds to a step function, and the problem of edge detection can now be stated as finding the abrupt variations of this function. This can be easily accomplished by calculating the differences $g(q) - g(q - 1)$, which is shown in Figure 3.28 (d). The regions corresponding to uniform gray level correspond to the zeros of $g(q) - g(q - 1)$ (i.e., the black left portion and the white right portion). The points where the two regions meet yield $255 - 0 = 255$ (Figure 3.28 (d)), which is much greater than 0. Therefore, the edge pixels could be easily determined by searching for the largest values of $g(q) - g(q - 1)$. In case both regions were not perfectly uniform, but rather composed of pixels with similar but varying gray levels along each region, an analogous approach could be considered. In that case $g(q) - g(q - 1)$ would be small (but not 0) within each region but larger at the transition between the two regions.

This strategy for edge detection can be easily implemented as a linear filtering process, in analogous fashion to the previously discussed average filter. The simplest operator for detecting vertical edges is $[-1 \quad 1]$ with the reference point placed at the second element.

Note: *Numerical Differentiation*

The difference $g(q) - g(q - 1)$ is the simplest approximation to the numerical differentiation of $g(q)$. Although numerical differentiation is one of the most important mathematical operations in computer vision and image analysis [Weiss, 1994], it does not have a unique solution and can vary for different situations. In fact, the problem of numerical differentiation appears in different contexts, such as regularization and wavelets [Grossmann, 1988; Poggio et al., 1985]. The connection between the above simple approach to edge detection and differentiation can be generalized by considering the continuous case. A smoothed version of the edge profile of Figure 3.26, which is more likely to be found in real images, can be represented by a continuous signal $g(t)$, as well as its respective first and second derivatives $\dot{g}(t)$ and $\ddot{g}(t)$. The point of maximum gray level variation defining an edge is associated with a local maximum of $\dot{g}(t)$ and with a *zero-crossing* of $\ddot{g}(t)$. These considerations introduce the two main approaches for edge detection, based on first- and second-order image derivatives. The 2D extension of these operators for image processing, i.e., the *gradient* and the *Laplacian*, are presented in the following sections.

3.3.3 Gradient-Based Edge Detection

It is easy to see that the above discussed 1D edge detection equally applies to vertical edges. In fact, it can be generalized to edges of any orientation by using the important mathematical concept of *gradient* (see Section 2.4.3). As is well known from differential calculus, the gradient of a 2D scalar field (e.g., an image) indicates the direction along which the fastest increase of the function is verified. Therefore, the high contrasts in an image correspond to the edges, which can be immediately estimated by considering the image gradient. Let $g(x, y)$ be a 2D continuous function admitting partial derivatives at (x_0, y_0). The *gradient vector* of g at (x_0, y_0) is defined as follows:

$$\vec{\nabla} g\,(x_0, y_0) = \left(\frac{\partial g}{\partial x}\,(x_0, y_0), \frac{\partial g}{\partial y}\,(x_0, y_0) \right)$$

As an illustration, let $g(x, y) = x^2 + y^2$. Then

$$\vec{\nabla} g\,(x, y) = \left(\frac{\partial g}{\partial x}\,(x, y), \frac{\partial g}{\partial y}\,(x, y) \right) = (2x, 2y)$$

In this case, the gradient of g at the point $(1, 1)$ is given by

$$\vec{\nabla} g\,(1, 1) = (2, 2)$$

Observe that the gradient orientation can be used for the analysis of edge orientation, an important problem in texture and motion analysis [Zucker, 1985].

One of the simplest approaches to gradient estimation is in terms of the above explained linear filtering method (Section 3.2.3). In fact, linear filtering with the operator $[-1 \quad 1]$ is equivalent to estimating $\frac{\partial g}{\partial x}$. Similarly, if $[-1 \quad 1]^T$ is instead adopted as operator, an estimation of $\frac{\partial g}{\partial y}$ is obtained. The following algorithm implements this operation:

Algorithm: *Simple Edge Detection*

1. $hx \leftarrow [-1 \quad 1]$;
2. $hy \leftarrow [-1 \quad 1]^T$;
3. $sk \leftarrow \begin{bmatrix} 1/9 & 1/9 & 1/9 \\ 1/9 & 1/9 & 1/9 \\ 1/9 & 1/9 & 1/9 \end{bmatrix}$;
4. $sg \leftarrow \text{LINFILT}(g, sk)$;
5. $gx \leftarrow \text{LINFILT}(g, hx)$;
6. $gy \leftarrow \text{LINFILT}(g, hy)$;

In the above algorithm, LINFILT(g, h) indicates linear filtering the image g with a mask h. Furthermore, the input image is previously smoothed with the 3×3 average filter (leading to the sg image) before the differentiating filters are applied. The modulus of the gradient of the image g is defined as follows:

$$\left| \vec{\nabla} g \right| = \sqrt{\left(\frac{\partial g}{\partial x} \right)^2 + \left(\frac{\partial g}{\partial y} \right)^2}$$

The implementation of the above expression can be done straightforwardly by using the resulting matrices of the gradient algorithm above (the expression should be applied to each of the corresponding pixels of the matrices gx and gy). Many alternative edge detection methods can be found in the literature. Some of the most popular are discussed in the following sections.

3.3.4 Roberts Operator

Roberts gradient approximation, also known as Roberts cross operator, is based on the value of the pixel under consideration, and its 3 neighbors (p_0+1, q_0), (p_0, q_0+1) and $(p_0 + 1, q_0 + 1)$, as depicted in Figure 3.29 (a). In fact, Roberts operator is based on the differences $(g(p_0, q_0) - g(p_0 + 1, q_0 + 1))$ and $(g(p_0 + 1, q_0) - g(p_0, q_0 + 1))$, which explains the name "Roberts cross operator."

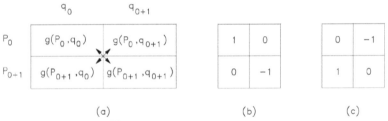

(a)   (b)   (c)

Figure 3.29: *The Roberts operator.*

Let g be a digital image. The value of the *Roberts operator* at (p_0, q_0) is typically defined as

$$r(p_0, q_0) = \sqrt{(g(p_0, q_0) - g(p_0 + 1, q_0 + 1))^2 + (g(p_0 + 1, q_0) - g(p_0, q_0 + 1))^2}$$

It is observed that a number of alternative definitions for this operator can be found in the related literature. Schalkoff [1989], for example, defines the Roberts operator as

$$r(p_0, q_0) = \max \{|g(p_0, q_0) - g(p_0 + 1, q_0 + 1)|, |g(p_0 + 1, q_0) - g(p_0, q_0 + 1)|\}$$

On the other hand, Angulo & Madrigal [1986] present the following definition:

$$r(p_0, q_0) = |g(p_0, q_0) - g(p_0 + 1, q_0 + 1)| + |g(p_0 + 1, q_0) - g(p_0, q_0 + 1)|$$

These two definitions may be preferable for real-time applications since they do not involve the square root operation. Finally, Castleman [1996] adopts

$$r = \sqrt{\left(\sqrt{g(p_0, q_0)} - \sqrt{g(p_0 + 1, q_0 + 1)}\right)^2 + \left(\sqrt{g(p_0 + 1, q_0)} - \sqrt{g(p_0, q_0 + 1)}\right)^2}$$

In the latter definition, the square root of each pixel is taken before the calculus of the differences. Castleman argues that this operation makes the Roberts operator more similar to edge detection in the human visual system.

The above definitions of Roberts operators can be implemented within the framework of the linear filtering algorithm. Both the involved cross differences, which are henceforth denoted as g_1 and g_2, can be easily calculated by applying the linear filtering algorithm with the operators shown in Figure 3.29 (b) and (c) (in the latter definition, presented by Castleman [1996], the square root of each pixel should be taken previously). The reference point of both operators is in the upper-left position. Any of the above definitions can be easily calculated in terms of g_1 and g_2.

3.3.5 Sobel, Prewitt, and Kirsch Operators

Another class of edge operators can be obtained by varying the linear operators to be applied to the image, before their combination (such as taking the maximum, or the sum of the modula, or the square root of the sum of the squared partial results). The edge operators known as Sobel, Prewitt and Kirsch employ larger masks in order to estimate the edges, i.e., they work at a larger spatial *scale*. As previously discussed, this implies a smoother, less noisy output, at the expense of losing some image detail. The Sobel edge operator is based on linear filtering the image with some special masks. Two implementations of the Sobel edge detector are possible after applying the linear filtering algorithm with the two masks below: (i) to take the largest partial value and (ii) to take the square root of the sum of the two squared partial values. Similar procedures can readily be implemented for the Prewitt Kirsh operator. The Sobel masks are defined as

$$\Delta x = \begin{bmatrix} -1 & 0 & 1 \\ -2 & 0 & 2 \\ -1 & 0 & 1 \end{bmatrix} \quad \Delta y = \begin{bmatrix} 1 & 2 & 1 \\ 0 & 0 & 0 \\ -1 & -2 & -1 \end{bmatrix}$$

The Prewitt masks are defined as

$$\Delta x = \begin{bmatrix} 1 & 0 & -1 \\ 1 & 0 & -1 \\ 1 & 0 & -1 \end{bmatrix} \quad \Delta y = \begin{bmatrix} -1 & -1 & -1 \\ 0 & 0 & 0 \\ 1 & 1 & 1 \end{bmatrix}$$

Finally, the Kirsch masks are defined as

$$\begin{bmatrix} 5 & 5 & 5 \\ -3 & 0 & -3 \\ -3 & -3 & -3 \end{bmatrix} \begin{bmatrix} -3 & 5 & 5 \\ -3 & 0 & 5 \\ -3 & -3 & -3 \end{bmatrix} \begin{bmatrix} -3 & -3 & 5 \\ -3 & 0 & 5 \\ -3 & -3 & 5 \end{bmatrix} \begin{bmatrix} -3 & -3 & -3 \\ -3 & 0 & 5 \\ -3 & 5 & 5 \end{bmatrix}$$

$$\begin{bmatrix} -3 & -3 & -3 \\ -3 & 0 & -3 \\ 5 & 5 & 5 \end{bmatrix} \begin{bmatrix} -3 & -3 & -3 \\ 5 & 0 & -3 \\ 5 & 5 & -3 \end{bmatrix} \begin{bmatrix} 5 & -3 & -3 \\ 5 & 0 & -3 \\ 5 & -3 & -3 \end{bmatrix} \begin{bmatrix} 5 & 5 & -3 \\ 5 & 0 & -3 \\ -3 & -3 & -3 \end{bmatrix}$$

In this case, the result for each pixel is taken as the maximum response for any of the above masks. The reader is referred to standard image processing textbooks for additional detail (e.g., Prewitt and Kirsch, refer to Castleman [1996]; Prewitt & Mendelsohn [1966]).

Fundamentally, all these edge operators vary on the weights and orientations, being more selective to a specific direction, giving more importance (or not) to the reference pixel, and so on. As discussed in the beginning of Section 3.3, the edges are detected by thresholding the edge enhanced image produced by the application of one of the above filters.

3.3.6 Fourier-Based Edge Detection

As far as edge detection through image differentiation is concerned, the Fourier transform provides an interesting and powerful approach. Consider the following 2D Fourier derivative property:

$$\left(\frac{\partial}{\partial x}\right)\left(\frac{\partial}{\partial y}\right) g(x, y) \leftrightarrow (j2\pi u)(j2\pi v) G(u, v)$$

Therefore, an image can be differentiated by taking its 2D Fourier transform, applying the above property, and taking the inverse transform. Figure 3.30 (a)

shows an image that has undergone this Fourier-based differentiation process, resulting in the edge enhanced image of Figure 3.30 (b). A low-pass Gaussian filtering has been applied to the original image prior to the application of the derivative property, in order to attenuate the high-pass nature of the derivative, which
tends to enhance high frequency noise in the image. A similar low-pass scheme is
also adopted in the Marr-Hildreth filter described in Section 3.3.8. The edges can
be detected by thresholding the edge-enhanced image, which is illustrated in Figure 3.30 (c). As an illustrative comparison, Sobel edge enhancement and detection
are shown in Figure 3.30 (d) and (e), respectively.

3.3.7 Second-Order Operators: Laplacian

As observed in Section 3.3, edges can be located based on local maxima of first-
order image differentiation or based on zero-crossings of second-order image differentiation. We have already argued that the gradient is the natural candidate for first-
order image differentiation. Second-order image differentiation is usually achieved
in terms of the *Laplacian*. Let $g(x, y)$ be a 2D continuous function possessing valid
partial derivatives in (x_0, y_0), and let $\vec{\nabla} g$ be the gradient of g. The *Laplacian* $\nabla^2 g$ of
g is defined as

$$\nabla^2 g = \vec{\nabla} \cdot \vec{\nabla} g = \frac{\partial^2 g}{\partial x^2}(x, y) + \frac{\partial^2 g}{\partial y^2}(x, y)$$

A simple comparison between the gradient and the Laplacian expressions reveals that, while the gradient is a vector, the Laplacian is a scalar quantity (see
Section 2.4.3). Among the different numerical expressions of the Laplacian, three
possible masks are presented, adapted from Jain [1989]:

$$L_1 = \begin{bmatrix} 0 & -1 & 0 \\ -1 & 4 & -1 \\ 0 & -1 & 0 \end{bmatrix} \quad L_2 = \begin{bmatrix} -1 & -1 & -1 \\ -1 & 8 & -1 \\ -1 & -1 & -1 \end{bmatrix} \quad L_3 = \begin{bmatrix} 1 & -2 & 1 \\ -2 & 4 & -2 \\ 1 & -2 & 1 \end{bmatrix}$$

It is also worth noting that, in practice, the Laplacian-of-Gaussian (Section 3.3.8)
is more frequently applied than the simple Laplacian.

3.3.8 Multiscale Edge Detection: The Marr-Hildreth Transform

The previous sections have discussed several image differentiation-based techniques
for edge detection. Image differentiation acts as a high-pass filter, which is a consequence of the fact that the filter function in the Fourier derivative property has a
high-pass nature. As a consequence, high-frequency image noise is generally enhanced by the differentiation process. Therefore, it is highly desirable to introduce a
mechanism to control this type of noise, which can also arise from the spatial quantization inherent to digital images (i.e., the stair-like patterns defined by the pixels
at the object boundaries). A simple and powerful technique to cope with this problem is to apply a low-pass filter to the image, thus compensating for the high-pass

238

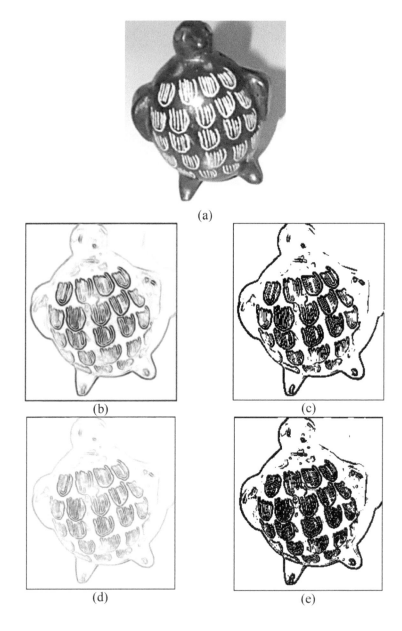

(a)

(b)

(c)

(d)

(e)

Figure 3.30: *Digital image **(a)** and respective Fourier-based edge enhancement **(b)**. The binary image containing the edges obtained by thresholding* (b) *are presented in **(c)**. Analogously, the Sobel edge enhancement and detection are shown in **(d)** and **(e)**, respectively.*

filter. In this context, a Gaussian filter can be used along with the differentiation process. Let $u(t)$ be a 1D signal and $\frac{d}{dt}(g * u)$ be the differentiated version of u, which has undergone a smoothing operation through a convolution with a Gaussian filter g. From the Fourier derivative property and the convolution theorem, we have

$$\frac{d}{dt}(g * u) = \left(\frac{d}{dt}g\right) * u = g * \frac{d}{dt}u$$

The above expression means that differentiating the smoothed version of u is equivalent to convolving u with the differentiated version of g. In other words, if the differentiated version $\frac{d}{dt}g$ of g is known *a priori*, then any signal u can be differentiated by convolving it with $\frac{d}{dt}g$, leading to a useful and versatile approach to numerical signal differentiation. This result can be generalized for the 2D case. Indeed, the introduction of this type of filter by Marr & Hildreth [1980] represented a fundamental tool for edge detection which has become a classical approach, having played a central role in the modern multiscale theories of scale-space and wavelets. The Marr-Hildreth operator $\nabla^2 g$, based on the *Laplacian of Gaussian (LoG)*, can be defined as follows:

$$\nabla^2 g = \left(1 - \frac{x^2 + y^2}{\sigma^2}\right)\exp\left(-\frac{x^2 + y^2}{2\sigma^2}\right)$$

where g is a 2D Gaussian. Considering the above discussion about the low-pass filtering step required for high-frequency noise attenuation, the Marr-Hildreth operator is equivalent to the Laplacian edge detection performed on a smoothed version of the original image. Therefore, the Marr-Hildreth edge detection approach is based on searching for zero-crossings of the scalar field $o = \nabla^2 g * f$, where $\nabla^2 g$ denotes the Laplacian of the Gaussian and f is the input image. A discrete mask which approximates the Laplacian of Gaussian follows [Jain et al., 1995]:

$$\nabla^2 g = \begin{bmatrix} 0 & 0 & -1 & 0 & 0 \\ 0 & -1 & -2 & -1 & 0 \\ -1 & -2 & 16 & -2 & -1 \\ 0 & -1 & -2 & -1 & 0 \\ 0 & 0 & -1 & 0 & 0 \end{bmatrix}$$

Finally, it is important to note that the above expression depends on a spatial scale parameter, defined by the Gaussian spread parameter σ. As the analyzing scale σ increases, the filtered image becomes more blurred, thus allowing small details to vanish. Figure 3.31 (a) shows an image of a leaf, with the respective edge detected images for two different scales shown in Figure 3.31 (b) (smaller scale) and (c) (larger scale). As it can be clearly seen, as the analyzing scale increases, the edge map becomes dominated by the largest structures and the small-scale detail is filtered out.

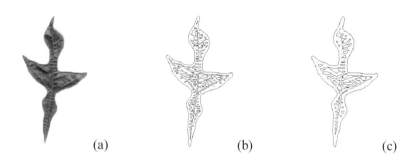

(a)   (b)   (c)

Figure 3.31: *Digital image* ***(a)*** *and the LoG-based detected edges for two increasing scales* ***(b-c).***

To probe further: *Edge and Other Feature Detection Filters*

The problem of performance assessment and comparison between image filters is very important and only recently has been addressed in the literature. So, it is difficult to decide *a priori* which filter is the best to use in a specific situation. The book by Jain et al. [1995] presents an introductory discussion and some simple results about performance assessment of edge detectors. Additional materials include Cho et al. [1997]; Demigny & Kamlé [1997].

There are many different approaches to the edge detection problem that complement those discussed in this book, and some of the most important include the Kirsch masks [Castleman, 1996; Kirsch, 1971] (a similar approach to Sobel), Canny operator [Canny, 1986], wavelets-based (Section 6.3 of Mallat [1998] and Mallat & Zhong [1992]) and Markov-based techniques [Geiger & Yuille, 1991; Li, 1995; Marroquin et al., 1986; Mumford & Shah, 1985]. On the other hand, the reader interested in numerical differentiation and regularization theory will find it useful to consider Bertero et al. [1988]; Grossmann [1988]; Poggio et al. [1985]; Weiss [1994]. Similar linear methods have been developed for the detection of other features besides edges, such as for generic differentiation [Koenderink & van Doorn, 1992], 2D isophote curvature [Estrozi et al., 2003; Kitchen & Rosenfeld, 1982; Koenderink & Richards, 1988], corner detection [Chen et al., 1995; Cooper et al., 1993], local frequency and local periodic patterns (e.g., texture) analysis [Bovik et al., 1990; Daugman, 1988; Dunn & Higgins, 1995; Freeman & Adelson, 1991; Reed & Weschsler, 1990; Simoncelli et al., 1992; Valois & Valois, 1990; Watson, 1987], and curves [Iverson & Zucker, 1995; Steger, 1998]. The papers Kubota & Alford [1995, 1996, 1997] may be of interest, as far as real-time implementations are concerned. Finally, gray level mathematical morphology also includes powerful tools in this context [Gonzalez & Woods, 1993].

3.4 Image Segmentation: Additional Algorithms

As already observed, a critical task in image analysis and vision corresponds to properly segmenting the image into meaningful regions to be subsequently analyzed. Image segmentation can also be thought of as the process of separating the object of interest in an image from the background and from other secondary entities. Figure 3.32 illustrates these two complementary views. Figure 3.32 (a) shows

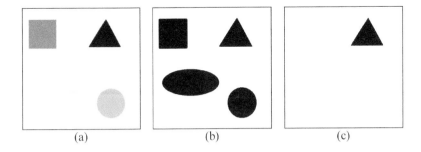

(a)   (b)   (c)

Figure 3.32: *Digital image **(a)**; binary image with all objects segmented **(b)**; binary image with a specific object segmented **(c)**.*

a typical image containing the background and some objects. A possible segmentation result is presented in Figure 3.32 (b), where the objects have been separated from the background. On the other hand, it would be possible that, in a particular application, only a specific object would be of interest. In this case, as shown in Figure 3.32 (c), the image segmentation result should be an image containing only one of the objects. From this discussion we have that the segmentation process presents two different aspects: it can be implemented either as a purely bottom-up, "blind" procedure whose result depends only on intrinsic image features (e.g., gray level or texture); or based on top-down approach guided by knowledge about the particular problem (e.g., "One wants to separate triangles from the rest"). These two approaches can also be integrated into a hybrid method involving both bottom-up and top-down information.

Segmentation is a particularly difficult problem and the success (or not) of any algorithm depends heavily on the image acquisition quality. It is therefore common to perform the segmentation in an interactive fashion, where the intermediate results are analyzed and improved through (human) expert supervision and intervention (see Section 9.7). The following sections review some of the classic approaches to digital image segmentation which, however basic, are useful in many shape analysis problems.

3.4.1 Image Thresholding

The simplest and most useful and often adopted methods for image segmentation are based on thresholding. The basic idea underlying this method is simple. Suppose that we want to segment a simple image containing dark objects over a light background (e.g., Figure 3.32 (a)). Observe that not all the background pixels have necessarily the same gray level (in fact, they generally do not), the same being verified for the object pixels. For instance, in the image in Figure 3.32 (a), the pixels of the background range from 250 to 255, while those belonging to the objects range from 0 to 240. A straightforward procedure to classify each element as an object pixel or as a background pixel is to set a *threshold* value in the intermediate range between the background and the foreground gray levels, and to consider as a foreground any point smaller or equal to this value. This is summarized by the following algorithm:

Algorithm: *Thresholding*

1. **for** each pixel in the image
2.   **do**
3.   **if** (pixel_value < threshold)
4.   **then**
5.   label the pixel as an object pixel;
6.   **else**
7.   label the pixel as a background pixel;

The result of the above procedure can be represented as a binary image in which the background pixels are represented by 0s, and the object pixels by 1s. The above presented image thresholding procedure has a direct relationship with the concept of histograms discussed in Section 3.2.1. The original image shown in Figure 3.32 (a) is shown again in Figure 3.33, together with its histogram. As it can be seen from Figure 3.33 (b), the histogram of this image presents modes around the concentration of the gray levels of each object and the background. The image histogram can provide a useful tool for setting the correct threshold value. For instance, an examination of the histogram presented in Figure 3.33 (b) shows that the thresholds 150 and 200 can be used for the segmentation of the circle, which can be easily accomplished through the following algorithm:

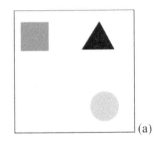

(a)

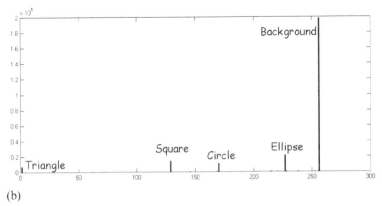

(b)

Figure 3.33: *Digital image (a) and corresponding histogram identifying the respective image structures (b).*

Algorithm: *Thresholding2*

1. **for** each pixel in the image
2.   **do**
3.   **if** (pixel_value $< t_1$ and pixel_value $> t_2$)
4.   **then**
5.   label the pixel as an object pixel;
6.   **else**
7.   label the pixel as a background pixel;

Note: *Improved Thresholding*

There are many different techniques that attempt to improve the thresholding algorithm. For instance, if the illumination is not homogeneous throughout the image, finding a single threshold that produces acceptable results may be impossible. In some of these situations, it is possible to find different suitable thresholds for different image regions, thus suggesting the use of *adaptive thresholding algorithms* [Fisher et al., 1996]. Another important problem is how to properly select the threshold value. Although this can be done manually by trial-and-error, it is desirable to have procedures for *automatic threshold selection* in many situations. Such procedures normally analyze the image histogram and try to set the threshold value in order to optimize some criterion function. An interesting approach is presented in Jansing et al. [1999], where the criterion function is defined from the entropy of the histogram portions defined by the threshold. This method has been applied to the top image shown in Figure 3.34: using four different criterion functions, and the results are shown in Figure 3.34. Thresholding can also be generalized as a problem of assigning the pixels to different classes, leading to a pattern recognition problem as pixel classification, an approach with many powerful tools to be explored (e.g., Fisher et al. [1996] and Chapter 8).

3.4.2 Region-Growing Algorithm

This method requires a seed pixel to be assigned inside the object to be segmented, which is followed by growing a region around that pixel until filling up the whole object. The growing process starts by testing the neighboring pixels of the seed in order to verify whether they belong to the object. The same test is subsequently applied to the neighbors of the neighbors (that have proven to belong to the object) and so on, until all such pixels are checked.

An important step of the region-growing algorithm is the test to verify whether a given pixel belongs to the region to be segmented. Such a test can be implemented in various ways, depending on each specific problem. The easiest and most straightforward option is to verify if the gray level of the pixel under consideration is not significantly different from the seed pixel, which may be done by checking whether the absolute difference between their gray levels does not exceed a given value. A computational strategy must also be devised in order to deal with the process of considering the neighbors, the neighbors of the neighbors, and so on. This can be easily implemented through the use of a *stack* [Langsam et al., 1996], which is a data structure into which each pixel to be considered (a neighbor in the aforementioned sequence) is *pushed*. In each iteration of the algorithm, a pixel is *popped* from the stack and verified to see whether it belongs to the region to be segmented. If it does not, it is simply discarded. Otherwise, its neighbors are pushed into the stack. An algorithm for region growing based on such principles is presented next.

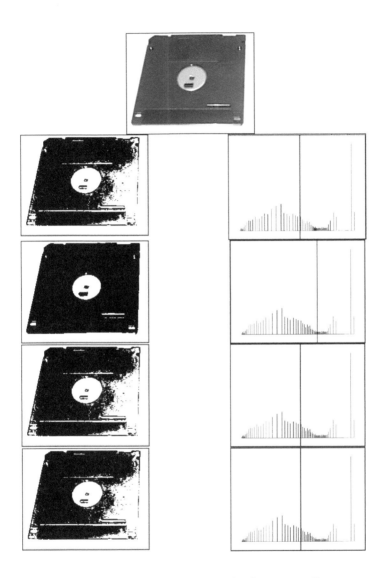

Figure 3.34: *The top image has been segmented with automatically set thresholds using four different criterion functions, with the results being shown in the four subsequent rows (refer to Jansing et al. [1999] for additional information).*

Algorithm: *Region-Growing*

REGION_GROWING(*seed, image*)

1.   INITIALIZE(stack);
2.   PUSH(stack, seed);
3.   seed_gl ← image[seed.p, seed.q];
4.   **while** (**NOT** (EMPTY(stack))
5.   **do**
6.   current_pixel ← POP(stack);
7.   **if** (BELONG(image, seed_gl, current_pixel))
8.   **then**
9.   LABEL(image, current_pixel);
10.   PUSH(stack, right (current_pixel));
11.   PUSH(stack, above (current_pixel));
12.   PUSH(stack, left (current_pixel));
13.   PUSH(stack, below (current_pixel));

A pixel is represented by a data structure such as a *record* (in Pascal) or a *struct* (in C or in MATLAB®) that contains two fields p and q. In the above algorithm, the variables *current_pixel* and *seed* are of this type, and *stack* is a stack of pixels. The above algorithm can also be easily implemented in programming languages not supporting such structures by treating the p and q coordinates separately and using two stacks instead of one, one for each coordinate. The functions *right, above, left* and *below* return the respective 4-neighbors of the current pixel. The function *belong*, which determines whether a given pixel belongs to the growing region, can be implemented as

---

Algorithm: *Belong*

BELONG(*image, seed_gl, current_pixel*)

1.   bel ← **FALSE**;
2.   **if** ((current_pixel.p >= 1) **AND** (current_pixel.p <= P)
          AND (current_pixel.q >= 1) **AND** (current_pixel.q <= Q))
3.   **then**
4.   **if** (abs (image[current_pixel.p, Current_pixel.q] - seed_gl) < ERROR)
5.   **then**
6.   bel ← **TRUE**;

The function *belong* is initialized with FALSE, so that if nothing happens in the following *if*, it returns FALSE, indicating that the pixel does not belong to the region. The first *if* verifies that the pixel coordinates are valid, assuming that the indexes p and q of the image vary from 1 to P and 1 to Q, respectively. The second *if* checks the difference between the gray levels of the current pixel and that of the seed, where ERROR is a constant that must be set *a priori*. Knowledge about each specific problem can be incorporated into the above procedure. For instance, if the objects to be segmented in the image are known not to present a diameter larger than a given value, this condition can be easily incorporated into the second *if* of the function *belong* by testing the distance between the current pixel and the seed (or between the current pixel and the farthest one still belonging to the object).

In addition, a mechanism has to be devised so that the already considered pixels are not revisited. This can be easily accomplished as follows. First, it is important to note that the output of the region growing procedure is a labeled image and, therefore, we must define some labeling representation. This can be achieved by marking the pixels that belong to the growing region with a special value, preferably one that is not a valid gray level. For instance, if the input gray levels vary from 0 and 255, we could mark the growing region with, say 1000, in such a way that a simple thresholding operation over the output image with a threshold of 300 would produce a binary image with the segmented region of interest. Furthermore, this labeling scheme also avoids the algorithm to reconsider already visited pixels, because if the label to be assigned to the pixels that belong to the region is large enough, the difference between the labeled pixels and the seed will always exceed the threshold ERROR. Therefore, the procedure *label (image, current_pixel)* only has to assign a (large enough) marking value to the *current_pixel* of the *image*.

The seed pixel, which is a parameter of the algorithm, may be set in different manners. For instance, in an interactive environment, a human expert could click the mouse at any pixel inside the image he or she wants to segment. Automatic procedures can also be implemented, such as to pick the first pixel of the image that does not (surely) belong to the background, or the foveated pixel of a selective attention-based system.

The region-growing algorithm can be adapted to different problems in a number of ways. For instance, it can be used for segmentation of images based on texture by first assigning to each pixel a specific texture measure, followed by the application of the region-growing algorithm. Straightforward procedures for other kinds of images (such as color and multispectral images) can also be easily obtained. In addition, it can be applied in order to label the different connected components in an image with different labels (it is sufficient to modify the algorithm so that different labels can be used, e.g., a different integer for each connected component).

Note: **Note:** *Image Processing Meets Computer Graphics*

The region growing algorithm for image segmentation can also be used to solve the image synthesis (computer graphics) problem known as *region filling*. Suppose that a 2D shape, like that in Figure 3.35 (a) has to be filled, as illustrated in Figure 3.35 (b)

(a)   (b)

Figure 3.35: *(a) Original contour and shape obtained by filling the contour in* (a) *by using a region-filling algorithm* **(b)**.

Clearly, an *ad hoc* approach would be to paint by hand the interior of the shape in a pixel-by-pixel fashion. A much better solution is simply to apply the region-growing algorithm using the desired color (or gray level) to label the object pixels. Most off-the-shelf imaging software include a flood fill tool in order to perform this task. Additional information about this operation can be found in Esakov & Weiss [1989].

To probe further: *Image Segmentation*

More sophisticated image segmentation techniques include level-sets, diffusion and variational methods. Good references on these topics are Morel & Solimini [1995] and Sethian [1999].

3.5 Binary Mathematical Morphology

The field of *mathematical morphology* originated from the work of the French mathematicians J. Serra and G. Matheron in the mid-1960s and has now become an important framework that provides many useful tools for image processing and analysis. In fact, mathematical morphology has been applied to a wide variety of practical problems such as image pre-processing (binary, gray level and color images), noise filtering, shape detection and decomposition, and pattern association, to name but a few. In this context, besides the theoretical aspects, mathematical morphology is an important approach to solve many practical image analysis problems. As far as the mathematical background is concerned, mathematical morphology differs from many of the previously discussed image processing techniques, mainly because it is a nonlinear approach, usually dealing with discrete data in terms of sets and set operations. The present section is aimed at presenting some useful basic concepts so that the reader will be able to probe further in the specialized literature as well as some particularly interesting possibilities for their applications in image and shape analysis. As a deep coverage of this subject is beyond the scope of the present book, the reader is directed to the rich mathematical morphology literature referenced at the end of this section.

Mathematical morphology algorithms are progressively created by combining a set of simple operations, allowing the creation of techniques capable of solving complex shape processing problems. Only binary images are considered in this section.

3.5.1 Image Dilation

Most mathematical morphology operations involve a binary image, f, to be processed and a *structuring element*, g. Different structuring elements define different outputs for the same processed shape. The first basic mathematical morphology operation is the so-called *dilation*, whose net effect is to expand or dilate the shapes in an input image. Figure 3.36 (a) illustrates the effect of dilating a continuous square having a circle as structuring element. In this case, the dilation process can

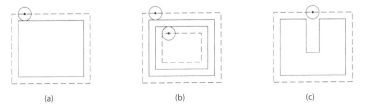

(a)   (b)   (c)

Figure 3.36: *Morphological dilation effects.*

be considered as if the circle were tracked along the square boundary while kept in touch with it. The expansion area is defined by the pixels between the square boundary and the circle center, i.e., the shaded region in Figure 3.36 (a). It is important to note that the structuring element has a reference point, or origin, defined *a priori*, which is assumed to be at the center of the circle in the previous example. Different reference points usually lead to different results. Two important effects of morphological dilation are illustrated in Figure 3.36 (b) and (c). Figure 3.36 (b) shows that dilation acts over both external and internal shape boundaries. Therefore, holes in objects are diminished, or even eliminated, depending on their size, type and reference point of the structuring element. Furthermore, Figure 3.36 (c) shows that concave regions narrower than the structuring element are also filled up by dilation.

Morphological dilation can be considered as an operation where the structuring element is translated over each pixel of the image. If the intersection between the translated structuring element and the image is non-empty, then the current pixel is considered valid. Let $p = (p_1, p_2)$ be the current pixel and g_p denote the translated structuring element, i.e., g with its origin centered at p. Therefore, image dilation can be defined as

$$\delta_g(f) = \left\{ p \mid g_p \cap f \neq \emptyset \right\}$$

In other words, the result of image dilation is the set of image pixels where the intersection between g_p and f is not empty.

Example: *Image Dilation*

An example of image dilation considering the structuring element in Figure 3.37 is presented in the following. Figure 3.38 illustrates the effect of discrete image dilation.

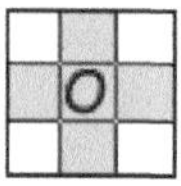

Figure 3.37: *Structuring element, with the origin indicated.*

As the introduced concepts can be applied to binary images, both image and structuring elements are represented by binary arrays, i.e., arrays containing 1s and 0s. Figure 3.38 (a) shows an image with a hypothetical binary object to be dilated with the structuring element of Figure 3.37. It is worth observing that this structuring element can be represented by the following binary array:

$$g = \begin{bmatrix} 0 & 1 & 0 \\ 1 & 1 & 1 \\ 0 & 1 & 0 \end{bmatrix}$$

The "1" at the second column, second line, represents the origin of the structuring element, which is denoted as an "o" in Figure 3.37. This origin can be defined arbitrarily. It is noted that the 1s of the structuring element are presented in a different gray level in this example only for didactic purposes.

Image dilation in a digital image can be understood in terms of the following conceptual procedure: the structuring element is shifted so that its origin passes through each pixel of the object. At each shifted position, all the image pixels covered by the 1-valued pixels of the structuring elements are set to 1. Figure 3.38 (b) illustrates this process by showing the structuring element shifted for the cases A, B and C. Some examples of such situations are shown in gray in Figure 3.38 (b),

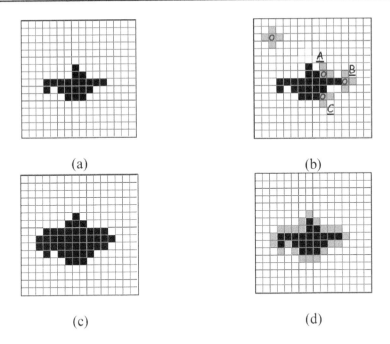

(a)   (b)

(c)   (d)

Figure 3.38: *Binary shape (a); situations found during image dilation (b);
the dilated shape represented in black (c); and showing the
dilated pixels in gray (d).*

while Figure 3.38 (c) presents the resulting dilated shape. The dilated pixels are
shown in gray in Figure 3.38 (d).

Morphological dilation is widely applied to many important problems in image
processing and shape analysis. As previously mentioned, mathematical morphology-
based algorithms are constructed by using simple operations as building blocks,
and dilation is frequently explored in this sense. In the specific case of dilation,
it can be used alone for reducing noisy structures such as small gaps and holes,
though morphological closing (see Section 3.5.2) is more often adopted for this
purpose. In addition, it should be observed that image dilation is an important tool
for biological shape analysis as it allows the implementation of a fractal dimension
estimation method called *Minkowsky sausages* and other interesting techniques, as
described in Section 6.3. Figure 3.39 presents an example of image dilation applied to a neural cell. A neural shape is illustrated in (a), and the resulting shapes
after successive dilations are respectively shown in (b), (c), and (d). As we can
see, the smallest scale structures (mainly those located on the neural boundary) are
merged as dilation proceeds. This strategy can be useful in order to expand narrow
structures, (see for instance Cesar-Jr. & Costa [1997]), where 1-pixel wide neural

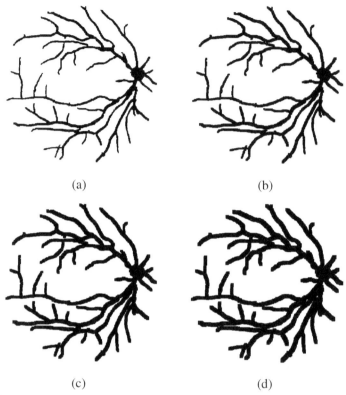

(a)   (b)

(c)   (d)

Figure 3.39: *Binary image of retina vessels (a) and results after 1 (b), 2 (c)*
and 3 (d) dilations. The structuring element used in this ex-
ample is a 3 × 3 square.

terminations tend to introduce instabilities in the numerical estimation of contour
curvature.

3.5.2 Image Erosion

As in the case of image dilation, morphological *erosion* involves a binary image,
f, to be processed and a *structuring element*, g. As before, it is observed that different structuring elements generally imply different results. The practical effect of
image erosion is to shrink the shapes in the image, hence the name. Two important
effects of morphological erosion are (i) morphological erosion acts over both external and internal shape boundaries; and (ii) thin regions like isthmuses can vanish,
thus substantially changing the shape and topology of the objects. Erosion is the
dual operation of dilation: eroding a shape is equivalent to dilating the background.

Morphological erosion can be considered as an operation where the structuring
element is translated over each pixel of the object: if the shifted structuring element

is completely contained in the object, then the current pixel is set to 1. Let $p = (p_1, p_2)$ be the current pixel and g_p denote the translated structuring element, i.e., g has its origin centered at p. Therefore, the image erosion operation is defined as follows:

$$\varepsilon_g(f) = \left\{ p \mid g_p \subset f \right\}$$

where $g_p \subset f$ denotes that g_p is a subset of f. In other words, the result of image erosion is the set of image pixels p where g_p is a subset of f.

Example: *Image Erosion*

An example, analogous to the previously discussed dilation example and with the same structuring element as in Figure 3.37, is presented in the following. Figure 3.40 (a) shows an image with a hypothetical binary object to be eroded. Image

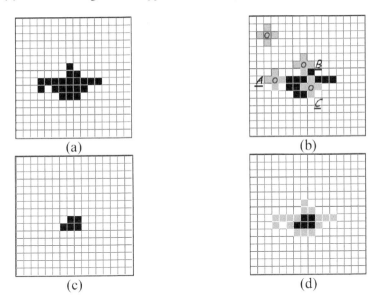

(a)

(b)

(c)

(d)

Figure 3.40: *Binary shape (a); examples of possible situations found during erosion (b); the eroded shape represented in black (c); and the eroded pixels shown in gray (d).*

erosion in a digital image can be understood in terms of the following intuitive procedure: the structuring element is shifted so that its origin is located over each pixel of the image. At each shifted position, if the whole set of 1-valued pixels of the structuring element are found to correspond to 1-valued pixels in the image, then the output image pixel corresponding to the structuring element origin is set to 1. Otherwise, it is set to 0. Figure 3.40 (b) illustrates this process by showing the structuring element shifted for the cases *A*, *B* and *C*. Since in the cases *A* and *B* the structuring element is not completely contained in the image foreground (i.e., the black object), the corresponding reference pixels are set to 0, i.e., eroded. On the other hand, in the case *C*, the 1-valued pixels of the structuring element are completely contained in the input object pixels, and the corresponding origin pixel is therefore not eroded. Figure 3.40 (c) presents the eroded shape, while Figure 3.40 (d) shows in gray the pixels that have been eroded from the original shape.

As with dilation, morphological erosion is widely applied to image processing and shape analysis problems, especially as a component of more complex algorithms. Although image erosion can be used for reducing noisy structures, such as small regions, morphological opening (see below) is more often adopted for this purpose. An interesting application of image erosion is to use it as a pre-processing tool before labeling (see Section 3.1.5). As an example, refer to Figure 3.41 and suppose that we want to count how many objects (i.e., leaves) there are in the image in Figure 3.41 (a), which can be done by identifying the connected components (refer to Section 3.1.5) in that image. A simple thresholding algorithm can be ap-

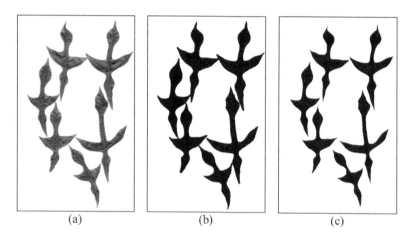

(a)   (b)   (c)

Figure 3.41: *Practical application of image erosion: original image (a), segmented image (b), and eroded image yielding the separated objects (c).*

plied to segment the initial image, which results in Figure 3.41 (b). Because the objects touch each other at different points, the correct number of objects cannot be inferred from the connected components. By suitably eroding the thresholded image, the thin contacts between the objects are removed, yielding the correct number of connected components (Figure 3.41 (c)).

Morphological erosion and dilation are used in more elaborated shape processing algorithms (see the box below for several related references). As an example of how such operations are used as intermediate steps for more complex procedures, suppose that we have a shape with several holes to be filled (see Figure 3.42 (a)). As it has been explained, a first solution to this problem is to apply morphologi-

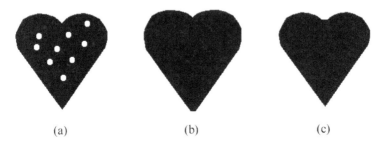

(a)   (b)   (c)

Figure 3.42: *Original shape (a), its dilated version (b), and shape obtained by morphological closing (c). The structuring element was a 3 × 3 square.*

cal dilation with a structuring element large enough to close the holes, resulting an image such as that illustrated in Figure 3.42 (b). The problem with this solution is clear: the dilation substantially affects all boundary image points, which may be undesirable, depending on the application. This problem can be circumvented by eroding the dilated shape with the same structuring element, which is shown in Figure 3.42 (c). In this case, the holes have been closed and the shape boundary has not been substantially modified. This operation, defined by dilation followed by erosion with the same structuring element, is called *morphological closing*. Its dual operation, i.e., erosion followed by dilation, is called *morphological opening*, which also exhibits analogous important properties.

To probe further: *Mathematical Morphology*

Most aforementioned books include some material about nonlinear filtering (such as median filters) and mathematical morphology. Chapter 8 of Gonzalez & Woods [1993] presents a nice and simple introduction to many useful mathematical morphology algorithms. The reader is also referred to Fisher et al. [1996] for a quick introduction to many useful algorithms. Additional references more specialized on these issues include Dougherty & Lotufo [2003]; Dougherty & Astola [1994, 1999]; Heijmans [1994]; Pitas & Venetsanopoulos [1990]; Soille [1999] as well as the classical books Matheron [1975]; Serra [1982]. The papers of Bloch & Maitre [1995]; Haralick et al. [1987]; Serra [1986, 1989] can also be of interest, as they develop in a survey fashion. As far as shape representation is concerned, the papers of Pitas & Venetsanopoulos [1990, 1992] present a morphological approach to shape decomposition, while Zois & Anastassopoulos [2000] explore morphology-based feature vectors.

It is worth noting that, although mathematical morphology is widely applied as a pre-processing tool, it is by no means limited to this kind of problem. In fact, mathematical morphology has been successfully applied to a wide range of problems, such as shape decomposition (as mentioned above), pattern association and recognition [Barrera et al., 1997], shape detection and spatial reasoning [Bloch, 1999], gray level image processing [Gonzalez & Woods, 1993], image segmentation (especially the watershed algorithm and the Beucher-Meyer paradigm) [Meyer & Beucher, 1990], multiscale analysis [Chen & Yan, 1989; Jackway & Deriche, 1996; Sapiro & Tannenbaum, 1993], and many others (see also [Goutsias & Batman, 2000]).

Additional resources: *Interactive image segmentation using watershed*

A watershed-based image segmentation tool is available at `http://watershed.sourceforge.net/`.

3.6 | Further Image Processing References

This chapter has presented an overview of the main approaches to image processing as used for image analysis. Most of the topics discussed are important and represent broad issues on their own, being covered in several books and in-depth tutorials in the scientific literature. Nevertheless, we hope that the present chapter, together with the mathematical introduction of Chapter 2, can provide the necessary background to the understanding of the shape analysis tools and applications discussed in the subsequent chapters. There are a great deal of books on image processing and computer vision, many of them being released or re-issued every year, therefore it is almost impossible to provide a complete list. Some of the classical and representative references are commented upon in the following box.

To probe further: *Image Processing and Correlated Areas*

Good introductory textbooks to image processing that have become classical references include the Gonzalez & Wintz [1987] and its respective revised edition Gonzalez & Woods [1993], as well as Castleman [1996]; Jain [1989]; Pitas [2000]; Pratt [1991]; Rosenfeld & Kak [1982]; Schalkoff [1989]. Additional texts that may be of interest are Baxes [1994]; Bracewell [1995]; Jähne [1997]; Lim [1990]; Pavlidis [1977]; Russ [1995]. The latter is a book that largely explores the signal processing approach applied to image processing, including the Z-transform, finite and infinite impulse response filters, and other potentially useful tools. In the context of signal processing and Fourier analysis, the reader can refer to Bracewell [1986]; Brigham [1988]; Morrison [1994]. The book by Glasbey & Horgan [1995], as well as van der Heijden [1994], introduce image processing concepts under an applied point-of-view with respect to the biological and natural sciences.

Image processing is twice a multidisciplinary field. Not only do its methods draw from several different areas (signal processing, statistics, differential geometry, numerical analysis, and so on), but its applications extend through a wide variety of problems, from telecommunications, medicine and biology to archeology and engineering, to name a few. As a consequence, non-experts in image processing techniques often experience difficulties while implementing some algorithms. With this respect, there are several good textbooks in the scientific literature that emphasize the algorithmic aspects of the techniques, being therefore useful for implementation, to which the reader can refer [Crane, 1996; Myler & Weeks, 1993; Parker, 1997; Phillips, 1997; Ritter & Wilson, 1996].

Some interesting books emphasize computer vision and image analysis-related issues, though with some intersection with the aforementioned texts on image processing. In this context, it is worth referring to Ballard & Brown [1982]; Haralick & Shapiro [1993]; Horn [1986]; Jain et al. [1995]; Levine [1985]; Marr [1982]; Trucco & Verri [1998]; Wechsler [1990]. More recently, the interface between computer graphics and computer vision has been narrowed, mainly because of some modern applications such as virtual reality and human–machine interaction. The paper by Lengyel [1998] provides a brief but compelling introduction to this topic. See also Velho et al. [2008] for image processing techniques applied to graphics and vision.

Additional resources: *Image databases*

Useful image databases may be found on the Web such as

- The Berkeley Segmentation Dataset and Benchmark:

  `www.eecs.berkeley.edu/Research/Projects/CS/vision/grouping/segbench/`

- Caltech 101:

  `www.vision.caltech.edu/Image_Datasets/Caltech101/`

- Caltech 256:

  `www.vision.caltech.edu/Image_Datasets/Caltech256/`

- The PASCAL Object Recognition Database Collection:

  `pascallin.ecs.soton.ac.uk/challenges/VOC/databases.html`

Shape Concepts

Chapter Overview

THE OBJECTIVE OF THE PRESENT CHAPTER is to identify and discuss in some detail the main concepts and issues in shape analysis, concentrating on continuous shapes and addressing shape characterization and representation, operations between shapes, shape metrics, position and size normalization, transformations, equivalence and similarity under transformations, besides shape recognition and classification. These concepts are extended to spatially quantized shapes in the subsequent chapters.

4.1 | Introduction to Two-Dimensional Shapes

However ordinary it could appear at first sight, the precise meaning of the word *shape*, as well as its synonym *form*, cannot be so easily formalized or translated into a mathematical concept. It is that kind of word that is promptly understood by many, but which can be hardly defined by any ("I do not know what a shape precisely is, but I know when a see one"). As well observed by Dryden & Mardia [1998], people tend to express and characterize shapes in a relative manner in terms of similarities and metaphors, bypassing a more formal definition. Therefore, the moon is round like a cheese, a doughnut is similar to a ring, and so on. At the same time, the importance of shapes to humans has been so great that it has been extended as a metaphor even for non-visual concepts, such as in the word trans*form* and the expression *to lick into shape* (see Section 1.1 for additional related examples). It is this broad and varied use of the term *shape* that makes it particularly difficult to be defined, for a good definition should be fully faithful to the human concept of shape.

But, why bother so much about defining shape? The point is that if we want to analyze shapes successfully, we must precisely know what they are, what properties they exhibit, and, often, to take into account the way they are perceived by humans. One of the most useful ideas in science has been the unification of several concepts. In physics, for instance, Newton's laws provide the key to understanding not only the fall of apples and other mechanical phenomena in our world, but also the rules governing the orbits in the solar system. Indeed, *all* formulations and concepts in classical mechanics (in opposition to quantum mechanics) can be subsumed to the three Newton's laws. To unify is important in science because it not only makes explicit the essence of the involved concepts, but also provides a nice means for generalizing over many different situations. Those interested in better understanding the importance of unification in science should refer to the interesting book [Weinberg, 1994].

The attempts at defining shapes usually found in the related literature are often based on the concept of invariant properties of an object or data set to translation, rotation and scaling (e.g., [Dryden & Mardia, 1998; Small, 1996]). While such definitions manage to capture an important property of shapes as perceived by humans, namely their invariance to basic geometric transformations (typically similarity transformations), they do not clearly specify what an "object" or "data set" is. Moreover, it is felt that invariance to transformations is not an essential feature in the definition of shape, but a property that may not even be limited to translation, rotation, and scaling. For instance, objects that have undergone other types of transformations (e.g., affine transformations), projections, deformations, occlusion, noise, and so on, are often recognized by humans as having the same shape. An object and its reflection by a mirror, for example, are usually understood as having the same shape. Moreover, a generic type of shape, such as a neural cell, can include an infinity of rather varying shapes that, however, present some common characteristic (in the case of neurons, many branches). An interesting approach to shape theory in terms of *pre-shapes* and transformations can be found in [Kendall et al., 1999]

In this book, we adopt a definition of shape which tries to take into account what the "object" is and to allow (or not) shape equivalence under a generic class of transformations. Regarding the former aspect, a shape is henceforth understood as any *single* visual entity, which is possibly the most essential feature underlying the concept of shape. In other words, whenever we talk about shapes, we are considering some whole entity or "object." Fortunately, the concept of "single," "whole" and "united" can be nicely formalized in terms of the mathematical concept of *connectivity* which leads to the following definition:

> **SHAPE** *is any connected set of points.*

Observe that the above definition includes shapes both in continuous and discrete (spatially sampled) spaces, provided suitable respective definitions of connectivity are considered. While the former types of shapes are addressed in this chapter, the discrete case is covered by the subsequent chapters in this book. By being quite generic, the above definition is compatible with many alternative approaches and still manages to capture the key concept underlying a shape. Moreover, it says nothing about invariance to specific transformations, since it is felt that such invariances are related to the *equivalence* and *similarity* between shapes, an issue that will be pursued separately in Section 4.6.

A possible problem with the above definition of shape is that it fails to identify some structures normally understood as shapes, such as the two examples in Figure 4.1. However, the face in 4.1 (a) should not be understood as a shape, but

(a)   (b)

Figure 4.1: *Two examples of possible "shapes" not satisfying the definition adopted in this book. See text for discussion.*

actually as a *set of shapes*, also referred to as a *composed shape* or a shape with multiple parts (see Chapter 9). As a matter of fact, a face is usually understood as a shape because its 3D version actually is a connected set of points (a surface). In Figure 4.1 (b), an originally connected set of points (and therefore a shape) has been reproduced with some graphical problem causing it to break apart into two connected sets of points. While some of us might still classify such a structure as a shape, it is actually more precise to identify it as a composed shape, which makes clear that it is broken. This latter example makes it clear that shapes, as defined in this book, are concepts corresponding to geometrical abstractions that may never be perfectly represented in the real world, a fact that had already been realized by Plato while proposing his world of ideas. The shape in Figure 4.1 (b), for instance, is usually understood as a shape because it is first identified in terms of the respective fully connected archetype, very likely according to perceptual principles (e.g., Gestalt). The same is true when shapes are represented into digital images (discrete shapes). In this case, as discussed in Section 3.1.4, since continuous shapes are transformed into finite sets of isolated points of the orthogonal grid, alternative concepts of connectivity are required (e.g., in terms of the 4- and 8-neighborhoods).

4.2 Continuous Two-Dimensional Shapes

Most shapes to be considered in the following chapters have in common the property of being *spatially quantized*, or *discrete*, as a consequence of the capture of continuous objects from our world by cameras, scanners and other acquisition devices (see Section 3.1.1). Therefore, a discrete shape will correspond to a finite set of points represented in the orthogonal lattice, such as is the case with most digital images (in some situations a hexagonal grid may be adopted instead of the orthogonal lattice). While we could have proceeded directly to discrete shapes, it is felt that a more solid and comprehensive understanding of the essence of shapes and shape-related issues including invariance, transformations and characterization can be achieved by starting with continuous shapes as connected sets of points in *continuous* spaces. Indeed, most discrete shapes can be understood as the result of sampling some continuous shape according to a pre-defined quantization scheme (see Section 3.1.4). Since the choice of such a scheme defines many of the properties of the respectively obtained discrete shapes and possibly affects the properties of continuous shapes (such as invariance to rotation) in particular ways, the continuous approach provides a more unified and uniform treatment of shapes and their properties.

4.2.1 Continuous Shapes and Their Types

As discussed above, connectivity is herein considered to be *the* essential feature characterizing shapes. As will soon become clear, connectivity corresponds to a well-defined, although not trivial, mathematical concept. Informally speaking, it indicates that any two points inside a given set can be reached through at least a path fully contained in that set. Figure 4.2 presents several examples of planar (i.e., 2D) sets of points that satisfy the shape definition in the previous section. The reader should have no difficulty verifying that all these shapes correspond to connected sets.

An interesting approach to a more formal treatment of connectivity involves the concept of *transformations*. Indeed, it can be shown (e.g., [Buck, 1978; Chinn & Steenrod, 1966; Kahn, 1995; Mendelson, 1990; Wall, 1972]) that a continuous transformation (see Sections 2.1.2 and 2.4.1) maps a connected set into a necessarily connected set. For instance, the transformation of scaling a shape by an isotropic factor a (i.e., the same scaling is observed in any orientation) is a continuous transformation that will always take a shape into another shape, not violating the connectivity constraint. However, it is important to bear in mind that this is also possible for discontinuous transformations. For instance, as shown in [Buck, 1978], the open unit square can be mapped by a discontinuous function, in a one-to-one fashion, into the interval [0, 1].

As hinted above, an alternative approach to the concept of connectivity can be developed in terms of the existence of *paths* connecting any two points of a set. Two types of paths are usually considered: polygonal and continuous. A *polygonal path* is defined as a sequence of connected straight line segments, i.e., straight segments sharing their extremities, as illustrated in Figure 4.3 (a). In case there is a polygonal path, with all its points contained in the set S, linking any two points P and Q of S, then S is said to be *polygon connected*. An open set S (see

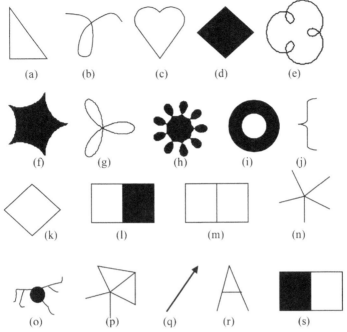

Figure 4.2: *Examples of planar continuous shapes.*

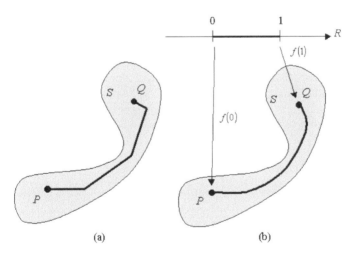

Figure 4.3: *A polygonal path (a) and a continuous path (b) connecting two points P and Q in a set S.*

the box entitled *Elementary Topological Concepts*) is *connected* if and only if it is polygon connected. A *continuous path* between two points P and Q is defined by a continuous mapping $f(t)$, $0 \leqslant t \leqslant 1$, such that $f(0) = P$ and $f(1) = Q$, as illustrated in Figure 4.3 (b). Observe that a polygonal path, including a straight line segment, corresponds to a particular case of a continuous path, and that two continuous paths can be merged at their extremities. A set S is *pathwise connected* in case any two of its points P and Q can be joined by a continuous path entirely contained in S. It can be shown (e.g., [Buck, 1978]) that any pathwise connected set is connected. Parametric curves, for instance, are connected sets. However, note that there are connected sets that are not pathwise connected (see [Buck, 1978] and [Wall, 1972] for examples). We also observe that a convex set[1] (including a single isolated point) is necessarily connected.

At this point we limit our study of thin shapes to those corresponding to pathwise connected sets. In addition, the following developments are observed *not to be completely formal*, although we try to keep the most important mathematical concepts and results in mind. Let us now return to Figure 4.2 and observe that generic shapes can be grouped into two main classes: those including or not including filled regions, which are henceforth called *thick* and *thin* shapes, respectively. Examples of the former include the shapes in Figure 4.2 (d, f, h, i, l, o, q, s), all the others being thin shapes. In addition, the trace (see Section 2.3.1) defined by any connected parametric curve can also be verified to be a thin shape. More formally, a thin shape can be defined to be a connected set that is equal to its boundary (see the box entitled *Elementary Topological Concepts*), which is never verified for a thick shape. Consequently, any thin shape has infinitesimal width (i.e., it is extremely thin).

Note: *Elementary Topological Concepts*

Let S be a subset of R^2, as shown in Figure 4.4. A generic point $P \in R^2$ can be an *interior, exterior* or *boundary* point of S. The point $P = (x_P, y_P)$ is said to be *interior to S* if there exists at least a real value $r > 0$ such that the open ball $B_r(P)$, i.e., the points (x, y) satisfying $\sqrt{(x - x_P)^2 + (y - y_P)^2} < r$, is entirely contained in S. This is illustrated in Figure 4.4. If all the points in a set $S \subset R^2$ are interior to S, this set is said to be *open*. A set S is said to be *closed* if its complement (i.e., the set of elements that are not in S) is open.

If a point Q is not an element of S, and there is at least an open ball centered at Q not contained in S, then Q is said to be an *exterior point to S*. A point T that is neither interior nor exterior is called a *boundary point of S*. In this case, any open ball centered at T must contain at least a point of S and a point not belonging to S. The points Q and T in Figure 4.4 are an exterior and boundary point of S, respectively. The *closure* of a set S is the set defined by the union of S with its boundary. Observe that the boundary of an open set is never contained in that set.

---

[1] A convex set is such that the straight line segment defined by any of its points is also contained in the set.

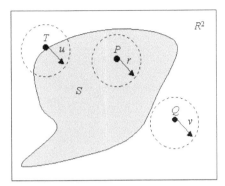

Figure 4.4: *A set S and examples of interior (P), exterior (Q) and bound-ary (T) points.*

Given a generic continuous thick shape, its *edge detected* version (i.e., its cor-responding thin shape) is defined as the set of its boundary points. For instance, the shape in Figure 4.2 (m) corresponds to the edge detected version of the shape in Figure 4.2 (l). It should be observed that edge detecting a thick continuous shape implies some of its information to be lost. For instance, the shape in Fig-ure 4.2 (l) cannot be completely recovered from the shape in Figure 4.2 (m), since it is impossible to know which hole to fill. As every point in a thin shape must be a boundary point, such shapes can be thought of as being defined by a number of connected continuous paths, as shown in Figure 4.5. It is also clear from this figure that a generic thin shape, in addition to a series of pathwise connected curves, typ-ically presents some special points, such as *extremities*—marked by small squares in Figure 4.5 (a), and *intersections* or *branches*—marked by small circles. A pos-sible suitable representation of such a shape can be achieved in terms of *graphs*, which are data structures involving *nodes* and *arcs* connecting such nodes [Bol-lobás, 1979; Bondy & Murty, 1976; Chartrand, 1985]. Figure 4.5 (c) illustrates the representation of the shape in (a) in terms of a non-oriented graph (i.e., the arcs have arrows at both their sides). Observe that the nodes in (c) correspond to branches or extremities in (b), enumerated in an arbitrary fashion, and the arcs correspond to the curve segments, labeled with arbitrary letters. Chapter 9 covers the topic of shape representation by graphs in more detail.

It is clear that such a kind of shape representation allows us to treat most shapes in terms of curve segments corresponding to the portions of the shape comprised between two nodes. Indeed, even thick shapes, after being edge detected, can be almost completely understood as a collection of connected curve segments. Given the special importance of connected curves, which can often be represented as para-metric curves, special attention is therefore paid to the characterization and analysis of curves and shape boundaries.

Figure 4.6 summarizes a possible classification of planar shapes. The shapes are first divided as thin and thick, the former being further subdivided into shapes

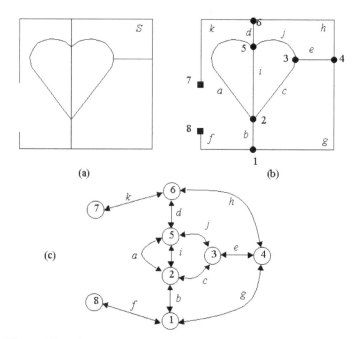

Figure 4.5: *A thin shape S **(a)**, its labeled branches and curves **(b)**, and the representation of S as a graph **(c)**, where each node corresponds to an extremity or branch point, and each arc corresponds to a parameterized curve.*

involving a *single* or *composed parametric curves*, the latter being characterized by containing several merged parametric curves, such as in Figure 4.5 (a). The single segment case is already familiar to the reader since, in practice, it corresponds to the parametric curves covered in Chapter 2. As shown in the hierarchical taxonomy in Figure 4.6, single curve shapes can also be classified as being *open* or *closed*, *Jordan* or not (a Jordan curve is such that it never intersects itself), *smooth* or not (a smooth curve is such that all its derivatives exist), and *regular* or not (see Section 2.3.2). Observe that thick shapes can be almost completely understood in terms of their thin versions, and composed parametric curves can be addressed in terms of single parametric curves.

4.3 | Planar Shape Transformations

Planar shapes, represented in terms of a connected set $S \subset R^2$, can be mapped by a generic transformation T into its respective image $Q = T(S)$, as illustrated in Figure 4.7.

In the most general case, a transformation can take the shape into virtually

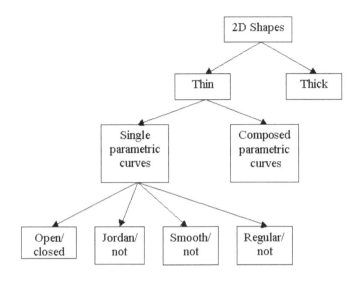

Figure 4.6: *Some of the principal types of shapes. See Section 2.3.2 for additional information about the subclasses of single parametric curves.*

Figure 4.7: *A shape S is mapped by the transformation T into its image Q = T (S).*

anything. Consider the following possibilities:

① $Q = \text{area}(S)$;

② $Q = (1, 1)$;

③ $Q = (\text{Rand}, \text{Rand})$, where Rand is the uniform random number generator;

④ $Q = 3$;

⑤ $Q = x^2, x \in R$;

⑥ $Q = S$ rotated of $30°$ clockwise with respect to the x-axis;

⑦ $Q = S$;

⑧ $Q = \begin{cases} \text{True if } S \text{ is a polygon} \\ \text{False otherwise} \end{cases}$;

⑨ $Q = (R, G, B)$, where R, G, and B are non-negative real values indicating the color of the shape.

It is clear that each of the above transformations has their intrinsic properties. For instance, both ② and ③ are from R^2 into R^2 and map the shape into a single point; transformations ① and ④ are from R^2 into R; ⑤ takes a shape in R^2 into a specific real function; and ⑧ takes S into a Boolean value indicating a property of the shape. The transformations ⑥ and ⑦ are special in the sense that they take a shape into another shape. As a consequence of such diverse properties, the useful-ness of the above transformations varies broadly. For instance, there is little use for transformations ② and ④, since they map *any* shape into constant values, thus hav-ing no discriminative properties. Although transformation ③ can (very probably) produce a different point every time it is applied, it is also of little interest, since it does not take into account any of the intrinsic properties of shape S. Therefore, let us now concentrate on those transformations that are particularly useful in shape analysis. As the reader shall agree, out of the above examples, transformations ①, ⑥, ⑦ and ⑧ can be included into this class of transformations. More specifically, two classes of transformations will receive special attention in this book: those mapping shapes into useful measures; i.e., $T : R^2 \rightarrow R^n$, henceforth called *feature transformations*; and those mapping shapes into shapes, i.e., $T : R^2 \rightarrow R^2$, which are henceforth called *morphic transformations*. Observe that such transformations can take a composed shape into composed shape, a single shape into a composed shape, and a composed shape into a single shape. Figure 4.8 illustrates a possible morphic transformation taking a shape into another.

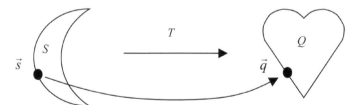

Figure 4.8: *The set of points composing shape S, including $\vec{s}$, are mapped by the morphic transformation T into a new shape $Q = T(S)$.*

Morphic transformations are particularly interesting because they allow us to relate different shapes. As a matter of fact, to identify the transformation taking one shape into another corresponds to learning almost everything about the relationship between both of these shapes. The transformation itself, as well as its properties (such as local magnification and phase, to be discussed in Section 4.9), often pro-vides important insights about the physical processes relating distinct versions of a shape. For instance, the uniform distortion of a square into a rhombus, under the action of some parallel but opposing forces applied at two opposite vertices, can be verified to be representable by an affine transformation. Observe, however, that the interpretation of the obtained transformation taking a shape into another shape can only be properly validated and understood when considered in the light of additional information about the possible processes acting over the shapes, and the physical properties of the latter. Morphic transformations are discussed in more depth in Section 4.9.

We conclude this section with an example of a morphic transformation, illustrated in Figure 4.9 in terms of its x- (a) and y- (b) scalar field components (refer to Section 2.4.1), that transform a single shape (c) into the composed shape in (d).

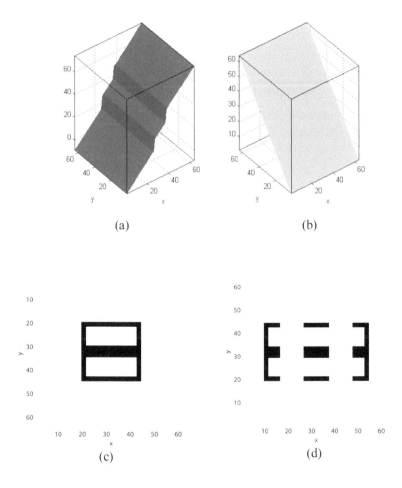

Figure 4.9: *The x- (a) and y-scalar (b) fields composing a discontinuous morphic transformation capable of slicing a shape, such as that in (c), into three parts (d).*

4.4 Characterizing 2D Shapes in Terms of Features

Given a specific shape, or a composed shape, it is often necessary to quantify some of its properties. This task, which constitutes one of the basic steps in shape classification, is henceforth called *shape characterization*. The respective *characterization* of a shape S can be made in terms of a series of its respective measures and properties, which are commonly referred to as *features*. For instance, a shape can be characterized in terms of its area, the total arc length of its boundary (i.e., its perimeter), its number of holes, its number of extremities, the fact that it belongs to some class (such as polygons) and so on. Most features can be thought of as corresponding to some real, complex, integer or Boolean scalar or vector. As a matter of fact, even properties exhibited by shapes, such as the fact of being a polygon, can be represented in terms of Boolean features. Such considerations lead us to the following definition of shape characterization:

Definition (Shape characterization). *To CHARACTERIZE A SHAPE = to measure a number of its features and properties.*

It follows from the above discussion that the process of shape characterization can be understood as involving a series of P transformations T_i taking the shape into a series of scalar measures or features F_k; $k = 1, 2, \ldots, M$; as illustrated in Figure 4.10, which can be combined into a *feature vector* $\vec{F} = (F_1, F_2, \ldots, F_M)$.

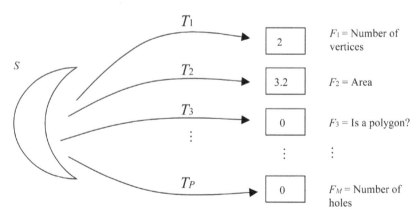

Figure 4.10: *A generic shape S can be transformed into a series of feature values F_i, which define a feature vector. Observe that the value 0 for the third feature is understood as "false."*

It should be observed that the number P of transformations is not necessarily the same as the number M of features, since a single transformation (for instance, the transformation that takes the shape into its center of mass) can produce more than one scalar feature. Thus, in general $M \geqslant P$. Chapters 6 and 7 present several possible shape features and the means for their respective estimation.

While no additional restriction could, at least in principle, be imposed onto a feature, except that it must assume some scalar or vector value, it is important that the considered features emphasize the property of interest and/or have good discriminative power, while not being too much sensitive to noise and artifacts. For instance, in case we are interested in characterizing polygons, the number of sides is a much more significant feature than, for instance, the number of holes, and it soon becomes clear that the choice of features has to take into account the types of shapes under consideration. It is important to observe that sometimes the features can also be defined by some specific properties required from the shapes, and not by the shapes themselves. For instance, one might be interested in obtaining measures so as to find out if a shape fits inside a square of side L. In this case, the relevant feature to be considered, i.e., its diameter, involves considering the distances between several of the points in the shape. Chapter 8 provides further discussion on how to select features for shape classification.

We have seen in the above discussion that features, represented in terms of a feature vector, can be associated with shapes. It is important to note that feature vectors live in a respective *feature space*. For instance, a feature vector composed of M real scalar measures is a vector in R^M. Although it is possible to consider heterogeneous feature spaces, defined by feature vectors involving more than one type of scalar (such as real, Boolean, etc.), most of the situations in this book will be restricted to real feature spaces, i.e., R^M. This important concept is illustrated in Figure 4.11, where a shape has been measured with respect to its area and perimeter, allowing it to be mapped into the feature space (perimeter) × (area), which is clearly a subset of R^2.

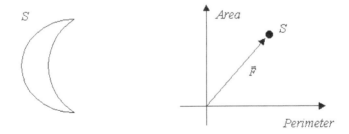

Figure 4.11: *The S shape mapped into the feature space defined by area and perimeter.*

It is generally expected that similar shapes will be mapped into feature vectors (i.e., points in the feature space) that are close to each other. Such a property is often implied by the continuity of the transformation taking the shape into its respective features. In case this transformation is continuous and the shapes define a continuum of variations, the respectively obtained representation in the feature space produced by a continuous feature extraction transformation will also be connected. The concept that similar shapes have similar features is illustrated in Figure 4.12,

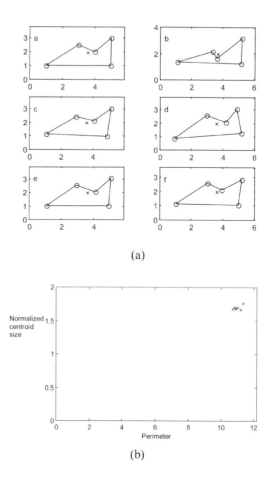

(a)

(b)

Figure 4.12: *A set of similar shapes **(a)** and their respective character-*
izations in the feature space (Normalized centroid size) ×
*(Perimeter) **(b)**. The similarity between shapes is nicely cap-*
tured by the fact that the respective feature vectors are close
to one another.

which shows a series of feature points (b), respective to the normalized centroid size (see Section 4.8.3) and perimeter of the similar polygonal shapes (a). It is clear that the feature points are close to each other, starting to form a *cluster* in the feature space. On the other hand, a discontinuous feature extraction transformation will tend to imply disconnected regions in the feature space.

When the features are properly chosen, the clusters respectively defined by the different types of shapes are expected to be well separated in the feature space. It should be also observed that the process of selecting adequate features is not easy, and should take into account not only each specific problem, but all the available information about the involved shapes' peculiarities and variability.

4.5 Classifying 2D Shapes

One of the central issues in shape analysis is the classification of shapes. More specifically, given a shape we have to find out to which class it belongs. While classification is often naturally performed by humans, it is in fact a rather difficult task, deserving special attention in the present book. While this issue is covered in detail in Chapter 8, an introduction to the basic concepts in shape classification are outlined in the present chapter in order to provide a sound overall idea of the most important shape-related issues.

Figure 4.13 presents the main elements involved in pattern classification. The

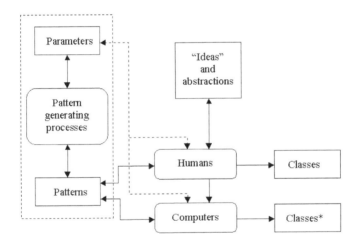

Figure 4.13: *Basic interactions in pattern recognition. See text for explanation.*

dashed box represents pattern generation in nature, which is controlled by a set of parameters. For instance, the length of a leaf is influenced by the duration of the summer season, the shape of the skull is influenced by genes, etc. Humans interact with patterns in nature, measuring their properties and identifying some of the parameters that produced them, which is used to organize the patterns into classes. This process is influenced by human abstractions such as mathematics and geometry, involving shapes that only exist in perfect fashion in Plato's world of ideas (such as a perfect triangle). The process of automated pattern classification involves the use of computers to process information about the patterns, which is typically done in terms of specific measures of their properties. Particularly, in the case of pattern classification based on visual information, an image of the object to be classified is fed as input to the computer, which derives suitable measures (features) and then classification methods are applied to such measures in order to determine the correct class of the original object. Since we do not yet know how classification is performed by humans, the classes obtained by computers may not agree with those produced by humans, which is indicated by the asterisk. Observe also in Figure 4.13 (dashed lines) that it is interesting that both humans and computers have access to the parameters controlling the generation of the shapes, since this allows more information about the shapes, therefore enhancing the chances of correct classification. For example, the knowledge that a particular leaf comes from a sunny region can help explain its length.

An important aspect related to shape classification regards the fact that classification tends to group together shapes that are similar to one another. Indeed, as discussed in the previous section, similar objects tend to yield feature vectors that are close to one another, leading to the formation of clusters in the feature space. Much of shape classification is performed by taking into account this tendency. However, as further discussed in Chapter 8, there are situations in which relatively dissimilar objects are grouped together as a consequence of an arbitrary definition of the classes. In addition, observe that the choice of poor features for shape characterization can lead to incorrect classifications.

4.6 Representing 2D Shapes

Henceforth in this book, *to represent* a shape shall mean to obtain a set of features characterizing it to such a degree that it becomes possible to reconstruct the original shape from the respective features. In other words, the representation of a shape can be understood as an invertible (or complete) transformation. This important concept is emphasized below.

Definition (Shape representation). *To* REPRESENT A SHAPE = *to characterize the shape in terms of a set of features in such a way that it becomes possible to reconstruct it, exactly or to a certain degree of precision, from its features.*

Observe that, according to the definitions in the previous and current sections, to represent a shape corresponds to a special case of shape characterization discussed in the previous section, but involving a more complete set of features. When a bijective correspondence is verified between the shapes and the respective feature vectors, the transformation taking the shape into the feature vector can be inverted. It should be borne in mind that this bijective mapping between objects and properties represents one of the most important basic concepts in the area of pattern analysis and recognition, ultimately allowing us to define *what an object is*. In other words, each object (including shapes) is fully characterized in terms of all its respective properties, possibly represented as a feature vector. An immediate important consequence of this principle is that two objects will be the same whenever all their properties are the same, providing a clear definition of *equivalence* between objects. Furthermore, it is also reasonable to expect that similar objects have similar properties, in such a way that the *similarity* between objects can be quantified in terms of the distances between their respective feature vectors, with respect to some adopted metrics. These important concepts are summarized in Table 4.1.

A *shape*	Defined bijectively by all its properties.
Equivalence of two shapes A and B	All their properties are equal. In other words, their respective complete feature vectors are identical, i.e., $\vec{F}_A = \vec{F}_B$.
Similarity between two shapes A and B	Their properties are similar, implying the distances between their feature vectors to be reasonably small, i.e., $\left\| \vec{F}_A - \vec{F}_B \right\| \leqslant \varepsilon$.

Table 4.1: *Shape equivalence and similarity.*

Although bijective correspondences between shapes and feature vectors constitute the ideal situation in shape representation, this is not always the case we encounter in practice. The problem is that, as already observed in the previous section, the process of selecting effective features for representation must rely on previous experience, intuition, the peculiarities and variability of the considered shapes, trial-and-error experiments and other inexact methods. As a matter of fact, there is no definite and automatic procedure for feature selection. In practice, the best alternative, when exact representation is not possible, is to seek features allowing the original shape to be reconstructed to a reasonable degree of precision. This situation, which is opposite to the ideal *exact representation*, will be called an *approximated representation*. More formally, we will say that a feature vector $\vec{F}$ provides an approximated representation of the shape S in case the respective reconstructed version $\tilde{S} = T^{-1}\left(\vec{F}\right)$ is such that $\mathrm{dist}\{S, \tilde{S}\} = \left\| S - \tilde{S} \right\| \leqslant \varepsilon$, where ε is the parameter representing the maximum allowed error in the reconstruction, which may vary from case to case. In case $\varepsilon = 0$ we say the representation is exact.

While a more careful discussion about how distances between shapes can be defined is postponed to the next section, the intuitive meaning of the above formulation as a measure of discrepancy between the original and reconstructed shapes should be clear enough for now. It is also clear that, typically, the larger the number of considered suitable features, the more complete the feature vector tends to be, allowing better and better characterization and reconstruction of the original shape. Informally speaking, this is analogous to a traditional crossword puzzle, where the chance of identifying the hidden word (i.e., the shape) increases with the number of questions that are answered about it (i.e., the number of features).

The main reason why one is typically interested in representing shapes is for compacting the amount of data needed for storage and/or analysis of the shapes. Since shapes are sets, the most natural and immediate way to represent them is in terms of their elements. Although an infinite amount of points is required to represent a continuous shape, its discrete version (see Section 3.1.4) can be immediately represented by listing all their constituent points. However, there are more effective alternatives for shape representation, which are outlined in the following with respect to continuous shapes; each of these strategies being easily extensible to discrete shapes (see Chapter 5). The first alternative is to recall from the previous section that thin shapes can be decomposed into a number of continuous paths corresponding to parametric curves, which would serve as an effective representation of the respective shapes. On the other hand, thick shapes can be represented in terms of their boundaries (which are thin shapes) plus an indication about the regions to be filled. For instance, we can represent a disk by specifying its boundary in terms of the respective circle equation and indicating that this circular region should be filled. Such possibilities, which have frequently been considered in computer graphics, especially in Constructive Solid Geometry—*CSG* [Foley et al., 1994], shall not be treated in greater depth in the present text. Indeed, it will suffice to observe that it is possible to represent most thin and thick continuous shapes in terms of connected parametric curves. Yet another possibility for shape representation is to use one of the many complete transformations (i.e., invertible), such as Fourier or Hadamard. Our interest here concentrates on two principal types of shape representations: (i) in terms of general morphometric measures (i.e., features) such as area, perimeter, complexity, number of vertices, etc.; and (ii) in terms of landmark points. These two particularly relevant situations are described in more detail in the following subsections.

4.6.1 General Shape Representations

A useful approach to understanding shape characterization, representation and classification is developed in terms of the mathematical concepts of *equivalence relations* and *partitions* (see the accompanying box entitled *Equivalence Relations and Partitions*).

Note: *Equivalence Relations and Partitions*

Given a set A, a *relation* between the elements in this set is any subset of $A \times A$, i.e., the Cartesian product between A and A, corresponding to all the possible ordered pairs (x, y) such that $x \in A$ and $y \in A$. For instance, if $A = \{a, d, f\}$, then $A \times B = \{(a, a); (a, d); (a, f); (d, a); (d, d); (d, f); (f, a); (f, d); (f, f)\}$, and some of the possible relations are $\{(a, f); (d, a)\}$ and $\{(a, a); (d, d); (f, f)\}$. An *equivalence relation* is a relation that is *reflexive* (x relates to x), *transitive* (if x relates to y and y relates to z, then x relates to z) and *symmetric* (x relates to y if and only y relates to x). Equivalent relations are particularly interesting because every such relation defines a *partition* of a set A, and every partition of this set is associated to an equivalence relation. A partition of a set A is a collection of subsets A_i of A which do not intersect each other and whose union produces A. For example, a possible partition of the set $A = \{1, 2, 3, 4, 7, 8, 9\}$ is defined by the subsets $A_1 = \{1, 3, 8\}$; $A_2 = \{2, 9\}$;

and $A_3 = \{4, 7\}$. The relation between the elements of A defined by the property "to be even," i.e., the subset $\{(2, 2); (2, 4); (2, 8); (4, 2); (4, 4); (4, 8); (8, 2); (8, 4); (8, 8)\}$ of A, can be easily verified as being an equivalence relation, partitioning A into $\{2, 4, 8\}$ and $\{1, 3, 7, 9\}$.

Interestingly, the fact that a subset of shapes have identical (or similar) features defines an equivalence relation between such shapes that induces a partitioning of the shape space. Let us clarify this important concept through the following specific example: Let our universe of shapes be restricted to squares having two internal line segments (or bars), one horizontal and the other vertical, both appearing at any of the possible internal positions, as illustrated in Figure 4.14. Let $\frac{a}{L} = T_1(S)$ and

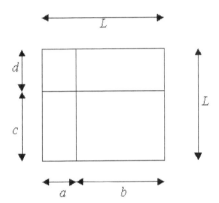

Figure 4.14: *A family of shapes defined by two parameters, namely the relative positions $0 \leqslant \frac{a}{L} \leqslant 1$ and $0 \leqslant \frac{c}{L} \leqslant 1$ of both a vertical and a horizontal internal segments contained in a square of side L.*

$\frac{c}{L} = T_2(S)$ be the transformations taking the shape S and producing as a result the relative position of the vertical and horizontal internal bars, i.e., $0 \leqslant \frac{a}{L} \leqslant 1$ and $0 \leqslant \frac{c}{L} \leqslant 1$. Figure 4.15 illustrates a series of shapes centered at the respective feature vectors in the $\frac{a}{L} \times \frac{c}{L}$ feature space.

Some important facts can be verified from Figure 4.15. First, it should be clear that in case only a single feature is taken into account, it will only be possible to distinguish between the shapes as far as the chosen feature is concerned. For instance, if we decide to characterize the shapes in terms of $T_2(S)$ only, we will not be able to distinguish between shapes differing by the position of the vertical bar. It is also clear from Figure 4.15 that as we reduce the region of interest in the feature space, the smaller the variation of the respective shapes will be. For instance, by restricting $T_2(S)$ to $0.25 \leqslant T_2(S) < 0.35$, as represented by the rectangle in Figure 4.15, we will be selecting only the shapes characterized by having the horizontal bar positioned inside this interval. In case the additional restriction

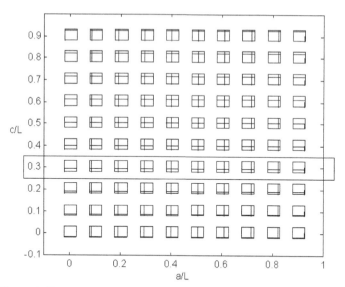

Figure 4.15: *Several instances of the family of shapes defined in Figure 4.14.*

$T_1(S) = 0.5$ is also imposed, the respective set of selected shapes will be further restricted and the representation will become more and more exact. In this simple example, exact representation will be achieved by specifying a single value for each of the two features. Observe, however, that such an exact representation accounts for the fact that the outer boundary of the shapes are always squares, in such a way that no additional features are necessary for representing the outline of the shapes. It should be noted that completely generic partitions of the feature space into regions are possible, including eventually disconnected regions.

4.6.2 Landmark Representations

One of the most natural and frequently used planar shape representations are known as *landmark points*. Although several types of landmarks have been proposed in the related literature (e.g., [Bookstein, 1991]), in this book we will distinguish only between the following two main types: (a) the intersection and extremity points of a shape (see Figure 4.5); and (b) salient points along parametric curves. Since the former are clearly defined, special attention is given to the second situation, namely the salient landmarks. More specifically, given a continuous planar shape S containing a single curve segment, which is an infinite set of points, a finite set of particularly meaningful and/or salient points can be identified and used for representation of the shape. As an interesting way to gain insight about what is meant by salience, suppose you had to select the smallest number of points along a curve so that afterwards you would be able to recover a reasonable reconstruction of the original shape [Fischler & Wolf, 1994]. It follows that landmark points are natural shape features, as they are related to shape reconstruction. Figure 4.16 illustrates landmark points $S_i = \left(S_{x,i}, S_{y,i} \right)$ allowing exact representation of a polygonal shape (a) and approximate representations of a closed parametric curve (b)–(d).

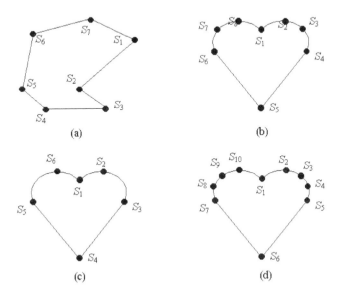

Figure 4.16: *Landmark points for two shapes. Observe that the represen-tation is exact in (a), but only approximate in (b–d).*

It is clear that it is often possible to choose several alternative landmarks for a given shape. For instance, more (or less) accurate representations can typically be obtained by considering more (or less) landmark points, as illustrated in Figures 4.16 (d) and (c), respectively. In the general case, the landmark approach implies some loss of information about the shape, in such a way that the transformation taking shapes to landmarks becomes degenerated and no longer invertible, conse-quently defining an *approximate representation*. However, it is always possible to obtain better representations by considering a larger number of landmarks and/or better placed landmarks. As a matter of fact, it is observed that the original shape is always perfectly represented when all its infinite points are taken into account as landmarks. The concept of landmarks can be immediately extended to discrete shapes (see Section 3.1.4). Although it is often difficult to identify good landmarks in an automated fashion, a particularly interesting alternative involving the calcula-tion of the shape curvature is described in Section 7.2. It is also possible, and even common, for the landmarks to correspond to some specific imposed constraint such as an anatomical reference point.

It is important to emphasize that proper landmark representations involve a careful consideration of the order of the landmark points. This can be achieved by storing the coordinates of each landmark point, starting from a fixed reference landmark, in sequential fashion into the feature vector, in such a way that the original shape can be recovered by joining each subsequent point, starting with the last and proceeding backwards to the first point, a process defining a *polygonal interpolation*. This type of strategy, henceforth called *ordered landmarks*, implies that the shape S is represented as the vector $\vec{S} = (S_{x,1}, S_{y,1}, S_{x,2}, S_{y,2}, \ldots, S_{x,n}, S_{y,n})$. This type of representation will be from now on called *2n-vector*.

An alternative approach where the order of the landmark points is not taken into account is henceforth called *free landmarks*, implying the shape to be represented in terms of the set $S = \{S_{x,1}, S_{y,1}, S_{x,2}, S_{y,2}, \ldots, S_{x,n}, S_{y,n}\}$, which not only does not allow exact representations, but also precludes any attempt to interpolate between the landmark points. Indeed, this less complete representation, which is henceforth called *2n-set*, should be considered only when the order of the landmarks is not necessary or available.

A third interesting approach, henceforth referred to as *planar-by-vector*, consists in representing any planar shape S in terms of the 2D vectors corresponding to each of its landmark points, which are stored in sequential fashion into an $n \times 2$ matrix, i.e.,

$$S = \begin{bmatrix} S_{x,1} & S_{y,1} \\ S_{x,2} & S_{y,2} \\ \vdots & \vdots \\ S_{x,n} & S_{y,n} \end{bmatrix}$$

Yet another possibility for shape representation, similar to the latter approach and called *planar-by-complex*, is achieved by using complex values to represent each landmark point, i.e., $S_i = S_{x,i} + jS_{y,i}$, in such a way that the whole set of landmarks can be stored into an $n \times 1$ vector $\vec{S} = (S_1, S_2, \ldots, S_n)$. Figure 4.17 summarizes the four above landmark representation schemes.

Although all the above schemes allow the exact representation of the landmarks (but not necessarily the shape), it is felt that the planar-by-vector and planar-by-complex alternatives are intuitively more satisfying since they do not involve mapping the shape into a $2n$-dimensional space. It is observed that the concept of landmark also provides a particularly suitable means for computational representation of continuous or discrete shapes, corresponding to a preliminary model of discrete shapes, an issue to be elaborated further in Chapter 5.

4.7 Shape Operations

Having discussed what a shape is and how it can be characterized and represented, we now address the interesting issue of defining operations between shapes. Since shapes are connected *sets*, it is natural to extend all the traditional operations between sets—including complementation, union, difference and intersection—in order to be applicable to shapes. Observe that the *difference between two sets* S and Q, represented as $S - Q$, is defined as the set of those elements of S that do not

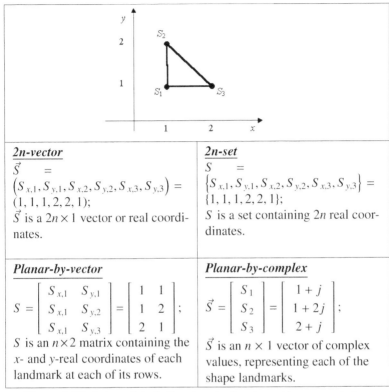

Figure 4.17: *A shape S including three landmarks and its representation in terms of 2n-vector, 2n-set, planar-by-vector and planar-by complex representations.*

belong to Q. These four intuitive and simple shape operations are illustrated in Figure 4.18.

In addition to such simple operations, it is interesting to define an *arithmetic of shapes*, hopefully leading to the concept of vector shape spaces. This is useful in practice because having shapes as elements in a vector space allows more structure and possibilities for shape analysis. As for the *2n-vector* shape representation, this can be straightforwardly done because here the shapes are associated with traditional vectors in R^{2N}, which is naturally a vector space endowed with addition and multiplication by a scalar, as well as norm, inner product and distance. The main problem with this approach, however, is that it involves shape representation in a high-dimensional space, which is counterintuitive. In this sense, both the planar-by-vector and the planar-by-complex representations provide a more interesting alternative to define shape arithmetics. Let two generic planar shapes S and Q be represented in terms of n respective landmark points S_i and Q_i; $i = 1, 2, \ldots, n$; as illustrated in Figure 4.16 (a) and (b), respectively. First, we define the product of

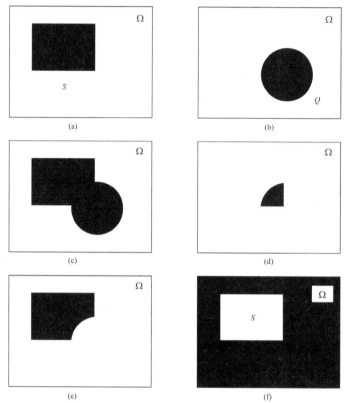

Figure 4.18: *Two shapes S and Q in a shape space Ω, respectively shown in (a) and (b), and their union (c), intersection (d), difference (e), and the complementation of S (f).*

the shape S by a scalar a as being the shape R, also containing n landmarks R_i, obtained as

$$R_{x,i} = aS_{x,i} \quad \text{and} \quad R_{y,i} = aS_{y,i}, \quad i = 1, 2, \ldots, n$$

Observe that all the landmark coordinates become zero for $a = 0$. The algebraic sum of S and Q can now be defined as the new shape R, also having n landmarks, given by

$$R_{x,i} = S_{x,i} + Q_{x,i} \quad \text{and} \quad R_{y,i} = S_{y,i} + Q_{y,i}, \quad i = 1, 2, \ldots, n$$

It is clear that both these operations are closed in R^2, producing as a result a planar shape also with n landmarks. Figure 4.19 illustrates two shapes S (a) and Q (b), and $\frac{Q}{2}$ (c), $-Q$ (d), $S + Q$ (e) and $S - Q$ (f). A similar approach can be easily developed for the case of planar-by-complex representation. Actually, in this case, the landmark points are all represented in the complex space (i.e., C), allowing even

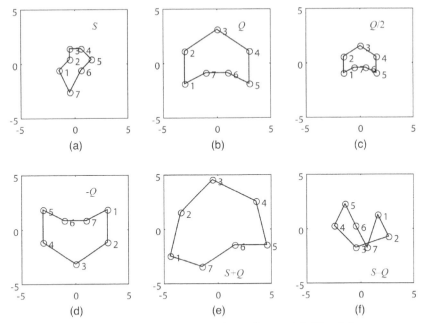

Figure 4.19: *Shape arithmetics: two shapes S (a) and Q (b) and the shapes $\frac{Q}{2}$ (c), $-Q$ (d), $S + Q$ (e) and $S - Q$ (f).*

more possibilities for shape arithmetics. For instance, it becomes possible to multiply two shapes, conjugate a shape and so on. A more comprehensive exploration of such possibilities is left as a project suggestion to the reader. Additional information about complex arithmetics, making intensive use of geometric constructions and visualizations, can be found in [Needham, 1997].

4.8 Shape Metrics

One of the most intuitive and simple attributes of a shape is its *size*. We have seen in Section 2.2 that the size of a vector is related to the concept of *norm*, which is a function (more precisely a functional) taking a vector into a scalar and obeying the following three conditions:

① $\|\vec{p}\| \geqslant 0$ for any $\vec{p} \in S$, and $\|\vec{p}\| = 0 \Leftrightarrow \vec{p} = \vec{0}$;

② $\|\alpha\vec{p}\| = |\alpha| \|\vec{p}\|$ for any $\vec{p} \in S$ and any real value α;

③ $\|\vec{p} + \vec{q}\| \leqslant \|\vec{p}\| + \|\vec{q}\|$ for any $\vec{p}, \vec{q} \in S$.

It is therefore natural to approach the problem of shape size in terms of norms, which can be understood as shape transformations mapping a shape into a scalar corresponding to its size. However, we will see that condition ③ is hard to meet in practice because the number n of landmarks frequently varies from shape to shape.

4.8.1 The 2n Euclidean Norm

Let us start our discussion of shape sizes by considering the *2n-vector* shape representation introduced in the previous section. In this case, it is only natural to consider the size of the plane shape S, represented in terms of n landmark points, as being equal to the Euclidean norm of $\vec{S}$, henceforth referred to as the *2n Euclidean norm* of the shape, i.e.,

$$\|S\| = \sqrt{\sum_{i=1}^{n} S_{x,i}^2 + \sum_{i=1}^{n} S_{y,i}^2} \tag{4.1}$$

While this definition is guaranteed, provided n is fixed, to satisfy all the three norm properties above, it will produce different values as the shape is shifted along the plane, producing larger values when the shape is further away from the coordinate system origin. Since it is clear that this behavior fails to capture our intuitive concept of shape size, which is invariant to translations, position normalization is required in order to achieve a more reasonable measure of size. The simplest solution is to move the shape S until its center of mass (or *centroid*), represented in terms of the vector $\vec{S}_0 = (S_{x,0}, S_{y,0})$, coincides with the origin of the coordinate system. Observe that the coordinates of the center of mass can be immediately calculated as

$$S_{x,0} = \frac{\sum_{i=1}^{n} S_{x,i}}{n} \tag{4.2}$$

and

$$S_{y,0} = \frac{\sum_{i=1}^{n} S_{y,i}}{n} \tag{4.3}$$

Now, the position of the shape S can be conveniently normalized by using the free-variable transformations $\tilde{S}_{x,i} = S_{x,i} - S_{x,0}$ and $\tilde{S}_{y,i} = S_{y,i} - S_{y,0}$. Figure 4.20 illustrates an original simple shape S (a) and its normalized version $\vec{\tilde{S}}$ (b).

Once the position of the shape is normalized, the $2n$ Euclidean norm (or size) of the shape S can be obtained as

$$\|S\| = \sqrt{\sum_{i=1}^{n} (S_{x,i} - S_{x,0})^2 + \sum_{i=1}^{n} (S_{y,i} - S_{y,0})^2} = \sqrt{\sum_{i=1}^{n} \tilde{S}_{x,i}^2 + \sum_{i=1}^{n} \tilde{S}_{y,i}^2} \tag{4.4}$$

Since the position normalization fixes the values of the coordinates of S whatever its original position, the above norm will now be invariant to translations of S. For simplicity's sake, it is henceforth considered that the shapes are position-normalized, i.e.,

$$\sum_{i=1}^{n} S_{x,i} = 0 \tag{4.5}$$

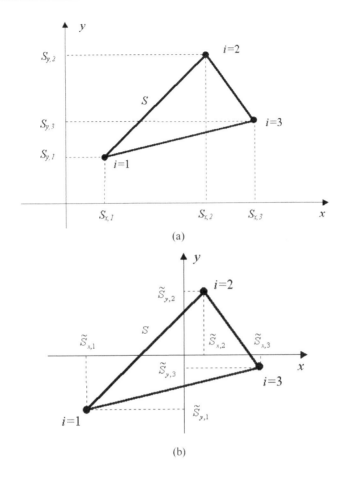

(a)

(b)

Figure 4.20: *A shape S **(a)** and its position-normalized version **(b)** obtained by translating the shape until its centroid coincides with the origin of the coordinate system.*

and

$$\sum_{i=1}^{n} S_{y,i} = 0 \tag{4.6}$$

Now, consider that shape S undergoes a rotation by an angle θ counterclockwise with respect to the x-axis, i.e.,

$$\begin{bmatrix} \hat{S}_{x,i} \\ \hat{S}_{y,i} \end{bmatrix} = \begin{bmatrix} \cos(\theta) & \sin(\theta) \\ -\sin(\theta) & \cos(\theta) \end{bmatrix} \begin{bmatrix} S_{x,i} \\ S_{y,i} \end{bmatrix} = \begin{bmatrix} S_{x,i}\cos(\theta) + S_{y,i}\sin(\theta) \\ -S_{x,i}\sin(\theta) + S_{y,i}\cos(\theta) \end{bmatrix}$$

By equation (4.2) the new coordinate $\hat{S}_{x,0}$ of the center of mass is

$$\hat{S}_{x,0} = \frac{\sum_{i=1}^{n} \hat{S}_{x,i}}{n} = \frac{\sum_{i=1}^{n} \left[S_{x,i} \cos(\theta) + S_{y,i} \sin(\theta)\right]}{n} = \frac{\cos(\theta) \sum_{i=1}^{n} S_{x,i} + \sin(\theta) \sum_{i=1}^{n} S_{y,i}}{n}$$

From equation (4.5) we have that

$$\hat{S}_{x,0} = \frac{\cos(\theta) \sum_{i=1}^{n} S_{x,i} + \sin(\theta) \sum_{i=1}^{n} S_{y,i}}{n} = \frac{\cos(\theta)(0) + \sin(\theta)(0)}{n} = 0$$

The reader is invited to verify that we also have $\hat{S}_{y,0} = 0$. Therefore, rotations do not displace the center of mass of the shape S from the origin of the coordinate system. In addition, we know from Section 2.2.5 that, since orthogonal transformations conserve the norm of vectors, the norm of the shape will be preserved by rotation. Therefore, the Euclidean norm of equation (4.1) has been verified to be invariant to translation and rotation, which constitutes an important practical property of this type of norm (for instance, the chessboard and city block distances are not rotation invariant). Observe that this type of shape size is consequently in full agreement with our intuitive concept of size.

Since size, as defined in equation (4.1), tends to become larger as the number n of landmark points increases, it is interesting to consider some type of normalization, which can be obtained by dividing the sums in equation (4.1) by the total number of coordinates, i.e., $2n$, yielding equation (4.7). Observe that this definition leads to the root mean square size, henceforth abbreviated as *RMS size*.

$$\left\| \vec{S} \right\| = \sqrt{\frac{\sum_{i=1}^{n} (S_{x,i} - S_{x,0})^2 + \sum_{i=1}^{n} \left(S_{y,i} - S_{y,0}\right)^2}{2n}} \tag{4.7}$$

4.8.2 The Mean Size

An alternative definition of the size of a shape S is the mean distance from each of the landmark points to its center of mass, i.e.,

$$\|S\| = \frac{\sum_{i=1}^{n} \sqrt{(S_{x,i} - S_{x,0})^2 + \left(S_{y,i} - S_{y,0}\right)^2}}{n} \tag{4.8}$$

This type of shape size can also be verified to be invariant to shape translation and rotation (the reader should verify this as an exercise). In addition, this definition is probably more intuitive than the $2n$ Euclidean norm, since it does not involve mapping the shape into a $2n$-dimensional space.

In the planar-by-complex shape representation, the mean size is given in equation (4.9).

$$\|S\| = \frac{\sqrt{Re^2\{S_1\} + Im^2\{S_1\}} + \sqrt{Re^2\{S_2\} + Im^2\{S_2\}} + \cdots + \sqrt{Re^2\{S_n\} + Im^2\{S_n\}}}{n}$$

$$(4.9)$$

4.8.3 Alternative Shape Sizes

Centroid Size This is defined as being the square root of the sum of the squared Euclidean distances between each landmark point and the shape center of mass, i.e.:

$$\|S\| = \sqrt{\sum_{i=1}^{n}\left[\left(S_{x,i} - S_{x,0}\right)^2 + \left(S_{y,i} - S_{y,0}\right)^2\right]}. \qquad (4.10)$$

Observe that, although defined in a different way, this size actually corresponds to the $2n$ Euclidean norm discussed in Section 4.8.1. As a consequence of its distinct definition, its normalized version becomes

$$\|S\| = \sqrt{\frac{\sum_{i=1}^{n}\left[\left(S_{x,i} - S_{x,0}\right)^2 + \left(S_{y,i} - S_{y,0}\right)^2\right]}{n}} \qquad (4.11)$$

which is slightly different from the RMS size (by the factor $\frac{1}{\sqrt{2}}$). This is one of the most frequently considered definitions of size in statistical shape analysis, being adopted, for instance, in [Bookstein, 1991; Dryden & Mardia, 1998; Kendall et al., 1999]. This shape size can be verified to be invariant to translations and rotations.

Baseline Distance In this case, which has been traditionally associated with the Bookstein coordinates (see, for instance [Dryden & Mardia, 1998]), the shape size is defined as being equal to the distance between the two initial landmark points. Despite its invariance to translation and rotation, and fulfillment of the norm conditions ① and ②, it is somewhat unstable since shape perturbations (such as noise or distortions) may imply a large variation of the overall shape size. In other words, it lacks the global nature of the previously discussed distances. In addition, the choice of the first two landmark points is also arbitrary, implying difficulties when comparing sizes of different shapes.

Shape Diameter This definition of size is similar to the baseline distance, except that the largest distance between landmark points is adopted instead. While the baseline distance avoids the arbitrary choice of the initial landmarks, it is also susceptible to perturbations.

4.8.4 Which Size?

Having discussed a series of possible alternative shape sizes, and considering that so many other sizes are possible (including chessboard and city block), we are left with the question of which size definition one should adopt for shape analysis. As it happens so often in applied sciences, there is no definitive option to be used in the generic case. Indeed, the choice of a suitable shape size definition usually must account for the specific demands imposed by each specific problem in shape analysis. However, Euclidean-based sizes are typically preferable because they are isotropic and consequently rotation invariant. Even so, all the alternative definitions of size discussed in Sections 4.8.1 and 4.8.3, which are based on Euclidean distances, satisfy this requirement. Considering that the baseline distance and the maximum diameter sizes are somewhat prone to disturbances, we are left with the $2n$, the RMS, the mean, the centroid and the normalized centroid sizes. However, these can be grouped into two main classes, since the $2n$, the RMS (i.e., a normalized version of the former), the centroid and its normalized version can be easily verified to be closely related, differing only by a multiplicative constant. Therefore, the choice reduces to using this large class of sizes or the mean size. However, this choice typically turns out to be not particularly critical, since these two types of distance differ little in practice, as illustrated in Figure 4.21, which shows a series of shapes and their respective mean and normalized centroid sizes. In addition to illustrating the fact that the two types of sizes are generally similar, this figure also illustrates the verification of the norm condition ② for the case of scaling the shape in (c) by 2, yielding (e), and scaling the shape (a) by 0.5, yielding (f). For such reasons, we henceforth consider the mean size, since this is arguably more compatible with our intuitive concept of shape size. Figure 4.22 illustrates the mean and centroid sizes for a series of 500 distorted versions of the shapes in Figure 4.21, obtained by adding uniformly distributed random noise with amplitude 10 to each coordinate of the landmark points. This graph clearly indicates that both these two size definitions are highly correlated and very similar.

4.8.5 Distances between Shapes

As observed in Section 4.1, humans frequently treat shapes in terms of similarities and metaphors. In biology, for instance, one is often interested in establishing correspondences between organs and bones from different animals or in relating the shapes obtained along either an evolutionary or a growing process, such as the development of neurons and other cells. Such shape correspondences have commonly been termed *homologies*. As we will soon see, the important issue of shape similarity involves the concept of distance, discussed in some detail in Chapter 2 (Section 2.2.4). It is therefore important to consider how a suitable distance between two shapes can be defined. This problem is simpler when the shapes have corresponding landmarks, which is the situation considered in the following.

It can be argued that one of the most natural definitions of distance between shapes, and that adopted in this book, takes into account the fact that a distance is

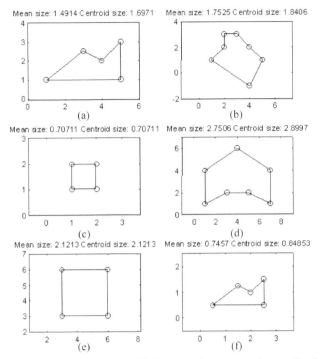

Figure 4.21: *Several shapes and their respective mean and normalized centroid sizes. Observe the good agreement between these two sizes.*

immediately induced by a norm, i.e., dist {shape 1, shape 2} = $\|$shape 1 − shape 2$\|$. Let the planar shapes S and Q to be compared be represented in terms of n respective landmark points S_i and Q_i; $i = 1, 2, \ldots, n$, as illustrated for the simple shape in Figure 4.23.

For simplicity's sake, let us define the quantities below, which will be referred to as the *x*- and *y*-coordinate differences:

$$\Delta_{x,i} = S_{x,i} - Q_{x,i} \quad \text{and} \quad \Delta_{y,i} = S_{y,i} - Q_{y,i}$$

The introduction of homologous landmarks allows distances between shapes to be defined by taking into account only the coordinates of the landmark points. As discussed in Section 4.6.2, there are two main possibilities for representing the landmarks: (i) as a set of n points in R^2, corresponding to the landmarks respective coordinates; and (ii) as a single vector in R^{2n} containing all the *x*- and *y*-coordinates of the landmarks. In the former case, a possible intuitive distance between the two shapes is obtained as the mean of the Euclidean distances between each respective landmark, as expressed in equation (4.12), which is henceforth called *mean*

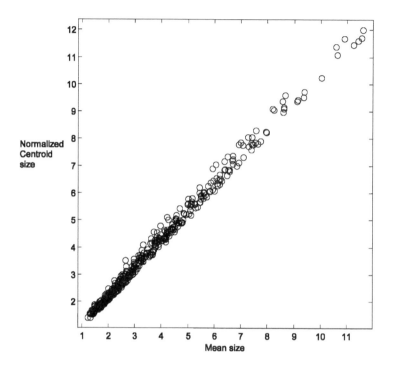

Figure 4.22: *The mean and centroid sizes for* 500 *distorted versions of the shapes in Figure 4.21. It is clearly seen that these two sizes are highly correlated.*

distance.

$$\varepsilon_a = \frac{\sum\limits_{i=1}^{n} \sqrt{(\Delta_{x,i})^2 + (\Delta_{y,i})^2}}{n} \qquad (4.12)$$

In the second case, i.e., the landmarks are represented as a single vector, we can define the *root mean square distance*, hence RMS distance, between the shapes as in equation (4.13).

$$\varepsilon_b = \sqrt{\frac{\sum\limits_{i=1}^{n} (\Delta_{x,i})^2 + \sum\limits_{i=1}^{n} (\Delta_{y,i})^2}{n}} \qquad (4.13)$$

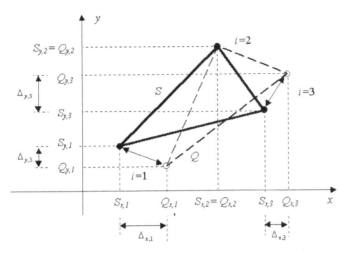

Figure 4.23: *The two shapes S and Q to be compared.*

As with the two main shape sizes discussed in the previous section, the mean and RMS distances between shapes tend to produce quite similar results, as illustrated in Figure 4.24. It also should be observed that the above distance definitions can be modified to be invariant to the position of the two shapes, which can be done by translating their centers of mass to the coordinate system origin. The choice of whether to normalize the shapes in this manner should be made by considering the specific requirements of each specific problem in shape analysis and classification.

4.9 | Morphic Transformations

As introduced in Section 4.3, morphic transformations are general transformations taking a shape into a shape. Figure 4.25 presents a possible hierarchical classification of morphic transformations. First, they are separated into two large classes corresponding to those transformations having functional analytic representation (e.g., $(x, y) = \vec{F}(x, y) = (xy, 2y^2)$), and those that only can be described in terms of tables, rules, etc. We concentrate our interest on the former type, which can be further divided as being continuous and discontinuous (see Section 2.4.1), the former appearing as smooth and non-smooth transformations. By smooth, it is meant that all the partial derivatives, up to infinity, of the vector field representing the transformation exist. Recall that a smooth function is necessarily continuous, but not vice versa. The several subclasses of smooth morphic transformations are discussed in more detail in the remainder of this chapter.

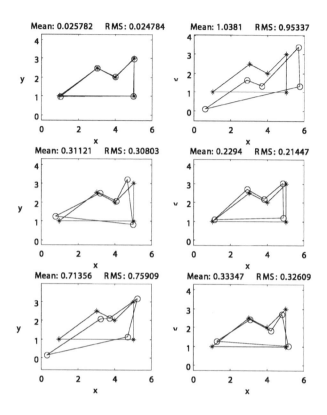

Figure 4.24: *The mean and RMS distances between a reference shape (represented by crosses) and several of its distorted versions (represented by circles). Observe the similarity between these two distance definitions.*

Let us now have a closer look at analytical morphic transformations. Let $\vec{F}$ be an analytic morphic transformation. By defining

$$\vec{s} = \begin{bmatrix} s_x \\ s_y \end{bmatrix} \quad \text{and} \quad \vec{q} = \begin{bmatrix} q_x \\ q_y \end{bmatrix},$$

we can represent the above transformation $\vec{F}$ of $\vec{s}$ into $\vec{q}$ in terms of the vector field $\vec{F}(s_x, s_y)$, i.e.,

$$\begin{bmatrix} q_x \\ q_y \end{bmatrix} = \vec{F}(s_x, s_y) = \begin{bmatrix} f_x(s_x, s_y) \\ f_y(s_x, s_y) \end{bmatrix} \tag{4.14}$$

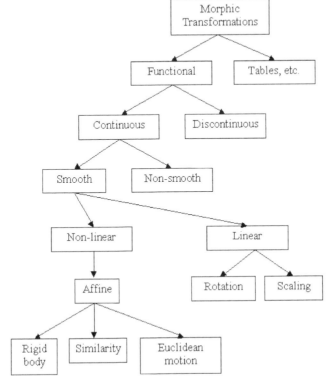

Figure 4.25: *A taxonomy of some of the principal classes of morphic transformations.*

Observe that the transformation $\vec{F}\left(s_x, s_y\right)$ is a 2×1 vector field having the scalar fields $f_x\left(s_x, s_y\right)$ and $f_y\left(s_x, s_y\right)$ as components. Given each input point $\vec{s}$, the transformation produces two scalar values corresponding to the x- and y-coordinates of the point $\vec{q}$, i.e., $q_x = f_x\left(s_x, s_y\right)$ and $q_y = f_y\left(s_x, s_y\right)$. Let $\vec{v}$ and $\vec{r}$ be two vectors representing two points of the original shape S. A transformation $\vec{F}$ such as $dist\{\vec{v}, \vec{r}\} = dist\{\vec{F}(\vec{v}), \vec{F}(\vec{v})\}$ is said to be *isometric*.

As an example of an isometric morphic transformation, consider the following transformation:

$$\begin{bmatrix} q_x \\ q_y \end{bmatrix} = \vec{F}\left(s_x, s_y\right) = \begin{bmatrix} f_x\left(s_x, s_y\right) \\ f_y\left(s_x, s_y\right) \end{bmatrix} = \begin{bmatrix} x\cos\left(30°\right) - y\sin\left(30°\right) \\ x\sin\left(30°\right) + y\cos\left(30°\right) \end{bmatrix}$$

As seen in Section 2.2.3, this is a linear transformation rotating a shape by $30°$ clockwise. Figure 4.26 presents the scalar fields $f_x\left(s_x, s_y\right)$ (a) and $f_y\left(s_x, s_y\right)$ (b) composing the above morphic transformation, whose effect is illustrated with respect to the simple shape in (c), producing as a result the shape in (d). Another

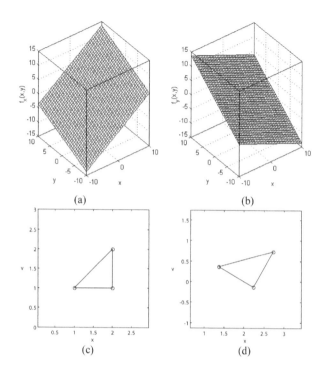

Figure 4.26: *The two scalar fields defining the transformation that rotates a shape by 30° clockwise, shown in (a) and (b), and the result (d) obtained when it is applied to the shape in (c).*

example of a morphic transformation can be found in the box entitled *Characterizing a Morphic Transformation.*

Analytic transformations are particularly interesting because they allow us to immediately apply the many useful tools and concepts from analysis (differential and integral calculus), differential geometry, linear algebra and so on, to shape analysis. One of the interesting possibilities of considering analytic morphic transformations, provided they are also smooth (or at least their first partial derivatives exist), consists in investigating the local properties of the transformation in terms of the Jacobian matrix and the Jacobian (i.e., the determinant of the Jacobian matrix). We have already seen in Section 2.4.2 that the Jacobian matrix plays a role in vector fields that is similar to that of the first derivative in univariate functions, especially regarding the approximation of the behavior of the transformation in a small neighborhood of a point. More specifically, the Jacobian matrix allows us to understand general transformations in terms of an approximate linear transformation. Therefore, provided the Jacobian matrix of the morphic transformation of interest exists, it can be used to characterize how the implemented mapping will affect small portions of the shape, in a small neighborhood. This situation is shown in Figure 4.27, illustrating how the small vector $\vec{\Delta}_{\vec{v}}$ placed at the position indicated by the vector $\vec{v}$, corresponding to a small portion of the shape S is mapped by the transformation into the vector $\vec{\Delta}_{\vec{q}}$, representing a small portion of the transformed shape Q. Provided these vectors are small enough, we have the linear approximation $\vec{F}\left(\vec{v}+\vec{\Delta}_{\vec{v}}\right) = \vec{F}(\vec{v}) + \vec{\Delta}_{\vec{q}} \cong \vec{q} + J\vec{\Delta}_{\vec{v}}$, and $\vec{\Delta}_{\vec{q}} \cong J\vec{\Delta}_{\vec{v}}$.

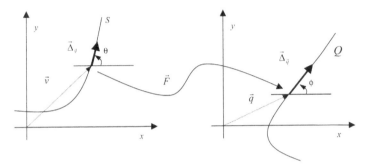

Figure 4.27: *A small portion of the original shape S, represented in terms of the vector $\vec{\Delta}_{\vec{v}}$, is mapped by the transformation $\vec{F}$ into the vector $\vec{\Delta}_{\vec{q}}$ corresponding to a small portion of the transformed shape Q.*

It is clear that the difference between the original and transformed vectors can be completely characterized in terms of changes in *magnitude* and *phase*. Let $\vec{\Delta}_{\vec{v}} = (\delta\cos(\theta), \delta\sin(\theta))$, where $\delta = \left\|\vec{\Delta}_{\vec{v}}\right\|$ is a small value. The *magnification ratio* $M(\vec{v}, \theta)$ of the transformation in a small neighborhood around the point $\vec{v}$ is henceforth defined as

$$M(\vec{v}, \theta) = \frac{\left\|\vec{\Delta}_{\vec{q}}\right\|}{\left\|\vec{\Delta}_{\vec{v}}\right\|} = \frac{\left\|J(\vec{v})\begin{bmatrix}\delta\cos(\theta)\\\delta\sin(\theta)\end{bmatrix}\right\|}{\delta} = \left\|J(\vec{v})\begin{bmatrix}\cos(\theta)\\\sin(\theta)\end{bmatrix}\right\|$$

and the *phase difference* of the transformation in a small neighborhood of the point $\vec{v}$ is defined as

$$\Phi(\vec{v}, \theta) = \theta - \phi(\vec{v}, \theta) = \theta - \arctan\left(\frac{\Delta_{q,y}}{\Delta_{q,x}}\right)$$

The above two measures allow us to completely and clearly characterize the local properties of each smooth morphic transformation. The box entitled *Characterizing a Morphic Transformation* presents a complete example illustrating such possibilities.

Example: *Characterizing a Morphic Transformation*

Characterize the local properties of the morphic transformation defined by the vector field $\vec{F}(S_x, S_y) = (S_x^3, S_y)$.

Solution:

First, it is interesting to visualize the effect of this transformation, as illustrated in Figure 4.28 for a simple shape. We have from Section 2.4.2 that the Jacobian

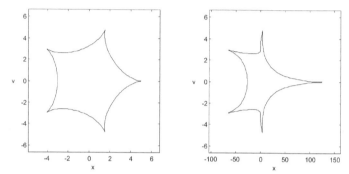

Figure 4.28: *A shape (a) and its transformation by the vector field* $\vec{F}(S_x, S_y) = ((S_x)^3, S_y)$ *(b).*

matrix of this transformation is

$$
J(S_x, S_y) = \begin{bmatrix} \dfrac{\partial F_x(S_x, S_y)}{\partial x} & \dfrac{\partial F_x(S_x, S_y)}{\partial y} \\ \dfrac{\partial F_y(S_x, S_y)}{\partial x} & \dfrac{\partial F_y(S_x, S_y)}{\partial y} \end{bmatrix} = \begin{bmatrix} 3(S_x)^2 & 0 \\ 0 & 1 \end{bmatrix}
$$

Therefore

$$
\det\{J\} = 3(S_x)^2; \quad \vec{\Delta}_{\vec{q}}(\vec{v}, \theta) = \delta J(\vec{v}) \begin{bmatrix} \cos(\theta) \\ \sin(\theta) \end{bmatrix} = \delta \begin{bmatrix} 3(S_x)^2 \cos(\theta) \\ \sin(\theta) \end{bmatrix}
$$

$$
M(\vec{v}, \theta) = \left\| J(\vec{v}) \begin{bmatrix} \cos(\theta) \\ \sin(\theta) \end{bmatrix} \right\| = \sqrt{9(S_x)^4 \cos^2(\theta) + \sin^2(\theta)}
$$

$$
\Phi(\vec{v}, \theta) = \theta - \arctan\left(\frac{\Delta_{\vec{q}, y}}{\Delta_{\vec{q}, x}}\right) = \theta - \arctan\left(\frac{\sin(\theta)}{3(S_x)^2 \cos(\theta)}\right) = \theta - \arctan\left(\frac{\tan(\theta)}{3(S_x)^2}\right)
$$

It thus becomes clear that the transformation is not isotropic regarding magnification or phase. Figure 4.29 presents the relative magnification (a) and phase difference (b) of the transformation in terms of θ for $S_x \in [0, 1]$. It is also clear from the above results and from Figure 4.28 that the magnification increases strongly with the absolute value of S_x. Observe that the relative area variation, given by $\det\{J\} = 3(S_x)^2$, also increases with S_x.

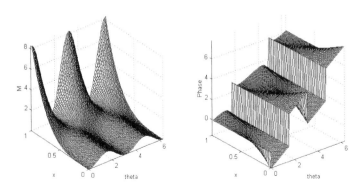

Figure 4.29: *The local relative magnification* **(a)** *and phase difference (in Radians)* **(b)** *of the transformation in terms of* θ *for* $S_x \in$ [0, 1].

An interesting possibility for graphically illustrating the local effect of morphic transformations as far as the magnification is concerned is to apply this transformation over a set of adjacent circles with fixed small radius, such as those in Figure 4.30 (a), in such a way that the obtained result indicates how these circles are locally affected. Figure 4.30 (b) illustrates this graphical visualization procedure

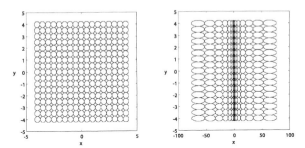

Figure 4.30: *The local effects of a morphic transformation can be graphically illustrated by applying the transformation onto a set of small circles* **(a)**. *The obtained result* **(b)** *indicates how the transformation will locally affect shapes at each specific position.*

with respect to the same transformation as in the box. It is clear from this result that the circles are not affected as far as their respective heights are concerned, but their widths are strongly amplified for larger absolute values of S_x. It is observed that a similar graphical alternative for visualizing the local shape alterations implemented by a transformation can be obtained by using an orthogonal grid as initial image.

The local properties of the shape changes implemented by a given analytic transformation can be used to further classify it. For instance, the case in which the transformation affects each small portion of the shape as if it were a similarity transformation (see Section 4.9.4) is called *conformal*, indicating that the shape is locally preserved. It can be verified (see, for instance, [Small, 1996]), that this situation is characterized by

$$\frac{\partial F_x\left(S_x, S_y\right)}{\partial x} = \frac{\partial F_y\left(S_x, S_y\right)}{\partial y} \quad \text{and} \quad \frac{\partial F_x\left(S_x, S_y\right)}{\partial y} = -\frac{\partial F_y\left(S_x, S_y\right)}{\partial x}$$

By being locally similar to similarity transformations, conformal transformations will also locally preserve the angles between curves. Observe that this situation is analogous to that encountered for conformal complex functions (Section 2.1.5).

Another interesting characterization of the local effect of a morphic transformation can be provided through the determinant of the Jacobian matrix which, as discussed in Section 2.4.2, provides an indication about the magnification of the area implemented by the transformation. For instance, the transformation addressed in the previous box has $\det\left\{J\left(\vec{F}\right)\right\} = 3\left(S_x\right)^2$, which indicates that the local area is amplified with the square of S_x. It is also possible to identify changes in the coordinate system orientation by observing the signal of $\det\left\{J\left(\vec{F}\right)\right\}$. The Jacobian of the transformation of interest also provides valuable information about the nature of the transformation and the possibility of defining its inverse. Indeed, it can be formally proven that a transformation $\vec{F}$ is bijective (and therefore invertible) in an open subset of S if and only $\det\{J\} \neq 0$ at each point of this subset. In other words, the fact that the Jacobian of the transformation is non-zero in a small neighborhood implies that the transformation is invertible in that neighborhood. Observe that in this case we have that the Jacobian matrix is also invertible, and $\vec{\Delta}_{\vec{v}} \cong J^{-1}\vec{\Delta}_{\vec{q}}$.

The following subsections illustrate some of the principal smooth analytical morphic transformations, including the affine, Euclidean motion, rigid body and similarity transformations. The latter three transformations can be verified to be special cases of the affine transformation:

$$T : S \rightarrow Q \mid \vec{q} = T\left(\vec{s}\right) = \alpha\left(\tilde{A}\vec{s} + \tilde{\vec{b}}\right)$$

where A is a general real matrix, $\vec{b}$ is a real vector, and α is a real scalar value. The subclassifications of this transformation are illustrated in the diagram in Figure 4.31.

4.9.1 | Affine Transformations

An affine transformation is any transformation that can be expressed in the following form:

$$\vec{q} = \alpha\left(\tilde{A}\vec{s} + \tilde{\vec{b}}\right) = A\vec{s} + \vec{b} \tag{4.15}$$

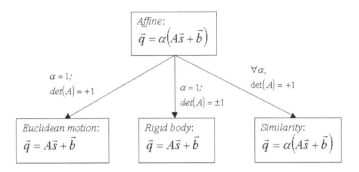

Figure 4.31: *The similarity, Euclidean motion and rigid body transforma-*
tions as particular cases of the affine transformation.

where $\tilde{A}$ is any 2×2 non-singular real matrix, $\tilde{b}$ is any 2×1 real vector, and α is any real scalar value. Observe that a generic affine transformation involves six parameters, four corresponding to the elements of the matrix A, plus the two elements of the vector $\vec{b}$. In other words, this transformation includes the linear transformation $\vec{q} = A\vec{s}$ plus the translation by $\vec{b}$. When applied to all points $\vec{s}$ of a shape S, the affine transformation produces a shape Q. Figure 4.32 illustrates the effect of the application of the same affine transformation defined by the parameters in equation (4.16) over two different shapes.

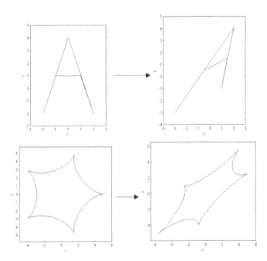

Figure 4.32: *The same affine transform applied over two different shapes.*

$$\alpha = 1; \quad A = \begin{bmatrix} 1 & 0.5 \\ 0.5 & 1 \end{bmatrix} \quad \text{and} \quad \vec{b} = \begin{bmatrix} 0 \\ 0 \end{bmatrix} \tag{4.16}$$

For affine transformations $\vec{q} = A\vec{s} + \vec{b}$, the Jacobian matrix becomes equal to the matrix A, and the area variation (see Section 2.2.5) given by $\det(A)$ is constant for any point in the plane. As shown in the following, *affine transformations take lines into lines*. Let $\vec{s}(t)$ be the parametric equation of a generic straight line (see Section 2.3.1), i.e.,

$$\vec{s}(t) = \begin{bmatrix} x(t) \\ y(t) \end{bmatrix} = \begin{bmatrix} bt + c \\ dt + e \end{bmatrix}$$

where b, c, d and e, with $b \neq 0$ and $d \neq 0$, are any real values. The straight line defined by $\vec{s}(t)$ is mapped by an affine transform into $\vec{q}(t)$ as follows:

$$\vec{q}(t) = A\vec{s}(t) + \vec{b} = \begin{bmatrix} a_{1,1} & a_{1,2} \\ a_{2,1} & a_{2,2} \end{bmatrix} \begin{bmatrix} bt + c \\ dt + e \end{bmatrix} + \begin{bmatrix} b_1 \\ b_2 \end{bmatrix} =$$

$$= \begin{bmatrix} a_{1,1}bt + a_{1,1}c + a_{1,2}dt + a_{1,2}e + b_1 \\ a_{2,1}bt + a_{2,1}c + a_{2,2}dt + a_{2,2}e + b_2 \end{bmatrix} =$$

$$= \begin{bmatrix} (a_{1,1}b + a_{1,2}d)t + (a_{1,1}c + a_{1,2}e + b_1) \\ (a_{2,1}b + a_{2,2}d)t + (a_{2,1}c + a_{2,2}e + b_2) \end{bmatrix} = \begin{bmatrix} \tilde{b}t + \tilde{c} \\ \tilde{d}t + \tilde{e} \end{bmatrix} = \tilde{\vec{s}}(t)$$

It is thus clear that the parametric straight line $\vec{s}(t)$ is transformed into the new straight line $\tilde{\vec{s}}(t)$. It is left as an exercise to the reader to verify that an affine transformation will also map parallel lines into parallel lines. Another interesting property of affine transformations is that, when S and Q are non-degenerated triangles (i.e., with non-null areas), it is possible to show that there is a unique affine transform A taking the shape S into Q, i.e., $Q = A(S)$. This transformation can be easily found, as illustrated in the following example.

Example: *Affine transforming*

Let the shapes S and Q be two triangles defined in terms of their vertices,

$$V_S = \left\{ \begin{bmatrix} 1 \\ 1 \end{bmatrix}; \begin{bmatrix} 1 \\ 4 \end{bmatrix}; \begin{bmatrix} 5 \\ 4 \end{bmatrix} \right\} \quad \text{and} \quad V_Q = \left\{ \begin{bmatrix} 2 \\ 1 \end{bmatrix}; \begin{bmatrix} 1 \\ 2 \end{bmatrix}; \begin{bmatrix} 3 \\ 4 \end{bmatrix} \right\}$$

respectively, as shown in Figure 4.33. Find the affine transformation A that takes the shape S into the shape Q.

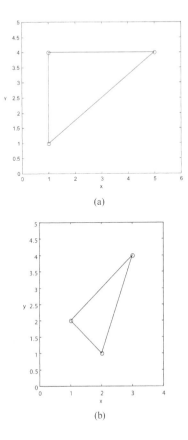

(a)

(b)

Figure 4.33: *The shapes S* **(a)** *and Q* **(b)**.

Solution:

Let us start by substituting the first vertex $\vec{s}_1 = \begin{bmatrix} s_{1,x} \\ s_{1,y} \end{bmatrix} = \begin{bmatrix} 1 \\ 1 \end{bmatrix}$ of S into the general affine transformation equation (4.15):

$$\vec{q}_1 = A\vec{s}_1 + \vec{b} = \begin{bmatrix} a_{1,1} & a_{1,2} \\ a_{2,1} & a_{2,2} \end{bmatrix} \begin{bmatrix} 1 \\ 1 \end{bmatrix} + \begin{bmatrix} b_1 \\ b_2 \end{bmatrix} = \begin{bmatrix} a_{1,1} + a_{1,2} + b_1 \\ a_{2,1} + a_{2,2} + b_2 \end{bmatrix} = \begin{bmatrix} q_1 \\ q_2 \end{bmatrix}$$

It is thus clear that each vertex yields two simple linear equations involving the sought parameters. The consideration of all vertices leads to the following matrix equation:

$$
\begin{bmatrix}
1 & 1 & 1 & 0 & 0 & 0 \\
1 & 4 & 1 & 0 & 0 & 0 \\
5 & 4 & 1 & 0 & 0 & 0 \\
0 & 0 & 0 & 1 & 1 & 1 \\
0 & 0 & 0 & 1 & 4 & 1 \\
0 & 0 & 0 & 5 & 4 & 1
\end{bmatrix}
\begin{bmatrix}
a_{1,1} \\
a_{1,2} \\
b_1 \\
a_{2,1} \\
a_{2,2} \\
b_2
\end{bmatrix}
=
\begin{bmatrix}
2 \\
1 \\
3 \\
1 \\
1 \\
4
\end{bmatrix}
$$

which defines a determined system of linear equations, whose solution supplies the sought parameters:

$$
A = \begin{bmatrix} 0.5 & -\frac{1}{3} \\ 0.5 & \frac{1}{3} \end{bmatrix} \quad \text{and} \quad \vec{b} = \begin{bmatrix} \frac{11}{6} \\ \frac{1}{3} \end{bmatrix}
$$

Observe that such an exact affine transformation is usually guaranteed only for triangles. For pairs of more complex shapes, involving several deformations such as those illustrated in Figure 4.34, the minimum squares approach described in Chapter 2 can still be used to obtain the affine transformation yielding the best approximation (in the least sum of squares sense) of the sought mapping between the two shapes, as illustrated in Figure 4.34.

The class of affine transformations of shapes includes several useful situations, such as the distortions you get while looking at shapes through a perfectly plane mirror. Another relevant case of affine transforms includes the rigid body transformations and similarity transformations, which are covered in the next sections.

4.9.2 Euclidean Motions

A *Euclidean motion* is any transformation *T* *involving a rotation (i.e., a special orthogonal transformation) followed by a translation*, having the following general form:

$$
T : S \to Q \mid \vec{q} = T(\vec{s}) = A\vec{s} + \vec{b}
$$

where A is an orthogonal matrix with $\det(A) = 1$ (i.e., A is a *special orthogonal matrix*). It is therefore clear that a Euclidean motion is a special case of an affine transformation. Figure 4.35 illustrates several Euclidean motion transformations of the shape S.

If two shapes S and Q are related by a Euclidean motion, i.e., if $\vec{q} = A\vec{s} + \vec{b}$ and $\vec{s} = A^{-1}(\vec{q} - \vec{b})$, where $\vec{s} \in S$ and $\vec{q} \in Q$, we say that these shapes are *congruent*. For instance, all shapes in Figure 4.35 can be said to be congruent. Observe, however, that the inverse transformation is unique only because all the landmarks in S have precise homologous landmarks in the other transformed shapes. Otherwise, a line segment, for instance, could be transformed into the same shape by rotating it

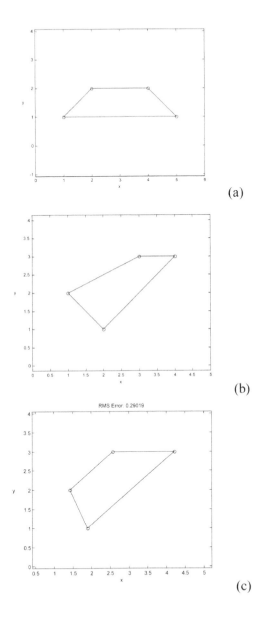

Figure 4.34: *Original shape (a) to be transformed into the shape in (b). The best solution, in the minimum squares sense, using an affine transformation is shown in (c) together with the respective RMS error.*

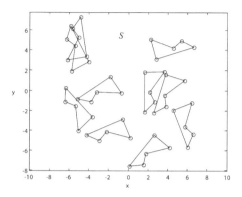

Figure 4.35: *Euclidean motions of the shape S.*

30° counterclockwise or 330° clockwise with respect to the x-axis. Congruence under Euclidean motion is one of the shape relationships closest to the human concept of shapes.

4.9.3 | Rigid Body Transformations

A rigid body transformation T *preserves all the distances between any two points of the shape*, i.e., given any two points $\vec{s}_1, \vec{s}_2 \in S$ with $d_{1,2} = \text{dist}\,(\vec{s}_1, \vec{s}_2) \in S$, we have necessarily that $\text{dist}\,(T\,(\vec{s}_1), T\,(\vec{s}_2)) = d_{1,2}$. Therefore, rigid body transformations can be verified to correspond to transformations having the following general form:

$$T : S \rightarrow Q \mid \vec{q} = T\,(\vec{s}) = A\vec{s} + \vec{b}$$

where A is a *general orthogonal matrix* (see Section 2.2.5), and consequently $\det\,(A) = \pm 1$. A rigid body transformation is therefore also a special case of the affine transformation. It should be observed that this type of transformation can be useful to treat situations where planar shapes can appear in reflected version, such as while scanning leaves by using a scanner. Figure 4.36 presents several examples of rigid body transformations of the shape S in Figure 4.35.

4.9.4 | Similarity Transformations

A similarity transformation T consists of a Euclidean motion followed by a scaling, as described below:

$$T : S \rightarrow Q \mid \vec{q} = T\,(\vec{s}) = \alpha\left(A\vec{s} + \vec{b}\right)$$

where $\alpha \in R$ and A is a *special orthogonal matrix* (corresponding to a rotation) implying $\det\,\{A\} = +1$. The similarity transformation is also a specific case of the affine transformation, where $\det\,\{A\} = +1$. It can easily be verified that a similarity

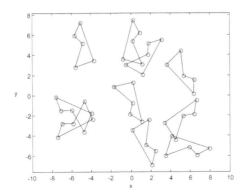

Figure 4.36: *Rigid body transformations of the shape S in Figure* 4.35.

transformation will preserve the angles between any two curves at their intersecting points, a property that is usually interpreted as guaranteeing *shape preservation*, or *conformality*. For instance, a similarity transformation of a circle will always produce a circle as result, although possibly in another position and at a different scale. Figure 4.37 illustrates the results obtained by random similarity transformations of the shape S in Figure 4.35. Observe that the similarity transform can be generalized by allowing $\det\{A\} = \pm 1$.

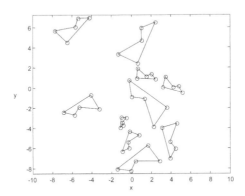

Figure 4.37: *Similarity transformations of the shape S in Figure* 4.35.

4.9.5 Other Transformations and Some Important Remarks

In addition to the above special cases of the affine transform, there is an infinity of other morphic transformations. For instance, pure rotation, scaling and translation are relatively simpler morphic transformations. Of utmost importance to shape analysis is the ample class of conformal transforms outlined in Section 2.1.5.

As indicated by their own name, conformal transformations *preserve* much of the original shape, but mostly at a *local level* (or small neighborhood), i.e., although the larger scale details of the shape are altered (i.e., distorted) by the transformation, its smaller details are preserved. This type of transformation has often been used in practice, for instance as a means of relating different species (see [Thompson, 1992]). Another important type of morphic transformations are defined by projections (see, for instance, the interesting book of [Dubery & Willats, 1972]).

The main importance of morphic transformations in shape analysis stems from the fact that the shapes of interest often appear in a transformed fashion. For instance, a shape submitted to a shearing process (see Figure 4.32) will be related to the original shape by an affine transform. In case both such types of shapes, i.e., the original and transformed versions, are to be understood as equivalent, the selected features have to be invariant to the transformation in question. This is an especially important concept in shape analysis and classification, deserving special attention. For instance, in case the original shape and its affine transformation in Figure 4.32 are to be considered as equivalent, a suitable feature would be the number of vertices or the number of holes, which is invariant to the transformation in question. In case a shape and one of its rotated versions are to be considered as equivalent, suitable features would be the area, perimeter, number of vertices, etc. Therefore, an important issue in shape analysis and classification concerns the identification of the transformations underlying the several observed instances of the same shape. Often, such transformations are a direct consequence of the natural processes producing the shapes. Unfortunately, it is not always easy to identify such transformations, except in the case of simple transformations such as the affine and its specific cases. To cope with this problem one should consider several transformations and find which better explains the observed shape variations. The remainder of this chapter presents the application of thin-plate splines as a reasonably general means to interpolate shape transformations. More specifically, given the original shape and its transformed version, both expressed in terms of landmark points, the thin-plate formulation allows us to obtain an interpolated approximation of the sought transformation.

4.9.6 | Thin-Plate Splines

The concept of thin-plate splines was first applied to the analysis of plane shapes by Bookstein [Bookstein, 1991]. In such approaches, a thin plate is understood as a thin sheet of some stiff material (e.g., steel) with infinite extension. When specific control points along the plate are displaced, the plate undergoes a deformation in such a way as to minimize the total bending energy E implied by the transformation. This formulation can be immediately extended to planar shapes by using pairs of thin-plates, represented in terms of landmark points. We start by presenting and illustrating the traditional thin-plate and proceed by discussing its extension as a means to interpolate morphic transformations.

Single Thin-Plate Splines

Consider the thin-plate *basic function* $g(\rho)$ defined by equation (4.17), where ρ is a non-negative real value (typically $\rho = \sqrt{x^2 + y^2}$). This function is shown in Figure 4.38 as $-g(\rho)$, for the sake of proper visualization. It should be observed

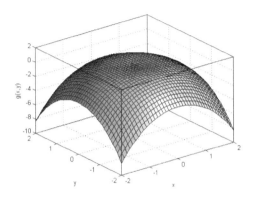

Figure 4.38: *The basic thin-plate function $g(\rho)$, shown as $-g(\rho)$ for the sake of better visualization.*

that it is also possible to use the square of the distance ρ as argument of the log function (see, for instance, [Bookstein, 1991]), which will however lead to the same thin-plate spline interpolation.

$$g(\rho) = \begin{cases} 0 & \text{if} \quad \rho = 0 \\ \rho^2 \log(\rho) & \text{otherwise} \end{cases} \qquad (4.17)$$

The *interpolating thin-plate spline* is the bivariate function $\Psi(x, y)$ defined as

$$\Psi(x, y) = a + b_x x + b_y y + \sum_{k=1}^{n} w_k g\left(\sqrt{(x - x_k)^2 + (y - y_k)^2} \right) \qquad (4.18)$$

or, in a more compact matrix form

$$\Psi(\vec{v}) = a + \vec{b}^T \vec{v} + W^T \vec{g}(\vec{v}) \qquad (4.19)$$

where

$$\vec{v} = \begin{bmatrix} x \\ y \end{bmatrix} ; \quad \vec{b} = \begin{bmatrix} b_x \\ b_y \end{bmatrix} ; \quad W = \begin{bmatrix} w_1 \\ w_2 \\ \vdots \\ w_n \end{bmatrix} \quad \text{and}$$

$$\vec{g}(x,y) = \begin{bmatrix} g\left(\sqrt{(x-x_1)^2 + (y-y_1)^2}\right) \\ g\left(\sqrt{(x-x_2)^2 + (y-y_2)^2}\right) \\ \vdots \\ g\left(\sqrt{(x-x_n)^2 + (y-y_n)^2}\right) \end{bmatrix}$$

In addition, the following constraints are imposed in order to minimize the bending energy:

$$\sum_{k=1}^{n} w_k = \sum_{k=1}^{n} w_k x_k = \sum_{k=1}^{n} w_k y_k = 0$$

It is clear from equations (4.18) and (4.19) that the thin-plate spline $\Psi(x,y)$ involves three main parts: (i) a constant term a_0; (ii) the internal product between the coefficient vector $\vec{b}$ and $\vec{v}$, i.e., $\vec{b}^T \vec{v}$; and (iii) a linear combination of the basic functions $g(x,y)$, centered at each of the n points $\vec{p}_k = (x_k, y_k)$. Each thin-plate fit to a set of control points is guaranteed to have *minimal bending energy E*, which is defined as

$$E = \sum_{k=1}^{2} \int_{x=-\infty}^{\infty} \int_{y=-\infty}^{\infty} \left[\left(\frac{\partial^2 \Psi_k}{\partial x^2}\right)^2 + 2\left(\frac{\partial^2 \Psi_k}{\partial x \partial y}\right)^2 + \left(\frac{\partial^2 \Psi_k}{\partial y^2}\right)^2 \right] dx dy$$

Having introduced the thin-plate function, it is time to address how it can be applied to interpolate over a set of control points, i.e., the set of n three-dimensional points $(x_k, y_k, h(x_k, y_k))$; $k = 1, 2, \ldots, n$; as illustrated in Figure 4.39. In other words, given n points, we wish to find the surface passing through them that exhibits minimal overall bending energy, i.e., which is "less bent." Such a solution can be verified to be given by the thin-plate spline $\Psi(x,y)$ whose coefficients are obtained as described below.

The following abbreviations are adopted henceforth:

① The $n \times n$ matrix T including the values of the basic function $g(x,y)$ evaluated at each of the distances defined between each pair of control points:

$$T = \left[g\left(\rho_{i,j}\right) \right]_{n \times n} = \begin{bmatrix} 0 & g(\rho_{1,2}) & \cdots & \cdots & g(\rho_{1,n-1}) & g(\rho 1, n) \\ g(\rho_{2,1}) & 0 & & & & \vdots \\ \vdots & & \ddots & & & \\ & & & 0 & & \\ \vdots & & & & \ddots & \vdots \\ g(\rho_{n-1,1}) & g(\rho_{n-1,2}) & & & 0 & g(\rho_{n-1,n}) \\ g(\rho_{n,1}) & g(\rho_{n,2}) & \cdots & \cdots & g(\rho_{n,n-1}) & 0 \end{bmatrix}$$

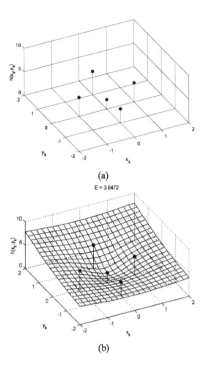

Figure 4.39: *A set of five control points* $(x_k, y_k, h(x_k, y_k))$, $k = 1, 2, \ldots, 5$ *(a)*
and respective interpolating thin-plate spline (b).

where $\rho_{i,j} = \sqrt{\left(x_i - x_j\right)^2 + \left(y_i - y_j\right)^2}$

② The $n \times 3$ matrix S_e, containing the *x*- and *y*-coordinates of the control points, plus an initial column of ones

$$S_e = \begin{bmatrix} 1 & x_1 & y_1 \\ 1 & x_2 & y_2 \\ \vdots & & \vdots \\ 1 & x_n & y_n \end{bmatrix}$$

③ The $(n + 3) \times (n + 3)$ matrix M, obtained by composing the matrices T and S_e as follows (see the accompanying box for an example)

$$M = \begin{bmatrix} T & S_e \\ S_e^T & 0 \end{bmatrix}$$

④ The $(n + 3) \times 1$ matrix H, containing the *z*-coordinates (i.e., the "heights"

$h(x_k, y_k))$ of the control points

$$H = \begin{bmatrix} h(x_1, y_1) & h(x_2, y_2) & \cdots & h(x_n, y_n) & 0 & 0 & 0 \end{bmatrix}^T$$

⑤ The $(n + 3) \times 1$ matrix C containing all the sought coefficients involved in equation (4.18) or (4.19):

$$C = \begin{bmatrix} w_1 & w_2 & \cdots & w_n & a & b_x & b_y \end{bmatrix}^T$$

Now, the sought coefficients required to interpolate through the n control points can be easily obtained by solving the following simple matrix equation, provided the matrix M is not singular:

$$C = M^{-1}H \tag{4.20}$$

The bending energy E can be immediately obtained as

$$E = W^T T W \tag{4.21}$$

where W is the $n \times 1$ vector containing the n first rows of C.

The accompanying box provides a complete example of the above described thin-plate spline interpolation.

Example: 1D Thin-Plate Spline Interpolation

Obtain the thin-plate spline passing through the following five control points $(-1, 0, 4)$; $(0, 1, 5)$; $(0, -1, 3)$; $(1, 0, 4)$; and $(0, 0, 2)$. These points are shown in Figure 4.39 (a).

Solution:

We have $n = 5$ and start by obtaining T, S_e, H, and M:

$$T = \begin{bmatrix} 0 & \alpha & \alpha & \beta & 0 \\ \alpha & 0 & \beta & \alpha & 0 \\ \alpha & \beta & 0 & \alpha & 0 \\ \beta & \alpha & \alpha & 0 & 0 \\ 0 & 0 & 0 & 0 & 0 \end{bmatrix}; \quad S_e = \begin{bmatrix} 1 & -1 & 0 \\ 1 & 0 & 1 \\ 1 & 0 & -1 \\ 1 & 1 & 0 \\ 1 & 0 & 0 \end{bmatrix}; \quad H = \begin{bmatrix} 4 \\ 5 \\ 3 \\ 4 \\ 2 \\ 0 \\ 0 \\ 0 \end{bmatrix}$$

and

$$M = \begin{bmatrix} 0 & \alpha & \alpha & \beta & \gamma & 1 & -1 & 0 \\ \alpha & 0 & \beta & \alpha & \gamma & 1 & 0 & 1 \\ \alpha & \beta & 0 & \alpha & \gamma & 1 & 0 & -1 \\ \beta & \alpha & \alpha & 0 & \gamma & 1 & 1 & 0 \\ \gamma & \gamma & \gamma & \gamma & 0 & 1 & 0 & 0 \\ 1 & 1 & 1 & 1 & 1 & 0 & 0 & 0 \\ -1 & 0 & 0 & 1 & 0 & 0 & 0 & 0 \\ 0 & 1 & -1 & 0 & 0 & 0 & 0 & 0 \end{bmatrix}$$

Note that values $\alpha = g\left(d_\alpha = \sqrt{2}\right) \cong 0.6931$, $\beta = g\left(d_\beta = 2\right) \cong 2.7726$ and $\gamma = g\left(d_\gamma = 1\right) \cong 0$ correspond to the respective distances between the x- and y-coordinates of the control points, as shown in Figure 4.40. Now, by applying

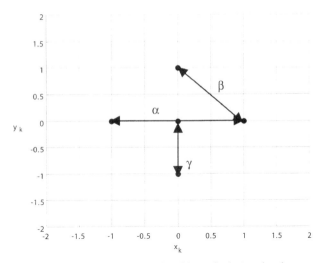

Figure 4.40: *The distances considered for calculating the elements of matrix M.*

equation (4.20) we get

$$C = M^{-1}H = \begin{bmatrix} 0.4809 & 0.4809 & 0.4809 & 0.4809 & -1.9236 & 2 & 0 & 1 \end{bmatrix}$$

and the bending energy is

$$E = W^T T W \approx 3.8472$$

The obtained interpolating spline has been used to transform a uniform grid, and the result is shown in Figure 4.39 (b).

Figure 4.41 illustrates a sequence of deformations of a thin-plate, starting with four control points, up to a highly deformed configuration involving seven control points. The configuration shown in Figure 4.41 (a) corresponds to a plane plate,

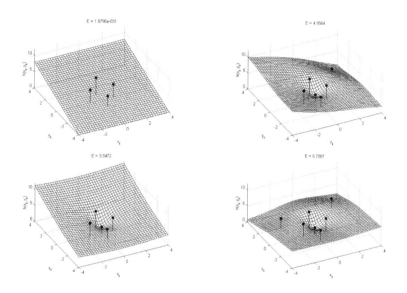

Figure 4.41: *Thin-plate splines obtained by progressive addition of control points: four points (a); five points (b); six points (c); seven points (d). Observe that the increasing of the foldings of the plate is expressed by higher bending energy values, shown above each respective configuration.*

implying zero bending energy (the extremely small values shown above each graph are a consequence of round-off noise during the numerical processing of the matrices). The addition of the fifth point in (b) implies the first bending of the plate, which is duly expressed by the higher bending energy, which continues to increase as more points are incorporated. Observe that each of these interpolating spline surfaces is characterized by minimum bending energies.

Pairs of Thin-Plate Splines   The thin-plate spline formulation can be immediately extended to address shape transformations. This is done by using a *pair of interpolating thin-plate splines*, i.e., the bivariate vector function $\vec{\Psi}(\vec{v} = (x, y)) = \left(\Psi_x(x,y), \Psi_y(x,y) \right)$ defined as

$$\vec{\Psi}_x(x,y) = a_x + b_{x,x}x + b_{x,y}y + \sum_{k=1}^{n} w_{x,k} g\left(\sqrt{(x - x_k)^2 + (y - y_k)^2} \right) \qquad (4.22)$$

and

$$\vec{\Psi}_y(x, y) = a_y + b_{y,x}x + b_{y,y}y + \sum_{k=1}^{n} w_{y,k}g\left(\sqrt{(x - x_k)^2 + (y - y_k)^2}\right) \tag{4.23}$$

which can be rewritten as the following matrix equation:

$$\vec{\Psi}(\vec{v}) = \vec{a} + B\vec{v} + W^T \vec{g}(\vec{v}) \tag{4.24}$$

where

$$\vec{v} = \begin{bmatrix} x \\ y \end{bmatrix}; \quad \vec{a}_0 = \begin{bmatrix} a_{x,0} \\ a_{y,0} \end{bmatrix}; \quad B = \begin{bmatrix} b_{x,x} & b_{x,y} \\ b_{y,x} & b_{y,y} \end{bmatrix}; \quad W = \begin{bmatrix} w_{x,1} & w_{y,1} \\ w_{x,2} & w_{y,2} \\ w_{x,3} & w_{y,3} \\ w_{x,4} & w_{y,4} \end{bmatrix}$$

$$\text{and} \quad \vec{g}(x, y) = \begin{bmatrix} g\left(\sqrt{(x - x_1)^2 + (y - y_1)^2}\right) \\ g\left(\sqrt{(x - x_2)^2 + (y - y_2)^2}\right) \\ \vdots \\ g\left(\sqrt{(x - x_n)^2 + (y - y_n)^2}\right) \end{bmatrix}$$

In the above formulation, the components $\Psi_x(x, y)$ and $\Psi_y(x, y)$, which are independent of one another, are used to represent the x- and y-coordinates of the transformed shape. More specifically, the original two-dimensional shape S is represented in terms of its n landmark points $\vec{s}_k = (x_k, y_k)$; and the shape Q, which constitutes either the transformation of S or another shape to which S is being compared, is also represented in terms of n landmark points, i.e., $\vec{q}_k = (\tilde{x}_k, \tilde{y}_k)$. Each of the thin-plate splines is then used to interpolate the bivariate functions (or scalar fields) defined by the x- and y-coordinates of S and the respective coordinates of Q, i.e.,

$$\Psi_x : S \to Q_x \mid \Psi_x(x_k, y_k) \to \tilde{x}_k \quad \text{and} \quad \Psi_y : S \to Q_y \mid \Psi_y(x_k, y_k) \to \tilde{y}_k$$

where $Q_x = \{\tilde{x}_1, \tilde{x}_2, \ldots, \tilde{x}_n\}$ and $Q_y = \{\tilde{y}_1, \tilde{y}_2, \ldots, \tilde{y}_n\}$.

The above concepts and representations are illustrated in Figure 4.42.

The sought interpolating thin-plate splines $\Psi_x(x, y)$ and $\Psi_y(x, y)$ can be obtained by applying the procedure described above separately, or through the integrated approach presented in the following. First, matrices T, Q_e and M are obtained by the method previously described (items A, B and C, respectively). The following two additional matrices are then constructed:

① The $(n + 3) \times 2$ matrix H_2 containing the coordinates of the landmark points of

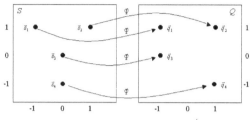

$$\vec{s}_1 = (x_1, y_1) = (-1, 1); \quad \vec{q}_1 = (\tilde{x}_1, \tilde{y}_1) = (-1, 1) = \left(\Psi_x(x_1, y_1), \Psi_y(x_1, y_1)\right)$$
$$\vec{s}_2 = (x_2, y_2) = (1, 1); \quad \vec{q}_2 = (\tilde{x}_2, \tilde{y}_2) = (1, 1) = \left(\Psi_x(x_2, y_2), \Psi_y(x_2, y_2)\right)$$
$$\vec{s}_3 = (x_3, y_3) = (0, 0); \quad \vec{q}_3 = (\tilde{x}_3, \tilde{y}_3) = (-1, 0) = \left(\Psi_x(x_3, y_3), \Psi_y(x_3, y_3)\right)$$
$$\vec{s}_4 = (x_4, y_4) = (0, -1); \quad \vec{q}_4 = (\tilde{x}_4, \tilde{y}_4) = (1, -1) = \left(\Psi_x(x_4, y_4), \Psi_y(x_4, y_4)\right)$$

Figure 4.42: *Concepts and representations adopted in the pair of thin-plate splines approximation to shape analysis. S is the original shape and Q its transformed version or another shape to which S is to be compared.*

shape Q is

$$H_2 = \left[\begin{array}{ccccccc} \tilde{x}_1 & \tilde{x}_2 & \cdots & \tilde{x}_n & 0 & 0 & 0 \\ \tilde{y}_1 & \tilde{y}_2 & \cdots & \tilde{y}_n & 0 & 0 & 0 \end{array}\right]^T$$

② The $(n + 3) \times 2$ matrix C_2, containing all the parameters in equations (4.22) or (4.23):

$$C_2 = \left[\begin{array}{ccccccc} w_{x,1} & w_{x,2} & \cdots & w_{x,n} & a_x & b_{x,x} & b_{x,y} \\ w_{y,1} & w_{y,2} & \cdots & w_{y,n} & a_y & b_{y,x} & b_{y,y} \end{array}\right]^T$$

The coefficients in equations (4.22) and (4.23) can now be obtained by solving the following matrix equation:

$$C_2 = M^{-1} H_2 \tag{4.25}$$

and the bending energy E can be obtained as

$$E = \text{Trace}\left\{W^T T W\right\} \tag{4.26}$$

Figure 4.43 illustrates the application of the thin-plate splines to planar shapes. S is the original shape, and the shape Q its transformation or the shape to which it is being compared. The obtained interpolating thin-plate spline has been used to transform the orthogonal grid of Figure 4.43 (a) into the deformed grid in (b). The bending energy E implied by this transformation is approximately 1.4779. It should be borne in mind that the bending energy implied by the inverse morphic transformation, as implemented by the thin-plate formulation, is not equal to the bending energy implied by the respective direct transformation. In addition, observe that,

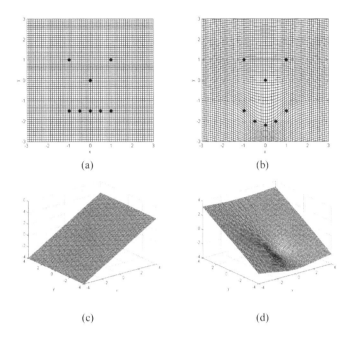

(a)　　　　　　　　　　　(b)

(c)　　　　　　　　　　　(d)

Figure 4.43: *Original* **(a)** *and transformed* **(b)** *shapes (the landmarks are indicated by the black dots); and the surfaces defined by the respective scalar fields* $\Psi_x(x, y)$ **(c)** *and* $\Psi_y(x, y)$ **(d)**.

as illustrated in Figure 4.44, excessive displacement of the control points can cause folding (or overlap) of the interpolating surface.

Figure 4.45 presents a more sophisticated example regarding the dynamic modification of a neural cell from the original configuration shown by circles into the new shape shown by asterisks. Observe that the thin-plate spline interpolation approach allows the comprehensive characterization of the spatial deformations related to the shape alterations, in this case the bending (identified by one asterisk) and growth (identified by two asterisks) of dendrites.

To probe further: *Shapes*

Additional material on shape related concepts can be found in the literature of the most diverse areas, from complex variables to biological shape analysis. A good reference on transformations underlying projective drawing systems can be found in [Dubery & Willats, 1972]. Relatively old but still interesting books are [Cook, 1979], which concentrates on spiral shapes, [Thompson, 1992], a traditional approach to biological shape, and [Ghyka, 1977], which provides a general perspective of shapes in art and life mostly in terms of proportions. More modern treat-

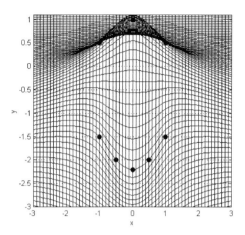

Figure 4.44: *Excessive displacement of the control points can cause over-lap of the interpolating surface.*

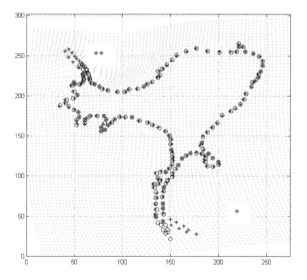

Figure 4.45: *The thin-plate formulation can be used to characterize the local effects of shape transformations, in this case, the bending (identified by one asterisk) and growth (identified by two asterisks) of dendrites. The original shape is represented by circles, and its new, deformed version, by crosses.*

ments of shapes include [Bookstein, 1991; Dryden & Mardia, 1998; Small, 1996; van Otterloo, 1991].

5

Shape Representation

Chapter Overview

THE IMPORTANT ISSUE OF SHAPE REPRESENTATION is covered in this chapter. First, the problem of contour extraction is characterized and respective algorithms are presented, followed by a practical discussion about interpolation of contours (including Ramer's and the split-and-merge algorithms). The chapter proceeds by introducing the concept and characteristics of digital straight lines and methods for their recognition in digital images, as well as a series of region-based representative concepts and techniques, including the distance transform, Voronoi tessellations, a powerful simple method for multiscale skeletonization and reconstruction, and the principal bounding region approaches.

5.1 Introduction

Shape analysis is generally carried out starting from an intermediate representation typically involving the segmented image (where the object shape has been located) and/or special shape descriptors. Since there are many approaches to shape representation, a classification of the main and more popular techniques is presented in this chapter in order to make their relationship clearer, and some of the more representative approaches are presented and discussed in some detail. Figure 5.1 shows the first level of a possible taxonomy of computational shape representation approaches, which can be broadly divided into contour-based, region-based and transform-based. Both contour-based and region-based approaches have already been discussed in Chapter 4, in such a way that only their respective computational aspects are covered in this chapter. The underlying idea behind the transform-based

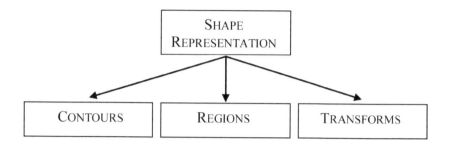

Figure 5.1: *A taxonomy of shape representation techniques.*

approach is the application of a transform, such as Fourier or wavelets, in order to represent the shape in terms of the transform coefficients. Nevertheless, it is important to note that the transform approach is frequently used both for shape representation and for description, in which case the transform coefficients are used as features for classification (see Section 6.5). Therefore, some of the transform-based techniques are discussed in more detail in Chapters 6 and 7.

A brief overview on the main representation methods, i.e., contours, regions and transforms, is presented in the following. Observe that the main boxes in Figure 5.1 have been expanded and detailed in Figures 5.2, 5.3 and 5.4.

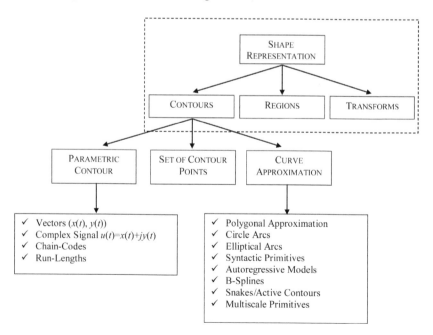

Figure 5.2: *A taxonomy of contour-based shape representation techniques.*

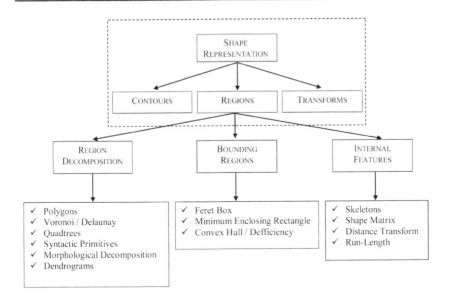

Figure 5.3: *A taxonomy of region-based shape representation techniques.*

As illustrated in Figure 5.2, the proposed taxonomy divides the contour-based approach into the following three classes:

1. **Parametric contours:** the shape outline is represented as a parametric curve, thus implying a sequential order along it;

2. **Set of contour points:** the shape outline is simply represented as a set of points, without any special order among them;

3. **Curve approximation:** a set of geometric primitives (e.g., straight line segments or splines) are fitted to the shape outline.

These three classes are discussed from Section 5.2 to Section 5.6 in this chapter. Figure 5.3 shows the subdivision of the region-based approach, which includes

Region decomposition: the shape region is partitioned into simpler forms (e.g., polygons) and represented by the set of such primitives;

Bounding regions: the shape is approximated by a special pre-defined geometric primitive (e.g., an enclosing rectangle) fitted to it;

Internal features: the shape is represented by a set of features related to its internal region (e.g., a skeleton).

Each class involves its own methods, and some of the most important are presented from Section 5.7 to Section 5.12 in this chapter. Figure 5.4 shows the subdivision of the transform approach, which includes the following categories:

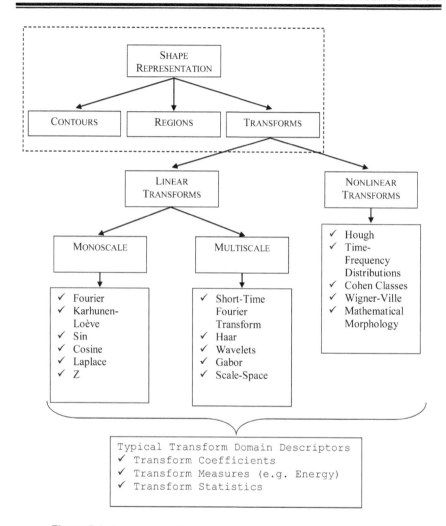

Figure 5.4: *A taxonomy of transform-based shape representation techniques.*

Linear: a linear transform T is such that, given two shapes A and B, and two scalars α and β, we have $T(\alpha A + \beta B) = \alpha T(A) + \beta T(B)$. A linear transform-based approach represents a shape A in terms of $T(A)$, e.g., the coefficients (or energies) of its transformation (e.g., Fourier);

Nonlinear: similar to the above, i.e., the shape is represented in terms of its transformation, with the difference that the transforms are nonlinear.

As mentioned above, transforms are used both for shape representation and description, and Figure 5.4 summarizes some of the main techniques that are usually applied for obtaining descriptors from transforms. Some transform-based shape representations and descriptors are discussed in Chapter 6.

To probe further: *Signal and Image Transforms for Shape Analysis*

Additional transforms to those listed in the diagram of Figure 5.4 include non-linear multiscale transforms, singular value decomposition, Hadamard, Slant, Hart-ley, Hilbert, Radon, Abel, Hankel and Mellin. The reader is referred to Bracewell [1986, 1995]; Brigham [1988]; Castleman [1996]; Flandrin [1999]; Gonzalez & Woods [1993]; Jain [1989]; Mallat [1998]; Morrison [1994]; Poularikas [2000]; Vetterli & Kovacevic [1995] for additional information.

5.2 Parametric Contours

5.2.1 Contour Extraction

This section discusses some methods for obtaining a representation of the contours of a binary object (and of curves in general) in terms of a list of points. It is important to note that the contours of spatially quantized shapes can be represented directly in a binary image, which is adopted by several algorithms (see Chapter 6). Nevertheless, representing shapes directly by binary images implies some drawbacks (e.g., demands large storage space and does not explicitly identify the shape elements) that motivate the creation of alternative schemes. An important way of representing a contour is the so-called *parameterized* or *parametric representation*, which is analogous to parametric representation of curves in differential geometry. The process of obtaining this type of parameterized representation is henceforth referred to as *contour extraction*, though the terms *contour linearization* and *contour following* are also used in the literature.

In order to better understand the concept of parametric contours, refer to Figure 5.5, which presents the contour of a square shape in a hypothetical simple image. The parametric representation of this contour is obtained by initially defining an arbitrary starting point and traversing the contour from this point onwards. The contour can be traversed clockwise or counterclockwise. If the curve is open, the contour is traversed from one extremity to the other, and if it is closed, the contour is traversed until the initial point is revisited, assuming the case of *simple curves* (see Section 2.3.2), i.e., curves devoid of intersections. In case intersections and bifurcations occur, some alternative mechanism to represent such structures has to be devised [Leandro et al., 2008] (see related comments in Section 4.2.1).

The sense adopted for traversing a contour defines important differences in many situations. For instance, if the order is inverted, then the sign of the curvature is also inverted. Let the lower-left corner of the contour of Figure 5.5 be the

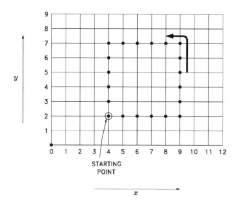

Figure 5.5: *Parametric contour representation on a discrete grid. In this case, the contour is followed arbitrarily in counterclockwise fashion.*

starting point and suppose that the contour is traversed in counterclockwise fashion. Therefore, the parametric representation of the contour begins with the initial point $(4, 2)$ (i.e., $x = 4$, $y = 2$), followed by the next point in the counterclockwise direction $(5, 2)$, which is followed by $(6, 2)$, and so on. The obtained complete ordered set of points is $(4, 2)$, $(5, 2)$, $(6, 2)$, $(7, 2)$, $(8, 2)$, $(9, 2)$, $(9, 3)$, $(9, 4)$, $(9, 5)$, $(9, 6)$, $(9, 7)$, $(8, 7)$, $(7, 7)$, $(6, 7)$, $(5, 7)$, $(4, 7)$, $(4, 6)$, $(4, 5)$, $(4, 4)$, $(4, 3)$.

Special attention should be taken since several applications and references consider the y-axis values increasing downward. The opposite sense, i.e., y-axis values increasing in the upward direction, is adopted throughout this book.

In case both open and closed curves can occur in a given application, it is useful to establish a convention in order to distinguish between them. For instance, it can be defined that closed curves are represented by letting the last point be equal to the first (e.g., as if $(4, 2)$ were the last point in the above example). Nevertheless, it is important to emphasize that this is just a convention. Moreover, in case closed curves are represented with the first point repeated at the end of the list, the latter has to be removed in order not to affect subsequent contour analysis techniques. Observe also that the contour in the above example can also be parametrically represented in terms of the following two lists of integer points representing the x and y coordinates:

$$x = \{4, 5, 6, 7, 8, 9, 9, 9, 9, 9, 9, 8, 7, 6, 5, 4, 4, 4, 4, 4\}$$

$$y = \{2, 2, 2, 2, 2, 2, 3, 4, 5, 6, 7, 7, 7, 7, 7, 7, 6, 5, 4, 3\}$$

The above two sequences of values can be thought of as two discrete signals and plotted as shown in Figure 5.6.

It is important to emphasize that the order of the points plays a fundamental role in several approaches. In addition, both the starting point and the tracking direction (clockwise or counterclockwise) should be defined by convention and used

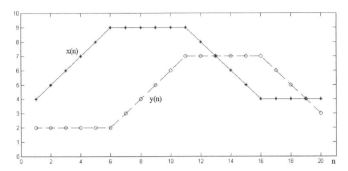

Figure 5.6: *Signals defined by the parametric contour representation of Figure 5.5.*

in a consistent fashion. This allows the set of points that compose the parametric contour (or the sets of x and y coordinates) to be indexed in terms of increasing parametric values. An example is the parameter n, in this case corresponding to the order of the contour elements along the shape contour, displayed in the horizontal axis of Figure 5.6.

It is worth noting that such signal representation of contours depends on the starting point in the sense that changing such point implies the whole signal to be shifted (to the left or to the right, depending on the situation) along the parameter (i.e., x-) axis.

Figure 5.7 shows four examples of shapes represented by parametric contours with the respective $x(t)$ and $y(t)$ signals. The 'o' symbol in each contour represents the corresponding starting points.

An important issue regards which contour is chosen in order to represent the object, and how open curves should be treated. More specifically, it should be first defined whether *external* or *internal contours* are to be taken into account. These concepts are illustrated in Figure 5.8 (a), which presents a filled rectangle as the object to be represented by a contour.

In this sense, depending on whether the contour points belong to the object, two different approaches are possible: if the contour points belong to the object, then it is called an *internal* contour; if the contour is connected to the object but does not belong to it, then it is called an *external* contour. The pixels that compose the external and the internal contours of the rectangle in Fgures 5.8 (a) and (b) are indicated by 'E' and 'I', respectively. Figures 5.8 (c) and (d) present the analogous situation where the object has one or more holes. Two contours are now required in order to represent such a shape, and it is again important to define what type of contours (i.e., internal or external) are expected. Both approaches, using internal or external contours, have their specific advantages and drawbacks, and the choice is often influenced by the adopted shape analysis methodology to be subsequently applied and by each particular application. It should be observed that it is also possible to have mixed schemes, where internal contours are adopted for holes and

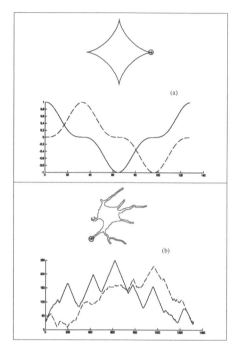

Figure 5.7: *Examples of shape contour and respective signals (solid plot represents x(t) while dashed plot corresponds to y(t)).*

external for outer boundaries, or vice versa. However, such schemes are rarely used.

One of the main differences between using internal or external contours is related to the fact that, if the considered shape has no holes, then the latter are always closed, which is not always the case with internal contours. A good example illustrating this difference is related to the representation of curve-like objects, i.e., thin objects that are one-pixel wide. An example is the shape of the character '1', which can be thought of as an open curve. If the shape is represented using the internal pixels, i.e., pixels that belong to the object, then the resulting representation is actually an open curve. On the other hand, if the external contour is taken instead, then a closed curve is obtained.

It is also important to observe that there are special situations occurring from time to time under both representations (i.e., by internal or external contours) that can adversely affect the measurement of the shape characteristics. The first situation is illustrated in Figure 5.9 (a), which presents a shape containing two parts joined by a one-pixel wide line. Clearly, no problem occurs if the external contour approach is adopted, since the contour would simply embrace the shape. However, if the internal approach were adopted instead, the contour would pass twice through the one-pixel wide linking segment. The problem is that this kind of

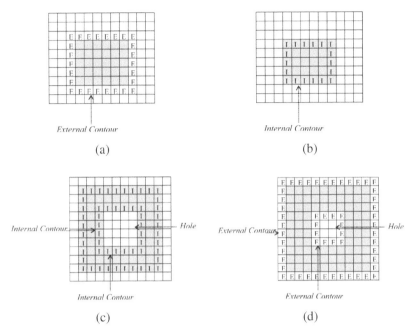

Figure 5.8: *Differences between external and internal contours obtained from discrete shapes. The internal contours are composed of pixels that belong to the shape, while the external contour pixels do not belong. See text for detail.*

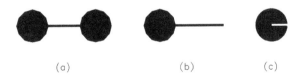

Figure 5.9: *Shape structures that imply special situations with respect to the obtained contours.*

resulting curve is no longer *simple*, i.e., it exhibits intersections. A more delicate situation arises in the case of a shape like that presented in Figure 5.9 (b). In this case, the shape presents one or more one-pixel wide appendix-like parts. While no problem occurs for the external contour approach, the internal contour representation would have to make a sharp turn at the appendix extremity, implying a pronounced singularity at that point. The curvature estimated by the majority of methods (see Section 6.4 for digital curvature estimation) is so high that it can undermine the result of curvature-based shape features, such as the bending energy. Special attention should be therefore taken in order to avoid such an effect.

A simple way to ameliorate the above problem is to search for such one-pixel wide structures and to introduce an extra point in order to eliminate the singularity, as illustrated in Figure 5.10. Suppose that the last three pixels near the extremity are

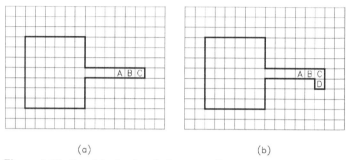

(a)   (b)

Figure 5.10: *Possible simple solution to ameliorate strong contour turns.*

labeled *A*, *B* and *C*, as shown in Figure 5.10 (a) and, therefore, the internal contour encoding this region is represented as ...*ABCBA*.... Clearly, the main problem is with the sequence *BCB*, corresponding to the strong turn. By introducing an extra pixel adjacent to the extremity, like pixel *D* in Figure 5.10 (b), the contour representation becomes ...*ABDCBA*... (assuming 8-neighborhood). The introduction of a pixel represents a relatively small price to pay in order to obtain more well-behaved contours, especially in situations involving one-pixel wide structures, such as in the analysis of neural cells or of line drawings.

Although both problems have been discussed with respect to the internal contour representation, similar problems can occur in the case of the external contours, an example of which is shown in Figure 5.9 (c). While in this case, which consists of a circle-like shape with a 1-pixel wide entrance, no problem occurs; in case of internal contours, a problem analogous to that in Figure 5.9 (b) arises while of external contour extraction, since that contour has to enter toward the shape and return through the same one-pixel wide path. A general technique to avoid the occurrence of all these problems is to apply morphological dilation (see Section 3.5.1) to the image objects before extracting the contours, since it allows one-pixel wide protuberances to be dilated and one-pixel wide entrances to be filled up. Nevertheless, this technique has the drawback of modifying the whole analyzed shape, which is undesirable in many situations.

The above considerations suggest a different method for representing open curves by allowing the internal contour to begin in one extremity, going along the curve until reaching the other extremity, and coming back through the same path. The resulting internal contour representation is periodic, but the obtained curve is not simple (in the mathematical sense).

5.2.2 A Contour Following Algorithm

There are many different algorithms for extracting parametric contours in a binary image, and some examples include the use of run-length codes [Kim et al., 1988] and chain codes (see Section 5.2.4) [Freeman, 1961, 1974]. A useful algorithm for extracting contours is explained in the following, which is usually applied to binary images such as previously segmented images and line drawings. This algorithm typically extracts the external contour of objects. Figure 5.11 illustrates the underlying idea behind the method.

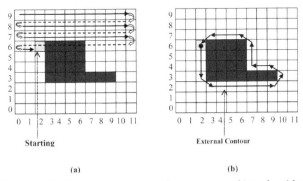

(a)   (b)

Figure 5.11: *Schemes illustrating the contour-tracking algorithm.*

First, an initial point belonging to the external contour of the object in hand must be selected. Assuming that the object pixels are black and that the background pixels are white, a simple approach to implement this initial step is to search line after line (i.e., from left to right, from up to bottom, as in Figure 5.11 (a)) until the first white pixel having a black pixel as a neighbor (at its right) is found. After identifying the starting pixel, the algorithm circumnavigates the connected object until the starting pixel is revisited, indicating the completion of the task (Figure 5.11 (b)). In fact, in order to obtain the parametric contour, it is sufficient to store the pixels' coordinates of the path tracked along the external contour.

It is assumed henceforth that the 8-neighbors of each pixel P are labeled 0, 1, 2, ..., 7, as shown in Figure 5.12 (as will be seen later, these numbers correspond to the *chain-code directions*). The resulting external contour will be 8-connected following the counterclockwise direction, as the example in Figure 5.11 (b). Once

3	2	1
4	P	0
5	6	7

Figure 5.12: *Labeled neighborhood used by the contour-tracking algorithm.*

the starting pixel has been found, the algorithm has to decide which is the next contour pixel. At each step, a search for the next neighbor pixel is carried out by placing the mask of Figure 5.12 in such a way that its central position P lies over the current contour pixel. The next pixel will be one of the neighbors labeled $0, 1, \ldots, 7$. In the initial case, where the current is the starting pixel, the algorithm tests the neighbors $4, 5, 6$ and 7. The other neighbors should not be tested since they have already been verified during the previous scanline search (Figure 5.13). Despite the fact that the neighbor labeled 4 has also already been visited, it has to

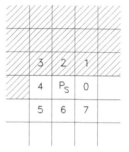

Figure 5.13: *Positions already verified by the initial scanline search of the tracking algorithm.*

be verified for an eventual object pixel immediately under it. Therefore, neighbor 4 is a candidate if it is white and neighbor 5 is black. Neighbor 5 is a candidate if it is white and neighbor 6 is black, and so forth. It is worth noting that neighbor 7 is a candidate if it is white and neighbor 0 is black. Figure 5.14 illustrates all these possibilities. In brief, at this initial step the method chooses the first candidate that it finds, searching from 4 to 7.

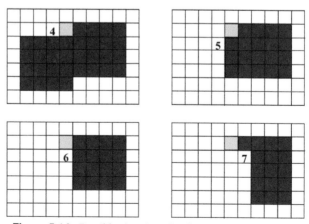

Figure 5.14: *Possible situations for the second contour point.*

After the second contour pixel has been found, the algorithm proceeds tracking along the external contour in an analogous, but slightly different manner. The algorithm needs 2 important variables at each step: the current contour pixel, denoted by $E[n]$ and the direction followed from the previous neighbor to the current pixel, denoted by d_{pc}. The direction from the current pixel to the previous, denoted as d_{cp}, can be calculated from d_{pc}, and we assume that there is a function invert(d) that implements this task, which is illustrated in Table 5.1.

D	invert(d)
0	4
1	5
2	6
3	7
4	0
5	1
6	2
7	3

Table 5.1: *The invert function.*

It is now desired to find the next pixel and the direction from the current pixel to it (denoted as d_{cn}). Starting from d_{cp}, the algorithm tests the neighbors (in counterclockwise direction) in order to find the contour pixel candidates, which is done in an analogous way to the above-explained initial step. Nevertheless, differently from that case, the last candidate (and not the first) is taken as the next contour pixel. Figure 5.15 illustrates schematically a typical situation. In that figure, the three pixels labeled P, C and N correspond to the previous, current and next pixels, respectively. Therefore, in the present example, the direction d_{pc} from the

Figure 5.15: *The tracking algorithm uses three pixel positions at each step, namely the previous (P), the current (C) and the next (N) contour pixels.*

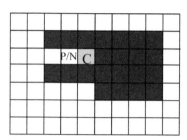

Figure 5.16: *An illustration of the special situation dealt with by the track-*
ing algorithm, identifying the previous (P), current (C) and
next (N) contour pixels.

previous to the current pixel is 6, implying $d_{cp} = 2$. The algorithm tests the neighbors from $d_{cp} = 2$ to $d_{cp} = 0$ (the neighbor corresponding to $d_{cp} = 1$ does not
need to be verified because its next neighbor is a contour pixel which is surely
white).

It is important to note that even the previous pixel can also be a candidate for
the next pixel, since this allows the algorithm to enter and to leave one pixel-wide
entrances into the contours, as illustrated in Figure 5.16. However, the previous
pixel is taken as the next one only when there are no more candidates, in order to
avoid a 'next-previous-next-previous. . .' loop situation.

The contour extraction algorithm may be summarized as follows:

Algorithm: *Contour Following*

    ▷ Starting pixel
1. find $E[1]$ in scanline;
2. $n \leftarrow 2$;
3. let next_pixel be the second contour pixel;
4. let d_{cn} be the direction from $E[1]$ to next_pixel;
5. **while** (next_pixel $\neq E[1]$)
6.   **do**
7. $E[n] \leftarrow$ next_pixel;
8. $d_{pc} \leftarrow d_{cn}$;
9.   FIND_NEXT($E[n]$, d_{pc}, next_pixel, d_{cn});
10. $n = n + 1$;

Algorithm: *Find Next Contour Pixel*

FIND_NEXT(P_c, d_{pc}, P_n, d_{cn})

1. $d_{cp} \leftarrow$ INVERT(d_{pc});
2.   **for** $r \leftarrow 0$ **to** 6
3.   **do**
4. $d_E \leftarrow$ MOD($d_{cp} + r, 8$);
5. $d_I \leftarrow$ MOD($d_{cp} + r + 1, 8$);
6. $P_E \leftarrow$ CHAINPOINT(P_c, d_E);
7. $P_I \leftarrow$ CHAINPOINT(P_c, d_I);
8.   **if** (ISBACKGROUND(P_E) and ISOBJECT(P_I))
9.   **then**
10. $P_n = P_E$;
11. $d_{cn} = d_E$;

After the completion of such a procedure, the vector $E[n]$ will store the external contour. The above algorithms use the modulus function $mod(x, y)$, which returns the remainder after the division of x by y, i.e.,

$$mod(x, y) = x - y * floor\left(\frac{x}{y}\right), y \neq 0$$

This function is present in the majority of programming languages. The above algorithm also assumes the existence of a function chainpoint(P, d) that simply returns the coordinates of the neighbor pixel of P in the direction d.

The above algorithm stops when it finds the starting point again, indicating that the object has been circumvented. Nevertheless, a potential problem can arise if an object of the type shown in Figure 5.17 appears.

Starting

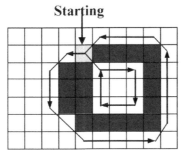

Figure 5.17: *This special shape structure can be addressed with a simple modification of the tracking algorithm. The same problem may also arise if the upper-left pixel in the above shape is not present, and this case should be addressed as well.*

The problem is that the starting point is also a point that the algorithm should traverse in order to trace the inside contour region (i.e., the curve is not Jordan). However, the above algorithm will stop when it finds the starting pixel and the internal contour region will not be traced. This problem can be solved by creating a flag that indicates that there are two different paths from the starting point. As explained, because the algorithm chooses the first candidate, the contour will be traced counterclockwise by the external region first (this flag can be set to 1 simply by verifying that the starting pixel has more than one contour neighbor candidate). Then, this flag should also be tested by the while loop of the above contour tracking algorithm, indicating that the algorithm should not stop if it is set on. In this case, the algorithm proceeds in the internal region and the flag is set off. Therefore, after the inside contour portion has been traced, the starting pixel is reached again and the algorithm stops.

It is also worth noting that the algorithm could be adapted in order to extract the internal shape contour simultaneously. On the other hand, the internal contour can also be easily obtained by inverting the binary image pixels (i.e., background pixels are set to 1 and object pixels to 0), followed by the application of the above algorithm. Nevertheless, it is important to note that this procedure leads to clockwise internal contour tracking. Finally, the algorithm assumes that there is only one object in the image. This is important to note because the test that verifies if a given neighbor is a contour candidate does not verify whether a possible black pixel actually belongs to the object corresponding to that contour. If more than one object can be present in the image, then a previous connected component labelling could be used in order to discriminate the different objects. In this case, it would be sufficient to adapt the isobject(P_I) function in the *find_next* procedure above.

Example: *Contour Following*

Consider the shape of Figure 5.18, which also indicates the path followed by the contour following algorithm. The obtained contour is

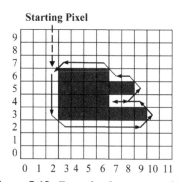

Figure 5.18: *Example of contour tracking.*

$(2, 6), (2, 5), (2, 4), (2, 3), (3, 2), (4, 2), (5, 2), (6, 2), (7, 2), (8, 2), (9, 2), (10, 3),$
$(9, 4), (8, 4), (7, 4), (8, 4), (9, 5), (8, 6), (7, 6), (6, 7), (5, 7), (4, 7), (3, 7).$

By convention, as the contour is closed, the first point could be repeated as the last point.

To probe further: *Contour Following*

There are many variations of the contour following algorithm which generate slightly different contours. Bennett & Donald [1975] have presented and evaluated a series of contour following algorithms and their implications with respect to digital curvature estimation, thus producing an important reference work. Additional references of interest are Chassery & Montanvert [1991]; Shahraray & Anderson [1985, 1989].

Additional resources: *Contour Following Software*

The open-source shape analysis software available at http://code.google.com/p/imagerecognitionsystem/ presents contour following and extraction functions available.

5.2.3 Contour Representation by Vectors and Complex Signals

Once the parametric contour has been extracted by a contour following algorithm (Section 5.2.2), it can be represented by different data structures. The simplest approach is to represent the x and y coordinates as ordinary vectors that are generally supported by the large majority of high-level programming languages, like C, Pascal, Delphi and MATLAB® scripts. Each contour pixel can be accessed by its parametric position, i.e., the n-th point along the contour has coordinates $x(n)$ and $y(n)$ (or $x[n]$ and $y[n]$, depending on the programming language). A structure can also be adopted, such as

```
struct {
   double x[MAXCONTOUR];
   double y[MAXCONTOUR];
} contour;
```

The variable *contour* in the above example is a structure in the C language, with the constant *MAXCONTOUR* denoting the maximum length allowed for the contour parameter value. The n-th contour point has coordinates (*contour.x[n]*, *contour.y[n]*). The above declaration also assumes that the contour coordinates can be expressed as real (floating points) numbers, rather than integers. Although the image coordinates are integers, contour manipulation can lead to real numbers, hence such a representation. Analogous representations can be defined in the majority of alternative programming languages (e.g., by using *record* in Pascal).

An alternative implementation for representing contours is by using linked lists as underlying data structure. For instance, a double-linked circular list composed of nodes with four fields could be used. In this case, two fields would be used for storing the x and y coordinates, and two pointer fields would be used to link the current node with its predecessor and successor, as depicted in Figure 5.19. Such dynamic structures can be straightforwardly implemented using structured

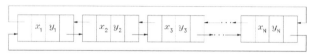

Figure 5.19: *Contour representation by a double-linked list.*

programming languages, which allow the use of flexible and efficient algorithms (Langsam et al. [1996] provides a good introductory textbook on this issue). One of the main advantages of adopting such powerful data structures is that they allow modifications, such as the inclusion or removal of additional cells and the incorporation of additional information. For instance, the data structure presented in Figure 5.19 could be easily modified in order to include additional fields to store the curvature along the contour or to indicate that the point is a vertex of a polygonal approximation (see Section 5.4.1).

In addition to using such interesting data structures, it is also useful to consider object-orientation paradigms. For instance, by creating a contour class, its objects can be endowed with interesting attributes and methods, such as contour characterization (curvature and derivatives), manipulation (Gaussian smoothing, zero-crossing detection, curvature thresholding, normalization), and measurements (perimeter, contour energy, bending energy, coefficient of variation, fractal dimension, among others). All these shape measures are discussed in detail in Chapter 6.

By assuming that the image is a complex plane, contours can also be represented in terms of complex-valued signals, whose real parts are represented in the x-axis and imaginary parts in the y-axis. In other words, an image point (x_0, y_0) is represented by the complex number $x_0 + jy_0$, with $j = \sqrt{-1}$. Therefore, a parametric contour $(x(n), y(n))$ defines a complex signal $u(n) = x(n) + jy(n)$, with $n = 0, \ldots, N - 1$. The complex signal representation is frequently adopted in signal processing methods (e.g., Fourier and wavelets, see Chapter 7), being particularly suitable for implementation in mathematical software, such as MATLAB®, Mathematica® and Scilab®, as well as for mathematical notation. In case of a programming language such as C, a record similar to the above contour data structure could be used, e.g.,

```
struct {
    double real_part[MAXCONTOUR];
    double imaginary_part[MAXCONTOUR];
} complex;
```

Obviously, all complex number operations (sum of two complex numbers, modulus, etc.) should be programmed as functions in the program.

To probe further: *Parametric Contour Representation*

It is worth noting that simple contour tracking algorithms do not lead to arc length parametric contours because the traditional square grid is not isotropic. The reader interested in contour interpolation and resampling is referred to Shahraray & Anderson [1985], which provides a good starting point into this area. In addition, there are alternative approaches to contour extraction based on edge and curve detection, such as through the Canny edge detector [Canny, 1986] (which includes contour following). Iverson & Zucker [1995] provide an example of a curve detection approach.

5.2.4 Contour Representation Based on the Chain Code

A simple and popular alternative to contour representation is based on the chain code proposed in Freeman [1961]. In order to understand the chain code scheme, refer to Figure 5.20, which represents a central pixel indicated over an image grid. In an 8-neighborhood, each pixel has 8 neighbors, which can be numbered from

3	2	1
4	●	0
5	6	7

Figure 5.20: *Chain code.*

0 to 7, as indicated in that figure. These numbers are the *chain codes* used by the representation. First, observe that the neighbors have been numbered counterclockwise starting from the right neighbor pixel. Now, refer to Figure 5.21 (the same contour of Figure 5.5), and let p be the starting point of that contour, i.e., $p = (4, 2)$. The next neighbor pixel in the counterclockwise direction is the pixel to the right, i.e., $(5, 2)$. If the mask of Figure 5.20 is superimposed onto the grid

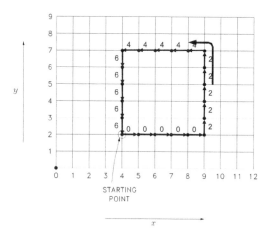

Figure 5.21: *Chain-code representation of a sample contour.*

in Figure 5.21 so that the mask central pixel matches the pixel p, then the pixel $(5, 2)$ will correspond to the chain code 0, as indicated in the figure. If the mask is now shifted over the new pixel $(5, 2)$, then the next neighbor is $(6, 2)$, whose chain code is also 0. Figure 5.21 illustrates the repetition of this process until the whole contour is traversed. It is easy to see that this contour is represented by the chain code 00000222224444466666 plus the starting point $(4, 2)$.

There are many interesting shape properties (see Box in Section 6.2.16) that can be extracted directly from the chain code representation [Angulo & Madrigal, 1986; Gonzalez & Woods, 1993]. Two important drawbacks with the chain code representation are that it depends on the starting point and that it varies as the shape is rotated. In order to be invariant to the choice of the starting point, the chain code should be circularly shifted (i.e., when shifting it in one position to the right, the last point becomes the first) a certain number of times in order to obtain the smallest integer formed by the chain code sequence (e.g., the chain code 7135 can be read as the integer number seven thousand, one hundred and thirty-five). The *first difference chain code* can be applied in order to circumvent the rotation problem, where the number of directions between each pair of consecutive codes is used. For instance, in the case of the above contour, whose chain code is 00000222224444466666, the first difference is 20000200002000020000, whose first element corresponds to the difference between the first and the last chain codes. The first difference representing the smallest integer defines a descriptor called *shape number*. The high frequency of repetitions of chain codes usually found in most contours also suggests the use of run-length coding [Gonzalez & Woods, 1993] for compression purposes. Basically, this technique involves replacing each run of repeated codes by a single code. The box in Section 6.2.16

discusses some interesting shape descriptors that can be obtained from the chain code.

5.3 Sets of Contour Points

Several of the approaches for representing the shape as a parametric contour discussed in Section 5.2 are based on tracking the shape boundary in a given order/sequence. A different approach to represent the shape boundary is to store the set of contour points without any given order, i.e., as a set. This representation is useful for obtaining several descriptors for shape characterization, mainly those presenting a global or statistic nature (such as the centroid and the major axis, see Chapter 6). The contour set of points could be represented by the following data structure:

```
struct {
    double x[MAXCONTOUR];
    double y[MAXCONTOUR];
} contourset;
```

Assuming that $g(p, q)$ is a binary image where $g(p, q) = 1$ for shape pixels and $g(p, q) = 0$ for background pixels, the algorithm below creates a set of contour points stored in the variable cs, which is of the above defined contourset type.

Algorithm: *Set of Contour Points*

1.   Calculate the edges of g and store them in *edges_g*;
2. $np \leftarrow 0$;
3.   **for** $p \leftarrow 1$ **to** P
4.   **do**
5.   **for** $q \leftarrow 1$ **to** Q
6.   **do**
7.   **if** *edges_g*$(p, q) = 1$
8.   **then**
9. $np \leftarrow np + 1$;
10. $cs.x[np] \leftarrow p$;
11. $cs.y[np] \leftarrow q$;

Refer to Medeiros et al. [2007] for additional information.

5.4 Curve Approximations

The third approach for representing shape boundaries is by approximating or interpolating them. Among the many techniques for curve approximation, vision researchers have recognized that rather than representing whole contours by a single function or series, it is generally more interesting to derive piecewise representations that approximate each contour portion by a geometric primitive such as a straight line segment [Pavlidis, 1986]. Therefore, this approach can be divided into the following two main steps: *contour segmentation* and *curve segment approximation*. The contour points that define the contour segmentation are called *dominant points*, and can be defined in a number of different ways. Chapter 7 discusses two different techniques for searching for dominant points based on curvature and on local periodic patterns, respectively. The following subsections present two approaches for polygonal approximation of contours, where the geometric primitives that are fitted to each contour segment are straight lines.

To probe further: *Dominant Points*

Although some specific references on contour segmentation and dominant points are discussed throughout the text, the reader is referred to the following papers for additional information: Fischler & Wolf [1994]; Held et al. [1994]; Medioni & Yasumoto [1987]; Stein & Medioni [1992]; Tsang et al. [1994].

5.4.1 Polygonal Approximation

A particularly important and difficult task in shape analysis is the segmentation or partitioning of a contour into meaningful parts [Fischler & Wolf, 1994; Garcia et al., 1995]. Such a decomposition of the contour in segments may be used in a number of different manners, such as in syntactical pattern recognition [Fu, 1982; Pavlidis, 1977] and in model-based recognition [Tsang et al., 1994]. Contour partitioning can be divided into two generic phases: the first phase determines the segmentation points along the contour, while the second phase represents each segment in terms of instances of a predefined geometric primitive. Since the simplest primitives, which are often adopted, are straight segment lines, the output of such process is a polygon-like representation of the original contour. This approach is called *polygonal approximation of contours* [Ramer, 1972]. An example is shown in Figure 5.22.

The problem of contour polygonal approximation can hence be understood as finding polygonal vertices along the contour in such a way that a good approximation of the original curve is obtained. The classical approach to this problem is taking as vertices the points with high modulus of curvature along the contour. A more in-depth discussion about this possibility is presented in Chapter 6, including some comments on the importance of high curvature points with respect to human perception and psychophysics. It is also observed that the use of high curvature points can be traced back to the early 1960s [Tsang et al., 1994].

The methods for vertex detection and polygonal approximation can be divided into two principal classes: global methods and local methods. Global methods

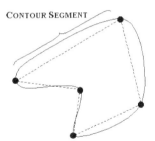

Figure 5.22: *Original shape contour and a possible representation by polygonal approximation.*

are generally based on the polygonal approximation of the contour in such a way that some error function is minimized. An example of such an error function is the larger distance between the contour points and the polygon edges. Additional examples include the following:

☞ maximal distance between the polygon line segments and the contour segments

☞ minimal polygon perimeter;

☞ maximal internal polygon area;

☞ minimal external polygon area.

Additional possible criteria for curve approximation can be found in the review of Loncaric [1998].

Some classical methods to solve this problem were developed during the 1970s, and the works of Pavlidis [1973]; Pavlidis & Horowitz [1974]; Ramer [1972] are particularly interesting. Ramer developed an elegant split-based algorithm, which has been extended by Horowitz and Pavlidis into a split-and-merge technique for contour segmentation. Such a method follows the same philosophy of the split-and-merge technique for image segmentation, presenting important connections with variational methods developed more recently [Morel & Solimini, 1995].

On the other hand, local methods are based on the idea of directly searching for high curvature points along the contour. Although the different techniques for vertex detection in contours include a large variety of approaches, such as neural networks [Sanchiz et al., 1996], electrical charge distributions [Wu & Levine, 1995], and non-linear algorithms [Zhu & Chirlian, 1995], the reality is that this apparently simple problem for human perception does not have an optimal unique solution, which accounts for the profusion of methods introduced in the literature (see for example Cornic [1997]). Chapter 7 illustrates the important fact that contour segmentation can actually depend on the spatial scale of analysis.

5.4.2 Ramer Algorithm for Polygonal Approximation

The polygonal approximation algorithm presented in this section has been introduced by Ramer [1972], being considered simple, elegant and with potential for
producing good results. This algorithm seeks a polygonal approximation of the
contour through an iterative mechanism with a small number of vertices, depending on a single predefined error parameter. The maximal distance between the
polygon line segments and the respective contour segments is used as the error
measure. The vertices of the polygonal approximation are the points of the original
contour identified by the iterative process as being the extremities of the fitted line
segments.

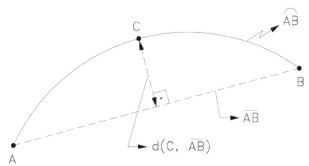

Figure 5.23: *Polygonal approximation of a contour segment.*

Figure 5.23 illustrates the typical situation in Ramer's algorithm. Contour segments are henceforth represented by a pair of contour points with a curve segment
as superscript (e.g., $\overset{\frown}{AB}$ denotes the contour segment between A and B), while line
segments adopt a straight line segment as superscript (e.g., $\overline{AB}$). Suppose that the
contour segment $\overset{\frown}{AB}$ is to be represented (or approximated) by the line segment $\overline{AB}$.
In order to determine whether this approximation is valid, the algorithm evaluates
the distance between every point of $\overset{\frown}{AB}$ and the line segment $\overline{AB}$ (recall that the
distance between a point and a line segment corresponds to the smallest distance
between the point and any of the points along the line segment). In the example
of Figure 5.23, the distance between point C of $\overset{\frown}{AB}$ and the line segment $\overline{AB}$ is
indicated as $d(C, \overline{AB})$. Suppose that C is the farthest point of $\overset{\frown}{AB}$ with respect to
$\overline{AB}$, i.e., $d(P, \overline{AB}) \leqslant d(C, \overline{AB}), \forall P \in \overset{\frown}{AB}, P \neq C$. In case $d(C, \overline{AB}) < \varepsilon$, where ε
is the maximal approximation error allowed, it is decided that $\overset{\frown}{AB}$ can be approxi-
mated by $\overline{AB}$. It is observed that the parameter ε corresponds to an input parameter
of the algorithm. Large values of ε imply coarse polygonal approximations, with
a smaller number of vertices and sides, whereas small values of ε tend to produce
polygonal approximations closer to the original curve but involving a larger number
of vertices and sides.

The algorithm starts with an initial solution and proceeds iteratively until the error measure is verified for every contour segment approximated by a straight line segment. In the case of Figure 5.23, if $d(C, \bar{AB}) > \varepsilon$ AB cannot be represented by $\bar{AB}$, since the largest distance between the points of AB to the line segment $\bar{AB}$ is larger than the maximum allowed approximation error (which corresponds to the distance between C and $\bar{AB}$). In this case, the contour segment $\widehat{AB}$ should be subdivided and approximated by two or more line segments. This is the *splitting* step of the algorithm, which explains why it is also known as the *split algorithm*.

Although there are many ways for dividing the curve segment $\widehat{AB}$ in two parts, a natural choice is to split it exactly at C, i.e., to take the farthest point with respect to $\bar{AB}$ as the splitting point. The obtained approximation after this splitting step is shown in Figure 5.24 (a), where the segment $\widehat{AB}$ is now represented by $\bar{AC}$ and $\bar{CB}$. It is worth noting that this splitting operation is equivalent to the introduction of a

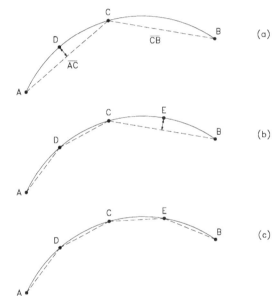

Figure 5.24: *Successive steps of the polygonal approximation by the Ramer algorithm.*

vertex in the polygon that is being created in order to represent the original contour. Furthermore, it is intuitive to think that in the case of contour segments like that in Figure 5.24 (a), the farthest (splitting) point should correspond to a high curvature point, which reflects the relationship between high curvature points and the vertices of contour polygonal approximations.

Once the splitting point has been defined and the contour segment divided, the algorithm proceeds iteratively (or recursively) by applying the same set of operations over each of the newly created contour segments. Suppose the contour segment $\widehat{AC}$ is the first to be considered. Therefore, the distances between all points of

$\bar{AC}$ and the line segment $\overline{AC}$ are calculated and the maximal distance is compared again to the maximal error ε. Let D be the farthest point of $\bar{AC}$ with respect to $\overline{AC}$ (see Figure 5.24 (a)) and suppose again that $d(D, \overline{AC}) > \varepsilon$. As before, the contour segment should be split into $\bar{AD}$ and $\bar{DC}$, which should be approximated by $\overline{AD}$ and $\overline{DC}$ respectively, as depicted in Figure 5.24 (b). Then, both newly created segments should undergo the same approximation tests. It is worth observing that as this iterative processing continues, the algorithm has to keep track of all segments that have not been tested yet (e.g., $\bar{CB}$ of Figure 5.24 (a)), so that it can come back to them later. This strategy can be easily implemented by using recursive procedures or, alternatively, by the adoption of stacks. The algorithm presented below adopts the stack-based solution.

Suppose that the approximation test over the segment $\bar{AD}$ succeeds, meaning that the maximal distance between the points of $\bar{AD}$ and the line segment $\overline{AD}$ is smaller than ε. Then, the algorithm considers the next segment in the stack, i.e., $\bar{DC}$. Let us also assume that $\bar{DC}$ can be successfully approximated by $\overline{DC}$. Then the algorithm proceeds by considering the next segment in the stack, which actually is $\bar{CB}$. Let E be the farthest point of $\bar{CB}$ with respect to $\overline{CB}$, in such a way that $d(E, \overline{CB}) > \varepsilon$. Therefore, the segment $\bar{CB}$ is split into $\bar{CE}$ and $\bar{EB}$, which should be approximated by $\overline{CE}$ and $\overline{EB}$, respectively. Supposing that the approximation tests succeed for both segments, the final polygonal approximation of the curve segment $\bar{AB}$ is given by the line segments $\overline{AD}$, $\overline{DC}$, $\overline{CE}$ and $\overline{EB}$, as shown in Figure 5.24 (c). It is worth emphasizing that, in order to store this polygonal approximation, it suffices to store the position of the vertices A, D, C, E and B.

The splitting procedure can be visualized as a binary tree-like hierarchical structure, which is shown in Figure 5.25. In this structure, each node corresponds to a

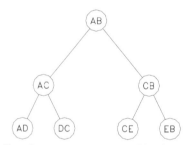

Figure 5.25: *Tree data structure generated by the Ramer algorithm.*

pair of polygonal vertices, being the final piecewise linear approximation defined by the leaves of the tree (i.e., the bottom nodes without sons). Each level of the tree corresponds to an analyzing scale, which means that the polygonal approximations defined by the nodes of the levels nearer to the root (the initial uppermost node) correspond to rough representations of the original contour. On the other hand, the polygonal approximations defined by the nodes of lower levels include progres-

sively small-scale details, leading to richer, although more sensitive to noise, representations. Such a tree representation is an example of a hierarchical data structure
for shape description (see Felzenszwalb & Schwartz [2007] for an application).

The algorithm depends on the specification of an initial solution (points A and
B in the case of the previous example), which can be done in many ways, with
open and closed curves being treated differently. In the case of open curves, the
easiest solution is to take the initial and final points as the initial solution (as in the
previous example, it is assumed that the segment $\overset{\frown}{AB}$ actually corresponds to the
whole curve). Before discussing the closed curves situation, it is worth noting that
the points of the initial solution will remain in the final polygonal approximation.
This implies that if a bad initial solution is specified, the final solution will be under-
mined. Refer to Figure 5.26 for an example, which shows a simple geometric figure
(a rectangle) as the original contour to be approximated (Figure 5.26 (a)). The sec-

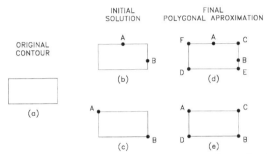

Figure 5.26: *A major problem with the Ramer algorithm is that the final
result depends on the initial solution.*

ond column of this figure illustrates two possible initial solutions (Figures 5.26 (b)
and (c)), while the third column presents the respective final solutions given by
the Ramer algorithm (respectively in Figures 5.26 (d) and (e)). In the case of Figure 5.26 (b), a bad initial solution has been provided (points A and B), and the final
solution, shown in Figure 5.26 (d), is a polygon with six vertices, instead of four
as would be desirable. On the other hand, the initial solution provided in the case
of Figure 5.26 (c) corresponds to two of the truly expected vertices, leading to the
4-vertices polygon in Figure 5.26 (e).

If the curve is closed, it suffices to provide two contour points as the initial
solution and to treat the contour as two open curves. For instance, in the case
of Figure 5.26 (c), the open curves algorithm is run on the segment $\overset{\frown}{AB}$ and, in a
second phase, $\overset{\frown}{BA}$. In addition to the above considerations about the initial solution,
it is important to carefully choose a good initial pair of points and, ideally, a pair
of vertices of the expected final polygonal approximation. Although there is no
optimal way of doing this, many different approaches can be adopted, including the
following:

Local maxima curvature points: The first approach is to use a curvature estimation technique (see Chapter 6) in order to detect a pair of points corresponding to local curvature maxima.

Upper-left and bottom-right points: A heuristic and simpler approach is to search for the upper-left and for the bottom-right points which, as extreme points, are usually good candidates for a polygonal representation of the original contour. An example of the application of this approach is presented in Figure 5.26 (c).

First and middle points: The simplest approach is to choose two points randomly or in a pre-defined manner. For instance, always choose the first and the middle point of the curve. Of course, this simple approach is more subject to sensitivity to the initial approximation, as discussed above.

The algorithm makes use of two stacks, *open* and *closed*. As the vertices are found (i.e., the splitting points), they are pushed into *open*. The contour segment defined by the vertices on the top of *open* and of *closed* is tested for approximation by a polygon side and, in case this test succeeds, the vertex is popped from *open* and pushed in *closed*. The algorithm proceeds iteratively until *open* is empty, which is the end condition. At the end of the algorithm execution, the *closed* stack contains the ordered vertices of the polygonal approximation. The implementation of the above algorithm by using stacks is presented below.

Algorithm: *Ramer for Open Curves*

1. PUSH(closed, first point)
2. PUSH(open, last point)
3. **while** (NOT(EMPTY(open)))
4.   **do**
5. $A \leftarrow$ STACKTOP(closed);
6. $B \leftarrow$ POP(open);
7.   Find $C \mid d(P, \bar{A}B) \leqslant d(C, \bar{A}B), \forall P \in \widehat{AB}, P \neq C$;
8.   **if** $d(C, \bar{A}B) < \varepsilon$
9.   **then**
10.   PUSH(closed, B);
11.   **else**
12.   PUSH(open, B);
13.   PUSH(open, C);

The reader should refer to textbooks on data structures for additional information on stacks [Langsam et al., 1996]. The above algorithm can be easily modified to treat closed curves:

Algorithm: *Ramer for Closed Curves*

1. PUSH(closed, first point);
2. PUSH(open, first point);
3. PUSH(open, last point);
4. (...)

Considering the curve in Figure 5.26 (c), this modified algorithm treats first the open curve defined as $\overset{\frown}{AB}$, and then $\overset{\frown}{BA}$. The reader is referred to the original work [Ramer, 1972] for a discussion on some methods for calculating the distance between points and lines.

5.4.3 Split-and-Merge Algorithm for Polygonal Approximation

Although the Ramer algorithm is simple and can produce good results depending on the contour and on the initial solution, it presents the following drawbacks:

☞ The points of the initial solution remain and will be present in the final solution. Therefore, even if a point of the initial solution does not actually correspond to an expected vertex, it is anyway taken (mistakenly) as a vertex.

☞ The Ramer algorithm does not include any special procedure to evaluate the segments between subsequent dominant points generated from the initial solution.

If these problems can be solved, the algorithm becomes independent of the initial solution, which defines the motivation for the split-and-merge algorithm discussed in this section.

We have already seen that the Ramer algorithm takes an initial solution in the form of a segment between two input dominant points and breaks it iteratively until all created segments can be approximated by line segments up to a pre-specified error measure. Now let us consider the situation depicted in Figure 5.27.

Figure 5.27: *Curve segment with some possible segmentation points.*

Assume that the dominant points $A, B, \ldots, I$ have been found as part of the polygonal approximation of the considered curve in the above figure. Observe that

G is the farthest point of the curve segment $\widehat{DC}$ with respect to the approximating segment $\overline{DC}$ and that F is the farthest point of $\widehat{DG}$ with respect to $\overline{DG}$. Therefore, if F and G are taken as dominant points, it is because the distance between these points and the respective approximating line segments is greater than the maximal allowed approximation error. Now, suppose that the distance between G and $\overline{FC}$ is smaller than the maximal approximation error. The line segment $\overline{FC}$ consequently becomes a good approximation for $\widehat{FC}$, and G can be eliminated from the representation. The main problem with the Ramer algorithm is that it is not able to eliminate vertices.

The above discussion can be immediately generalized. Let a be a curve segment that has been partitioned into the two segments b and c. Suppose that segment b has been further partitioned into the segments d and e, and that segment c has yielded segments f and g. In order to reach this situation, the curve segments a, b and c have had to be tested. Nevertheless, the curve segment composed by the union of the segments e and f is not evaluated. The split-and-merge algorithm incorporates a mechanism to evaluate such segments and *merge* them if necessary. This algorithm searches for the dominant points that define the polygonal approximation of a curve by using the following two steps:

① A curve segment is analyzed and if it cannot be approximated by a straight line segment then it is split in two parts.

② The union of adjacent segments are evaluated and eventually merged into a single segment.

The procedures that implement each of the above steps are respectively discussed below.

The Split Step

The Ramer algorithm uses two stacks, *open* and *closed*, and the current segment is defined by the dominant points on the top of both stacks. The split step can be easily adapted from the Ramer algorithm as follows. Instead of initializing the stacks in every iteration of the algorithm, it suffices to call the split procedure with the currently found dominant points stored in the stack *open*. This split procedure then pops the first point of *open* and pushes it into *closed*. From this point onward, the split procedure follows in analogous fashion to the above-presented Ramer's procedure. The output of the split step is a stack containing all the input plus the newly found dominant points (if there are any).

The Merge Step

Before introducing this algorithm, it is interesting to consider the following example. Refer to the contour illustrated in Figure 5.28, which also presents some previously found polygonal approximation dominant points. In this representation,

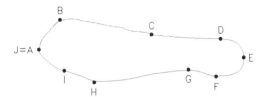

Figure 5.28: *Contour with possible segmentation points obtained by splitting.*

it is assumed that the contour is closed, which is indicated by the fact that the first point is equal to the last, i.e., $A = J$. As in the case of the split procedure, the input is defined by a stack storing the current dominant points. The order of these points in the stack obeys the distribution of the points along the contour. Let M_0 be this stack representing the points as $M_0 = \{A, B, C, D, E, F, G, H, I, J\}$.

As already observed, adjacent segments should be tested in order to verify whether they can be joined (merged) and approximated by a single line segment extending from the first dominant point of the first segment to the last dominant point of the second segment. In case $\overline{AB}$ and $\overline{BC}$ are the two adjacent segments to be considered, the algorithm should test the distance between the farthest point of $\overline{AC}$ and the line segment $\overline{AC}$. If this distance is larger than the maximal allowed error, the two contour segments cannot be merged.

Now, assume that the adjacent segments $\overline{CD}$ and $\overline{DE}$ have been tested and that the distance between the farthest point of $\overline{CE}$ and the line segment $\overline{CE}$ is smaller than the maximal allowed error. In this fortunate case, the two segments can be merged yielding the new segment $\overline{CE}$, which is represented by the polygonal side $\overline{CE}$. The algorithm proceeds analogously by verifying whether the newly defined segment $\overline{CE}$ and the subsequent segment $\overline{EF}$ can be merged.

These steps proceed iteratively until the last two segments, such as $\overline{HI}$ and $\overline{IJ}$ in Figure 5.28, are tested. Then, the segments $\overline{IJ}$ and $\overline{AB}$ can finally be tested. It is worth noting that it is just at this point that the point J $(= A)$ is tested. There are some important related issues that should be carefully considered:

☞ As with Ramer's algorithm (and the split procedure), the process can be controlled by considering each segment defined in terms of its respective dominant points.

☞ While testing each pair of adjacent segments, the algorithm manipulates three dominant points (the first, the middle, which is also the second extremity of the first segment and the first extremity of the second segment, and the last).

☞ The dominant points are already stored in M_0.

☞ Special care should be taken when testing for the union between the last and the first segments.

In practice, merging two segments is equivalent to discarding the middle corresponding dominant point from the list of valid vertices. Therefore, the algorithm uses an additional stack M_1 to store the valid dominant points, and the middle points of the merged segments are simply not included in this stack.

It is also important to note that the test for merging two segments takes into account only the middle point between these two segments. This is a consequence of the fact that even if there is a farther point with respect to the new approximation line segment, the middle point between the current pair of contour segments can be eliminated, and the farthest point will be subsequently located by the split procedure.

Controlling the Split and the Merge Procedures

The main algorithm should provide the initial solution to be used by the split and the merge algorithms. As observed above, the split-and-merge algorithm does not depend on the initial solution because if it includes non-critical points they can be eliminated by the merging procedure. Therefore, any pair of points can be adopted for the initial solution and any of the strategies discussed in the context of the Ramer algorithm can be applied. In the case of closed curves, by letting A and B be two points of the initial solution, the algorithm can initiate execution with *open* = $\{A, B, A\}$.

The execution stops when the dominant points stack has not changed (i.e., new dominant points have not been created and segments have not been merged) after the procedures *split* and *merge* have been called.

Therefore, the split-and-merge main algorithm can be defined as

---

Algorithm: *Split-and-Merge*

1.   PUSH(open, A);
2.   PUSH(open, B);
3.   PUSH(open, A);
4.   **while** (NOT(EMPTY(open)) AND MODIFIED(open))
5.   **do**
6.   SPLIT(open);
7.   MERGE(open);
8.   RETURN(open);

The above algorithm assumes the Boolean function *modified (open)*, which returns *true* if and only if the open stack is modified during the last iteration. For proper operation, this function should return true at the first time it is used. There are many different approaches to implement the function *modified*. A possible solution is to create a flag variable that can be set by the split and the merge procedures in case the *open* stack is modified. This solution is simple and efficient, and can be easily implemented by slightly modifying the split and the merge procedures.

The above algorithm can be immediately modified in order to cope with open curves. The first and the last points are assumed to be dominant by definition, and can always be taken as the initial solution (as with the Ramer algorithm). In addition, the condition for merging the last and the first segments of the curve (the last steps of the merge procedure) should not be carried because these segments should never be merged.

Note: *Representing Contour Segments by Alternative Primitives*

Other representative geometric primitives can also be used, such as circles and ellipses [Castleman, 1996; Rosin & West, 1995; West & Rosin, 1991] and other primitives [Fischler & Wolf, 1994; Rosin, 1993; Wuescher & Boyer, 1991]. Alternatively, syntactical primitives may also be adopted, mainly for syntactical pattern recognition [Fu, 1982]. A more versatile but somehow more complex approach is based on representing each contour segment by parametric curves of the type

$$P(t) = \sum_{j=1}^{n} V_j B_j(t)$$

where t is the parameter, B_j is the j-th basis function and V_j is the j-th control point. Although parametric curves are very popular within the computer graphics community, they have been less explored in imaging problems, except for applications such as fingerprint representation [Chong et al., 1992], recognition of partially occluded objects [Salari & Balaji, 1991] and image segmentation and curve representation using rational Gaussian curves [Goshtasby, 1992, 1993].

To probe further: *Contour Detection and Representation*

There are a large number of relevant alternative methods related to the topics discussed in the preceding sections, such as snakes and dynamic contours, which are covered, for example, in Blake & Isard [1998]; Cohen [1991]; Cootes et al. [1995]; Kass et al. [1988]; Morel & Solimini [1995]; Staib & Duncan [1992].

5.5 Digital Straight Lines

After the point, straight lines and straight line segments are the simplest geometric elements in continuous spaces. A straight line segment in the plane (and also in higher dimensional spaces) can be understood as the geometric place of the infinite set of points linking two points along their shortest Euclidean distance. Despite the triviality of straight lines in continuous spaces, their counterparts in spatially sampled spaces such as the orthogonal lattice underlying digital images, hence *digital straight lines*, are by no means simple. As described in Section 5.7, distances, a trivial concept in continuous spaces, exhibit a complex behavior when considered in the orthogonal lattice. Similarly, straight lines also become much more complicated when mapped into spatially sampled spaces. A poignant example of the intricacies underlying digital straight lines is that, given any two points in the orthogonal lattice, there is more than a single digital straight line segment passing through these two points. Even simple facts such as the number of possible distinct digital lines in an $N{\times}N$ image are not yet known. Being fundamental representative elements in shape processing and analysis, it is important to properly understand the properties exhibited by digital straight lines, so they can be properly represented and detected. This is the objective of the present section.

5.5.1 Straight Lines and Segments

Before discussing how digital straight lines (and segments) can be generated and recognized, it is opportune to review the two principal straight line parametric equations, i.e., the *slope-intercept parametrization* given by the equation $y = mx + c$, and the *normal parametrization* given by the equation $\rho = x\cos(\theta) + y\sin(\theta)$. Figure 5.29 illustrates both of these parametrizations.

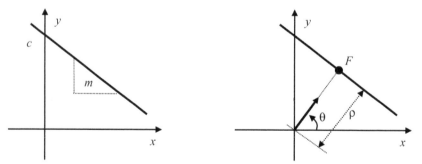

Figure 5.29: *The slope-intercept (a) and normal (b) parametrizations of straight lines.*

Both equations are called *parametric* because they involve parameters. As they define straight lines, which are specified by their position and orientation, both the slope-intercept and normal parametrizations require two parameters each. In the case of the slope-intercept, the parameters are the *slope m*, representing the line inclination (or tangent) and the *intercept* c, corresponding to the point where the line crosses the y-axis. The two parameters in the normal parametrization are the *normal angle* θ, corresponding to the angle between the x-axis and the normal to the line (anti-clockwise sense), and ρ, the *normal distance*, which is equal to the distance between the line and the coordinates system origin. As a matter of fact, the point of the straight line that is the closest to the coordinate system origin is called the *foot of the normal*, represented as F in Figure 5.29. This point can be represented by the position vector $\vec{F} = \rho\cos(\theta)\,\hat{\imath} + \rho\sin(\theta)\,\hat{\jmath}$, which defines the unit vector $\vec{u} = \cos(\theta)\,\hat{\imath} + \sin(\theta)\,\hat{\jmath}$ indicating the normal orientation of the straight line. It is now clear how the normal parametric equation arises: it corresponds to the points represented by the position vector $\vec{p} = x\hat{\imath} + y\hat{\jmath}$ whose projection over $\vec{u}$ is equal to ρ, i.e., $\langle \vec{u}, \vec{p} \rangle = x\cos(\theta) + y\sin(\theta) = \rho$.

It is important to note that each of such parameter pairs, i.e., (m, c) or (ρ, θ), defines a unique straight line in the plane, and vice versa. In addition, it should be observed that the slope-intercept parametrization is defective in the sense that it cannot represent vertical lines, i.e., when $m = \infty$ the parameter c cannot be specified. Table 5.2 presents the conversions between the slope-intercept and normal parametrizations, except for the situation $m = \infty$. *Straight line segments* can be

Slope-intercept to normal	$\theta = \pi + a\tan(m)$
	$\rho = \frac{c}{\sin(\theta)}$
Normal to	$m = \tan(\theta - \pi)$
slope-intercept	$c = \rho\sin(\theta)$

Table 5.2: *Conversions between the slope-intercept and normal parametrizations, except for m = ∞.*

immediately represented in terms of the coordinates of their respective two extremities, or by combining their slope-intercept of normal parameters with just one of their coordinates. For instance, we can say a straight line segment has parameters (ρ, θ), which corresponds to saying that the segment is part of the line defined by those parameters, and extends along an interval $[x_i, x_f]$.

5.5.2 | Generating Digital Straight Lines and Segments

The best way to start understanding digital straight lines (hence DSL) and digital straight line segments (hence DSLS) is by considering both these elements as the result of the spatial quantization of their continuous counterparts (sometimes called *preimages*). In other words, given a continuous straight line, a digital straight line can be obtained by spatially sampling this line with respect to a specific spatial resolution Δ. Although any of the quantization schemes discussed in Section 3.1.4 can in principle be considered for this task, we shall be limited—for simplicity's sake—to the grid intersect quantization scheme (see Section 3.1.4). Figure 5.30 illustrates a generic straight line segment and the respective digital straight line segment generated by its grid-intersect quantization in an orthogonal lattice with spatial sampling Δ.

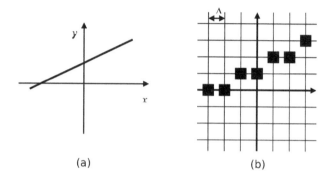

(a)   (b)

Figure 5.30: *A continuous straight line segment L **(a)** and the respective line segment **(b)** obtained by spatially sampling L with a specific spatial sampling Δ.*

For simplicity's sake, it is henceforth assumed that $\Delta = 1$.

It is therefore clear that, while a DSL is an infinite set of lattice points, a DSLS is a finite set of lattice points, which are henceforth represented in terms of their coordinates (x, y). In case the straight line is defined in terms of one of its possible parametric equations, such as the slope-intercept equation $y = mx + c$, the respective digital straight line can be obtained by varying either x or y while obtaining the other variable. There are several approaches to spatially sample the straight line given by equation $y = mx + c$, including the following:

$$y = \text{round}\,(mx + c), \quad x = \dots, -2, -1, 0, 1, 2, \dots \tag{5.1}$$

$$y = \text{trunc}\,(mx + c), \quad x = \dots, -2, -1, 0, 1, 2, \dots \tag{5.2}$$

$$y = \text{floor}\,(mx + c), \quad x = \dots, -2, -1, 0, 1, 2, \cdots \tag{5.3}$$

$$y = \text{ceil}\,(mx + c), \quad x = \dots, -2, -1, 0, 1, 2, \dots \tag{5.4}$$

Figures 5.31 (a) and (b) present the spatial sampling of the function $y = x - 0.5$ by using equations (5.1) and (5.2), respectively.

It is clear from this illustration that the *round* and *trunc* functions imply the quantized points behave differently for negative and positive values of y, which is undesirable since it enhances the DSLS discontinuity at $y = 0$. A more consistent approach would be to use the *floor* or *ceil* functions (see Section 2.1.4), but these functions would imply a shift of the quantized straight line towards the left and right, respectively. A more balanced approach consists in shifting the floor function by 0.5 to the left, yielding the following quantization scheme:

$$y_i = \text{floor}\,(mx + c + 0.5), \quad x = \dots, -2, -1, 0, 1, 2, \dots \tag{5.5}$$

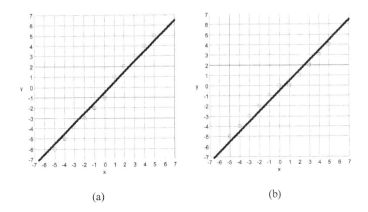

(a)   (b)

Figure 5.31: *The spatial sampling of y = x − 0.5 (represented by the continuous line); for y = −5,..., −1, 0, 1, 2,..., 5; obtained by using equation (5.1) in (a) and equation (5.2) in (b). The sampled values are represented by squares.*

Figure 5.32 illustrates the quantization of the lines $y = x - 0.5$ by using this scheme. The above characterized additional discontinuity at the transition of y from negative to positive is now clearly avoided.

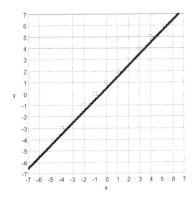

Figure 5.32: *The spatial sampling of the straight line defined by the equation y = x − 0.5; for x = −5,..., −1, 0, 1, 2,..., 5; by using equation (5.5).*

Therefore, equation (5.5) is henceforth adopted for obtaining DSLSs from straight line equations such as $y = mx + c$. It is interesting to have a closer look at the sampling implemented by this equation. In order to do so, let us consider that $b = \text{floor}(a) \Leftrightarrow b \leqslant a < b + 1$ and b is an integer value, and rewrite equation (5.5) as below. The reader is referred to the box entitled *Inequalities* for a brief review about inequalities and their algebraic handling.

$$y_i = \text{floor}(mx + c + 0.5) \Leftrightarrow y_i \leqslant mx + c + 0.5 < y_i + 1 \Leftrightarrow$$
$$\Leftrightarrow 0 \leqslant mx + c + 0.5 - y_i < 1 \Leftrightarrow -0.5 - mx - c \leqslant -y_i < -mx - c + 0.5 \Leftrightarrow$$
$$\Leftrightarrow mx + c - 0.5 < y_i \leqslant mx + c + 0.5$$

Note: *Inequalities*

An expression such as $a < x \leqslant b$ is called an *inequality*. As with equalities, inequalities can also be transformed into equivalent forms. For instance, an inequality does not change in case we add (or subtract) the same constant k to all its terms, e.g., $a < x \leqslant b \Leftrightarrow a + k < x + k \leqslant b + k$. A strictly positive constant k (i.e., $k > 0$) can also be multiplied to all terms, i.e., $a < x \leqslant b \Leftrightarrow ak < xk \leqslant bk$. Observe that in case k is strictly negative (i.e., $k < 0$) we have $a < x \leqslant b \Leftrightarrow bk \leqslant xk < ak$. For instance, if $k = -1$ we have $1 < x \leqslant 2 \Leftrightarrow -2 \leqslant -x < -1$.

Therefore, given a specific pair or parameters $(\tilde{m}, \tilde{c})$, the set of sampled values y_i satisfying $y_i = \text{floor}(\tilde{m}x + \tilde{c} + 0.5)$ must fall within the straight region defined by $\tilde{m}x + \tilde{c} - 0.5 < y_i \leqslant \tilde{m}x + \tilde{c} + 0.5$, which is henceforth denominated *upper bound straight band*. In other words, the set of points corresponding to the DSLS defined parameters $(\tilde{m}, \tilde{c})$ by using the quantization scheme of equation (5.5) comprehends all lattice points inside the above band. This concept is illustrated in Figure 5.33 with respect to $(\tilde{m} = 0.4, \tilde{c} = 0.2)$ (a) and $(\tilde{m} = -0.2, \tilde{c} = 0.5)$ (b).

It can be easily verified that the above-adopted sampling scheme produces the same quantization that would be obtained by using the grid-intersect quantization (see Section 3.1.4). However, it should be observed that, as the absolute value of the slope parameter m increases, steeper lines are obtained by equation (5.5), and ultimately gaps start appearing along the DSLS, as shown in Figure 5.34.

It can be verified that this occurs whenever $|m| > 1$. In this case, the solution is to vary y, in such a way that x is obtained as a function of y, which can be done by using the following criteria for generating the DSL respective to the continuous straight line defined by the slope-intercept equation:

$$\begin{cases} \text{if} \quad |m| \leqslant 1 \quad \text{then:} \quad y = \text{floor}(mi + c + 0.5) \\ \qquad\qquad \text{Otherwise:} \quad x = \text{floor}\left(\frac{i-c}{m} + 0.5\right) \end{cases} \quad \text{for} \quad i = \ldots, -2, -1, 0, 1, 2, \ldots$$

In case the normal parametric equation $\rho = x \cos(\theta) + y \sin(\theta)$ is adopted, the

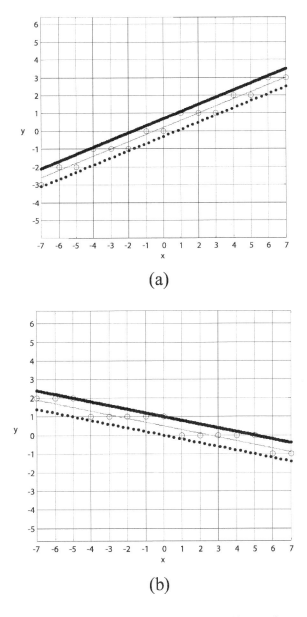

(a)

(b)

Figure 5.33: *The DSLSs defined by ($\tilde{m} = 0.4, \tilde{c} = 0.2$) **(a)** and ($\tilde{m} = -0.2, \tilde{c} = 0.5$) **(b)** correspond to the lattice points (indicated by circles) inside the respective upper-bound straight bands. The thin lines in the middle of the bands represent the continuous lines $y = 0.4x + 0.2$ (a) and $y = -0.2x + 0.5$ (b).*

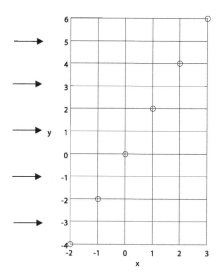

Figure 5.34: *The DSLSs defined by* $(\tilde{m} = 2, \tilde{c} = 0.2)$ *for* $x = -2, -1, 0, 1, 2,$
by using the sampling scheme $y_i = $ floor $(\tilde{m}x + \tilde{c} + 0.5)$.

following strategy should be applied:

$$
\begin{cases}
\text{if} \quad 45° \leqslant \theta \leqslant 135° \quad \text{then} \quad y = \text{floor}\left(\frac{\rho - x\cos(\theta)}{\sin(\theta)} + 0.5\right) \\
\text{for} \quad x = \ldots, -2, -1, 0, 1, 2, \ldots \\
\qquad \text{Otherwise} \quad x = \text{floor}\left(\frac{\rho - y\sin(\theta)}{\cos(\theta)} + 0.5\right) \\
\qquad \text{for} \quad y = \ldots, -2, -1, 0, 1, 2, \ldots
\end{cases}
$$

A similar development in terms of bands can also be obtained for the normal parametrization approach to DSLS generation. For instance, for $45° \leqslant \theta \leqslant 135°$ we have

$$
y_i = \text{floor}\left(\frac{\rho - x\cos(\theta)}{\sin(\theta)} + 0.5\right) \Leftrightarrow y_i \leqslant \frac{\rho - x\cos(\theta)}{\sin(\theta)} + 0.5 < y_i + 1 \Leftrightarrow
$$
$$
\Leftrightarrow 0 \leqslant \frac{\rho - x\cos(\theta)}{\sin(\theta)} + 0.5 - y_i < 1 \Leftrightarrow
$$
$$
\Leftrightarrow -\frac{\rho - x\cos(\theta)}{\sin(\theta)} - 0.5 \leqslant -y_i < -\frac{\rho - x\cos(\theta)}{\sin(\theta)} + 0.5 \Leftrightarrow
$$
$$
\Leftrightarrow \frac{\rho - x\cos(\theta)}{\sin(\theta)} - 0.5 < y_i \leqslant \frac{\rho - x\cos(\theta)}{\sin(\theta)} + 0.5
$$

This can be easily verified to correspond exactly to the band defined by the slope-intercept parametrization, except that the parameters m and c are replaced by ρ and θ.

In case $0° \leqslant \theta < 45°$ and $135° < \theta < 180°$, we have

$$x_i = \text{floor2}\left(\frac{\rho - y\sin(\theta)}{\cos(\theta)} + 0.5\right) \Leftrightarrow x_i < \frac{\rho - y\sin(\theta)}{\cos(\theta)} + 0.5 \leqslant x_i + 1 \Leftrightarrow$$
$$\Leftrightarrow 0 < \frac{\rho - y\sin(\theta)}{\cos(\theta)} + 0.5 - x_i \leqslant 1 \Leftrightarrow -\frac{\rho - y\sin(\theta)}{\cos(\theta)} - 0.5 < -x_i \leqslant -\frac{\rho - y\sin(\theta)}{\cos(\theta)} + 0.5 \Leftrightarrow$$
$$\Leftrightarrow \frac{\rho - y\sin(\theta)}{\cos(\theta)} - 0.5 \leqslant x_i < \frac{\rho - y\sin(\theta)}{\cos(\theta)} + 0.5$$

Observe that a variation of the floor function (see the accompanying box), namely *floor2*, is required in order to guarantee that the straight bands are bound at their superior portion when seen at the conventional axis position (i.e., without exchanging x by y).

Note: *The floor2 and ceil2 functions*

The floor and ceil functions presented in Section 2.1.4 can be modified in order to exclude the "equal" in their respective definitions. The floor(x) function, defined as the greatest integer smaller or equal than x, therefore becomes floor2 (x), meaning the greatest integer smaller than x. Similarly, we have the function ceil2 (x), defined as the smallest integer greater than x. The floor2 (x) and ceil2 (x) functions are illustrated in Figure 5.35. Since most languages and environments do not include

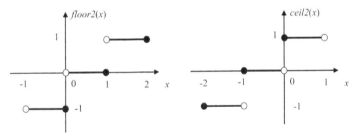

Figure 5.35: *Modified floor and ceil functions (see text for details).*

these functions, it is important to express them in terms of the more usual *floor* and *ceil* functions, i.e., floor2 (x) = ceil ($x - 1$) and ceil2 (x) = floor ($x + 1$).

It is interesting to observe that the spatial sampling of a straight line or line segment corresponds to a degenerate mapping of the original geometric entity (i.e., the line). In other words, as information is lost in the sampling process, the original line cannot be exactly recovered from its spatially sampled version. As a consequence, there is more than a continuous single line (actually an infinite amount) defining straight lines passing through any two distinct points in the orthogonal lattice. This phenomenon is discussed in greater detail in Section 5.6.

5.5.3 Recognizing an Isolated Digital Straight Line Segment

The problem of recognizing an isolated digital straight line segment consists in, given a set of lattice points, verifying whether they correspond to the spatial sampling of some continuous straight line segment. The first method for recognizing a DSLS, proposed in Freeman & Davis [1977], involves representing the set of points in terms of the chain code (see Section 5.2.4) and applying the following criteria:

① only two chain codes are verified,

② one of the chain codes always appears isolated, and

③ this isolated code should appear as evenly spaced as possible.

In case each of these three criteria is verified, the specific set of points is confirmed to be a DSLS. However, while criteria ① and ② still provide useful necessary conditions that can be eventually used to discard some situations, the third criterion is fuzzy and cannot be applied in a fully objective fashion. It should be observed that, even if Freeman's criteria were completely sound, they would say nothing about the continuous line (or lines) yielding the DSLS in question, but only that the set of points correspond to spatial quantization of *some* straight line.

While several alternative criteria have been proposed since Freeman [1974], we shall be limited to the recent criterion advanced by Melter et al. [1993], which can be summarized as:

> A given set S of lattice points corresponds to a DSLS if the grid-intersect quantization of the line equation obtained by their least square fitting reproduces the set S.

Let us clarify this simple criterion in terms of the following example:

Example: *Checking whether a Set of Points Corresponds to a DSLS*

Verify if

$$S_1 = \{(0, 1); (1, 1); (2, 2); (3, 2); (4, 3)\}$$

and

$$S_2 = \{(0, 1); (1, 1); (2, 3); (3, 2); (4, 3)\}$$

correspond or not to DSLSs.

Solution:

① For $S_1 = \{(0, 1); (1, 1); (2, 2); (3, 2); (4, 3)\}$: The first step consists in obtaining the linear regression of these points. By using the linear least squares equations provided in Section 2.2.5, we get the continuous straight line L defined by $m = 0.5$ and $c = 0.8$. Now, if we spatially sample L, by using equation (5.5) for $x = 0$

to 4, we get the set $T = \{(0, 1) ; (1, 1) ; (2, 2) ; (3, 2) ; (4, 3)\}$, which is readily verified to be equal to the above original set. Therefore, the set S_1 is verified to correspond to the grid intersect quantization of some continuous straight line, consequently being a DSLS.

② For $S_2 = \{(0, 1) ; (1, 1) ; (2, 3) ; (3, 2) ; (4, 3)\}$: Now we get the continuous straight line L defined by $m = 0.5$ and $c = 1.0$ from which, when sampled for $x = 0$ to 4, we get the set $T_2 = \{(0, 1) ; (1, 1) ; (2, 2) ; (3, 2) ; (4, 3)\}$, which is not equal to the original set. Therefore, the set S_2 is not verified to be a DSLS.

It should be remarked that the continuous line obtained in ① is only one of the infinite continuous straight lines whose spatially sampled version reproduces the original set of points.

Melter's approach to DSLS identification can be extended to cope with images containing several isolated DSLSs. In this case, a preliminary segmentation is required in order to identify each of the isolated lines, to be subsequently processed one-by-one using the above-described methodology.

5.6 Hough Transforms

While the above-described criterion allows an exact and simple means for verifying whether a set of lattice points is a DSLS, also identifying the parameters of one of the possible continuous lines yielding the DSLS, it does not work when the DSLSs cross each other, or are connected to other elements in the image. A classical methodology to effectively cope with such situations is the *Hough transform*, hence HT, which is capable of identifying DSLSs in cluttered and/or noisy images, and also to give an estimate of their respective parameters. Hough transforms are the object of the present section. Although there are many HT variations, we shall concentrate on the traditional approach, first reported in Hough's patent [Hough, 1962], in which the Hough transform acts as a mapping from the image space into a parameter space. In addition, we supply a step-by-step development of an accurate HT algorithm and a scheme for parameter space voting, both following the methodology described in Costa [1992, 1995].

5.6.1 Continuous Hough Transforms

The basic principle underlying the HT can be easily understood by considering both the image and parameter spaces as being continuous. Under such conditions, one of the many possible HT approaches would be defined by considering the above slope-intercept parametrization. However, since the alternative representation of vertical lines implies some difficulties (infinite slope), the normal parametrization of a straight line is often adopted. This parametrization, which is illustrated in Figure 5.29 (b), involves as parameters the normal distance between the straight line and the coordinate origin (ρ) and the normal angle (θ) defined between the line normal and the x-axis. In such a parametrization, a straight line with the above pair of parameters corresponds to the set of points in the plane satisfying the equation $\rho = \tilde{x}\cos(\theta) + \tilde{y}\sin(\theta)$.

The normal HT maps each of the foreground points $(\tilde{x}, \tilde{y})$ in the continuous image space into the sinusoidal $\rho = \tilde{x}\cos(\theta) + \tilde{y}\sin(\theta)$ in the normal parameter space, i.e., the space defined by the parameters θ and ρ. Figure 5.36 illustrates such a mapping with respect to an image containing a single point.

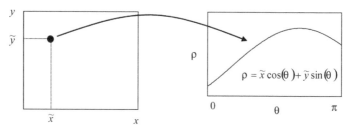

Figure 5.36: *A single point $(\tilde{x}, \tilde{y})$ in the continuous image space **(a)** is mapped into the sinusoidal $\rho = \tilde{x}\cos(\theta) + \tilde{y}\sin(\theta)$ in the continuous normal parameter space **(b)**.*

The important property allowed by such a mapping is that aligned points in the image space will produce sinusoidals that intercept, all of them, at a single point. The coordinates (θ, ρ) of this intersection point correspond to the parameters of the original line. Figure 5.37 illustrates such concepts.

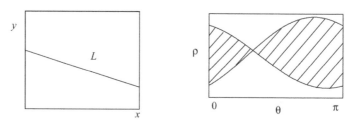

Figure 5.37: *A straight line L in the continuous image space **(a)** is mapped into a continuous collection of sinusoidals, sometimes called a butterfly, in the continuous normal parameter space **(b)**. All such sinusoidals intercept at the single point (θ, ρ) corresponding to the parameters of the original line L.*

Although alternative intervals can be considered, it is important to note that it is enough to have $0 \leqslant \theta < \pi$, which implies negative values of the normal distance. It is thus clear that the HT maps straight lines into crossing points in the parameter space, in such a way that we can identify a line in the image by searching crossing points in the respectively obtained parameter space. One of the useful HT properties consists of the fact that it can be applied to cluttered/noisy images containing many straight lines, curves and shapes, without requiring preliminary segmentation. As a matter of fact, the HT is capable of simultaneously performing segmentation and straight line identification.

5.6.2 | Discrete Image and Continuous Parameter Space

We saw in Section 5.5.2 that a DSLS generated by using the normal parametrization corresponds to the set of image points inside the respective upper-bound straight band. Let us now develop a formula expressing the inverse mapping which, given a specific image point $(\tilde{x}, \tilde{y})$, produces a region in the parameter space. This can be done as follows:

① If $45° \leqslant \theta \leqslant 135°$,

$$\frac{\rho - \tilde{x}\cos(\theta)}{\sin(\theta)} - 0.5 < \tilde{y} \leqslant \frac{\rho - \tilde{x}\cos(\theta)}{\sin(\theta)} + 0.5 \Leftrightarrow$$
$$\Leftrightarrow \rho - \tilde{x}\cos(\theta) - 0.5\sin(\theta) < \tilde{y}\sin(\theta) \leqslant \rho - \tilde{x}\cos(\theta) - 0.5\sin(\theta) \Leftrightarrow$$
$$\Leftrightarrow -\tilde{x}\cos(\theta) - 0.5\sin(\theta) < -\rho + \tilde{y}\sin(\theta) \leqslant -\tilde{x}\cos(\theta) + 0.5\sin(\theta) \Leftrightarrow$$
$$\Leftrightarrow -\tilde{x}\cos(\theta) - \tilde{y}\sin(\theta) - 0.5\sin(\theta) < -\rho \leqslant -\tilde{x}\cos(\theta) - \tilde{y}\sin(\theta) + 0.5\sin(\theta) \Leftrightarrow$$
$$\Leftrightarrow \tilde{x}\cos(\theta) + \tilde{y}\sin(\theta) - 0.5\sin(\theta) \leqslant \rho < \tilde{x}\cos(\theta) + \tilde{y}\sin(\theta) + 0.5\sin(\theta)$$

② If $0° \leqslant \theta < 45°$ (observe that $\cos(\theta) > 0$),

$$\frac{\rho - \tilde{y}\sin(\theta)}{\cos(\theta)} - 0.5 \leqslant \tilde{x} < \frac{\rho - \tilde{y}\sin(\theta)}{\cos(\theta)} + 0.5 \Leftrightarrow$$
$$\Leftrightarrow \rho - \tilde{y}\sin(\theta) - 0.5\cos(\theta) \leqslant \tilde{x}\cos(\theta) < \rho - \tilde{y}\sin(\theta) + 0.5\cos(\theta) \Leftrightarrow$$
$$\Leftrightarrow -\tilde{y}\sin(\theta) - 0.5\cos(\theta) \leqslant -\rho + \tilde{x}\cos(\theta) < -\tilde{y}\sin(\theta) + 0.5\cos(\theta) \Leftrightarrow$$
$$\Leftrightarrow -\tilde{y}\sin(\theta) - \tilde{x}\cos(\theta) - 0.5\cos(\theta) \leqslant -\rho < -\tilde{y}\sin(\theta) - \tilde{x}\cos(\theta) + 0.5\cos(\theta) \Leftrightarrow$$
$$\Leftrightarrow \tilde{x}\cos(\theta) + \tilde{y}\sin(\theta) - 0.5\cos(\theta) < \rho \leqslant \tilde{x}\cos(\theta) + \tilde{y}\sin(\theta) + 0.5\cos(\theta)$$

③ If $135° < \theta < 180°$ (observe that $\cos(\theta) < 0$),

$$\frac{\rho - \tilde{y}\sin(\theta)}{\cos(\theta)} - 0.5 \leqslant \tilde{x} < \frac{\rho - \tilde{y}\sin(\theta)}{\cos(\theta)} + 0.5 \Leftrightarrow$$
$$\Leftrightarrow \rho - \tilde{y}\sin(\theta) - 0.5\cos(\theta) < \tilde{x}\cos(\theta) \leqslant \rho - \tilde{y}\sin(\theta) + 0.5\cos(\theta) \Leftrightarrow$$
$$\Leftrightarrow -\tilde{y}\sin(\theta) + 0.5\cos(\theta) < -\rho + \tilde{x}\cos(\theta) \leqslant -\tilde{y}\sin(\theta) - 0.5\cos(\theta) \Leftrightarrow$$
$$\Leftrightarrow -\tilde{y}\sin(\theta) - \tilde{x}\cos(\theta) + 0.5\cos(\theta) < -\rho \leqslant -\tilde{y}\sin(\theta) - \tilde{x}\cos(\theta) - 0.5\cos(\theta) \Leftrightarrow$$
$$\Leftrightarrow \tilde{x}\cos(\theta) + \tilde{y}\sin(\theta) + 0.5\cos(\theta) \leqslant \rho < \tilde{x}\cos(\theta) + \tilde{y}\sin(\theta) - 0.5\cos(\theta)$$

Observe that the situation in ② implies an upper-bound sinusoidal band instead of the lower-bound bands defined by ① and ③. While each of these three situations can be treated separately in case maximum precision is required, situation ② is henceforth modified in order to produce a lower-bound band, in such a way that the whole band in the parameter space becomes upper bound. Although this can eventually imply a small bias for the normal distance with respect to normal angles in the interval $0° \leqslant \theta < 45°$, this alteration allows a more simplified development of the Hough transform. Therefore, the above results can be conveniently combined into the following equation:

$$\tilde{y}\sin(\theta) + \tilde{x}\cos(\theta) - 0.5\mu(\theta) \leqslant \rho < \tilde{y}\sin(\theta) + \tilde{x}\cos(\theta) + 0.5\mu(\theta) \qquad (5.6)$$

where

$$\mu(\theta) = \begin{cases} \sin(\theta) & \text{if } 45° \leqslant \theta \leqslant 135° \\ |\cos(\theta)| & \text{otherwise} \end{cases}$$

Thus, as illustrated in Figure 5.38, each specific image point $(\tilde{x}, \tilde{y})$ defines a sinusoidal *lower-bound straight band* in the normal parameter space. At the same

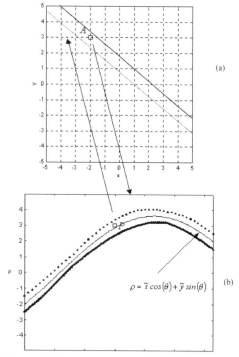

Figure 5.38: *A specific image point $A = (\tilde{x}, \tilde{y})$ (a) defines a sinusoidal lower-bound band in the normal parameter space (b). Any point P inside the sinusoidal band defines an upper-bound straight band in the image that contains the point A.*

time, any point inside this band defines an *upper-bound straight band* in the image space in such a way that this band necessarily contains the point $A = (\tilde{x}, \tilde{y})$. Although frequently overlooked in the related literature, this important relationship, actually an equivalence relation between image points and regions in the parameter space, provides the underlying concept for properly understanding and applying HTs.

As a consequence of the above-characterized mapping, each of the two points A and B in the image (see Figure 5.39) will define a lower-bound sinusoidal band in the parameter space, in such a way that any of the points inside the intersect region between these two bands will define the upper-bound straight band in the image that necessarily contains the points A and B.

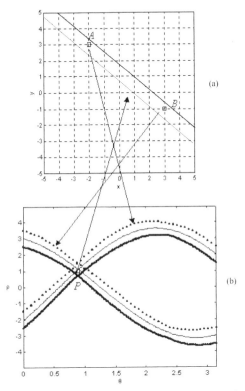

Figure 5.39: *Points A and B in the image **(a)** define a respective lower-bound sinusoidal band in the parameter space **(b)**. Any point P inside the intersection region between these two sinusoidal bands define an upper-bound straight band necessarily containing the points A and B.*

When considered at its most general form, the above-discussed mapping relationship provides a clear indication of how DSLSs are organized in the normal parameter space. More specifically, each of the DSLS constituent points will produce a lower-bound sinusoidal band in the parameter space, and any point inside the region defined by the intersection of all these bands defines an upper-bound straight band containing the DSLS. Consequently, each DSLS is mapped into a respective connected region in the parameter space, which typically has the shape

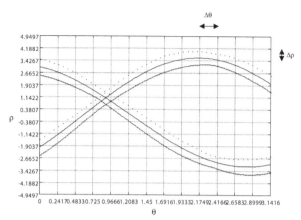

Figure 5.40: *The normal parameter space sampled by an orthogonal lattice with resolutions $\Delta\theta$ and $\Delta\rho$.*

of a diamond or rhombus (see Figure 5.40). Generally, the longer the DSLS, the smaller this intersection region.

5.6.3 Discrete Image and Parameter Space

Having so far dealt with the mapping of DSLSs into the continuous parameter space, it is time to consider how this space can be sampled in order to allow numeric implementations of the HT.[1]

The discrete image space is represented by an $N \times N$ matrix, and the parameter space, sampled with resolutions $\Delta\theta$ and $\Delta\rho$, by an $N_\theta \times N_\rho$ matrix, generally known as *accumulator array*. Special care must now be taken while defining the relationship between the continuous parameter values θ and ρ and their respective indexing values ith and irh in the matrix representing the discrete normal parameter space. We start by superimposing an orthogonal lattice with $\Delta\theta$ and $\Delta\rho$ onto the normal parameter space, as illustrated in Figure 5.40.

Since $0 \leqslant \theta < \pi$, we have that $\Delta\theta$ and N_θ are related as

$$\Delta\theta = \frac{\theta_{\max} - \theta_{\min}}{N_\theta - 1} = \frac{\pi - 0}{N_\theta - 1} = \frac{\pi}{N_\theta - 1}$$

For the sake of enhanced effectiveness (i.e., in order to allow the usage of a larger portion of the accumulator array), the center of the image coordinates should be

---

[1] As a matter of fact, it is indeed possible to implement Hough transforms with continuous parameter spaces [Cygansky et al., 1990], but such possibilities will not be explored in the present book.

placed at the center of the image, implying $\rho_{min} = -\sqrt{2}\frac{N}{2}$ and $\rho_{max} = \sqrt{2}\frac{N}{2}$. Thus, $\Delta\rho$ and N_ρ are related as

$$\Delta\rho = \frac{\rho_{max} - \rho_{min}}{N_\rho - 1} = \frac{\sqrt{2}N}{N_\rho - 1}$$

Therefore, the following two possible practical situations arise: (1) N_θ and N_ρ are given (for instance, limited by the available memory and processing speed), defining $\Delta\theta$ and $\Delta\rho$, respectively; and (2) $\Delta\theta$ and $\Delta\rho$ are given (for instance, chosen in terms of the expected resolutions), defining N_θ and N_ρ, respectively. It should be observed that the values of $\Delta\theta$ and $\Delta\rho$ are critical for the proper DSLS representation.

Now that we have spatially sampled the accumulator array, we need an effective means to represent the intersections between the lower-bound sinusoidal bands. This can be done by using a *voting scheme*. More specifically, every time one of the sampled points is found to be contained into one of such bands, the bin associated with the parameters defining this band is incremented. Such bins can be represented in terms of the $N_\theta \times N_\rho$ *accumulator array*. Therefore, the accumulator array will store the votes defined every time a sinusoidal band passes over one of the superimposed grids, in such a way that a DSLS will produce a peak in the accumulator array at the position where the respectively mapped sinusoidal bands intersect each other. This is the main principle underlying the Hough transform for discrete image and parameter array.

Although we are now close to a computational implementation of the normal HT, we still have to devise a means to map the discretized normal parameter space into the accumulator array, which can be done in terms of integer indexes ith and irh and the affine mapping function described in Section 2.1.4. For the *normal angle*, as we want to map $0 \leqslant \theta < \pi$ into $ith_{min} \leqslant ith < ith_{max}$, the proportional mapping is given by

$$ith = \frac{ith_{max} - ith_{min}}{\theta_{max} - \theta_{min}}\theta + \frac{\theta_{max}\,ith_{min} - \theta_{min}\,ith_{max}}{\theta_{max} - \theta_{min}} =$$
$$= \frac{ith_{max} - ith_{min}}{\pi}\theta + ith_{min}$$

For the *normal distance*, as we want to map $-\sqrt{2}\frac{N}{2} \leqslant \rho \leqslant \sqrt{2}\frac{N}{2}$ into $irh_{min} \leqslant irh < irh_{max}$, the proportional mapping is given by

$$irh = \frac{irh_{max} - irh_{min}}{\rho_{max} - \rho_{min}}\rho + \frac{\rho_{max}\,irh_{min} - \rho_{min}\,irh_{max}}{\rho_{max} - \rho_{min}} =$$
$$= \frac{irh_{max} - irh_{min}}{\sqrt{2}N}\rho + irh_{min} + irh_{max}$$

As the integer values ith and irh are intended to be used as indexes of the accumulator array, it is normally expected that they vary from 1 to the size of the respective size of the matrix, i.e., $1 \leqslant \text{ith} < N_\theta$ and $1 \leqslant \text{irh} \leqslant N_\rho$. This requirement is henceforth adopted, leading to ith $= \frac{N_\theta - 1}{\pi}\theta + 1$ and

$$\text{irh} = \frac{\text{irh}_{\max} - \text{irh}_{\min}}{\rho_{\max} - \rho_{\min}}\rho + \frac{\rho_{\max}\,\text{irh}_{\min} - \rho_{\min}\,\text{irh}_{\max}}{\rho_{\max} - \rho_{\min}} =$$

$$= \frac{\text{irh}_{\max} - \text{irh}_{\min}}{\sqrt{2}N}\rho + \frac{\sqrt{2}N\,\text{irh}_{\min} + \sqrt{2}N\,\text{irh}_{\max}}{2\sqrt{2}N} =$$

$$= \frac{\rho(N_\rho - 1)}{\sqrt{2}N} + \frac{\sqrt{2}N + \sqrt{2}NN_\rho}{2\sqrt{2}N} = \frac{\rho(N_\rho - 1)}{\sqrt{2}N} + \frac{\sqrt{2}\frac{N}{2}(N_\rho - 1)}{\sqrt{2}N} + \frac{\sqrt{2}N}{\sqrt{2}N} =$$

$$= \frac{\sqrt{2}\frac{N}{2}}{\Delta\rho} + 1 + \frac{\rho}{\Delta\rho} = 1 + \frac{\rho - \rho_{\min}}{\Delta\rho}$$

The expressions relating the integer indexing values, i.e., $1 \leqslant \text{ith} < N_\theta$ and $1 \leqslant \text{irh} \leqslant N_\rho$, and their respective real normal counterparts θ and ρ are summarized in Table 5.3.

	$\theta = (\text{ith} - 1)\Delta\theta$
ith $\rightarrow \theta$	ith $= 1, 2, \ldots, N_\theta - 1$
	$\Delta\theta = \frac{\pi}{N_\theta - 1}$
	$\rho = (\text{irh} - 1)\Delta\rho + \rho_{\min}$
irh $\rightarrow \rho$	irh $= 1, 2, \ldots, N_\rho$
	$\Delta\rho = \frac{\sqrt{2}N}{N_\rho - 1}$

Table 5.3: *Expressions relating the normal parameters and their respective indexes.*

A possible discrete implementation of the normal HT consists in incrementing all the accumulator array cells inside the lower-bound sinusoidal band defined by each image foreground point $(\tilde{x}, \tilde{y})$. Each band can be obtained by varying the normal angle ith from 1 to N_θ while updating the respective irh values, as illustrated in Figure 5.41.

We have from equation (5.6) that

$$rh\_ini = \tilde{y}\sin(\theta) + \tilde{x}\cos(\theta) - \frac{\mu(\theta)}{2}$$

and

$$rh\_fin = \tilde{y}\sin(\theta) + \tilde{x}\cos(\theta) + \frac{\mu(\theta)}{2}$$

where

$$\mu(\theta) = \begin{cases} \sin(\theta) & \text{if } 45° \leqslant \theta \leqslant 135° \\ |\cos(\theta)| & \text{otherwise} \end{cases}$$

Since irh_*ini* is the smallest integer greater than or equal to *rh_ini*, and considering the equations in Table 5.3, we have

$$\text{irh\_}ini = \text{ceil}\left\{\frac{(rh\_ini - rh_{\min})}{\Delta\rho + 1}\right\}$$

Now, as irh_*ini* is the greatest integer strictly smaller than *rh_ini*, we have to use the floor2 function (see the box in Section 5.5.2), yielding

$$\text{irh\_}end = \text{floor2}\left\{\frac{(rh\_fin - rh_{\min})}{\Delta\rho + 1}\right\}$$
$$= \text{ceil}\left\{\frac{(rh\_fin - rh_{\min})}{\Delta\rho + 1 - 1}\right\} =$$
$$= \text{ceil}\left\{\frac{(rh\_fin - rh_{\min})}{\Delta\rho}\right\}$$

Having revised the above concepts, we are now in a position to present a complete and accurate algorithm for discrete execution of the normal HT, which is given in below, where *Acc* is the accumulator array.

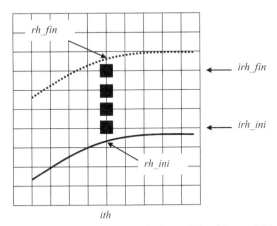

Figure 5.41: *A range rh_ini $\leqslant \rho <$ rh_fin is defined for each* ith, *and all the cells within this range should be incremented.*

Algorithm: *Normal Hough Transform*

1. $rh_{min} \leftarrow -\sqrt{2}N/2$;
2. Clear the accumulator array Acc;
3. **for** each foreground point $(\tilde{x}, \tilde{y})$ in the image
4.   **do**
5.   **for** ith $\leftarrow 1$ **to** $N_\theta - 1$
6.   **do**
7. $th \leftarrow (\text{ith} - 1)\,\Delta\theta$;
8. $rh\_ini \leftarrow \tilde{x}\cos(th) + \tilde{y}\sin(th) - \frac{\mu(th)}{2}$;
9. $rh\_fin \leftarrow rh\_ini + \mu(th)$;
10. $\text{irh}\_ini \leftarrow \text{ceil}\left\{\frac{(rh\_ini - rh_{min})}{\Delta\rho + 1}\right\}$;
11. $\text{irh}\_fin \leftarrow \text{ceil}\left\{\frac{(rh\_fin - rh_{min})}{\Delta\rho}\right\}$;
12.   **for** irh $\leftarrow \text{irh}\_ini$ **to** irh_fin
13.   **do**
14. $Acc(\text{ith}, \text{irh}) \leftarrow Acc(\text{ith}, \text{irh}) + 1$;

As we can see, this algorithm incorporates three loops: the first for covering the foreground (usually the edge) image elements, the second for covering ith, and the third for covering irh within the band. The step $Acc(\text{ith}, \text{irh}) = Acc(\text{ith}, \text{irh}) + 1$ is usually referred to as "*voting*" in the accumulator array. Figure 5.42 illustrates the application of the above HT algorithm with respect to a 33×33 noisy (uniformly distributed salt-and-pepper noise) binary image containing two DSLSs.

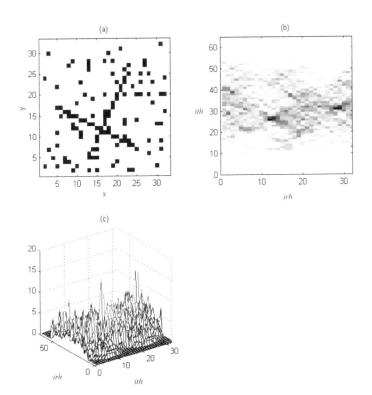

Figure 5.42: *A 33 × 33 noisy image containing two DSLSs **(a)**, the accumulator array (with dimension 33 × 66) obtained by the above normal HT algorithm shown as a gray-level image **(b)** and as a 3D plot **(c)**. It should be observed that higher accumulator values are represented by darker gray levels in* (b).

It is clear from the previous example that the crossings of the sinusoidals defined by each DSLS in the image produce a respective peak in the accumulator array. Therefore, DSLSs in the image can be identified, except for their terminations (see Section 5.6.6), by looking for peaks in the discrete parameter space represented in the accumulator array. Moreover, the parametric mapping implemented by the HT is reasonably robust to noise, since the two peaks corresponding to the two DSLSs in (a) are still identifiable in (b) and (c) despite the substantial noise in the image. However, it is at the same time verified that the two peaks appear in the middle of a background distribution of votes. Since the typical and simplest approach to peak detection consists in thresholding the accumulator array, i.e., every cell in the accumulator array exceeding a specific value thr is considered as a peak, this threshold value becomes critical for proper DSLS detection. The DSLS corresponding to each accumulator array peak $(\tilde{\theta}, \tilde{\rho})$ can be obtained by using the DSLS generation procedure described in Section 5.5.2, i.e.,

$$\left\{ \begin{array}{l} \text{if} \quad 45° \leqslant \theta \leqslant 135° \quad \text{then} \quad y = \text{floor}\left\{ \frac{\tilde{\rho} - x\cos(\tilde{\theta})}{\sin(\tilde{\theta})} + 0.5 \right\} \\ \text{for} \quad x = \ldots, -2, -1, 0, 1, 2, \ldots \\ \qquad \text{Otherwise} \quad x = \text{floor}\left\{ \frac{\tilde{\rho} - y\sin(\tilde{\theta})}{\cos(\tilde{\theta})} + 0.5 \right\} \\ \qquad \text{for} \quad y = \ldots, -2, -1, 0, 1, 2, \ldots \end{array} \right.$$

The binary image containing all the detected and reconstructed DSLSs is henceforth called a *reconstruction* of the original image. Figure 5.43 illustrates the reconstructions of the image in Figure 5.42 considering a series of successive threshold values.

Since the unwanted background of votes can conceal peaks, especially those corresponding to shorter DSLSs, it is important to consider means to attenuate this *background noise*. A simple and effective technique for this purpose is described in the following section.

5.6.4 | Backmapping

Introduced in Gerig & Klein [1986], the technique known as *backmapping* provides a simple and effective means for reducing the background noise generally obtained in discrete HTs. This technique consists in performing a standard HT, yielding Acc, and then repeating the HT calculation over the same input image. However, a search is performed in *Acc* for the peak along each sinusoidal path defined by each foreground pixel in the image, and only the cell in a secondary accumulator array Acc2 (initially cleared) having the same coordinates as that peak is incremented. The backmapping technique for the normal HT (the backmapping can easily be extended to other parametrizations) is described by the pseudo-code below. Figure 5.44 presents the normal HT over the image in Figure 5.42 (a), but now using backmapping. The effectiveness of this post-processing in filtering the background noise is evident, making the reconstruction of the DSLSs much more robust with respect to the choice of the threshold values.

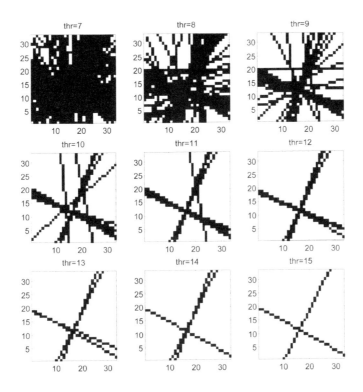

Figure 5.43: *Reconstructions of the image in Figure 5.42 obtained by using successive threshold values* (thr). *Observe the elimination of the surrounding noise allowed by the HT.*

Algorithm: *Backmapping for Normal HT*

1. Perform the normal HT, yielding Acc;
2. Clear the secondary accumulator array Acc2;
3. **for** each foreground point $(\tilde{x}, \tilde{y})$ in the image
4.   **do**
5.   max $\leftarrow -10000$;
6.   **for** ith $\leftarrow 1$ **to** $N_\theta - 1$
7.   **do**
8.   th $\leftarrow$ (ith $-1) \Delta\theta$;
9.   rh_ini $\leftarrow \tilde{x}\cos(\text{th}) + \tilde{y}\sin(\text{th}) - \frac{\mu(\text{th})}{2}$;
10.   rh_fin $\leftarrow$ rh_ini $+\mu(\text{th})$;
11.   irh_ini $\leftarrow$ ceil $\left\{ \frac{(\text{rh\_ini} - \text{rh}_{\min})}{\Delta\rho + 1} \right\}$;
12.   irh_fin $\leftarrow$ ceil $\left\{ \frac{(\text{rh\_fin} - \text{rh}_{\min})}{\Delta\rho} \right\}$;
13.   **for** irh $\leftarrow$ irh_ini **to** irh_fin
14.   **do**
15.   **if** Acc $(ith, irh) >$ max
16.   **then**
17.   max $\leftarrow$ Acc (ith, irh);
18.   ith_ max $\leftarrow$ ith;
19.   irh_ max $\leftarrow$ irh;
20.   Acc2 (ith_ max, irh_ max) $\leftarrow$ Acc2 (ith_ max, irh_ max) $+ 1$;

5.6.5 | Problems with the Hough Transform

In order to fully understand the HT underlying principles and problems, it is important to pay special attention to the way DSLSs are mapped into the discrete parameter space by each specific HT algorithm. Although this important issue is pursued in this section with respect to the normal HT, the obtained results can be easily extended to other HT parametrizations.

We have already seen that the HT involves mapping image pixels into a spatially quantized parameter space (i.e., the accumulator array). More specifically, each image pixel defines a sinusoidal band in the parameter space, and all accumulator array points within this band are incremented. Consequently, intersections between such bands appear as peaks in the accumulator array, thus indicating possible instances of DSLSs in the image. The main problem with the HT is that this mapping is degenerated, in the sense that it is impossible to recover a DSLS from the parameters of the respectively detected peak. The main problem is that longer DSLSs define smaller diamonds that fall inside larger diamonds defined by shorter DSLSs, as illustrated in Figure 5.45. Therefore, and indeed as could be expected,

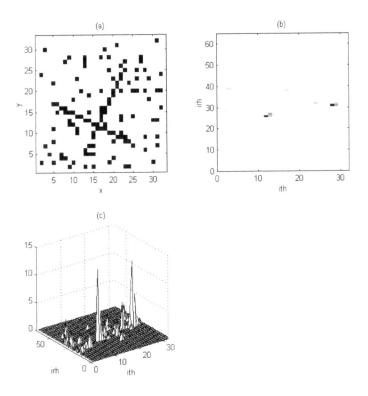

Figure 5.44: *A noisy image containing two DSLSs (a) and the accumulator array obtained by the normal HT algorithm with backmapping shown as a gray level image (b) and as a 3D plot (c). It should be observed that higher accumulator values are represented by darker gray levels in (b).*

the diamonds in the accumulator array are not in a one-to-one correspondence with the DSLSs. In addition, it is impossible to recover the end points of the DSLSs by considering only the peaks in the accumulator array.

The problem is aggravated by the fact that the diamonds are usually mapped into peaks in the accumulator array. In other words, it can be verified that, given the irregular shape of the diamonds, no spatial sampling can be obtained by using an orthogonal lattice in such a way that each diamond is sampled only once. While a coarse grid implies that some of the diamonds will not be sampled, characterizing the *undersampling* of the parameter space, a fine grid will tend to sample some of the diamonds more than once (*oversampling*). While irregular lattices could be used to enforce bijective sampling of the diamonds, this would imply sophisticated

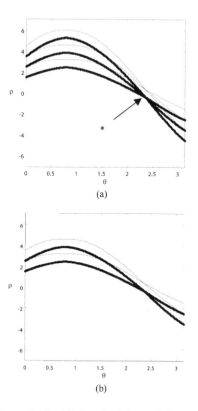

Figure 5.45: *The diamond, identified by the * in (a), defined by the intersection of the points* $(3, 3)$, $(4, 4)$ *and* $(5, 5)$, *is contained inside the diamond defined by the points* $(3, 3)$ *and* $(4, 4)$, *shown in (b).*

data structures and HT algorithms. In practice, it is recommended that you use an orthogonal grid with $N_\theta = N$ and $N_\rho = 2N$ and, if required to use some additional strategy to cope with under- and oversampling, some of which are described in the remainder of this section.

5.6.6 Improving the Hough Transform

Connectivity Analysis

It is clear from the above sections that the peaks resulting in the accumulator array have intensity equal to the number of image foreground points falling inside the upper-bound straight band defined by the respective peak parameters. Since this global value does not express the connectivity (a local property) of the image

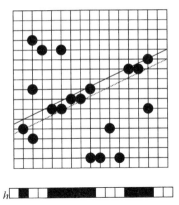

Figure 5.46: *The accumulator array cell defining the upper-bound straight band in this image will contain a peak with intensity equal to the foreground points inside the band, i.e., 9, no matter if they are connected or not. It is possible to consider the projection of the points inside the band in order to obtain information about the connectivity and end points of DSLSs.*

foreground peaks, a set of sparse foreground points, or a set of collinear DSLSs (see Figure 5.46) will map into a peak in the accumulator array. In addition, even if the peak does correspond to a single DSLS in the image, it provides no information about its end points. In this section we present a simple approach, namely a connectivity analysis of the foreground peaks covered by the upper-bound straight band, capable of minimizing all these shortcomings.

The connectivity analysis algorithm involves projecting the points inside each of the upper-bound straight bands defined by each peak in the accumulator array into a vector, as illustrated in Figure 5.46. This projection has to be done in terms of the normal angle of the straight band, more specifically:

if $45° \leqslant \theta \leqslant 135°$ project onto the x-axis

Otherwise project onto the y-axis

These projections are done while pursuing the points inside each upper-bound straight band. Once a projected vector, hence h, is obtained, it is checked for maximal gaps and minimal lengths. In other words, if the vector h contains a gap smaller or equal to a critical value max_gap, specified by the user, the gap is filled with ones. In addition, if a connected run of ones smaller than a critical minimal value min_length is found, it is not considered. A possible implementation of the gap-filling procedure is summarized by the following pseudo-code:

Algorithm: *Gap Filling*

   1.   cnt ← 0; fl ← 0;
   2. **for** i ← 2 **to** N
   3.   **do**
   4.   **if** $(h(i-1) = 1)$ **AND** $(h(i) = 0)$
   5.   **then**
   6.   fl ← 1;
   7.   **if** $(h(i-1) = 0)$ **AND** $(h(i) = 1)$
   8.   **then**
   9.   fl ← 0;
 10.   **if** cnt ⩽ max_gap
 11.   **then**
 12.   **for** k ← i − cnt **to** i − 1
 13.   **do**
 14. $h(k)$ ← 1;
 15.   cnt ← 0;
 16.   **if** fl = 1
 17.   **then** cnt ← cnt +1;

Observe that although the above algorithm assumes that DSLSs never start or terminate at the borders of the image, it can be easily extended to cope with situations where this is not verified. In addition, for more uniform results, the gap threshold should take into account the normal angle so that the gap has the same length for any inclination.

A possible implementation of the procedure to remove connected groups of pixels shorter than min_length is presented in the following:

Algorithm: *Short Group Elimination*

```
1.   cnt ← 0;
2.   fl ← 0;
3.   for i ← 2 to N
4.        do
5.             if (h(i − 1) = 0) AND (h(i) = 1)
6.                 then
7.                     fl ← 1;
8.             if (h(i − 1) = 1) AND (h(i) = 0)
9.                 then
10.                    fl ← 0;
11.                    if cnt ⩽ min_length
12.                        then
13.                            for k ← i − cnt to i − 1
14.                                do
15.                                    h(k) ← 0;
16.                    cnt ← 0;
17.             if fl = 1
18.                 then cnt ← cnt + 1;
```

Observe that the above two algorithms have to be applied in this respective order, i.e., the gap filling should precede the short group elimination. Once a vector h has been processed, the upper-bound straight band is followed in the new image, marking as foreground elements only those points set as one in the respective projection vector. Figure 5.47 presents an example of the application of the above procedure, including the original image (a), the straightforward reconstruction (as described in Section 5.6.3) (b), and the reconstruction after the connectivity analysis (c).

It should be observed that the values of the parameters max_gap and min_length must be set for each specific application, by taking into account the type of lines and noise present in the image. The above example adopted max_gap = 4 and min_length = 5.

Once a projection vector h has been checked for minimal gap and short group extraction, the extremity of the defined DSLS can be easily determined, as described in the following pseudo-code:

Algorithm: *DSLS Extremity Detection*

1.   **for** $i \leftarrow 1$ to N
2.   **do**
3.   **if** $(h(i-1) = 1)$ **AND** $(h(i) = 0)$
4.   **then**
5.   sl_start $\leftarrow i$;
6.   **if** $(h(i-1) = 0)$ **AND** $(h(i) = 1)$
7.   **then**
8.   sl_end $\leftarrow i$;
9.   STORE (sl_start, sl_end, th, rh)

where the procedure *Store* stores the obtained DSLS into a list of detected DSLSs. Observe that it is enough to store only the projected coordinates of the extremities, since the other coordinates can be obtained by considering the stored normal parameters th and rh of the respective upper-bound straight band. Indeed, since two extremity points do not define a single DSLS in the orthogonal lattice, it becomes more precise to represent a DSLS in terms of the projected coordinates of its extremities and its normal parameters. Therefore, a DSLS is henceforth represented and stored in terms of the vector DSLS = $(\theta, \rho, \text{sl\_start}, \text{sl\_end})$. Observe that the type of projection (onto the x– or y-axis) can be immediately given by inspecting the value of θ.

Dilated Straight Bands, Progressive Elimination and DSLS Merging

We have seen in the previous section that some relatively simple post-HT processing, in the case connectivity analysis, can substantially improve the quality and information provided by the HT about the DSLSs in the original image. In this section we present three additional strategies useful for improving the normal HT in some specific situations, namely the use of wider straight bands during connectivity analysis, the progressive elimination of the detected DSLSs from the image during connectivity analysis, and the merging of detected DSLSs.

One of the principal problems with the HT is that, because of the spatially sampled nature of the accumulator array, the obtained peaks define upper-bound straight bands in the image that do not exactly fit the original DSLS, in the sense that some of the reconstructed points in the DSLS do not match the original DSLS pixels. This is clearly illustrated in Figure 5.47 (c), which shows the reconstructed DSLSs corresponding to the two detected peaks superimposed on the original image. The main implication of such an effect is that the projection vectors corresponding to these reconstructed DSLSs will present many small gaps. Although this can be dealt with by using larger values for max_gap, such an approach affects the quality of the DSLS detection since it will also allow noise points to merge. A possible alternative to reduce the number and extent of the gaps in the projection vectors consists in using a wider, dilated version of the upper-bound straight band while constructing these vectors. This process is illustrated in Figure 5.48.

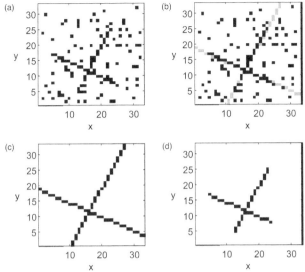

Figure 5.47: *The original image (a), its straightforward reconstruction after the standard normal HT shown superimposed to the original image (b) and separately (c), and the reconstruction after the connectivity analysis (d) considering* max_gap = 4 *and* min_length = 5.

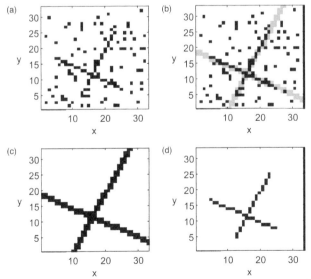

Figure 5.48: *The original image (a), its reconstruction (with doubled width) after the normal HT shown superimposed on the original image (b) and separately (c), and the reconstruction after the connectivity analysis (d) considering* max_gap = 2 *and* min_length = 5.

It is clear from Figure 5.48 that the use of dilated upper bound straight bands (width 2 in this case) has allowed the parameter max_gap to be reduced to half the value used in Figure 5.47. The connectivity analysis considering upper bound straight bands with generic width w is illustrated by the following algorithm, assuming $45° \leqslant \theta \leqslant 135°$ (a similar approach can be applied otherwise), which assumes N to be odd:

Algorithm: *Connectivity Analysis with w Larger than* 1

1. $N2 \leftarrow (N - 1)/2$;
2. $N2p1 \leftarrow N2 + 1$;
3. **for** $x \leftarrow -N2$ **to** $N2$
4.   **do**
5. $y \leftarrow (\text{rh} - x * \cos(\text{th}))/ \sin(\text{th})$;
6. $iy\_ini \leftarrow \text{floor}(y - w)$;
7. $iy\_fin \leftarrow \text{floor}(y + w + 1)$;
8.   **for** $y \leftarrow iy\_ini$ **to** iy_fin
9.   **do**
10.   **if** $(y > -N2)$ **AND** $(y < N2)$
11.   **then**
12.   **if** $(\text{img}(x + N2p1, y + N2p1) = 1$
13.   **then**
14. $h(x + N2p1) \leftarrow 1$;

The second strategy for improving DSLS detection by using the normal HT, backmapping and connectivity analysis involves the elimination of each detected DSLS from the image during the connectivity analysis. More specifically, once a DSLS indicated by a specific peak in the accumulator array is confirmed (i.e., it presents a connected extension larger or equal to the parameter min_length), the points corresponding to its reconstruction are removed from the image. Better results are usually obtained by processing the peaks in decreasing order of their intensity. This strategy is particularly useful for images involving too many lines or too much noise. Observe that although the detected DSLSs' removal may cause some gaps in the remaining DSLSs, this can be compensated by the connectivity analysis gap.

The third approach that can improve HT-based DSLS detection consists in merging the DSLSs detected during the connectivity analysis [Costa, 1992; Costa & Sandler, 1993]. Since, as a consequence of undersampling the parameter space, DSLSs can be detected as a set of broken pieces (which, by the way, are also DSLSs), special strategies may be required in order to enhance the integrity of the detected DSLSs. Two simple possibilities are described in the following. The first involves considering a dilated band while performing the connectivity analysis, in such a way that the projected vector reflects a broader neighborhood around the detected DSLSs. The second possibility consists of detecting the pieces of larger DSLSs and merging them according to their position and the similarity between their parameters (see Costa [1992]; Costa & Sandler [1993]).

Finally, it should be observed that the above described additional care and processing is not necessarily required in every HT application. As a matter of fact, as is so common in computer vision and shape analysis, it is the need implied by each specific problem and application that defines the required sophistication and accuracy. Simple HT applications, such as those aimed at identifying the main orientations of straight elements of the image (see, for instance, Costa et al. [1991]), can involve just the standard HT eventually followed by the backmapping methodology.

5.6.7 General Remarks on the Hough Transform

It should be clear by now that the Hough transform is an interesting technique for pattern recognition that, although seemingly simple at first sight, becomes relatively complex when one is interested in its full implementation. This has implied that the coverage of the HT in the present book (including some original results) had to be limited to some of its more popular aspects, namely the normal parametrization and straight line detection. In order to present a reasonably in-depth and sound treatment of the related issues, especially the understanding of the basic HT mapping mechanism in terms of bands, several of the alternative approaches reported in the rich HT literature, have not been presented here and are left as suggestions for further reading, as provided in the accompanying box.

To probe further: *Hough Transform*

Although the basic idea underlying the Hough transform is described in some of the traditional books on computer vision and image analysis, including Gonzalez & Woods [1993]; Jain [1989]; Schalkoff [1989]; Wechsler [1990], the more specific HT aspects, especially the many HT variations and additional processing aimed at improving some of the HT deficiencies are mostly found in specialized papers. A good overview of the HT state-of-the-art up to 1988 can be found in Illingworth & Kittler [1988], while the book of Leavers [1992] provides up-to-date material. Recently published, the special issue Costa & Wechsler [2000] addresses several important HT aspects, including its application and implementations in hardware.

5.7 Exact Dilations

We discussed in Section 3.5.1 the concept of shape dilations from the perspective of mathematical morphology. In the present section we present the concept of *exact dilations*, introduced in Costa et al. [1999]; Costa & Consularo [1999], as well as a simple algorithm for its implementation.

Let S be a two-dimensional binary shape. The exact dilations of this shape correspond to the sequence of *all* successive dilations, without repetition, by using circles with increasing radii as structuring elements. The important point to be taken into account is that, given the spatially sampled nature of digital images, only a few of the infinite distances that would be allowed in a continuous space are observed between any two points in the orthogonal lattice typically underlying digital images. Let us clarify this fact with a simple example. Consider the shape containing only an isolated point, as shown in Figure 5.49 (a). If we start dilating

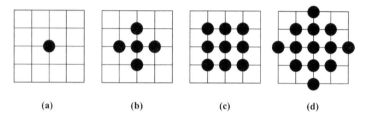

(a)   (b)   (c)   (d)

Figure 5.49: *The first four exact dilations around an isolated point.*

this single point with disks of increasing radii R, the first observed dilation would be verified for $R = 1$, which is represented in Figure 5.49 (b). In other words, all the dilations with $0 < R < 1$ produce the same original isolated point. As we continue to increase R, the next change will be verified only for $R = \sqrt{2}$, producing the dilated shape shown in Figure 5.49 (c).

It is clear that, because of the discrete nature of the lattice, only a few of the infinite radii will imply a new dilated shape. The distances defined by such critical radii are henceforth called *exact distances*, and the order in which a specific exact distance occurs is henceforth called its respective *distance index*, k. For instance, starting with $k = 0$ for the null distance, we have $k = 1$ for $d = 1$, $k = 2$ for $d = \sqrt{2}$, and so on, until $k = N_M$ for $d = d_M$. The set of all dilated shapes for each of the possible exact distances, i.e., for k varying from 0 to some maximum value N_M, corresponds to the *exact dilations* of the original shape. As a consequence, there are neither overlooked intermediate dilations between any two successive exact dilations, nor replicated dilations.

Exact dilations, considering Euclidean metric, can be performed by using the simple algorithm described in the following. The first step to calculate the exact dilations of any binary shape consists in obtaining an auxiliary data structure, henceforth called *sorted Euclidean distance representation* (*SEDR*). This data structure includes each of the possible Euclidean distances in the lattice and the relative position of each of the points presenting that distance. Figure 5.50 illustrates the general organization of the SEDR considering the first four exact distances.

A simple form to implement the SEDR is as a standard matrix, reserving a line for each of the successive exact distances. Given the k-th exact distance, the

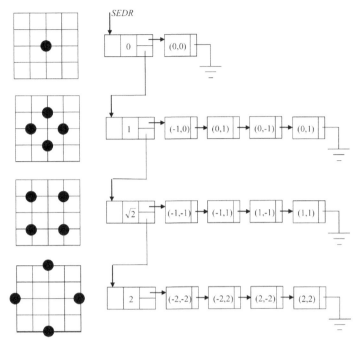

Figure 5.50: *The SEDR for the first four exact distances (i.e., $k_M = 3$).*

element $SEDR$ $(k, 1)$ will contain the value of d (this is optional, since it can be easily calculated from the additional information in the SEDR line), the element $SEDR$ $(k, 2)$ will indicate the number $M(k)$ of points presenting distance d, and $SEDR$ $(k, 3)$ to $SEDR$ $(k, 3 + M(k))$ will contain the x and y coordinates of these points, stored sequentially as pairs. This simple implementation of the SEDR as a matrix is illustrated below for $N_M = 4$.

$$SEDR = \begin{bmatrix} 0 & 1 & 0 & 0 \\ 1 & 4 & -1 & 0 & 1 & 0 & 0 & -1 & 0 & 1 \\ \sqrt{2} & 4 & -1 & -1 & -1 & 1 & 1 & -1 & 1 & 1 \\ 2 & 4 & -2 & 0 & 0 & 2 & 2 & 0 & 0 & -2 \\ \sqrt{5} & 8 & -2 & 1 & -1 & 2 & 1 & 2 & 2 & 1 & 2 & -1 & \cdots \end{bmatrix}$$

It is clear that this implementation of the SEDR is not particularly efficient, (especially when compared to the linked list structure in Figure 5.50) as far as storage is concerned, since the number of columns has to be defined in terms of the maximum number of relative positions (in the above example, eight positions for $d = \sqrt{5}$). However, provided that N_M is not too large, the implied additional amount of memory is a small sacrifice to be made in exchange for increased simplicity. It should also be borne in mind that the SEDR can be calculated only once, and stored for afterward use.

Having discussed how the SEDR can be represented, it is time to describe a process for obtaining this structure. A simple approach is outlined in the following:

① Calculate the distances between *all* the points inside a digital ball of radius d_M centered at $(0, 0)$, i.e., $\sqrt{x^2 + y^2} \leqslant d_M$.

② Sort the above distances in increasing order, eliminating repeated distances, and recording the respective relative positions of the points presenting each specific distance.

Figure 5.51 presents all the Euclidean exact distances up to the maximum distance of 30 pixels. The distances are represented by their respective gray level, smaller distances being represented by darker gray levels.

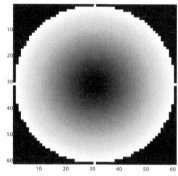

Figure 5.51: *The Euclidean exact distances up to the maximum distance of 30 pixels, represented by their respective gray level.*

Now, let S be a binary shape represented in terms of two lists L_x and L_y containing the coordinates x and y, respectively, of all its N elements. The exact dilations of S can be obtained by scanning all the elements of S for each of the successive possible distances, and marking all the positions exhibiting each considered distance from each of the shape elements. This process is summarized by the following pseudo-code:

Algorithm: *Exact Euclidean Dilations*

1. CLEAR dil_im;
2. **for** $k = 1$ **to** N_M
3.   **do**
4.   **for** $j = 1$ **to** N
5.   **do**
6.   **for** $i = 1$ **to** $SEDR$ $(k, 2)$
7.   **do**
8. $x = L\_x(j) + SEDR(k, 2i + 1)$;
9. $y = L\_y(j) + SEDR(k, 2i + 2)$;
10.   dil_im$(x, y) = 1$;

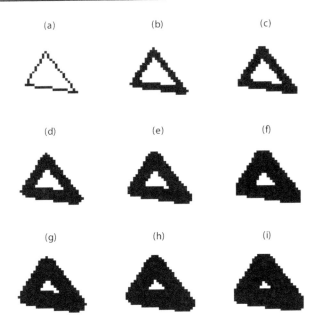

Figure 5.52: *The first eight exact dilations* **(b)** *through* **(i)** *of the binary shape in* **(a)**.

Each of the subsequent exact dilations of the original image is obtained in the image dil_im for each instance of i. Figures 5.52 (b) through (i) present the first nine exact dilations of the shape in (a). Since the number of distinct distances tends to increase as N_M increases, the performance of this algorithm tends to degrade for larger values of d_M. However, reasonably fast execution can be obtained for values of d_M up to about 20 or 30 pixels, depending on the specific hardware.

It should be observed that the above algorithm can be adapted to generate the exact dilations considering any of the possible distances presented in Section 2.2.4. Particularly fast schemes can be obtained for special cases such as the chessboard and city-block distances, although at the expense of the important isotropy property allowed by the Euclidean distance, which makes it largely invariant to rotations (perfect invariance is not possible because of the anisotropic nature of the orthogonal grid).

5.8 Distance Transforms

Given a two-dimensional binary shape S (continuous or discrete) and one of its external points P, the *distance* $d(P, S)$ between the point P and the shape S is defined to be the smallest distance between P and any of the elements of the set

S (recall that a shape is a set of connected points). It should be observed that $d(P,S) = d(S,P)$ and that several alternative types of distances can be considered, including chessboard, city-block and Euclidean (see Section 2.2.4). The important concept of distance between a point and a shape (alternatively the set of all the shape elements) is illustrated in Figure 5.53 with respect to the Euclidean distance.

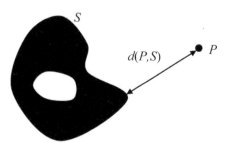

Figure 5.53: *The distance $d(P,S)$ between a point P, external to a shape S, and this shape is defined as the smallest distance between P and any of the elements of S. The Euclidean metrics has been considered.*

The *distance transform* [Fabbri et al., 2008] of the shape S can now be defined as the process that assigns to every external point P of S the respective distance $d(P,S)$. Figure 5.54 presents the distance transform of the shape in Figure 5.52 (a) with respect to the Euclidean metrics (a). The distance can also be visualized as a 3D surface, such as illustrated in Figure 5.54 (b).

While the distance transform poses no serious conceptual problems, its calculation in spatially sampled spaces, such as the orthogonal grid usually underlying digital images, deserves special attention. The critical point is that only a few of the infinite continuous distances are possible in a digital image. For instance, a 3×3 image allows only six different distances, as illustrated in Figure 5.55 (i.e., the five distances represented by arrows plus the zero distance, not shown in the figure). Observe that, because of the inherent symmetry of the orthogonal grid, each possible distance in a digital number occurs for a number of points that is always a multiple of 4.

While the number of possible distances, hence N_d, increases substantially for larger image sizes, it will always be finite (as long as the image size is finite). The complexity exhibited by the Euclidean distances in the orthogonal lattice, a consequence of the anisotropy of the latter, has implied a series of practical difficulties that, ultimately, led to many simplified schemes for the calculation of *approximated* Euclidean distance transforms (see, for instance Borgefors [1984]; Hilditch [1969]; Lee & Horng [1999]). On the other hand, recent advancements have allowed exact Euclidean distance calculation by using vector schemes (e.g., Cuisenaire [1999]; Vincent [1991]) and Voronoi diagrams (e.g., Ogniewicz [1992]; Sherbrooke et al. [1996]) which, although highly effective, imply relatively more

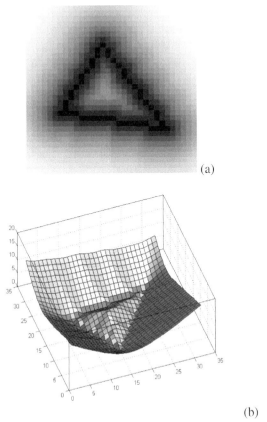

(a)

(b)

Figure 5.54: *The exact Euclidean distance transform of the shape in Figure 5.52 (a) shown as a gray-level image* **(a)** *and as a surface* **(b)**.

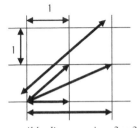

Figure 5.55: *The possible distances in a 3 × 3 orthogonal grid.*

complex algorithms. Another related reference can be found in Ragnemalm [1993]. In the present book, we describe an approach to exact Euclidean transform not as simple as the approximated schemes, but also not as complex as the fastest approaches. Therefore, it represents an alternative approach that is easy to implement and whose speed, although not optimal, should be enough to deal with a series of problems in shape analysis, especially when the maximum distance is not too large. This technique, based on the concept of exact dilations described in Section 5.7, is described in the following.

5.9 Exact Distance Transform through Exact Dilations

One of the simplest approaches to obtaining the exact Euclidean distance transform consists in slightly modifying the exact dilation algorithm presented in Section 5.7. This can be done simply by replacing the line dil_im$(x, y) := 1$ by dst_tr$(x, y) := $ SEDR$(k, 1)$ and including a test in order not to rewrite already calculated distances. Observe that the matrix dst_tr, used to store the distance transform, is now accordingly initiated with -1. The complete pseudo-code for exact Euclidean distance transform is

---

Algorithm: *Exact Distance Transform through Exact Dilations*

1.   Initiate the whole matrix dst_tr with -1;
2.   **for** $k = 1$ **to** N_M
3.   **do**
4.   **for** $j = 1$ **to** N
5.   **do**
6.   **for** $i = 1$ **to** SEDR$(k, 2)$
7.   **do**
8. $x = L_x(j) + $ SEDR$(k, 2i + 1)$;
9. $y = L_y(j) + $ SEDR$(k, 2i + 2)$;
10.   **if** dst_tr$(x, y) = -1$
11.   **then**
12.   dst_tr$(x, y) = $ SEDR$(k, 1)$;

5.10 Voronoi Tessellations

Any set of isolated points P_i, $i = 1, 2, \ldots, N$ in $\mathbb{R}^2$ defines a respective *Voronoi tessellation* of $\mathbb{R}^2$. This tessellation corresponds to partitioning the $\mathbb{R}^2$ space into N regions, henceforth called $\mathbb{R}_i$, associated with each of the points P_i, in such a way any point inside the region $\mathbb{R}_i$ is closest to the point P_i. The points that are equidistant to two or more points P_i define the separating boundaries. Figure 5.56 (b) illustrates the Voronoi diagram obtained for the set of isolated points in (a).

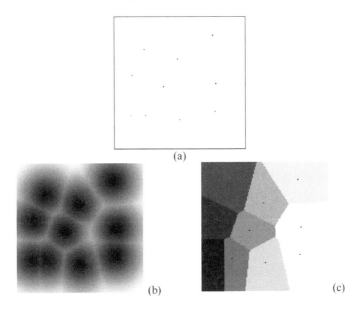

(a)

(b)					(c)

Figure 5.56: *A set of isolated points **(a)** and the respective distance transform **(b)** and Voronoi diagram **(c)**. Each influence region has been represented by a distinct gray level.*

The Voronoi tessellation provides a logical and powerful way to partition the $\mathbb{R}^2$ space in the sense that each region $\mathbb{R}_i$ can be understood as the *region of influence*, as far as Euclidean distance is concerned, of the point P_i. Of course, Voronoi tessellations considering other distances are also possible. The Voronoi tessellation concept can be generalized to a set of isolated *shapes*, instead of points, by considering the concept of distance between a point and a set discussed in Section 5.8. Such a *generalized Voronoi tessellation* is illustrated in Figure 5.57.

Although a series of highly effective algorithms for calculating Voronoi tessellations have been reported in the literature (e.g., Ogniewicz [1992]), these algorithms are relatively complex. Here, we present a simple approximated approach to obtaining generalized Voronoi tessellation. This approach is based on the concept of label propagation, also known by the name of SKIZ [Lantuèjoul, 1980; Vincent, 1991], performed by using a simple modification of the above exact dilation algorithm. In this approach, each of the N isolated shapes in the image is labeled with a subsequent integer value (i.e., $L = 1, 2, \ldots, N$) and these labels are propagated around the surrounding space, for instance by using exact dilations. A possible label propagation algorithm is given by the following pseudo-code:

Figure 5.57: *A set of isolated shapes and its respective generalized*
Voronoi tessellation.

Algorithm: *Label Propagation through Exact Dilations*

1.   **Initiate** img_lbl with -1;
2.   **for** $k \leftarrow 1$ **to** N_M
3.   **do**
4.   **for** $j \leftarrow 1$ **to** N
5.   **do**
6.   **for** $i \leftarrow 1$ **to** $\text{SEDR}(k, 2)$
7.   **do**
8. $x \leftarrow L\_x(j) + \text{SEDR}(k, 2i + 1)$;
9. $y \leftarrow L\_y(j) + \text{SEDR}(k, 2i + 2)$;
10.   **if** img_lbl$(x, y) \neq -1$
11.   **then**
12.   img_lbl$(x, y) \leftarrow L\_lb(j)$;

The labels are propagated through the image img_lbl. The list L_lb contains
the labels of each of the image elements, assigned as described above. Observe
that all the pixels in each isolated shape have their coordinates stored in the same
lists L_x and L_y. Although the order in which the isolated shapes are labeled is

unimportant, except for a few one-pixel displacements of the separating frontiers, all the pixels in each of the shapes should be stored subsequently into the lists L_x, L_y and L_lb. Figure 5.58 illustrates this fact.

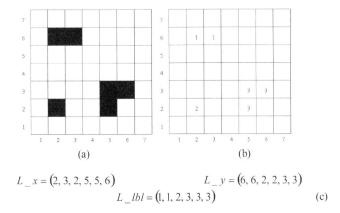

(a)   (b)

$$L\_x = (2, 3, 2, 5, 5, 6) \qquad\qquad L\_y = (6, 6, 2, 2, 3, 3)$$
$$L\_lbl = (1, 1, 2, 3, 3, 3) \qquad\qquad\qquad \text{(c)}$$

Figure 5.58: *A simple image containing three isolated shapes **(a)** one of its possible labelings **(b)**, and the lists L_x, L_y and L_lb **(c)**. Observe that the elements of each isolated shape are stored consecutively into the lists.*

Once the labels have been propagated by using the above-described procedure, a good approximation of the Voronoi tessellation is obtained. This is illustrated in Figure 5.59, with respect to the Voronoi tessellation (b) defined by the isolated shapes in (a).

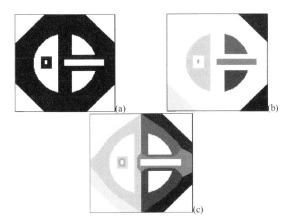

Figure 5.59: *A binary image containing a set of isolated shapes represented in white **(a)**, the labeled shapes with each label represented as a gray-level **(b)**, and the respective Voronoi tessellation obtained by the above-described label propagation algorithm **(c)**.*

It should be observed that a point that is equidistant to two isolated shapes will be marked as belonging to the shape that comes first in the lists L_x, L_y, and L_lb, hence the above-commented small variations induced by the order in which the shapes are labeled. For this reason, the above presented Voronoi diagram algorithm is not exact, but varies little with the order of labeling. Exact algorithms for obtaining Voronoi diagrams can be found in the literature, but involve more complex structures (for additional information on Voronoi algorithms, see Aurenhammer [1991]; Fortune [1987]; Preparata & Shamos [1985]).

5.11 Scale-Space Skeletonization

Given a specific shape, what are its most essential features? What makes a square a square and a circle a circle? Humans have been trying to answer such questions, although in a subjective way, since their earliest times, and with considerable success, since we are able to distinguish a myriad of distinct shapes. Besides the special relevance of curvature for shape characterization and analysis in Section 6.4, it is also important to concentrate on an especially relevant alternative, namely the *skeletons* of shapes. As implied by the name, skeletons are related to the essential structure of each shape. While no satisfactory formal definition of a shape skeleton has been thus far advanced, we shall understand skeletons in an informal way as providing a representation of a shape that is as thin as possible and that lies near the middle of the many portions of a shape. One of the formal mathematical concepts underlying skeletons is known as the *medial axis transform —MAT*, also-called *symmetry axis transform—SAT*, described by Blum (see, for instance, Blum [1967]). While alternative symmetry representations can be found in the literature, including the SLS and PISA skeletons (see, for instance, Wright [1995]), we shall in this book be limited to the SAT as an interesting approach to shape skeletons.

Fundamentally, the SAT of a specific continuous shape corresponds to all possible positions of the center of circles satisfying the following two conditions: ① are bitangent to the shape, i.e., touch the curve at two distinct points; and ② are completely enclosed inside the shape. Figure 5.60 presents a simple shape (a rectangle) and its respective SAT. This type of skeleton is particularly interesting since it is

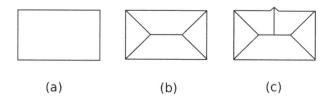

(a)   (b)   (c)

Figure 5.60: *A simple shape (a) and its respective SAT (b). Small disturbances in the shape outlined typically add new branches to the SAT skeleton (c).*

closely related to the *grass-fire model* (e.g., Blum [1967]) and the Eikonal equation [Sethian, 1999]. Think of the shape as being a grass field. If a fire is simultaneously started at every point along the borders of the shape, and assuming that this fire propagates with constant speed, the positions where the fire extinguishes itself will correspond to the SAT of the shape. More generally speaking, the SAT elements can be thought of as being the positions where waves propagating with constant speed from the shape outline points shock each other, a concept that is directly related to the exact dilation approach described in Section 5.7.

As is clear from Figure 5.60, the SAT of a shape obeys the two above identified criteria for being a good skeleton, namely it is thin and located at the middle of the shape. While the SAT provides a particularly interesting possibility for many applications, such as in dendrogram extraction (see Section 1.3.2), its standard version suffers from the fact of being too susceptible to noise at small spatial scales. Figure 5.60 (c) illustrates how a small noise at the outline of the shape in Figure 5.60 (a) can cause a new branch to appear in the skeleton.

Fortunately, it is possible to develop a version of the SAT that controls this effect. In this section a simple and relatively effective approach to SAT of binary shapes is described that incorporates a scale-space control of the SAT representation. This technique is based, again, on the concept of exact dilations presented in Section 5.7. More specifically, we start by labeling the shape outline elements with successive integer values, such as those illustrated in Figure 5.61.

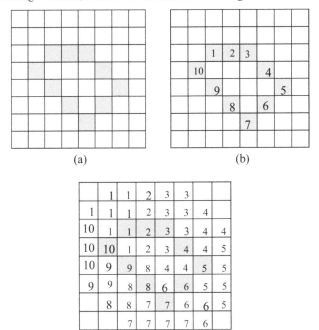

Figure 5.61: *Original shape outline (a), contour labeling (b) and propagated labels up to the maximum distance (c).*

It is observed that the choice of the initial point is irrelevant and that both clockwise and counterclockwise senses can be adopted with slightly different results.
The contour-following algorithm described in Section 5.2.2 can be applied in order
to obtain this successive labeling. Next, the labels are propagated through exact dilations as described in Section 5.10, which is illustrated in Figure 5.61 (c), yielding
the label image img_lbl. Once this image has been obtained, the maximum difference between each of its values and the respective four neighbors is determined,
and updated into a difference image img_dif. This procedure is summarized in
terms of the following pseudo-code, where Nx and Ny indicate the image size. Observe that line 7 means that max is chosen as the maximum difference between the
central label value and its four immediate neighbors.

Algorithm: *Scale-Space Skeletonization*

---

1. Label the shape outline elements with consecutive integer values;
2. Propagate these labels;
3. **for** $x \leftarrow 2$ **to** $Nx - 1$
4.   **do**
5.   **for** $y \leftarrow 2$ **to** $Ny - 1$
6.   **do**
                    ▷ Max below is taken for $i, j = -1, 0, 1$ and $|i| + |j| = 1$
7.   max $\leftarrow$ Max $\{\text{img\_lbl}\,(x, y) - \text{img\_lbl}\,(x + i, y + j)\}$
8.   **if** max $< \frac{N}{2}$
9.   **then**
10.   img_dif$(x, y) \leftarrow$ max;
11.   **else**
12.   img_dif$(x, y) \leftarrow N -$ max;

The test inside the loop is required in order to cope with the shocks defined
between elements near the label extremities, i.e., 0 and N. A scale-space family
of SATs can now be obtained simply by thresholding img_dif at different integer
values T_i. Figure 5.62 presents an original shape (a), its distance transform (b), the
propagated labels (c), and the difference image (d).

It should be noted that the difference values are proportional to the arc length
along the shape contour between the two shape elements causing the respective
shock. In other words, high difference values are obtained whenever two distant
shape elements (arc length along the contour) shock each other. This indicates that
smaller details, corresponding to shocks between close shape elements, can be removed by thresholding the difference image. Some of the skeletons and respective
reconstructions for specific thresholds obtained for the shape in the previous figure
are presented in Figure 5.63.

While both the *internal* and *external* skeletons of the shape are simultaneously
obtained, they can be immediately separated by using the filled original shape as

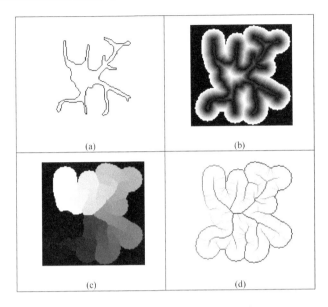

Figure 5.62: *An original shape **(a)** and its respective distance transform **(b)**, label propagation **(c)**, and difference image **(d)**.*

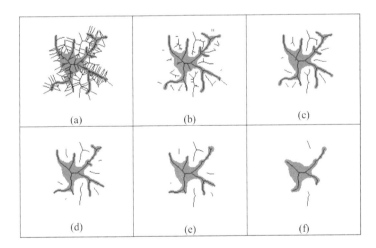

Figure 5.63: *In black: Multiscale skeletons (internal and external) of the
shape in Figure 5.62 (a) obtained by using the proposed
methodology considering threshold values of 2 **(a)**, 5 **(b)**,
10 **(c)**, 20 **(d)**, 30 **(e)** and 60 **(f)**. In gray: Respective reconstructions considering only the internal skeletons.*

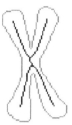

Figure 5.64: *The internal skeleton (in black) of a chromosome obtained*
for threshold 25. The outline of the original shape is also
shown (in gray).

a mask. An example of the internal skeleton of a chromosome shape is shown
in Figure 5.64. The obtained skeletons can be verified to be always 8-connected
and one-pixel wide. Since each obtained skeleton should correspond to the SAT of
some shape, an interesting possibility is to obtain reconstructions of such shapes,
which can be readily verified to represent filtered reconstructions of the original
shape. This reconstruction can be achieved in a straightforward manner, involving
the union of the digital balls centered along the specific skeleton and having as radius the respective distance transform value, as shown in gray in Figure 5.63. It is
also clear from the above example that skeletons and respective reconstructions obtained for larger threshold values tend to be smoothed out, while small scale details
are eliminated. Conversely, skeletons defined for smaller threshold values tend to
incorporate more and more small-scale information. In other words, the threshold
value acts as a *spatial scale parameter*, allowing us to choose particular values for
defining suitable skeletons. For instance, the choice of a very small threshold value
(near 1), tends to produce skeletons that are affected by every small detail along
their outlines, including the noise caused by their spatial sampling. By adopting a slightly larger threshold value, such interferences can be nicely eliminated.
More importantly, such an elimination of smaller scale detail does not substantially change the larger details in the shape, which would be impossible to achieve
by using standard linear scale-space filtering such as those implemented by using
Gaussian smoothing (see Section 3.2.4), which are known to displace the borders
of the original shape as the smaller details are removed. As the reader may have
already anticipated, the choice of a suitable threshold depends on each specific application, being typically defined by the spatial scale of the information of interest
in the image. It is also observed that a slightly distinct version of the skeletons,
with maximum displacement of one pixel, is obtained in case the labels are initially
assigned along the opposing sense. It is possible to improve the invariance of the
obtained skeletons with respect to rotations of the original shapes by using the arc
lengths along the shape border as labels (see [Costa, 2003]).

Although we have been limited to images containing single shapes, the above scale space skeletonization algorithm can be immediately extended to multiple shape images. This can be done by applying the above described generalized Voronoi tessellation in order to define the regions of influence of each shape, and then performing the scale space skeletonization separately inside each of these regions. Observe that the skeletonization inside each region of influence is completely independent of the other areas, allowing parallel execution. As far as the algorithm complexity is concerned, it is largely defined by the exact dilation process used to propagate the labels which, as already observed, is more effective for relatively small values of the maximum distance. In case larger distances have to be considered, the use of the vector approach described in Cuisenaire [1999] to propagate the labels, followed by the above-described label differences and thresholding, can yield scale space skeletonization in a fraction of the time demanded by the exact dilation procedure.

As an example of the many possible applications of scale-space skeletonization, we describe in the remainder of this section how to obtain fully automated extraction of dendrograms from shapes [Costa & Wechsler, 2000]. A dendrogram is a data structure, more specifically a tree, defining the hierarchical structure of the shape branches, as well as their respective lengths (additional information such as the respective width, roughness, etc., can also be incorporated into dendrograms). Basically, once a skeleton has been defined by selecting a suitable scale (i.e., threshold), it is tracked by using some traditional contour-following technique (see Section 5.2.2), which builds the tree as it identifies the branching points and extremities. Such an identification is facilitated because the obtained skeletons are 8-connected and exhibit one-pixel wide branches, in such a way that branching points correspond to skeleton points having more than two neighbors (considering 8-neighborhood), and extremities correspond to skeleton points having a single neighbor. We illustrate the above outlined possibility in Figure 5.65 with respect to the neural cell in Figure 5.62 (a).

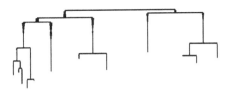

Figure 5.65: *The dendrogram of the neural cell in Figure* 5.62 (a).

Observe that the above application illustrates one of the most important properties of skeletons, namely the identification of the hierarchical structure not only of neurons, but also of generic shapes. The scale-space nature of the above described skeletons can also be used in order to identify particularly salient points of the shape by identifying those branches and segments exhibiting longer lifetimes, establishing an interesting analogy with the curvegram and scalegrams discussed

in Chapter 7. Another interesting possibility allowed by skeletonization is the logical decomposition of generic shapes in terms of balls (as above) or in terms of
sausages. The latter can be obtained by partitioning the skeleton into its basic
segments, which can be done by cutting all the branch points [Costa & Wechsler,
2000]. Observe that each sausage will therefore correspond to a skeleton segment.

5.12 Bounding Regions

An important class of shape representation methods is based on defining a bounding
region that encloses the shape of interest. Some of the most popular approaches
include the following:

☞ The *Feret Box*, which is defined as the smallest rectangle (oriented according
to a specific reference) that encloses the shape [Levine, 1985]. Usually, the
rectangle is oriented with respect to the coordinate axis, as in Figure 5.66 (a).

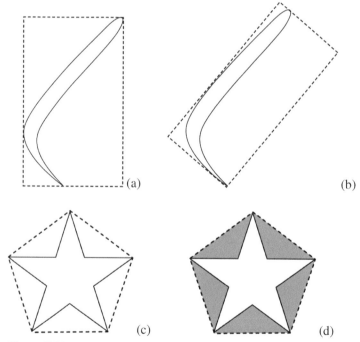

Figure 5.66: *The Feret box **(a)**, the minimum enclosing rectangle **(b)**,
the convex hull **(c)** and the convex deficiency (gray) **(d)**.*

☞ The *Minimum Enclosing Rectangle* (*MER*), also known as bounding box or
bounding rectangle [Castleman, 1996; Schalkoff, 1989], which is defined as the
smallest rectangle (oriented along any direction) that encloses the shape (see
Figure 5.66 (b)).

☞ The *Convex Hull*, which is defined as the smallest convex set that contains the shape [Costa, 1995; Duda & Hart, 1973; Gonzalez & Woods, 1993] (see Figure 5.66 (c)). A related concept is that of *convex deficiency*, defined as the difference between the convex hull and the shape (see Figure 5.66 (d)).

Shape Characterization

Chapter Overview

THIS CHAPTER STARTS by reviewing a number of shape measures (or features) that are usually adopted for shape characterization and classification and proceeds by covering several complexity descriptors, including the fractal dimension and curvature-based measures. The chapter concludes by presenting Fourier descriptors, their alternative definitions and properties. Once the shape of interest has been detected and properly represented (Chapter 5), a number of approaches can be adopted to extract important information regarding its properties. As we have said before, many approaches to shape characterization are based on obtaining shape measures to be used as features for classification and pattern recognition. This is the subject of the present chapter, which covers several shape measures from simple alternatives such as area and perimeter to more sophisticated features. A discussion about the desirable properties of classification features can be found in Chapter 8.

6.1 Statistics for Shape Descriptors

Before introducing shape descriptors, it is worth noting that a natural means to produce interesting features is by obtaining new measures from previously calculated ones. For example, once the convex hull or skeleton of a shape (Chapter 5), which are features by themselves, have been obtained it is possible to consider their perimeter or area as new features. Therefore, the reader should bear in mind that combinations of the representations discussed in Chapter 5 and the shape measures discussed in Chapters 6 and 7 are also potentially useful tools for solving specific problems.

At the same time, many shape measures covered in this chapter, and in the scientific literature in general, are based on statistics of some shape feature, which provides an interesting source for the definition of new descriptors. Let the random variable X be some shape feature, for instance, the curvature or the tangent orientation along the shape contour. The probability $P(X = x)$ can be estimated, for instance, by calculating the histogram of the observed values x (see Chapter 2). In this context, the following measures can be calculated:

☞ the mean value of X;

☞ the median value of X;

☞ the value of x_{max} so that $P(X = x_{max})$ is maximum;

☞ the value of x_{min} so that $P(X = x_{min})$ is minimum;

☞ ratios such as $\frac{x_{max}}{x_{min}}$;

☞ statistical moments of $P(X = x)$.

As the reader can verify from this book and related literature, such statistical features are commonly used in shape analysis.

6.2 Some General Descriptors

This section presents a series of simple measures, many of them related to metric aspects of shapes. These features are useful whenever the size of the shape is important (e.g., to quantify the size of a tumor in biomedical applications), and provide important resources for comparing and classifying shapes. All metric features derived in this section are expressed with respect to the *pixel dimension*. As such measures often have to be converted to the scene original units, such as meters or centimeters, proper conversion is required, which should take into account the resolution adopted for the image sampling. It is also important to note that metric features can be affected by the choice of neighborhood and distance definition.

Among the many heuristic shape measures that have been proposed in the literature, most present good performance only in specific situations, which is typically a consequence of the degenerated mapping that may result from the description of the shape in terms of features (implying that many shapes can generate the same feature values). This fact can be immediately illustrated by considering, for instance, the area as a shape measure: all shapes with the same area generate the same feature value, despite their great variability in form. Because of such a behavior, this class of features (global nature) has often been criticized in the literature. Nevertheless, it is important to note that even some simple shape measures can be applied successfully to specific situations, although it is difficult to choose which are the most suitable set of features for each application (see Chapter 8). A possible approach is to test a large set of possible features and to apply automatic feature selection algorithms in order to define a proper set of features with respect to a given training set.

6.2.1 Perimeter

The arc length of a spatially sampled curve (or contour) can be estimated by using many approaches. Supposing the contour is represented isolated in a binary image, the simplest approach to estimate its perimeter is given by the following algorithm:

Algorithm: *Region-Based Perimeter Estimation*

1. Read the input binary image;
2. Apply the binary image edge detection algorithm;
3. Perimeter ← number of detected edge points;

The above algorithm provides good results when 4-connected edges are produced by step ②, but a problem arises in the case of 8-neighborhood. This is because the latter implies the distances between consecutive pixels are not constant, i.e., it is $\sqrt{2}$, if the pixels are diagonal neighbors, and 1 otherwise. As an example, consider Figure 6.1, which shows a contour portion with three consecutive points, namely A, B and C. In this example, the distance $d(B,A)d(B,A) = \sqrt{2}$, while

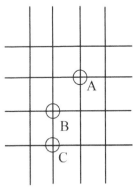

Figure 6.1: *Distance differences between consecutive points in a grid: the distance between A and B is larger than the distance between B and C.*

the distance $d(B,C) = 1$. Clearly, these local differences in the distance between consecutive points affect the total arc length, which is not taken into account by the above algorithm. The easiest way to circumvent this problem is by estimating the arc length from a contour-based representation (see Section 2.3.2). For instance, if the 8-neighborhood chain code is used to represent the contour, then the arc length can be estimated as follows [Castleman, 1996]:

$$P = N_e + N_o \sqrt{2},$$

where N_e and N_o denote, respectively, the number of even and odd codes in the chain-coded contour representation (see Section 5.2.4). It is worth emphasizing that the perimeter is calculated for the discrete curve representation, and not for the original contour (which is sampled in order to be digitally represented; see Section 3.1.4).

If the contour is represented as a complex-valued signal $u(n) = x(n) + jy(n)$, with $j = \sqrt{-1}$ and $n = 0, \ldots, N - 1$, its arc length P can be estimated as:

$$P = \sum_{n=0}^{N-1} |u(n) - u(n-1)|,$$

where $u(-1) = u(N-1)$, for closed contours, and $|u(n)|$ denotes the complex modulus.

Observe that the perimeter of a shape can be defined as the arc length of its internal contour. The reader interested in additional methods for arc length estimation should refer to [Jain et al., 1995].

6.2.2 Area

The simplest approach to estimate the area of an object is to count the number of pixels representing that shape. Supposing that g is a binary image where $g(p,q) = 1$ for shape pixels and $g(p,q) = 0$ for background pixels, the algorithm below provides an estimation of the area of the object.

Algorithm: *Region-Based Area Estimation*

1. *area* $\leftarrow$ 0;
2. **for** $p \leftarrow 1$ **to** MAX_P
3.   **do**
4.   **for** $q \leftarrow 1$ **to** MAX_Q
5.   **do**
6.   *area* $\leftarrow$ *area* $+g(p,q)$;

This algorithm assumes that the shape has been isolated into a binary image. In case the latter presents more than one shape (i.e., connected component), it is necessary to separate each object and to apply the above algorithm to each of the connected components in order to estimate the respective areas. A variation of this simple method takes into account the fact that, on the average, the pixels on the

shape boundary are half on the object and half on the background, and is defined
as follows [Castleman, 1996]:

$$A = N_o - \left[\frac{N_b}{2} + 1 \right],$$

where N_o and N_b denote, respectively, the number of interior and of boundary pixels. The area feature, together with two previously discussed image processing
techniques (i.e., labeling and histogram), can be useful as a means to solve the following interesting problem[1]. In case a given image presents several shapes to be
sorted according to their sizes, the following algorithm can be used:

Algorithm: *Area-Based Object Sorting*

1. Label each connected component in the image;
2. Calculate the histogram of the labeled image;
3. Sort the connected components using the histogram;

The above algorithm explores the idea that the histogram of the connected component labeled image actually calculates how many pixels constitute each labeled
object. The largest and the smallest objects can also be easily detected by analyzing
the above-defined histogram.

To probe further: *Alternative Approaches to Perimeter and Area Estimation*

The book of van der Heijden [1994] presents some expressions for estimating the
perimeter from its Fourier descriptors, an issue also discussed in Chapter 7 of the
present book. Additional works of interest include [Jain et al., 1995; Kiryati &
Maydan, 1989], as well as approaches based on computational geometry [Castle-
man, 1996; Chassery & Montanvert, 1991].

6.2.3 Centroid (Center of Mass)

The easiest way to estimate the shape centroid is as the average values of the shape
points' coordinates. Suppose that g is a binary image containing only the shape of
interest, where $g(p, q) = 1$ for shape pixels and $g(p, q) = 0$ for background pixels.
The next algorithm can be used to calculate the centroid of the object represented
in g.

Algorithm: *Region-Based Centroid Estimation*

---

[1] Suggested to the authors by F. C. Flores.

1. $centroid_p \leftarrow 0$;
2. $centroid_q \leftarrow 0$;
3. $area \leftarrow 0$;
4. **for** $p \leftarrow 1$ **to** P
5.   **do**
6.   **for** $q \leftarrow 1$ **to** Q
7.   **do**
8.   **if** $(g(p, q) = 1)$
9.   **then**
10. $centroid_p \leftarrow centroid_p + p$;
11. $centroid_q \leftarrow centroid_q + q$;
12. $area \leftarrow area + 1$;
13. $centroid_p \leftarrow centroid_p \,/\, area$;
14. $centroid_q \leftarrow centroid_q \,/\, area$;

The centroid can also be obtained for a shape represented by its contour. For instance, the center of mass M of a contour represented by a complex signal $u(n)$ can be easily calculated as the average value of all points of $u(n)$, i.e.,

$$M = \frac{\sum_{n=0}^{N-1} u(n)}{N},$$

where $M = z_1 + jz_2$ is a complex number and (z_1, z_2) are the centroid coordinates. Figure 6.2 presents some shapes and their respective centers of mass. It is interesting to note that the shape centroid may be located outside the shape.

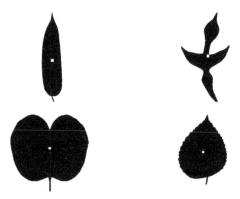

Figure 6.2: *Shapes and respective centroids, marked as small white squares.*

6.2.4 | Maximum and Minimum Distance to Centroid

Once the shape centroid has been obtained, the following additional interesting metric features can be obtained from it:

☞ maximum distance D_{max} between the centroid and the boundary points;

☞ minimum distance D_{min} between the centroid and the boundary points;

☞ mean distance D_{mean} between the centroid and the boundary points;

☞ histogram of the distances between the centroid and the boundary points;

☞ $\frac{D_{max}}{D_{min}}, \frac{D_{max}}{D_{mean}}, \frac{D_{min}}{D_{mean}}$.

The ratio-based features above are dimensionless and, therefore, more suitable when size independence is required. They can be used, for instance, in order to discriminate between circle-like shapes (where $D_{max} \cong D_{min}$) from elongated shapes. Furthermore, the features involving D_{mean} are more tolerant to local modifications along the shape.

6.2.5 | Mean Distance to the Boundary

The mean distance between the shape internal points and the shape boundary points can also be used as a shape feature. Let g be the shape of interest composed of N points, $r \in g$ be a point of g and $d(r, \text{boundary}(g))$ be the smallest distance between r and all boundary points. Therefore, the mean distance can be calculated as

$$\beta = \frac{1}{N} \sum d(r, \text{boundary}(g)).$$

It is easy to see that β can also be easily calculated from the distance transform of the shape (Chapter 5). The mean distance to the boundary can be used to define the shape complexity measure f below:

$$f = \frac{A}{\beta^2},$$

where A is the area of the shape represented by g.

6.2.6 | Diameter

The *diameter* is defined as the largest distance between any two points of a shape. A brute force method to compute the shape diameter is to search for the maximum distance between every pair of points that constitutes the shape. It is easy to see that it is sufficient to test only the points on the shape boundary, i.e., the distances between every pair of boundary points. Assuming that g is a binary image containing only the shape of interest, where $g(p, q) = 1$ for shape pixels and $g(p, q) = 0$ for background pixels, the algorithm below calculates the shape diameter and the corresponding coordinates of the farthest points of the object represented in $g(p, q)$.

Algorithm: *Estimation of the Shape Diameter*

1. Calculate the edges of g;
   ▷ b and c are the coordinates of the edge points;
2. $u \leftarrow b + cj$;
3. $n \leftarrow$ LENGTH(u);
4. $dmax \leftarrow 0$;
5. **for** $i1 \leftarrow 1$ **to** $(n - 1)$
6.   **do**
7.   **for** $i2 = (i1 + 1)$ **to** n
8.   **do**
9.   **if** (ABS($u(i1) - u(i2)$) > $dmax$)
10.   **then**
11. $dmax \leftarrow$ ABS($u(i1) - u(i2)$);
12. $lm \leftarrow i1$;
13. $cm \leftarrow i2$;

In the above algorithm, lm and cm indicate the farthest points (i.e., u (lm) and u (cm)), while dmax is the shape diameter. Figure 6.3 presents four shape examples with the corresponding farthest points indicated as "■". The respective diameters are the distances between the corresponding farthest points.

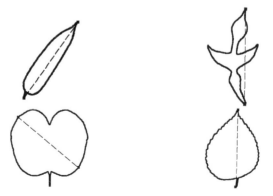

Figure 6.3: *Shapes and respective diameters, indicated by the maximum chords.*

The line segment joining the farthest points of a shape is known as the *maximum chord*. It is easy to verify that the shape diameter, as defined above, is actually the length of the maximum chord, as illustrated in Figure 6.3.

6.2.7 Norm Features

Additional shape measures that quantify shape sizes are discussed in Section 4.8 and include the following:

☞ $2n$ Euclidean norm;

☞ RMS size;

☞ mean size;

☞ centroid size;

☞ normalized centroid size;

☞ baseline distance;

☞ landmark-based shape diameter.

All the above shape features can be straightforwardly implemented from the respective discussion in Chapter 4.

6.2.8 Maximum Arc Length

Recall from Chapter 4 that some generic shapes can be represented as a set of pathwise connected curves organized as a graph (Section 4.2.1). Important features can be estimated from such shapes by statistical analysis of the composing curves, e.g., the maximum, the minimum and the mean arc length of the curves. This kind of graph can be used in syntactic pattern recognition [Fu, 1982].

6.2.9 Major and Minor Axes

Important approaches to extract shape features are based on the concept of eigenvalues. In order to understand this concept, consider the shapes in Figure 6.4. The direction along which each shape is more elongated (i.e., the direction along which the shape points are more dispersed) is known as the *major axis*. The corresponding major axes of Figure 6.4 are indicated as a longer axis within each respective shape. Perpendicular to the major axis we have the *minor axis*, which is shown superimposed onto the contours in Figure 6.4. The major and minor axes are referred to as the *principal axes*. It is important to note that only the line orientations and length (related to the eigenvalues, see below) are important and that the axes senses in Figure 6.4 are arbitrary. In addition, observe that some shapes, such as a square, can have interchangeable minor and major axes.

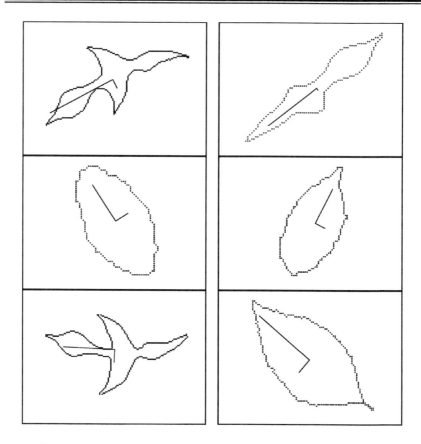

Figure 6.4: *Several shapes and their respective major and minor axes (i.e.,*
the eigenaxes). The axes lengths are represented proportion-
ally to the respective eigenvalues. It is worth emphasizing that
only the axes' orientations are important, and that their senses
are arbitrary.

These axes bear a close relationship with the *eigenvectors* of covariance matrices in multivariate probability theory (see Section 2.6.3), but now the random
vectors are understood as being defined by the coordinates of the shape pixels. A
technique for obtaining the "eigenaxes" of shapes, which can be applied to thin or
thick shapes, is described in the following. Assume that a shape is represented by
a set of points:

$$\{(x_1, y_1), (x_2, y_2), (x_3, y_3), \ldots, (x_n, y_n)\}.$$

Now, suppose that the point coordinates are represented by a random vector
$S = [x, y]$, and let C be the covariance matrix considering all such vectors. The
eigenaxes of S (i.e., of the shape) are defined as the eigenvectors of C. The eigenvector associated with the larger eigenvalue is the shape major axis, while the sec-

ond eigenvector is the *minor* axis. Suppose g is a binary image containing only the shape of interest, where $g(p,q) = 1$ for shape pixels and $g(p,q) = 0$ for background pixels. Therefore, the algorithm below calculates the major and minor axes of the shape represented in $g(p,q)$ and stores it in the array *evectors*, where each column is an eigenvector associated with respective eigenvalues stored in the vector *evalues*.

Algorithm: *Eigenaxes*

1. Calculate the edges of g;
2. Store the coordinates of the edge points into the vectors b and c;
3. $n \leftarrow$ LENGTH(b);
4. **for** $i \leftarrow 1$ **to** n
5.   **do**
6. $X(i, 1) = b(i)$;
7. $X(i, 2) = c(i)$;
8. Calculate the covariance matrix K of X;
9. Calculate the eigenvectors and the eigenvalues of K;

The results in Figure 6.4 have been calculated from the shape boundaries, while Figure 6.5 illustrates the major and minor axes obtained for a region-based shape. It is worth noting that most mathematical software, such as MATLAB®, has pre-

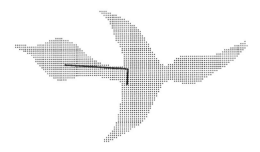

Figure 6.5: *Shape eigenaxes calculated from a region-based representation.*

defined functions that can be used to calculate the eigenvectors and eigenvalues of matrices.

Observe that the concepts of major and minor axes discussed in the present section can be related to the minimum enclosing rectangle (MER) of a shape [Castle-

man, 1996]. In fact, the major and minimum axes could be alternatively defined as
the major and minor sides of the MER. Nevertheless, it is important to emphasize
that these approaches are not equivalent and may lead to different results depending
on the shapes under consideration. Some of the features that can be obtained from
the major and minor axes include the following:

☞ the lengths of the principal axes, which can be defined as the associated eigenvalues;

☞ the *aspect ratio*, also known as *elongation*, defined by the ratio between the
major and the minor axes' sizes;

☞ $\dfrac{\text{length}(\text{major axis})}{\text{perimeter}(\text{shape})}$;

☞ rectangularity, defined as $\dfrac{\text{area}(\text{shape})}{\text{area}(\text{MER})}$.

The rectangularity can be thought of as a measure of how well the shape can
be approximated by the MER, while the aspect ratio discriminates between non-
elongated and elongated shapes. Both measures are dimensionless.

6.2.10 Thickness

The shape *thickness* is defined as the number of erosions, performed with the same
structuring element, that are necessary to completely erode the object. Observe that
such a definition of shape thickness therefore depends on the adopted structuring
element. For instance, a one-pixel wide curve has unit thickness, as one erosion is
sufficient to completely eliminate it. It is also worth noting that the thickness can
be directly related to the distance transform of a shape (see Chapter 5).

A slightly modified shape feature can be defined as the number of erosions
needed to break the shape connectivity. For example, in case the shape includes an
isthmus portion, it would break after a small number of erosions (see Figure 6.6).
The evolution of the number of connected components that are created as the shape

Figure 6.6: *Illustration of shape thickness: this shape is broken in two
after a small number of erosions.*

is eroded can also be statistically analyzed in order to generate shape features reflecting the different width isthmuses.

6.2.11 Hole-Based Shape Features

In case the analyzing shapes present holes, these can be considered as the reference for extracting several shapes. Actually, in such a situation each hole can be treated as if it were a shape. Examples of hole-based features therefore include area, perimeter, principal axes, etc. Figure 6.7 shows three shapes of characters with the respective hole shapes indicated.

Figure 6.7: *Hole-based features: the holes are themselves shapes.*

6.2.12 Statistical Moments

Statistical moments, together with the Fourier descriptors and curvature, are among the most classical and popular shape descriptors. Statistical moments descriptors can be obtained both from shape boundaries and from 2D regions.

Let g denote the shape image. The standard 2D moments are defined as

$$m_{r,s} = \sum_{p=0}^{P-1} \sum_{q=0}^{Q-1} p^r q^s \, g(p,q) \, .$$

In order to obtain translation invariance, the central moments should be applied:

$$\mu_{r,s} = \sum_{p=0}^{P-1} \sum_{q=0}^{Q-1} (p - \bar{p})^r (q - \bar{q})^s \, g(p,q) \, ,$$

where

$$\bar{p} = \frac{m_{1,0}}{m_{0,0}} \quad \text{and} \quad \bar{q} = \frac{m_{0,1}}{m_{0,0}}$$

To probe further: *Statistical Moments for Shape Analysis*

One of the nice features exhibited by these shape descriptors, which are often adopted in shape analysis, is that the first moments have appealing geometrical interpretations, being even possible to reconstruct the shape from the complete set of moments (i.e., the moment representation is invertible). Additional information about this approach can be found in [Jiang & Bunke, 1991; Kiryati & Maydan, 1989; Li, 1992; Loncaric, 1998; Marshall, 1989; Safaee-Rad et al., 1992; Schalkoff, 1989; Trier et al., 1996; Wood, 1996].

6.2.13 Symmetry

Symmetry represents an important feature that may be decisive for the effective solution of several problems in shape characterization and classification. The first important fact to be noted is that there are several types of symmetries (see, for instance, [Weyl, 1980]). While a comprehensive discussion and presentation of techniques to cope with each possible symmetry type is beyond the scope of the present book, we present a complete example of how the degree of *bilateral symmetry* can be numerically estimated from binary image representations. Consider the shape in Figure 6.8 (a). The first step consists of reflecting the shape with respect to the line having orientation defined by its first principal component or major axis (see Section 6.2.9) and passing through the respective center of mass. After filling the eventual holes caused by the reflection, which can be done by using

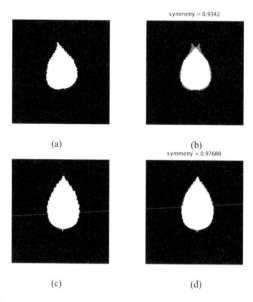

Figure 6.8: *Original shape (a) and its reflected superimposition and bilateral symmetry estimation (b). A more symmetric shape (c), and its reflected version and symmetry estimation (d).*

the closing operator from mathematical morphology (see Chapter 3), this reflected version is superimposed onto the original shape, yielding the gray level image in Figure 6.8 (b). Observe that the elements in this image are limited to having gray levels 0 (background), 1 (portions of the shapes that are asymmetric) and 2 (symmetric pixels). Let N be the number of foreground elements in the superimposed image (i.e., the number of pixels having as gray level 1 or 2), and N_2 the number of pixels having 2 as gray level. The bilateral symmetry degree can therefore be estimated by the ratio $\frac{N_2}{N}$.

Observe that this index assumes positive values up to 1, achieving its maximum value when all the pixels in the original shape have symmetric counterparts. Figures 6.8 (c) and (d) present a similar example, but now for a more symmetric shape. Observe that higher value of the symmetry index is obtained for this second shape.

6.2.14 Shape Signatures

The underlying idea of shape representation by signals or signatures is to generate one or more 1D signals that somehow describe the 2D shape [Gonzalez & Wintz, 1987]. Observe that 1D signatures can be obtained from both contour-based and region-based shape representations, and that there are many different ways of defining these signatures. It is important to emphasize that, as shape signatures describe shapes in terms of a 1D signal, they allow the application of 1D signal processing techniques (e.g., scale-space and wavelets; refer to Chapter 7) for shape analysis.

In general, the contour-based signatures are created by starting from an initial point of the contour and traversing it either in a clockwise or counterclockwise sense. A simple signature illustrating this concept is to plot the distance between each contour point and the shape centroid in terms of the sequence of the former, which acts as the parameter. Figure 6.9 (b) shows an example of such a signature $d(n)$ obtained for the contour of Figure 6.9 (a). Note that lines joining the centroid and some sample points along the contour are shown, together with an indication of the respective parameter.

It is important to note that such signals are periodic for closed curves, as one can traverse the contour an indefinite number of times. An alternative and very popular approach to define the coordinate axis of the signatures is to take the angle between the line joining the current point and the centroid, and a reference a vector or axis (e.g., the image horizontal axis). This approach presents the problem that more than a single intersection point can be obtained for a given angle. In order to circumvent this problem, a more widely adopted angle-based signature parameterization is defined by continuously varying the angle between the reference and the rotating line. In case the current line intersects the shape contour in more than one point, some function of the intersected points can be taken as the respective signature value. For instance, in the case of the above-described distance to the centroid, the signature can be calculated as the maximum, the minimum or the mean distance to the centroid with respect to all intersected points defined by each angle. The resulting signatures are periodic in the case of closed curves. Examples

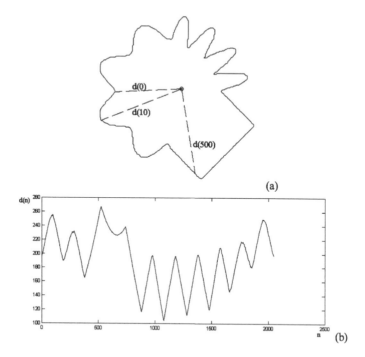

Figure 6.9: *Original contour (a) and distance to centroid-based signature (b).*

of contour-based signatures follow:

x **and** *y* **signals:** the parametric coordinates obtained by the contour extraction algorithm can be separately used as two 1D signals. A signature can also be defined in terms of the complex signal $x + jy$ or by some function of each such complex values, such as magnitude or phase.

Chain-code and shape number: the chain-code and the shape number presented in Section 5.2.4, can also be adopted as a 1D signature.

Curvature: the curvature can also be adopted as a 1D signature. Refer to Section 6.4.

Distance to the centroid: as explained above.

Number of intersections: this signature is possible only for the above-described angle-based parameterization, being defined by the number of times that the current line intersects the shape boundary.

Angle with an axis: the angle-based parameterization can be inverted in order to produce an interesting signature. In this approach, the angle between the reference vector and the line joining each point along the contour to the centroid is plotted as a function of the contour parameter. It is worth noting that no problems are verified in this case if, for a given angle, the current line intersects the contour in more than one point.

Affine signatures: affine curvature and affine parameterization can be defined as important signatures, mainly if the 2D shapes are formed by affine projections of 3D objects. The reader is referred to [Guggenheimer, 1977; Sapiro & Tannenbaum, 1993] for more detail.

Region-based signatures can also be obtained from 1D signals, but they use the whole shape information, and not the shape boundary as in the contour-based approach. An example of such a possibility is the use of the projection-based signatures, which are defined by projecting (actually integrating) the image pixel values along lines perpendicular to a reference orientation (e.g., projection signatures defined along the x and y axes). This concept is closely related to the Hough transform discussed in Section 5.6.

To probe further: *Shape Signatures*

Additional boundary-based signatures include the slope-density function [Gonzalez & Woods, 1993], the arc-height function [Lin et al., 1992] and the orientation along the contour [Lee et al., 1993].

Additional region-based signatures include the Quench function [Duda & Hart, 1973], the shape signal obtained by aligning the columns of the shape matrix [Goshtasby, 1985; Loncaric, 1998], integral geometric descriptions [Duda & Hart, 1973], line intersection based descriptions [Duda & Hart, 1973], the shape context [Belongie et al., 2001] and Sholl diagrams [Sholl, 1953], which have been proposed in the context of neuromorphology, though they can be easily applied to any 2D shape. The shape matrix, the shape context and the Sholl diagrams present interesting similarities, being useful alternatives for shape characterization. Further references on this issue can be found in [Loncaric, 1998; Pavlidis, 1980; van Otterloo, 1991].

6.2.15 Topological Descriptors

Shapes can also be analyzed with respect to their structural aspects not related to metric features. An example of such a structural aspect is the number of holes present in a shape. Some of the most important topological features for shape analysis are

☞ the *number of holes* N_H;

☞ the *number of connected components* N_C. It is important to note that this feature applies to composed shapes (e.g., Arabic character recognition),

☞ the *Euler number*, which is defined as $E = N_C - N_H$.

6.2.16 Polygonal Approximation-Based Shape Descriptors

Chapter 5 discussed the importance of polygonal approximations of shape contours, including some algorithms for obtaining them. Once a shape outline has been properly represented by a polygon, the following shape features can be extracted by straightforward procedures:

☞ *number of corners or vertices.*

☞ *angle and sides statistics*: such as mean, median, variance and moments, to name a few.

☞ *major and minor sides lengths.*

☞ *major and minor sides ratio.*

☞ *major and minor angles ratio.*

☞ *ratio between the major angle and the sum of all angles.*

☞ *ratio between the standard deviations of sides and angles.*

☞ *mean absolute difference of adjacent angles.*

☞ *symmetry measure*: a symmetry measure for polygonal segments (i.e., curve segments) is defined in [You & Fu, 1979] as

$$S = \int_0^L \left(\int_0^t k(l)\, dl - \frac{A}{2} \right) dt,$$

where t is the parameter along the curve, $\int_0^t k(l)\, dl$ is a measure of angular change until t, A is the total angular change of the curve segment and L is the length of the curve segment. It is observed that $k(l)$ can be alternatively taken as the curvature along the contour. Refer also to Section 6.2.13.

Note: *Chain-Code Based Descriptors*

As explained in Chapter 5, the chain code is a popular contour representation that has been applied to a large number of practical problems. An example of a shape measure that can be extracted from this representation is the chain cross correlation [Duda & Hart, 1973]. The reader interested to probe further on chain-code based shape descriptors should refer to [Freeman, 1961, 1970, 1974; Freeman & Davis, 1977; Jiang & Bunke, 1991; Kaneko & Okudaira, 1985; Li, 1995; Liu & Srinath, 1990; Sarkar, 1993; Wu, 1982].

6.2.17 Shape Descriptors Based on Regions and Graphs

Chapters 4 and 5 have discussed how shapes can be represented based on their regions, as well as in terms of graphs. In such situations, the following shape features can be easily extracted:

☞ *number of constituent parts*;

☞ *number of junctions*;

☞ *number of extremities*;

☞ *number of branches*;

☞ *branch size*;

☞ *convex hull and convex deficiency measures* such as area, perimeter, etc.;

☞ *distance transform statistics*;

☞ *number of self-intersections*;

☞ *geodesics statistics*.

6.2.18 Simple Complexity Descriptors

An important property of shapes is their *complexity*. Indeed, classification of objects based on shape complexity arises in many different situations. For instance, neurons have been organized into morphological classes by taking into account the complexity of their shapes (especially the dendritic tree). While complexity is a somewhat ambiguous concept, it is interesting to relate it to other geometric properties, such as *spatial coverage*. This concept, also known as space-filling capability, indicates the capacity of a biological entity for sampling or interacting/filling the surrounding space. In other words, spatial coverage defines the shape interface with the external medium, determining important capabilities of the biological entity. For instance, a bacterium with a more complex shape (and hence higher spatial coverage) will be more likely to find food. At a larger spatial scale, the amount of water a tree root can drain is related to its respective spatial coverage of the surrounding soil. Yet another example of spatial coverage concerns the power of a neural cell to make synaptic contacts (see, for instance, [Murray, 1995]). In brief, shape complexity is commonly related to spatial coverage in the sense that the more complex the shape is, the larger its spatial covering capacity.

Unfortunately, although scientists and other professionals have been widely using the term *shape complexity*, its precise definition, as yet, does not exist. Instead, there are many different useful shape measures that attempt to capture some aspects related to complexity, the most important of which are reviewed in the following, as well as in Section 6.3 and Chapter 7. The last two approaches deserve to be considered separately because of their versatility and potentiality.

Let P and A denote the shape perimeter and area, respectively. Some simple complexity measures include the following:

Circularity: defined as $\frac{P^2}{A}$.

Thinness ratio: it is inversely proportional of the circularity and defined as $4\pi\left(\frac{A}{P^2}\right)$. The 4π multiplying constant is a normalization factor.

Area to perimeter ratio: defined as $\frac{A}{P}$.

$\left(\frac{P-\sqrt{P^2-4\pi A}}{P+\sqrt{P^2-4\pi A}}\right)$: related to the thinness ratio and to circularity (see [O'Higgins, 1997]).

Rectangularity: the rectangularity has been introduced in Section 6.2.9, and is defined as $\frac{A}{\text{area(MER)}}$, where MER stands for minimum enclosing rectangle. Some of the other shape features discussed in Section 6.2.9 can be used as alternative and equivalent complexity measures.

Temperature: the temperature of a contour is defined as $T = \left(\log_2\left(\frac{2P}{P-H}\right)\right)^{-1}$, where H is the perimeter of the shape convex hull. The contour temperature is defined based on a thermodynamics formalism. The authors that proposed this feature conjecture that it bears strong relationships with the fractal dimension [DuPain et al., 1986].

Texture: the analysis of texture in images is an important topic *on its own*, but it can also be used as a means to define shape features. Texture analysis techniques generally produce a pixel-by-pixel output, where the value of the output pixel is related to the texture around it. Simple statistics about the output, such as mean texture value, can be used to define shape features. Among the many different approaches to texture analysis, it is worth remembering the co-occurrence matrix, and the linear transform-based approaches, including Fourier, Gabor and the wavelet transforms. The reader interested in more detailed related information is referred to [Gonzalez & Woods, 1993; Pratt et al., 1981; Randen & Husoy, 1999].

An interesting set of complexity measures can be defined by histogram analysis of blurred images. Recall from Chapter 3 that simple histogram measures cannot be used as shape features because the histogram does not take into account the spatial distribution of the pixels, but only the gray-level frequencies. For instance,

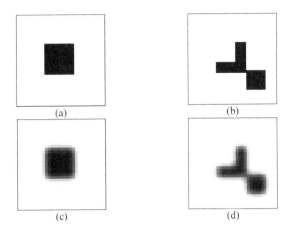

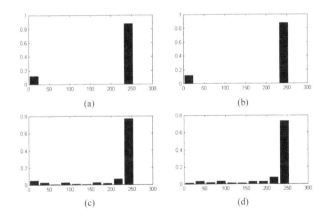

Figure 6.10: *Two different shapes (a) and (b), and respective blurred versions (c) and (d).*

refer to Figures 6.10 (a) and (b), where two different shapes are presented. The two shapes have the same area and, therefore, the same number of black pixels, thus implying identical histograms, which are shown in Figure 6.11 (a) and (b), respectively. Consider now the two shape images shown in Figures 6.10 (c) and (d),

Figure 6.11: *Respective histograms of the shapes in Figure 6.10.*

which have undergone a blurring process by mean filtering. It is observed that the output blurred images are influenced by the original shapes, thus leading to different image histograms, which is easily verified from Figure 6.11 (c) and (d). Therefore, shape features can be defined by extracting information from the latter type of histograms, i.e., those obtained after image blurring. A straightforward generalization of this idea, described in [Bruno et al., 1998], is the definition of multiscale families of features, which can be calculated based on the convolution of the input image with a set of multiscale Gaussians $g(x, y, a) = \exp \frac{-(x^2+y^2)}{2a^2}$. Let $f(x, y, a) = f(x, y) * g(x, y, a)$ be the family of multiscale (blurred) images indexed by the scale parameter a and obtained by 2D convolutions between f and g for each fixed value of a. Examples of multiscale histogram-based complexity features are

Multiscale entropy: defined as $E(a) = -\sum_i p_i(a) \ln p_i(a)$ where $p_i(a)$ is the relative frequency of the i-th gray level in the blurred image $f(x, y, a)$ for each value of a.

Multiscale standard deviation: defined as the square root of the variance of $f(x, y, a)$ for each scale a. The variance itself can also be used as a feature.

6.3 Fractal Geometry and Complexity Descriptors

6.3.1 Preliminary Considerations and Definitions

Extending its roots to the works of the nineteenth century, mathematicians Georg Cantor and Giuseppe Peano, who were interested in understanding highly irregular and fragmented sets [Mandelbrot, 1982], fractal measures have been widely applied to different problems in image processing, analysis, vision and pattern recognition. As far as shape analysis is concerned, fractal measures are useful in problems that require complexity analysis of self-similar structures across different scales. In this section we present a conceptual introduction to the underlying ideas of fractal shapes before discussing two approaches for fractal dimension estimation (box counting and Minkowsky sausage). The reader is referred to specific works in order to find more in-depth and formal related treatments [Falconer, 1990; Peitgen et al., 1992; Peitgen & Saupe, 1998], as well as to [Gleick, 1987] which provides a nice informal introduction to chaos and fractal science. Therefore, we start by introducing two possible distinct definitions of the *dimension* of a set in the Euclidean space $\mathbb{R}^N$:

☞ The *topological dimension* coincides with the number of degrees of freedom that characterize a point position in the set, denoted by d_T. Therefore, the topological dimension of a point is 0, of a curve is 1, of a plane is 2, and so on.

☞ The *Hausdorff-Besicovitch* dimension $d \geq d_T$ was formulated in 1919 by the German mathematician Felix Hausdorff. In $\mathbb{R}^N$ sets, both dimensions are comprised within 0 and N and the topological dimension assumes integer values, which is not necessarily the case for the Hausdorff-Besicovitch dimension. In the case of planar curves (such as image contours), the Hausdorff-Besicovitch dimension is an important concept that can be applied to complexity analysis, where the closer d is to 2, the more the curve fills the plane into which it belongs. Therefore, this dimension can be used as a curve complexity measure.

The mathematician Benoit Mandelbrot has been one of the main responsibles for the popularization of fractal theory [Mandelbrot, 1982]. Mandelbrot has defined the notion of a *fractal set*, i.e., a set whose Hausdorff-Besicovitch is larger than its topological dimension. Since d can assume non-integer values, he coined the name *fractal dimension*. It is nevertheless important to say that the formal definition of the Hausdorff-Besicovitch dimension is difficult to introduce and hard to calculate in practice. Therefore, it is often the case that alternative definitions for the fractal dimension are taken into account, such as the Minkowsky-Bouligand dimension [Arnèodo et al., 1995]. Frequently, these alternative definitions lead to practical methods to estimate the fractal dimensions of experimental data. In the next sections, we present two of the most popular fractal dimension estimation methods, namely the box-counting and the Minkowsky sausage methods.

6.3.2 The Box-Counting Approach

In this section, we briefly review a popular method for defining and estimating the fractal dimension of sets (and shapes). This alternative definition is based on the *box-counting dimension*, one of the most classical approaches to fractal analysis in image processing. Let S be a set of $\mathbb{R}^2$, and $M(\varepsilon)$ the number of open balls of radius ε that are necessary to cover S. An open ball of radius ε and centered at (x_0, y_0), in $\mathbb{R}^2$, can be defined as the set $\left\{(x, y) \in \mathbb{R}^2 \mid \left((x - x_0)^2 + (y - y_0)^2\right)^{\frac{1}{2}} < \varepsilon\right\}$. The box-counting fractal dimension d is defined as

$$M(\varepsilon) \sim \varepsilon^{-d}. \tag{6.1}$$

Therefore, the box-counting dimension is defined in terms of how the number of open balls necessary to cover S varies as a function of the size of the balls (i.e., the radius ε, which defines the *scale*). This concept is illustrated in a more conceptual way in the following section.

6.3.3 Case Study: The Classical Koch Curve

In this section we show how the box-counting dimension can be used in order to define the fractal dimension of Koch's triadic curve illustrated in Figure 6.12. As shown in that figure, a segment of this curve can be constructed from a straight line segment by dividing it into three identical portions, which is followed by substituting the intermediary portion by two segments having the same size and shape as the three initial segments. The complete fractal curve is obtained by recursively applying *ad infinitum* the above construction rule to each of the four resulting segments. Figure 6.12 shows four subsequent generations along the Koch curve construction process. The self-similarity property of fractal structures stems from the fact that these are similar whatever the spatial scale used for their visualization.

Recall that the definition of the box-counting dimension involves covering the set with $M(\varepsilon)$ balls for each radius ε. While analyzing the Koch curve, we take

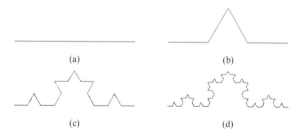

Figure 6.12: *The initial steps in constructing Koch's triadic curve.*

1D balls that cover line segments (such as rule pieces, i.e., a "ball" here means just a 1D line segment). For simplicity's sake, it is also assumed unit length for the initial straight line segment. Consequently, we need a single ball of radius $\varepsilon = \frac{1}{2}$ in order to cover the Koch curve segment. Recall that a line segment of radius $\varepsilon = \frac{1}{2}$ has as total length 1 (see Figure 6.13 (a)). In case smaller balls are used, say

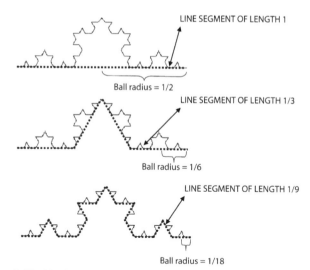

Figure 6.13: *The length of a fractal curve depends on the measuring device.*

$\varepsilon = \frac{1}{6}$, then $M(\varepsilon) = 4$ balls are needed to cover the Koch curve while, for $\varepsilon = \frac{1}{18}$, we have a total of $M(\varepsilon) = 16$ balls, as illustrated in Figures 6.13 (b) and (c). An interesting conceptual interpretation of this effect is as follows: suppose the line segments of Figures 6.13 (a), (b) and (c) represent measuring devices with the smallest measuring units. This means that, if we choose to measure the Koch curve of Figure 6.13 with the measuring device of length 1, we would conclude that the curve has length 1. This is because we cannot take into account details smaller than 1. On the other hand, if a measuring device with length $\frac{1}{3}$ were used instead, 4 segments would be needed, yielding a total length of $4\frac{1}{3} = 1.333...$

ε	$M(\varepsilon)$	Measured curve length
$\frac{1}{2} = \left(\frac{1}{2}\right)(1) = \left(\frac{1}{2}\right)\left(\frac{1}{3}\right)^0$	$1 = 4^0$	1
$\frac{1}{6} = \left(\frac{1}{2}\right)\left(\frac{1}{3}\right) = \left(\frac{1}{2}\right)\left(\frac{1}{3}\right)^1$	$4 = 4^1$	1.33
$\frac{1}{18} = \left(\frac{1}{2}\right)\left(\frac{1}{9}\right) = \left(\frac{1}{2}\right)\left(\frac{1}{3}\right)^2$	$16 = 4^2$	1.78
$\vdots$	$\vdots$	$\vdots$

Table 6.1: *The number $M(\varepsilon)$ of balls of radius ε necessary to cover the Koch curve in Figure 6.12.*

Table 6.1 summarizes this measuring process of $M(\varepsilon)$ as a function of ε. This analysis indicates that when ε is decreased by a factor of $\frac{1}{3}$, $M(\varepsilon)$ is increased by a factor of 4. From equation (6.1), this means that

$$4 \sim \left(\frac{1}{3}\right)^{-d}.$$

Therefore, we have $d = \frac{\log(4)}{\log(3)} \cong 1.26$, which actually is the fractal dimension of the Koch curve. The Koch curve is *exactly self-similar*, as it can be constructed recursively by applying the generating rule, which substitutes each segment by the basic element in Figure 6.12 (b), as explained above.

6.3.4 Implementing the Box-Counting Method

The basic algorithm for estimating the box-counting dimension is based on partitioning the image into square boxes of size $L \times L$ and counting the number $N(L)$ of boxes containing at least a portion (whatever small) of the shape. By varying L, it is possible to create the plot representing $\log(N(L)) \times \log(L)$. Figures 6.14 (a) and (b) show a retina vessels image and two superimposed grids whose non-empty boxes are represented as shaded. As implied by the discussion in the previous section, the fractal dimension can be calculated as the absolute value of the slope of the line interpolated to the $\log(N(L)) \times \log(L)$ plot, as illustrated in Figure 6.15 (a).

The sequence of box sizes, starting from the whole image, is usually reduced by $\frac{1}{2}$ from one level to the next [Jelinek & Fernandez, 1998]. Nevertheless, it is important to note that truly fractal structures are idealizations that can neither exist in nature nor be completely represented by computers. The main reasons for this are ① *ad infinitum* self-similarity is never verified for shapes in nature (both at the microscopic scale, where the atomic dimension is always eventually achieved; and the macroscopic scale, since the shape has a finite size); and ② the limited resolution adopted in digital imaging tends to filter the smallest shape detail. Therefore, it is necessary to assume that "fractal" shapes present limited fractality, i.e., they are fractal with respect to a limited scale interval (e.g., [Coelho & Costa, 1996]).

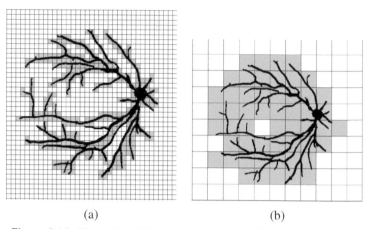

(a)   (b)

Figure 6.14: *Illustration of the box-counting method for fractal dimension.*

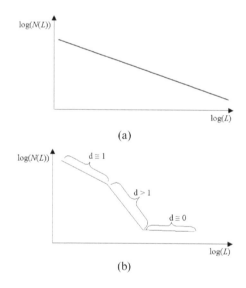

Figure 6.15: *Expected **(a)** and experimentally obtained **(b)** diagrams for the box-counting method.*

This is reflected in the $\log(N(L)) \times \log(L)$ plot in Figure 6.15 (b) by the presence of three distinct regions, i.e., a non-fractal region ($d \cong 1$), a fractal region ($d > 1$) and a region with dimension zero. This general organization of the curve can easily be understood by taking into account the relative sizes of the boxes and the shape. When the boxes are too small, they tend to "see" the shape portions as straight lines, yielding dimension 1. As the boxes' size increases, the complex details of the shape start to be perceived, yielding fractal dimension possibly larger than 1 and smaller or equal to 2. Finally, when the boxes become too large, the whole shape can be comprised within a single box, implying dimension zero. As a consequence, the line should be fitted to the fractal portion of the $\log(N(L)) \times \log(L)$ plot in order to obtain a more accurate estimation of the limited fractal behavior of the shape under analysis.

6.3.5 The Minkowsky Sausage or Dilation Method

An alternative approach for estimating the fractal dimension can be understood from a key concept related to 2D shape, namely the *area of influence* or *spatial* (or *territorial*) *coverage*. In spite of its importance and intuitive appeal, it is particularly difficult to define precisely. For instance, in shape analysis applications of neural cells, the area of influence of a neuron has been frequently associated with the polygon obtained by joining the cell extremities [Vaney, 1994]. The convex hull has been proposed as more stable alternative approach [Costa, 1995]. The problem with both approaches is that they do not consider the shape structure within the influence region, which can be a serious drawback in different situations. In order to circumvent this problem, the area of influence of a shape can be defined as the set of points located up to a maximum distance D from that shape. Therefore, a straightforward technique to calculate the area of influence is by dilating the shape by a disk of diameter D (see Figure 6.16). Such inflated shape representations are called Minkowsky's sausages [Jelinek & Fernandez, 1998; Tricot, 1995]. It is worth noting that the exact dilations approach of Section 5.7 can also be applied to obtain the Minkowsky's sausages.

The Bouligand-Minkowsky fractal dimension [Falconer, 1990; Peitgen & Saupe, 1988] is defined from the Minkowsky sausages by analyzing how the area of influence grows as the diameter D increases. It is easy to see that, if the shape is a single point, then the area of influence steadily grows. On the other hand, the area of influence tends to saturate for dense shapes, such as nearly filled regions. Therefore, similarly to the box-counting approach, the fractal dimension is obtained by analyzing the log-log plot of the area of influence *versus* D: curves with higher slopes are obtained for simple shapes, and the Bouligand-Minkowsky fractal dimension is defined as $d = 2 - \text{slope}$, where slope is the slope of the just mentioned log-log plot. A technique for fractal dimension estimation based on exact dilations (see Section 5.7) has been described [Costa et al., 1999] that allows the consideration of every possible Minkowsky sausage in an orthogonal grid. Another extension reported is the multiscale fractal dimension, which expresses the fractal behaviour of the shape in terms of the spatial scale [Costa et al., 2001].

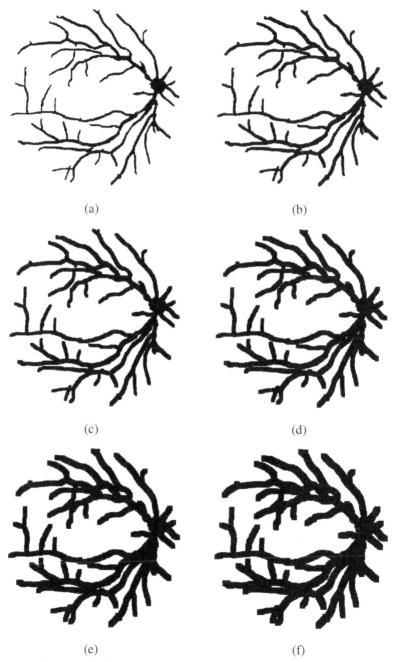

(a)   (b)

(c)   (d)

(e)   (f)

Figure 6.16: *The Minkowsky sausage method for fractal dimension.*

6.4 Curvature

6.4.1 Biological Motivation

The curvature is one of the most important features that can be extracted from contours. Indeed, strong biological motivation has been identified for studying curvature, which is apparently an important clue explored by the human visual system. In this section we discuss some important related aspects, concentrating on the developments by Fred Attneave, which have largely influenced many developments in shape analysis by computers. In one of his seminal papers [Attneave, 1954], Attneave emphasized the importance that transient events and asymmetries have in human visual perception. In his work, which considered psychological issues in terms of information processing, it was claimed that the visual information is highly redundant, both in space and time. In one of the experiments, Attneave explored the fact that the points in a scene about which the subjects tend to make more mistakes while trying to predict continuation (see below) are precisely those carrying more information. In the experiment, an image containing a small black bottle laid on a brown table in front of a white wall was gradually shown to the subject. This image was divided into 4000 cells, organized as an array of 50 rows and 80 columns and presented to subjects in a cell-by-cell fashion, from left to right and from up to bottom. That experiment was actually a kind of game in which the subject was requested to predict the color of the next hidden cell to be shown (black, brown or white). In the beginning of the experiment, after some mistakes, the subject perceived the background homogeneity and started to correctly guess the white cells until the bottle top has been reached. In fact, this special cell presented a high error rate. After a few more mistakes, the subject perceived the bottle homogeneity, returning to correct guesses. This kind of psychophysical result illustrates the importance of edges or outline contours in images, which are associated with the physical boundaries between the scene objects.

Another type of redundancy explored by the human subjects was related to the straight line segments along the shape edges. More specifically, while the first cell containing the top of the bottle concentrated a large amount of information (i.e., presented a high error rate), the next cells in the following rows presented less errors, since the subjects started to correctly infer the cells' color along the prolongation of the line segment along the following cells. This experiment demonstrated that since *corners* and *high curvature points* tended to induce higher error probability, they should *concentrate more information*.

Finally, the subjects were verified to be fully capable of correctly extrapolating the bottle *symmetry*, which implied the right side of the bottle to be perceived with fewer errors. Based on these considerations, it is possible to compare the information content of several shape elements, as shown in Table 6.2.

Attneave remarked that even Gestalt principles [Rock & Palmer, 1990] can be interpreted under the viewpoint of information redundancy exploration. Since typical Gestalt images present a high degree of some kind of redundancy, they induce

Less information	More information
homogeneous color region	edges
straight lines (null curvature) and constant curvature segments	corners and high curvature points
(local) periodic patterns	extremities of periodic patterns
symmetries	asymmetries

Table 6.2: *Some interesting relationships identified in Attneave's work.*

our perceptual system to explore such redundancies in order to achieve enhanced perception.

We conclude the discussion about Attneave's work by briefly describing two other of his psychophysical experiments. In the first, a series of 2D shapes was presented to the subjects, who were requested to represent each contour in terms of a set of 10 points. The results were presented as a histogram superimposed onto the shapes contours, from which it was clear that most subjects preferred to use high curvature points to represent each contour. In another experiment, Attneave created the now-classical picture of a cat by identifying high curvature points in an ordinary snapshot and linking them by line segments, showing that most of the information in the original image concentrated at the curvature extrema. Similar experiments and results have been reported in [Fischler & Wolf, 1994].

To probe further: *Biological Curvature Analysis*

The reader interested in further material on the biological importance of curvature analysis is referred to [Biederman, 1985; Blakemore & Over, 1974; Dobbins et al., 1987; Gibson, 1933; Guez et al., 1994; Levine, 1985; Perret & Oram, 1993; Timney & Macdonald, 1978].

As far as computational shape analysis is concerned, curvature plays an important role in the identification of many different geometric shape primitives. Table 6.3, adapted from [Levine, 1985], summarizes some important geometrical aspects that can be characterized by considering curvature.

6.4.2 Simple Approaches to Curvature

Recall from Chapter 2 that the curvature $k(t)$ of a parametric curve $c(t) = (x(t), y(t))$ is defined as

$$k(t) = \frac{\dot{x}(t)\ddot{y}(t) - \ddot{x}(t)\dot{y}(t)}{\left(\dot{x}(t)^2 + \dot{y}(t)^2\right)^{\frac{3}{2}}}. \tag{6.2}$$

Curvature	Geometrical aspect (assuming counterclockwise parameterization)
curvature local absolute value maximum	generic corner
curvature local positive maximum	convex corner
curvature local negative minimum	concave corner
constant zero curvature	straight line segment
constant non-zero curvature	circle segment
zero crossing	inflection point
average high curvature in absolute or squared values	shape complexity, related to the bending energy (Chapter 7)

Table 6.3: *Summary of some important geometrical properties that can be extracted by considering curvature.*

It is clear from this equation that estimating the curvature involves the derivatives of $x(t)$ and $y(t)$, which is a problem in the case of computational shape analysis where the contour is represented in digital (i.e., spatially sampled) form. There are two basic approaches to circumvent this problem:

① definition of alternative curvature measures based on angles between vectors defined in terms of the discrete contour elements;

② interpolation or approximation of $x(t)$ and $y(t)$ and differentiation of the fitted curves.

Figure 6.17 illustrates the two above-mentioned approaches.

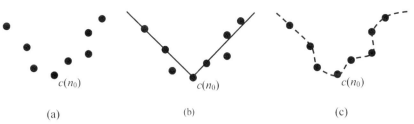

      (a)   (b)   (c)

Figure 6.17: *Approaches for estimating the curvature of a discrete curve: discrete contour portion (**a**), angle-based curvature measure (**b**) and interpolation-based curvature estimation (**c**).*

Figure 6.17 (a) shows a digital contour segment composed of a series of points. Suppose the curvature at $c(n_0)$ is to be estimated. The former approach involves fitting line segments intersecting at $c(n_0)$ along portions of the contour, as illustrated in Figure 6.17 (b). On the other hand, by interpolating the neighbor points around $c(n_0)$, such as in Figure 6.17 (c), it is possible to calculate the corresponding derivatives from the interpolated function. Some possible implementations of both such approaches are discussed below. More powerful techniques based on multiscale methods are addressed in Chapter 7.

Case I: Definition of alternative curvature measures based on angles between vectors

The approaches that have been proposed to estimate the angle defined at $c(n_0)$ by vectors along the contour differ in the way these vectors are fitted or in the method applied for estimating the angles. Let $c(n) = (x(n), y(n))$ be a discrete curve. The following vectors can be defined:

$$\begin{aligned}
v_i(n) &= (x(n) - x(n-i), y(n) - y(n-i)), \\
w_i(n) &= (x(n) - x(n+i), y(n) - y(n+i)).
\end{aligned} \tag{6.3}$$

These vectors are defined between $c(n)$ (the current point) and the i-th neighbors of $c(n)$ to the left and to the right (see Figure 6.18).

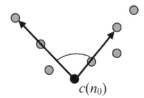

$c(n_0)$

Figure 6.18: *Angle-based curvature indication: the left and the right vectors represent $v_i(n)$ and $w_i(n)$, respectively.*

A digital model of high curvature points proposed in [Johnston & Rosenfeld, 1973] is defined by the following equation:

$$r_i(n) = \frac{v_i(n) \, w_i(n)}{\|v_i(n)\| \, \|w_i(n)\|}, \tag{6.4}$$

where $r_i(n)$ is the cosine of the angle between the vectors $v_i(n)$ and $w_i(n)$ (see Figure 6.18). Therefore, we have that $-1 \leqslant r_i(n) \leqslant 1$, with $r_i(n) = -1$ for straight lines and $r_i(n) = 1$ when the angle becomes $0°$ (the smallest possible angle). In this sense, $r_i(n)$ can be used as a measure capable of locating high curvature points.

Note: *Corner Detection from Curvature*

As commented in Chapter 5, there are two main approaches to corner detection: by polygonal approximation of 2D contours and by contour curvature analysis. Since, as indicated in Table 6.3, corners are normally associated with high absolute curvature values, a straightforward approach to corner detection is simply to threshold the curvature and to define as corners those points whose curvature absolute value is larger than a given threshold T. This simple approach has two main drawbacks, besides the fact of being global, which are explained in the following. Chapter 7 discusses the problem of searching for local maxima points.

First, groups of contour points near high curvature points tend to present curvature also exceeding the threshold, which produces clusters of corner points instead of the typically desired single corner points. This problem can be addressed by searching for curvature local maxima points and defining a minimum neighborhood around each maximum where only a single corner is allowed.

The second problem arises because, while true corners are often defined considering the global behavior of the curve, the curvature is itself a local measure. Refer to Figure 6.19, which shows an outline with five high curvature points indicated. The middle high curvature point in the upper vertex of the square is produced by the

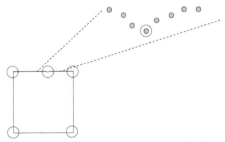

Figure 6.19: *A possibly spurious detected corner, which is shown expanded.*

contour structure shown expanded. Nevertheless, this middle point should probably not be selected as a true corner, since it belongs to an approximate straight line joining the two neighboring corners. This problem can be circumvented by calculating the angle formed by the line segment uniting each corner candidate to its two neighboring corner points (e.g., analogously to the equation (6.4)). The point should be considered as a corner only if this angle is smaller than a given threshold. Actually, this problem is also closely related to the spatial scale of the contour variations.

More detail about post-processing for corner detection can be found in [Pavlidis, 1977], and an alternative approach to corner detection based on multiscale curvature zero-crossings is described in [Cesar-Jr. & Costa, 1995]. In that work, a contour point P is considered a corner whenever the following conditions are simultaneously verified:

☞ the first derivative of the absolute curvature (i.e., the curvature taken in modulus, $|k(t)|$) is a zero-crossing at P;

☞ the absolute curvature at P exceeds the threshold T;

☞ the second derivative of the absolute curvature at P is negative.

Case II: Interpolation of $x(t)$ and $y(t)$ followed by differentiation

There are different approaches to approximating or interpolating the contour in order to analytically derive the curvature. The simplest approach is to approximate the derivatives of $x(n)$ and of $y(n)$ in terms of finite differences, i.e.,

$$\dot{x}(n) = x(n) - x(n-1),$$
$$\dot{y}(n) = y(n) - y(n-1),$$
$$\ddot{x}(n) = \dot{x}(n) - \dot{x}(n-1),$$
$$\ddot{y}(n) = \dot{y}(n) - \dot{y}(n-1).$$

The curvature can then be estimated by substituting the above-calculated values in the curvature equation (6.2). Although this simple approach can be efficiently implemented, it is rather sensitive to noise.

A more elaborate technique has been proposed by Medioni & Yasumoto [1987], which is based on approximating the contour in a piecewise fashion in terms of cubic B-splines. The Fourier-based approach, to be discussed in Chapter 7, can also be considered an interpolation-based scheme. Suppose that we want to approximate a contour segment by a cubic parametric polynomial in t, with $t \in [0, 1]$. The curve approximating the contour between two consecutive contour points A $(t = 0)$ and B $(t = 1)$ is defined as:

$$x(t) = a_1 t^3 + b_1 t^2 + c_1 t + d_1,$$
$$y(t) = a_2 t^3 + b_2 t^2 + c_2 t + d_2,$$

where a, b, c and d are taken as parameters of the approximating polynomial. By substituting the respective derivatives of the above parametric curve into equation (6.2) (for A $(t = 0)$, i.e., supposing that the curvature is to be estimated around the point A (0)) we get:

$$k = 2\frac{c_1 b_2 - c_2 b_1}{\left(c_1^2 + c_2^2\right)^{\frac{3}{2}}}. \tag{6.5}$$

The parametric curve segments can be adjusted by using cubic B-splines with equally spaced nodes, with the curve segments given by

$$x(t) = TMP_x \quad \text{and}$$
$$y(t) = TMP_y,$$

with

$$T = \begin{bmatrix} t^3 & t^2 & t & 1 \end{bmatrix},$$

$$M = \begin{bmatrix} -1 & 3 & -3 & 1 \\ 3 & -6 & 3 & 0 \\ -3 & 0 & 3 & 0 \\ 1 & 4 & 1 & 0 \end{bmatrix}$$

and

$$P_x = \left[\begin{array}{cccc} x_{n-1} & x_n & x_{n+1} & x_{n+2} \end{array} \right]^T ,$$

$$P_y = \left[\begin{array}{cccc} y_{n-1} & y_n & y_{n+1} & y_{n+2} \end{array} \right]^T ,$$

where $(x_n, y_n) = (x(n), y(n))$ are the coordinates of the n-th contour point. The above coefficients b_1, b_2, c_1 and c_2 can be calculated as follows (see [Medioni & Yasumoto, 1987] for further detail):

$$b_1 = \frac{1}{12} \left((x_{n-2} + x_{i+2}) + 2(x_{n-1} + x_{n+1}) - 6x_n \right),$$

$$b_2 = \frac{1}{12} \left((y_{n-2} + y_{n+2}) + 2(y_{n-1} + y_{n+1}) - y_n \right),$$

$$c_1 = \frac{1}{12} \left((x_{n+2} - x_{n-2}) + 4(x_{n+1} + x_{n-1}) \right),$$

$$c_2 = \frac{1}{12} \left((y_{n+2} - y_{n-2}) + 4(y_{n+1} + y_{n-1}) \right).$$

The curvature is calculated by substituting the above coefficients into the curvature equation (6.5).

To probe further: *Alternative Approaches to Digital Curvature Estimation*

Some classical works on angle-based digital curvature indication include [Beus & Tiu, 1987; Freeman & Davis, 1977; Liu & Srinath, 1990; Rosenfeld & Weszka, 1975]. Levine's book also discusses a method for curvature evaluation based on the chain-code (page 500 of [Levine, 1985]). On the other hand, [Asada & Brady, 1986; Baroni & Barletta, 1992; Mokhtarian & Mackworth, 1992] are examples of interpolation-based curvature estimators. The reader interested in works that deal with the important issue of precision evaluation of digital curvature estimators should refer to the works of [Bennett & Donald, 1975; Fairney & Fairney, 1994; Worring & Smeulders, 1993]. Finally, both affine parameterization and affine curvature are important concepts in different situations (e.g., refer to [Alferez & Wang, 1999; Guggenheimer, 1977; Sapiro & Tannenbaum, 1993]).

6.4.3 c-Measure

The fact that the important structures in an image generally occur at different spatial scales has been taken into account by several vision and image processing researchers [Koenderink, 1984; Marr, 1982; Witkin, 1983], which has paved the way for the development of multiscale shape analysis [Asada & Brady, 1986; Lowe, 1989; Mokhtarian & Mackworth, 1992; Rattarangsi & Chin, 1992]. The importance of multiscale shape analysis was recognized by vision researchers at an early stage, and the c-measure proposed by Larry Davis in 1977 is one of the first important works related to this paradigm [Davis, 1977]. Davis started from a digital model of high curvature points proposed in [Johnston & Rosenfeld, 1973]. As previously explained, the vectors $\underline{v}_i(n)$ and $\underline{w}_i(n)$ are defined between the n-th point of $c(n)$ and the i-th point $c(n-i)$ to one side, and $c(n+i)$ to the other side (equation (6.3)). As i increases, a larger neighborhood is taken into account in the estimation of $k_i(n)$, i.e., the angle is estimated at a *larger spatial scale*. Larger scales make the curvature more tolerant to noise, but also less precise with respect to small structures in the contour. In this context, Davis developed a hierarchical description method based on the evolution of $k_i(n)$ as i varies. Figure 6.20 (a) shows a filled *astroid*[1], with the respective x and y signals being presented in Figure 6.20 (b). This curve is defined as:

$$x(t) = a\cos^3(t) \quad \text{and}$$
$$y(t) = a\sin^3(t).$$

The c-measure is presented in Figure 6.20 (c), where the horizontal axis represents the parameter n along the contour, while the vertical axis, increasing downward, represents the scale parameter i. As can be seen, as the scale i decreases, the c-measure diagram high values makes a zoom around the four astroid corners (see Figure 6.20 (c)). As previously discussed, this method of curvature measuring may be alternatively implemented by taking into account not only the extremities $c(n-i)$ and $c(n+i)$, but also all the points within these neighborhoods. This can be easily accomplished by calculating the angle between the resulting vectors obtained by adding the vectors $c(n \pm m) - c(n)$, $m = 1, 2, \ldots, i$. The resulting c-measure diagram is shown in Figure 6.20 (d).

6.4.4 Curvature-Based Shape Descriptors

While the curvature itself can be used as a feature vector, this approach presents some serious drawbacks including the fact that the curvature signal can be too long (involving thousands of points, depending on the contour) and highly redundant.

---

[1] A selection of interesting curves (including the astroid) and their respective properties can be found in [Lawrence, 1972] and [Yates, 1952].

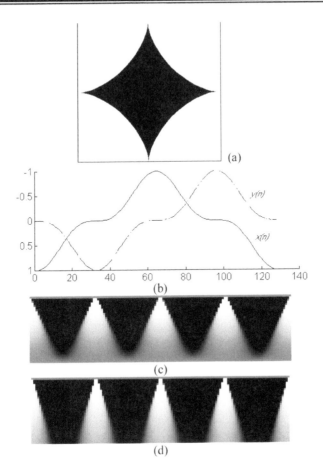

Figure 6.20: *Shape defined by a filled astroid, exhibiting four corners (**a**), x and y signals, extracted from the astroid (**b**), c-measure (**c**) and c-measure estimated by using the alternative implementation discussed in the text (**d**). Observe that higher values of the c-measure are represented with brighter gray levels in* (c) *and* (d).

Once the curvature has been estimated, the following shape measures can be calculated in order to circumvent these problems:

Sampled curvature: Instead of using all the curvature values along the contour, it is possible to sample the curvature signal in order to obtain a smaller feature set. For instance, Figure 6.21 (a) shows a map contour originally with 220 points that has been resampled at intervals of 10 points. Each sample is indicated by a "•". The curvature estimated by using the Fourier approach (Section 7.2) considering all the original 220 contour points, is presented in

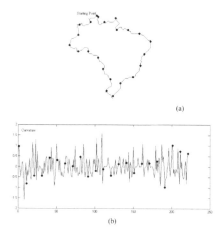

(a)

(b)

Figure 6.21: *Sampling the curvature along the contour in order to create a shorter feature vector.*

Figure 6.21 (b). The sampled curvature points are also indicated by "•". It is worth noting that, actually, the arc tangent of the curvature values are plotted in Figure 6.21 (b) for visualization's sake.

Curvature statistics: The histogram of curvatures can provide a series of useful global measures, such as the mean curvature, median, variance, standard deviation, entropy, moments, etc.

Maxima, minima, inflection points: The fact that not all points along a contour are equally important (in the sense of conveying information about the shape) motivates the analysis of dominant points such as those points where the curvature is either a positive maximum, or a negative minimum, or an inflection point. The number of such points, their position along the contour and the curvature values of them (in case of maxima and minima curvature points) can be used as shape measures.

Bending energy: In the continuous case, the mean bending energy, also known as *boundary energy* is defined as:

$$B = \frac{1}{P} \int k(t)^2 \, dt,$$

where P is the contour perimeter. The above equation means that the bending energy is obtained by integrating the squared curvature values along the contour and dividing the result by the curve perimeter. As one of the most important global measures related to shape complexity, the bending energy is discussed in more detail, including its multiscale versions, in Chapter 7.

$\boxed{6.5}$ **Fourier Descriptors**

Fourier Descriptors (FDs) is one of the most popular shape representation methods for vision and pattern recognition applications. It should be borne in mind that FDs refer to a *class* of methods, not a single one, since there are many different ways in which the FDs of a shape can be defined. The basic idea underlying this approach consists in representing the shape of interest in terms of a 1D or 2D signal, followed by taking the Fourier transform of this signal and calculating the FDs from this Fourier representation. FDs were originally proposed by R. L. Cosgriff in 1960 [Zahn & Roskies, 1972], and thereafter became popular among the pattern recognition community through the papers of Gösta H. Granlund [Granlund, 1972], Charles T. Zahn and Ralph Z. Roskies [Zahn & Roskies, 1972] and by Eric Persoon and King-Sun Fu [Persoon & Fu, 1977]. Particularly, the papers by Granlund and by Zahn and Roskies present two different approaches for obtaining the signal, which describes the shape. For a review on shape representation by signals (and signatures), refer to Section 6.2.14.

As in [Zahn & Roskies, 1972], let Γ be a curve defined by the shape contour, which may be parametrically represented as $c(t) = (x(t), y(t))$, where t is the arc length parameterization, $0 \leqslant t \leqslant L$, and L is the curve perimeter. Let $\theta(t)$ be the angular direction of Γ, $\delta_0 = \theta(0)$ be the angular direction at the starting point $c(0)$, and $\Phi(t)$ be the cumulative angular function obtained from the accumulative integral of $\theta(t)$ from 0 to t. The creation of FDs set invariant to translation, rotation and scaling requires the definition of a normalized function $\Phi_N(t)$:

$$\Phi_N(t) = \Phi\left(\frac{Lt}{2\pi}\right) + t.$$

$\Phi_N(t)$ can be represented as a Fourier series in polar form:

$$\Phi_N(t) = \mu_0 + \sum_{k=1}^{\infty} A_k \cos(kt - \alpha_k).$$

The coefficients A_k and α_k represent, respectively, the *harmonic modulus and phase of the FDs* of Γ.

The alternative approach proposed in [Granlund, 1972] for the set of FDs is as follows: initially, the complex signal $u(t) = x(t) + jy(t)$ should be constructed. It is observed that this is a complex periodic signal (for closed contours) having as period the perimeter of Γ, i.e., $u(t + iL) = u(t)$, $i = 0, 1, 2, \ldots$ The complex signal $u(t)$ may be expanded in a complex Fourier series defined as

$$\text{FD}(s) = \frac{1}{L} \int_0^L u(t) \, e^{\frac{-j2\pi st}{L}} \, dt.$$

Therefore

$$u(t) = \sum_{-\infty}^{\infty} \text{FD}(s) \, e^{\frac{j2\pi st}{L}}.$$

The coefficients FD (s) correspond to the Fourier descriptors proposed by Granlund. Special care is deserved when obtaining Fourier descriptors from an open curve, since the standard approach for extracting its parameterized representation leads to a non-periodic signal. More specifically, the signal discontinuity at its extremities typically implies instability in Fourier-based approaches because of Gibbs phenomenon. There are specific strategies that can be considered in such situations. For instance, an open curve can be treated as if it were closed if the contour is allowed to go from an extremity to the other and come back over the same path, which allows the generation of the above defined FDs [Mokhtarian & Mackworth, 1986; Persoon & Fu, 1977]. However, since this approach ensures only continuity of the signal, and not its derivatives, numerical instabilities can still arise from such a representation, especially at the abrupt turns at the extremities of the original open curve. An alternative approach is to treat the open curve as a one-pixel thick object and to extract its outer closed contour instead.

Some properties of the FDs directly follow from the underlying theory of the Fourier transforms and series. For instance, the 0-th component of the FDs obtained from the complex contour representation $u(n)$ is associated with the centroid of the original shape. The invariances to geometric transformations are also a direct consequence of the properties of the Fourier series. Such properties have helped to popularize this shape representation scheme, which has been adopted in a number of applications along the last four decades. Some of these properties and applications are presented in the next subsections.

6.5.1 Some Useful Properties

Some useful properties exhibited by FDs shape representation and characterization are discussed below.

Simplicity of implementation and efficient computation: Any FD definition can be computationally estimated by the application of the discrete Fourier transform:

$$\text{FD}(s) = \sum_{t=0}^{N-1} u(n) \, e^{\frac{-j2\pi ns}{N}}, \quad s = 0, \ldots, N-1.$$

Consequently, the FDs can be efficiently implemented by using FFT algorithms [Brigham, 1988]. For instance, a simple FD algorithm to be applied over a contour represented by a complex signal follows:

Algorithm: *Simple Fourier Descriptors*

1. Extract the shape contour and represent it as the complex signal $u(n)$;
2. $FD = \text{FFT}(u)$;

It is worth noting that the above approach can be adapted to the case in which the contour coordinates are represented by the real signals $x(n)$ and $y(n)$, implying the FFT will be calculated twice (over each signal).

Contour information is concentrated on the low frequencies: It is a known fact that most of the energy of contours obtained from typical (i.e., correlated) objects in image analysis applications concentrated in the low-frequency coefficients (see Section 3.2.5). For instance, refer to the neural cell presented in Figure 6.22, which was created by using a stochastic formal grammar-based technique [Costa et al., 1999]. The complex signal $u(n)$ and the respec-

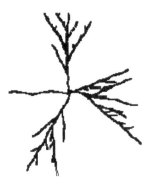

Figure 6.22: *An artificial neural cell.*

tive FDs have been calculated, and the parameterized signals $x(n)$, $y(n)$, the modulus $|FD(s)|$ (log-plot) and the phase $\arg(FD(s))$ are shown in Figure 6.23.

The *accumulated energy function* of the FDs is generally defined as

$$E(s) = \sum_{i=0}^{s} |FD(i)|^2 .$$

The function $E(s)$ describes how the FDs energy increases as more Fourier coefficients are considered (observe that this function takes into account only positive frequencies). The accumulated energy function of the neural cell in Figure 6.22 is shown in Figure 6.24. As can be observed, the energy steadily

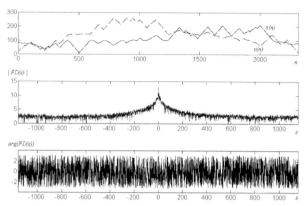

Figure 6.23: *Signals and Fourier descriptors (modulus and phase) obtained from the contour of the shape in Figure 6.22.*

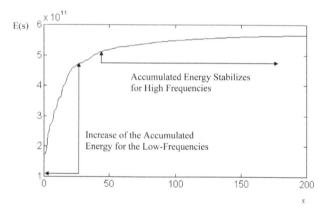

Figure 6.24: *The Fourier energy of typical contours concentrates on the low-frequency coefficients. The above diagram illustrates the case for the contour of Figure 6.22.*

increases for the first 30 coefficients, and stabilizes at higher frequencies, which indicates that most of the signal energy concentrates on the lowest frequencies. This is a consequence of the fact that, by being highly correlated (redundant), the Fourier kernel functions provide a good basis for representing the typically correlated elements in images.

In order to gain additional insight about shape analysis, consider the example in Figure 6.25. Here, the contour in Figure 6.22 has been reconstructed by considering four sets of FDs, where each set is composed of the i first and the i last Fourier coefficients, resulting $2i$ coefficients, since the discrete Fourier transform represents the negative low-frequency coefficients at the final portion of the vector. Figure 6.25 presents the reconstructions obtained by considering each of these four sets of coefficients, defined by $i = 10, 30, 60$

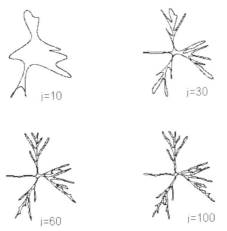

Figure 6.25: *Reconstruction of the neural shape in Figure 6.22 by using 2i coefficients, for i = 10, 30, 60 and 100.*

and 100. As we can see, the reconstructed shape for $i = 10$ is just a rough caricature of the original shape (compare it with Figure 6.22). On the other hand, the set for $i = 30$ allows a reconstruction where the main structural characteristics of the dendrites can be perceived. For $i = 60$, small-scale detail starts to appear. Finally, the reconstruction for $i = 100$ provides an almost exact reproduction of the original shape. In this example, the original contour has 2319 points, implying the same number of coefficients for the complete Fourier representation. Fortunately, the Fourier energy concentration property allows the FDs to be used effectively as a means to obtain reasonably small feature vectors to be used in pattern recognition. Indeed, such a possibility has motivated a series of applications reported in the related literature, such as recognizing calcifications in mammography [Shen et al., 1994] and Arabic character recognition [Mahmoud, 1994].

Several additional features can be obtained from Fourier descriptors: The mathematical background underlying the Fourier analysis theory provides a large set of useful tools for shape analysis. For instance, it is possible to estimate several shape measures directly from the respective FDs. In this context, Kiryati and Maydan have derived formulas to calculate the area and 2nd order moments of a contour from its FDs [Kiryati & Maydan, 1989]. Expressions for estimating the shape perimeter and area have been proposed in [van der Heijden, 1994] and [Baroni & Barletta, 1992] have estimated the digital curvature by using the Fourier series. An important possibility that is extensively used in this book is to use the convolution theorem to obtain different multiscale contour representations (see Section 7.2), mainly when used together with the Fourier derivative property.

Note: *2D Fourier Descriptors*

If the shape is represented in a binary image g, then analogous 2D Fourier descriptors can be obtained from the 2D transform:

$$G(r, s) = \mathfrak{F}\{g(p, q)\} = \frac{1}{MN} \sum_{p=0}^{M-1} \sum_{q=0}^{N-1} g(p, q) \exp\left\{-j2\pi\left(\frac{pr}{M} + \frac{qs}{N}\right)\right\}$$

6.5.2 Alternative Fourier Descriptors

Many different approaches for deriving shape measures based on Fourier techniques have been proposed in the literature, which can all be called Fourier descriptors. This section presents examples that can be useful for some applications.

Let $u(n)$ be a digital contour whose discrete Fourier transform $U(s)$ is defined as

$$U(s) = \frac{1}{N} \sum_{n=0}^{N-1} u(n) \, e^{-j\frac{2\pi}{N} sn}, \quad s = -\frac{N}{2} + 1, \ldots, \frac{N}{2}.$$

This Fourier series representation of $u(n)$ can be used to define a set of FDs invariant to translation, rotation, parameter shifting and scaling. These Fourier descriptors, represented as FD(s), can be defined as

$$FD(s) = |NFD(s)|,$$

with

$$NFD(s) = \begin{cases} 0 & \text{for} \quad s = 0 \\ \frac{U(s)}{U(1)} & \text{for} \quad s \neq 0 \end{cases}.$$

The normalized FDs, NFD(s), have been applied for shape characterization in mammographic images [Shen et al., 1994]. In that work, NFDs are used to define the following descriptor:

$$FF = \frac{\left[\sum_{s=-\frac{N}{2}+1}^{\frac{N}{2}} \frac{|NFD(s)|}{|s|}\right]}{\sum_{s=-\frac{N}{2}+1}^{\frac{N}{2}} |NFD(s)|}.$$

The authors of [Shen et al., 1994] claim that the above FDs are less sensitive to high-frequency noise. Furthermore, its values are limited to the range $[0, 1]$.

We can also easily derive a simple shape measure defined as the NFD energy, i.e.,

$$E_{FD} = \sum_{s=-\frac{N}{2}+1}^{\frac{N}{2}} |NFD(s)|^2 .$$

To probe further: *Fourier Descriptors*

As it has been said, one of the main reasons for the popularity of FDs is that the Fourier theory endows many useful mathematical tools for shape analysis. For instance, some of its properties can be used for multiscale curvature estimation by using the Fourier transform, as described in Chapter 7. Another important application of FDs is for shape normalization with respect to scale, rotation, translation and starting point, as described in [Kashi et al., 1996] (also [Pavlidis, 1977]). Otterloo's book [van Otterloo, 1991] is largely devoted to contour analysis using FD, as well as [Lestrel, 1997], the latter being dedicated to applications in biomedical sciences. An important aspect of FDs is their capability for shape generation [Zahn & Roskies, 1972], which can be used to assess the precision of contour analysis methods (e.g., curvature estimation, refer to [Estrozi et al., 2003]). Further references of interest are [Aguado et al., 1998; Arbter et al., 1990; Chellappa & Bagdazian, 1984; Kimia et al., 1995; Richard & Hemani, 1974; Rosin & Venkatesh, 1993; Staib & Duncan, 1992; Wallace & Wintz, 1980]. Other transforms may also be used for contour analysis, such as Hadamard [Ma et al., 1986], Gabor and wavelet transforms [Chuang & Kuo, 1996; Eichmann et al., 1990; Kashi et al., 1996]. The proponents of the use of the Hadamard transform claim that it allows the creation of shape measures invariant to the main geometric transformations while being more efficient for implementation than the FFT. Nevertheless, it is important to emphasize that no other transform has achieved the popularity and importance of the Fourier transform with respect to contour analysis.

Additional resources: *Shape characterization software*

An open-source software that includes many shape features discussed in this chapter is available at code.google.com/p/imagerecognitionsystem/. The reader is invited to contribute to this project by incorporating new tools to the software!

Multiscale Shape Characterization

Chapter Overview

THIS CHAPTER INTRODUCES a series of shape characterization tools and features based on multiscale transforms. Three types of multiscale transforms, namely the scale-space, Gabor and wavelets, are discussed in an
intuitive manner, including comments on how these transforms can be analyzed and used to produce shape features. Next, an approach for estimating the multiscale curvature of a contour based on the Fourier transform is
discussed, followed by a similar contour-based description calculated by
using the wavelet transform. Several algorithms for shape characterization
based on these descriptors are discussed. Finally, the multiscale energies
are presented as an interesting approach to obtaining global shape features.

7.1 Multiscale Transforms

This section presents a brief introduction to a series of *multiscale transforms* that
are widely applied to signal and image processing. Such transforms define a set
of basic mathematical tools underlying many of the methods covered in this book.
Playing a central role in these transforms, the concepts of *scale-space, time-
frequency* and *time-scale* planes are introduced in the initial portion of this chapter.
In fact, the multiscale transforms for signal analysis can be classified in different ways, especially in terms of the above concepts, yielding the following three
groups: the *scale-space* methods, the *time-scale transforms* and the *time-frequency
transforms*. This classification scheme is not rigid, though it is didactic and reflects
some important trends in computer vision research.

449

In general, signal analysis by using a multiscale transform is characterized by the following elements. Frequently, a signal $u(t)$ presents a set of features and structures occurring at different spatial scales. This signal is to be analyzed by a multiscale transform $U(b, a)$ involving two parameters: b, associated with the *time* variable t of $u(t)$, and a, associated with the *analyzing scale*. The scale parameter a is usually related to the inverse of the *frequency* f, i.e., $\frac{1}{a} \cong f$, leading to a dual interpretation of these transforms and suggesting the terms *time-scale* and *time-frequency*. Therefore, a 1D signal $u(t)$ is said to be *unfolded* by the 2D transform $U(b, a)$, with the purpose of *making explicit* the information underlying its structures at different scales.

On the other hand, it is worth emphasizing that the above transform process implies a substantial increase of data redundancy[1], because a 1D signal is represented by a 2D transform. As seen in Chapter 2, the complete representation of a 1D signal can be achieved by using a 1D-to-1D transformation. Nevertheless, redundancy increase is the price to be paid in order to make explicit important signal information, provided algorithms are available that can analyze the transform $U(b, a)$ in order to automatically detect and extract the relevant scale information. As a matter of fact, as it will be discussed in this chapter, multiscale approaches can also express the hierarchical relationship between noticeable points (e.g., singularities) in shapes and contours.

Therefore, the many different multiscale-based signal analysis frameworks can be summarized by the following algorithm:

Algorithm: *Multiscale Signal Analysis*

1. Obtain the signal to be analyzed $u(t)$;
2. Calculate the multiscale transform $U(b, a)$ of $u(t)$;
3. Extract the important scale characteristics of $u(t)$ from $U(b, a)$;

Such characteristics are expected to help several approaches in image analysis and classification. This chapter discusses some important multiscale transforms and strategies for automatic multiscale feature extraction.

7.1.1 Scale-Space

The scale-space approach to multiscale signal analysis is one of the most popular multiscale methods, which has been the object of growing attention since the publication of the seminal work [Witkin, 1983]. A key concept in Witkin's work is that important features of a signal are generally associated with extreme points such as local maxima or local minima points. Therefore, it is important to devise tools

---

[1]This is particularly true for continuous and redundant transforms, e.g., continuous wavelet transforms, which are generally more suitable for signal analysis. Transforms such as the discrete wavelet transform can avoid this problem of data redundancy.

capable of detecting and locating these points by taking into account the signal derivatives. The local maxima points of $u(t)$ correspond to *zero-crossings* of $u^{(1)}(t)$, the first derivative of $u(t)$. These concepts have been discussed in the context of edge detection in Chapter 3. The zero-crossings of $u^{(1)}(t)$ are henceforth denoted by $\{u^{(1)}(t)\}_{zc}$.

In practical image analysis situations, the analytical signal derivatives are unknown, requiring numerical differentiation methods. As these methods depend on a neighborhood around each point, the definition of the neighborhood length indeed corresponds to defining the *analyzing scale*. Basically, Witkin has developed and generalized some ideas implicit to the *Marr-Hildreth* or *Laplacian-of-Gaussian* operator, covered in some detail in Chapter 3. An important fact about many of the numerical differentiation methods is that this process tends to enhance high frequency noise, therefore requiring a filtering process to compensate for such an effect. Gaussian filtering is one of the most widely used approaches to this problem. In this case, the desired differentiated signal can be obtained by convolving $u^{(1)}(t)$ with a Gaussian $g(t)$, and, by applying the well-known property of convolution:

$$u^{(1)}(t) * g(t) = \left(u(t) * g(t)\right)^{(1)} = u(t) * g^{(1)}(t).$$

The above expression means that the signal $u(t)$ can be differentiated by convolving it with the first derivative of the Gaussian. The extreme points of $u(t)$ can now be located by searching for the zero-crossings of the convolution output. It is important to note that the Gaussian filter has a parameter (its standard deviation) which controls its width, thus controlling the smoothing degree of the filter. Marr-Hildreth's work had already paid great attention to the relationship between the Gaussian standard deviation and the scale of analysis, where the larger the standard deviation, the larger the analyzing scale and, consequently, less small scale details will remain in the signal. In his work, Witkin generalized this concept by defining the scale parameter (the Gaussian standard deviation) as a *continuous variable*. In this framework, for each scale, a set of extreme points of a smoothed version of $u(t)$ can be found at that scale. The *scale-space* of $u(t)$ is defined as the evolution of the set of these extreme points along the scale-parameter.

Let $u(t)$ be the input signal to be analyzed and $g_a(t)$ be a Gaussian with standard deviation $a > 0$:

$$g_a(t) = \frac{1}{a\sqrt{2\pi}} \exp\left(-\frac{t^2}{2a^2}\right).$$

The convolution between $u(t)$ and $g_a(t)$ is defined as

$$U(t, a) = u(t) * g_a(t).$$

At a specific observed scale a, and for a given a_0, $U(t, a_0)$ is the smoothed version of $u(t)$. The extreme points of $U(t, a)$ can be defined as the zero-crossings of $U^{(1)}(t, a)$, with

$$U^{(1)}(t, a) = u(t) * g_a^{(1)}(t),$$

where $g_a^{(1)}(t)$ denotes the first derivative of $g_a(t)$. In this context, the scale-space of $u(t)$ can now be defined.

Definition (Scale-Space). *Let $u(t)$ be the signal to be analyzed, $g_a^{(1)}(t)$ be the first derivative of the Gaussian function $g_a(t)$ and $U^{(1)}(t, a) = u(t)*g_a^{(1)}(t)$. Let $\left\{U^{(1)}(t, a_0)\right\}_{zc}$ denote the zero-crossings set of $U^{(1)}(t, a_0)$. The scale-space of $u(t)$ is defined as the set of zero-crossings of $U^{(1)}(t, a)$, i.e.,*

$$\left\{(b_0, a_0) \mid a_0, b_0 \in \mathbb{R}, a_0 > 0, \quad and \quad b_0 \in \left\{U^{(1)}(t, a_0)\right\}_{zc}\right\}.$$

Observe that the term scale-space is sometimes also used for $U(t, a)$, obtained by the convolution of the signal with a series of Gaussians. In general, the extrema of $U^{(n)}(t, a)$ can be defined from the zero-crossings of the scale-space generated by $g_a^{n+1}(t)$. The scale-space of a signal allows tracking its singularities (or of one of its derivatives) through the scale dimension, thus allowing the verification of a conjecture by Marr that the perceptually important dominant signal points correspond to singularities remaining along longer scale intervals. This idea defines the basis of the wavelet-based algorithm for dominant point detection discussed in Section 7.3.5.

As observed above, the simple introduction of a scale parameter, i.e., an additional dimension to the data, does not present an *a priori* advantage for shape analysis, unless effective algorithms are available that can take advantage of such a redundancy. Witkin, in the aforementioned seminal paper, presented an algorithm for the generation of a scale-space tree (a *dendrogram*) that organizes the important signal features in a hierarchical fashion as a function of the scale-space lifetime for each maxima line in the parameter plane (see Section 7.3.5).

Another seminal contribution to scale-space theory has been provided by the work of Jan Koenderink [Koenderink, 1984], a biophysics professor at the University of Utrecht. After a discussion with Andrew Witkin during a conference in honor of Marr in 1983, Koenderink perceived that the signal scale-space $U(t, a)$ defined by a series of Gaussian convolutions could be thought of as the solution of the 1D *diffusion equation* [Widder, 1975]:

$$\frac{\partial U}{\partial a} = k \frac{\partial^2 U}{\partial t^2},$$

where k is a real constant.

Therefore, the previously defined scale-space can be obtained from the solution of the above partial derivatives equation (PDE), taking $U(t, 0) = u(t)$ as the initial solution. Using this alternative definition, Koenderink was able to show that the Gaussian kernel is unique with respect to the restriction that new structures cannot be created during the filtering process as the analyzing scale increases [Koenderink, 1984].

7.1.2 Time-Frequency Transforms

The time-frequency transforms have originated as an alternative to Fourier analysis capable of *signal local analysis*. Actually, the fact that the Fourier transform cannot satisfactorily treat local transient events has been known for a long time, and attempts at defining a local frequency measure go back to Sommerfeld's Ph.D. thesis of 1890 [Flandrin, 1999]. Let us recall the Fourier transform:

$$U(f) = F\{u(t)\} = \int_{-\infty}^{\infty} u(t)\, e^{-i2\pi ft}\, dt.$$

It can be observed that this transform is based on integrating the whole signal for each frequency f, implying that events occurring at different instants t contribute in the same global manner, affecting the whole representation. For instance, if a short (along the time) disturbance is added to a signal, this will affect its *whole* Fourier representation. The *short-time Fourier transform* (also known as *windowed Fourier transform*) has been defined as an attempt to circumvent this problem by introducing an *observation window* aimed at selecting a signal portion during the transform, thus disallowing (or attenuating) signal events outside this window. Consider the following transform, defined from the Fourier equation:

$$U(b, f) = \int_{-\infty}^{\infty} g^*(t - b)\, u(t)\, e^{-i2\pi ft}\, dt, \qquad (7.1)$$

where $g(t)$ is the window function that *slides* over the signal $u(t)$[1]. Time-frequency analysis is based on the above considerations, and one of its most important tools is the *Gabor transform*, i.e., the short-time Fourier transform where $g(t)$ is a Gaussian window. This important transform has been developed from the seminal work by D. Gabor on signal decompositions using well-localized functions in the time and in the frequency domains *simultaneously* [Gabor, 1946].

The Gabor transform can be analyzed under two points of view. First, for each b, the window is shifted around b, supposing that $g(t)$ is concentrated around the time origin ($t = 0$). Therefore, $u(t)$ is multiplied by $g(t - b)$ (the complex conjugated of equation (7.1) can be ignored because the Gaussian is a real function), which means that only the portion of $u(t)$ that is "seen" through $g(t - b)$ will survive the multiplication, i.e., the signal $u(t)$ vanishes outside $g(t - b)$. Secondly, the Fourier transform of the resulting signal $u(t)\, g(t - b)$ is taken. In this context, the Fourier transform simply takes into account only (or mostly, since windowing functions such as the Gaussian can exhibit infinite decaying tails) the signal information available through $g(t - b)$, which explains the *time localization* property of the Gabor transform $U(b, f)$.

---

[1] The French name for this transform is "*Fourier à fenêtre glissante*," i.e., "sliding window Fourier transform."

The Gabor transform can be rewritten in terms of the Fourier transforms of the signal $U(v)$ and of the Gaussian window $G(v)$, which can be seen as the inverse Fourier transform, modulated by a complex exponential, of the product between the signal Fourier transform and a shifted version of the Fourier transform of the Gaussian. The fact that the Fourier transform of the Gaussian is also well localized in the frequency domain implies the product $G^*(v - f) U(v)$ to select a window around the frequency f. In other words, the Gabor transform also exhibits local analysis in the frequency domain.

7.1.3 Gabor Filters

The concept of Gabor filters is very important in computer vision, and can be understood from the discussion of the Gabor transform in the previous section. In this context, instead of the product $g^*(t - b) u(t)$ in equation (7.1), the kernel $h_f(t) = g^*(t) e^{-i2\pi ft}$ can be considered for each fixed value of f. The kernels $h_f(t)$ are functions defined as the Gaussian $g(t)$ modulated by the complex exponential $e^{-i2\pi ft}$ tuned to frequency f. Therefore, the Gabor transform can be defined as [Carmona et al., 1998]:

$$U(b, f) = \int_{-\infty}^{\infty} h_f(t - b) u(t) \, dt.$$

The above equation shows that $U(b, f)$ can be considered as a convolution between $u(t)$ and a *well-localized kernel around the frequency f*, i.e., $h_f(-t)$, which is an alternative manner of explaining the localization property of the Gabor transform $U(b, f)$.

To probe further: *Time-Frequency Analysis*

Many important concepts, such as the notion of *instantaneous frequency*, follow from the considerations in the previous section, and the interested reader can obtain in-depth information in the textbooks of Patrick Flandrin [Flandrin, 1999], or in the review papers of [Cohen, 1989] and [Hlwatsch & Bordreaux-Bartels, 1992]. An alternative approach to time-frequency analysis involves nonlinear transforms, such as Wigner-Ville distributions, based on the work of J. Ville in 1948 (therefore, a contemporary of Gabor's) and on the quantum mechanics developments by E. P. Wigner in 1932. Further details about this approach can be obtained from the aforementioned time-frequency analysis references.

7.1.4 Time-Scale Transforms or Wavelets

The wavelet transform, as it is modernly known, was introduced by J. Morlet, being subsequently formalized by him and the physicist A. Grossmann [Grossmann

& Morlet, 1984]. Morlet worked with signals presenting transient events such as local frequencies and exhibiting an important feature: high frequency events frequently occurred along short time intervals while low-frequency components remained longer during the signal evolution. The application of the traditional short-time Fourier transform had the drawback that the analyzing window size was the same for all frequencies. In order to circumvent this problem, Morlet introduced a transform where the kernel size varies with the frequency, allowing high frequency events to be localized with better time resolution, while low-frequency components are analyzed with better frequency resolution. This property is known as *relative bandwidth* or *constant-Q*, being one of the main differences between the wavelet transform and the short-time Fourier transform [Rioul & Vetterli, 1991][1]. The continuous wavelet transform is defined as:

$$U[\Psi, u](b, a) = U_\Psi(b, a) = \frac{1}{\sqrt{a}} \int_R \Psi^* \left(\frac{t - b}{a} \right) u(t)\, dt.$$

Section 7.3 discusses the above transform in detail.

Note: *Filter Bandwidth*

A better understanding of one of the main differences between the wavelet and the Gabor transform, namely the aforementioned constant-Q property, can be obtained by taking into account the concept of *bandwidth*. Given a window $g(t)$ and its Fourier transform $G(f)$, the *bandwidth* Δf of $g(t)$ is defined as

$$\Delta f^2 = \frac{\int f^2 |G(f)|^2\, df}{\int |G(f)|^2\, df}.$$

Analogously, the width of $g(t)$ can be defined in the time domain as:

$$\Delta t^2 = \frac{\int t^2 |g(t)|^2\, dt}{\int |g(t)|^2\, dt}.$$

While Δf is a measure of frequency resolution of $g(t)$, Δt is a measure of time resolution. Although it would intuitively be desirable to have a filter with the best resolution in both the time and in the frequency domains, this is not possible. In fact, the product of the two above-introduced resolution measures can be verified

---

[1] Ingrid Daubechies, in her paper about the history of wavelets [Daubechies, 1996], tells an anecdote that, despite all the possibilities opened by this idea, and the theoretical foundation created in collaboration with Grossmann, Morlet met some resistance among his geophysicist colleagues, who argued that "If it were true, then it would be in math books. Since it isn't in there, it is probably worthless."

to have a lower bound, defined as:

$$\Delta t \, \Delta f \geq \frac{1}{4\pi}.$$

The above property is known as *uncertainty* or *Heisenberg principle*, implying that the better the resolution is in one domain, the worse it is in the other. One of the main differences between the short-time Fourier transform and the wavelet transforms is that Δt and Δf are constant with respect to all frequencies in the case of the short-time Fourier transform. On the other hand, in the case of the wavelet transform, these parameters vary so that Δf is proportional to f, i.e.,

$$\frac{\Delta f}{f} = Q,$$

where Q is a constant. Figure 7.1 shows the two types of filter banks (i.e., a set of filters tuned to different scales), illustrating their differences as the analyzing scale varies.

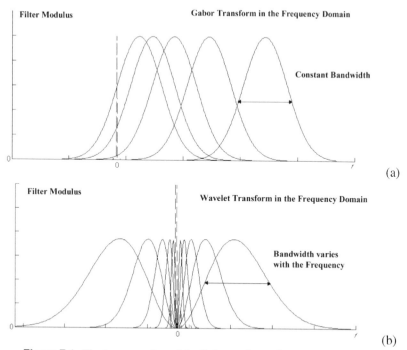

(a)

(b)

Figure 7.1: *The interpretation of the Gabor and wavelet transforms as filters illustrates that in the former the filter bandwidth remains the same for different frequencies, while the latter applies larger filters for higher frequencies.*

To probe further: *The Continuous and the Discrete Wavelet Transforms*

Among the important developments that followed from the wavelet theory, it is worth noting its connection to several related developments, such as the *multiresolution analysis*, developed in the works of Stéphane Mallat and Yves Meyer [Mallat, 1989; Meyer, 1993]; sub-band filter banks; and the orthonormal wavelets developed by Ingrid Daubechies (a former Ph.D. student of Grossmann) [Daubechies, 1992]. These issues, related to the *discrete wavelet transform*, have been successfully applied to many problems of signal coding, compression and transmission, among others. Some comments about the continuous and the discrete wavelet transform are given in the following.

Firstly, what are the main differences between the continuous and the discrete wavelet transforms? This is a difficult and possibly tricky question to answer, mainly because both approaches include a wide and versatile set of theoretical and practical tools. The continuous wavelet transform is highly redundant and generally makes use of nice analyzing wavelets, two features that are desirable for *signal analysis*. Furthermore, in the case of the Morlet wavelet, the Gabor transform and the so-called *Gabor wavelets* [Daugman, 1988; Kruger & Sommer, 1999], the transform presents a strong and interesting biological inspiration (i.e., the receptive fields of neural cells involved in visual processing). On the other hand, the discrete wavelet transform can be computed by fast algorithms, besides being capable of reaching high data compression rates when used together with vector coding techniques, thus explaining its popularity for *signal coding, compression* and *synthesis*. Additionally, discrete wavelets have led to successful results in numerical analysis, e.g., by allowing the development of efficient algorithms for matrix multiplication, as well as in mathematics, e.g., being used for building functional spaces, in the resolution of PDEs (partial differential equations), in statistical analysis of time series and many others.

7.1.5 Interpreting the Transforms

This section discusses some topics related to the visualization and visual analysis of a specific multiscale transform $U(b, a)$. There are interesting facts about visualizing these transforms to the extent that Morlet and Grossmann (in collaboration with R. Kronland-Martinet) have written a paper exclusively devoted to visualization and interpretation of the resulting data [Grossmann et al., 1989].

First, it is worth noting that the transform $U(b, a)$ can be seen as a function from $\mathbb{R}^2$ assuming real or complex values, depending on the transform and on the signal, i.e.:

$$U(b, a): \mathbb{R}^2 \to \mathbb{R} \quad \text{or} \quad U(b, a): \mathbb{R}^2 \to \mathbb{C}.$$

For instance, in the case of the scale-space, the transform is obtained by a se-

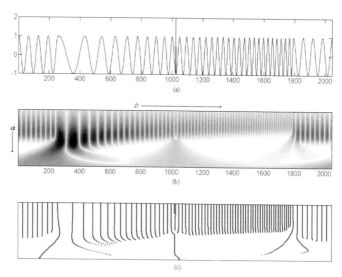

Figure 7.2: *Scale-space transform: the signal is shown in* **(a)***; the result of the convolution between the original signal and the set of Gaussian kernels is shown as a gray-level image in* **(b)***; and the maxima lines of the scale-space are shown in* **(c)***.*

ries of convolutions between the signal and a Gaussian derivative. Therefore, if the signal is real-valued, the transform will be real-valued; otherwise, both the input signal and the resulting transform are complex-valued. On the other hand, the Gabor transform involves inner products between the signal and a Gaussian modulated by the complex exponential function. In this case, the resulting transform will generally be complex-valued. As far as the wavelet transform is concerned, the resulting transform will be real or complex-valued depending on the mother wavelet and on the signal itself.

If $U(b, a)$ is real-valued, the transform can be visualized either as a surface, where (b, a) gives the coordinates on the xy plane and the real value $U(b, a)$ defines the height in the z direction or as a gray-level image, with (b, a) being the pixel coordinates and $U(b, a)$ being the gray level. On the other hand, if $U(b, a)$ is complex-valued, then the transform must be visualized by two real-valued components, i.e., modulus and phase or real and imaginary parts of $U(b, a)$ for each (b, a).

Next is presented an example comparing the three previously introduced approaches that clarifies the kind of analysis allowed by each transform. Figure 7.2 (a) presents the scale-space transform of a signal composed of three distinct perceptual parts: two portions, at each of the function extremities, consisting of a constant frequency sinusoidal wave, with a linear *chirp* function in the middle.

This latter type of function is of the type $\cos((mt + n)t)$, with the frequency

m linearly increasing with time [Torrésani, 1995]. A spike $\delta(t - 1024)$ has also been added in the middle of the signal, in order to produce the singularity in Figure 7.2 (a). A straightforward analysis indicates the main problem of global transforms, such as the Fourier transform, when dealing with this type of signal: although different features occur at different time locations, they globally affect the whole transform. It is therefore impossible to locate these signal localized events along the time domain by considering frequency domain representations.

Figure 7.2 (b) presents, as a gray-level image, the scale-space transform of the signal of Figure 7.2 (a) using a Gaussian function as the kernel. In this image, the horizontal axis represents the space (or time) parameter b, while the vertical axis represents the scale parameter a. The vertical axis increases downward, and the gray level at each position (b, a) represents the value $U(b, a)$. In the examples in this section, darker gray levels are associated with higher values of $U(b, a)$, i.e., $U(b, a)$ increases from white to black. The plane (b, a) that represents the support of $U(b, a)$ is called the *scale-space plane*. As can be seen from Figure 7.2 (b), small scale signal details, associated with higher frequency signal portions, are filtered and vanish as the analyzing scale is increased, i.e., the signal is convolved with Gaussians of larger standard deviation. Furthermore, the transform detects three singular points, corresponding to the two transition points between the three signal portions, and to the intermediate delta. These singularities are identified by stronger responses of the transform that survive as the analyzing scale increases. Figure 7.2 (c) shows the local maxima lines (or crests) of $U(b, a)$. Each maxima line is composed of points that are local maximum along each row of the representation (these points are called vertical maximum points; refer to Section 7.3.5). Suppose that $U(b, a)$ is discretized and represented by the array $U(p, q)$, where p and q are integer indexing values associated with b and a, respectively. Then, the simplest way to detect the local maximum points is to search for points that are larger than their neighbors to the left and to the right along each row. The algorithm below can be used in order to obtain the local maxima representation (also known as vertical skeleton) of $U(b, a)$.

Algorithm: *Vertical Maxima Detection*

1. **for** each q
2.   **do**
3.   **for** each p
4.   **do**
5.   **if** $(U(p, q) > U(p - 1, q))$ and $(U(p, q) > U(p + 1, q))$
6.   **then**
7.   *vertical_skeleton*$(p, q) \leftarrow 1$;
8.   **else**
9.   *vertical_skeleton*$(p, q) \leftarrow 0$;

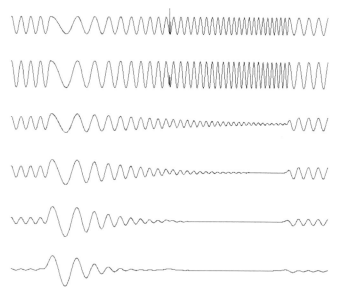

Figure 7.3: *Signal and some of its smoothed versions composing the respective scale-space representation.*

The above algorithm produces an output array *vertical_skeleton* whose each
element corresponding to a vertical maximum is labeled 1; otherwise the element
receives value 0. A typical output of this algorithm is shown in Figure 7.2 (c).

In order to better understand the construction of the above scale-space, refer
to Figure 7.3, which presents the input signal in the first row, and its smoothed
versions in the other rows. Each of these smoothed versions corresponds to a row
of $U(b, a)$, i.e., they are signals $U(b, a_0)$ for a series of five fixed values of a_0 in
increasing order downward. As can be noted, as the analyzing scale increases, the
high frequency details become weaker, while the structures related to low frequency
are maintained.

Figure 7.4 presents analogous results obtained by using the Gabor transform.

Because this transform is complex-valued, the modulus $|U(b, a)|$ is considered
in Figure 7.4 (b) in order to visualize the transform. As can be seen from this
example, the Gabor transform presents accentuated responses when it is correctly
tuned to the local frequencies of the signal, as well as when a signal singularity
occurs. Therefore, stronger responses appear in two horizontal regions at the first
and at the last portions of the *time-frequency plane*, corresponding to the sinusoidal
patterns at the beginning and at the end of the signal. The Gabor transform also
responds strongly along an inclined (or diagonal) stripe at the central portion of the
time-frequency plane, which corresponds to the increasing frequency portion of the
linear chirp. The time-frequency plane is represented with the same conventions as
the scale-space, i.e., the horizontal axis is associated with the time dimension while

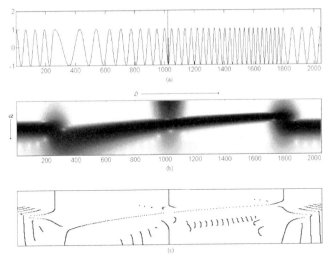

Figure 7.4: *The Gabor transform of the signal in **(a)** is shown in modulus in **(b)**. The vertical maxima lines of the Gabor transform are shown in **(c)**.*

the vertical axis is associated with the scale dimension, increasing downward. It is worth noting that, as the frequency can be considered inversely proportional to scale, i.e., $a = \frac{1}{f}$, the frequency actually increases upward.

Figure 7.5 illustrates the results of analyzing the same above signal by applying the wavelet transform with a Morlet mother wavelet (see Section 7.3.3 for additional detail about this type of wavelet).

Since this mother wavelet is also complex-valued (i.e., a Gaussian modulated by a complex exponential, similar to the Gabor functions), the visual representation is also shown in terms of the modulus of the transform. Although a result to some extent similar to the Gabor transform can be perceived, the wavelets' relative bandwidth property implies an interesting effect illustrated by this example. More specifically, the wavelet transform defines a stronger response with a funnel-like shape pointing towards the singular points as the scale decreases. This property is explored by the dominant point detection algorithm presented in Section 7.3.5.

7.1.6 Analyzing Multiscale Transforms

Once a signal has been represented by a multiscale transform, different approaches may be considered in order to extract information from it. Some of the most popular general techniques that can be applied to analyze the signal representation in the scale-space, time-frequency and time-scale planes are discussed in the following:

Local Maxima, Minima and Zeros: Since the work of David Marr [Marr, 1982], this has been one of the most explored approaches to analyze multiscale

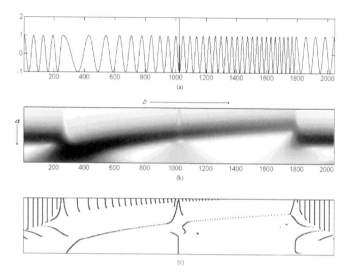

Figure 7.5: *The wavelet transform (using Morlet) of the signal in **(a)** is shown in modulus in **(b)**. The vertical maxima lines of the wavelet transform are shown in **(c)**.*

transforms. This approach is based on the fact that some signal points are perceptually more important than others, which also applies to shapes (e.g., corner points) and images (e.g., edge points). Normally, a suitable multiscale transform responds more strongly to this kind of singular points, suggesting a search for local maxima (or minima, or zeros, depending on the transform) points. Examples of this approach are the Marr-Hildreth zero-crossings based edge detector and the curvature scale-space [Mokhtarian & Mackworth, 1992]. The choice whether to search for local maxima, minima or zero points depends both on the transform behavior and on the features of interest in each particular problem. For instance, in the case of curvature-based shape analysis, W. Richards and collaborators have searched for curvature local negative minima, associated with concave regions, based on criteria associated with human perception [Richards et al., 1986]. On the other hand, the curvature scale-space developed by Mokhtarian and Mackworth is based on searching for inflection points along the contour, which correspond to curvature zero-crossings [Mokhtarian & Mackworth, 1992].

Frequency Tuning or Scale Selection: A second approach, explored in the literature under different denominations (frequency tuning, scale selection, natural scale detection, transform or wavelet calibration), is based on analyzing not the complete transform, but an *a priori* fixed scale a_0 (or a set of scales a_i, $i = 0, 1, \ldots, N$), which is equivalent to analyzing the response of a linear filter (or a set of filters) to the input signal. In this case, the linear filter is defined by the transform kernel tuned to the fixed scale (frequency). In the case of the example discussed in Figure 7.3, each filtered signal is actually the response of a Gaussian filter, tuned to five different scales, to the input signal shown in Figure 7.2 (a). This approach is adopted, for instance, in texture segmentation problems [Bovik et al., 1990; Dunn & Higgins, 1995; Reed & Weschsler, 1990]. The most important problem to be solved before the effective implementation of this approach is related to the difficulty of correctly tuning the transform parameters, i.e., the analyzing scale. The ganglion cell classification experiments shown in Section 8.5.1 illustrate that the multiscale bending energy (which is a shape feature obtained from a multiscale transform) can lead to both the best and worst recognition rates, depending on the chosen analyzing scale, thus illustrating the importance of correctly choosing the scale. A frequently adopted simple technique is to choose the scales with higher filter response (e.g., in terms of response energy) [Reed & Weschsler, 1990]. This idea has been used in the characterization of *natural scales*, as explained in Section 7.3.6, which consists in a generalization of the Fourier-based approach described in [Rosin & Venkatesh, 1993]. A second problem that must be addressed is that the magnitude of the response to each filter can vary widely, mainly because the energy of the Fourier transform of typical shapes is generally concentrated along the low-frequency coefficients. The difficulty is that such magnitude variations may imply misleading results while comparing the response of different filters. Therefore, it is sometimes desirable to normalize the energy of the different filter responses before further analysis. This normalization is particularly important in the case of the projection-based approach, which is explained in the following.

Projections onto the Time and onto the Scale (or Frequency) Axis: A simple and straightforward manner of extracting information from a multiscale transform is to project its coefficients onto the time or onto the scale parameter axis. In the case of projecting onto the scale axis, the resulting projection is directly related to the Fourier transform, assuming that no normalization like those previously explained have been applied. The relationship with the Fourier transform is due to the fact that time information is lost during the projection. Nevertheless, the projection of local maxima can be used to detect global natural scales, defining an approach analogous to [Rosin & Venkatesh, 1993]. On the other hand, projection onto the time axis can reveal which time regions carry more signal energy throughout the scales, which can be used to identify the contour dominant points [Cesar-Jr. & Costa, 1996].

We conclude this section by noting that algorithms involving combinations of these techniques can be devised in order to more properly meet the application requirements.

Note: *Wavelets, Vision and Image Processing*

Wavelet theory has been built by unifying a series of results developed in most diverse research fields such as filter bank theory, physics, time series, signal and image processing and computer vision, to name a few. As far as computer vision is concerned, it is worth noting some concepts developed in the context of vision research that have become precursors of wavelet theory and that have had great influence in the work of S. Mallat, who also acts in the computer vision field.

The first concept is that of an *image pyramid*, a data structure that represents an image at different resolution levels. This kind of data structure has been researched since the early 1970s by vision pioneers like S. Tanimoto, T. Pavlidis and A. Rosenfeld [Jolion & Rosenfeld, 1994; Tanimoto & Pavlidis, 1975]. The development of such multiscale structures has culminated in the works [Burt & Adelson, 1983; Crowley & Parker, 1984; Crowley & Sanderson, 1987; Lindeberg, 1994]. The creation of an image pyramid is based on a filtering step (e.g., Gaussian low-pass filtering), followed by subsampling, thus generating an image with less resolution and larger scale. The recursive application of these two rules allows the creation of the pyramid. More recently, nonlinear mathematical morphology-based approaches for image pyramid definition have also been addressed in the literature [Heijmans & Goutsias, 1999; Vaquero et al., 2005].

The second important early link between wavelet theory and computer vision has its origin in the edge detection technique introduced by D. Marr and E. Hildreth [Marr, 1982; Marr & Hildreth, 1980]. While interested in image edges, these researchers perceived that they tend to occur at different scales. Therefore, they introduced a mathematical tool that is capable of finding edge points at different spatial scales through the detection of the *zero-crossings* of the image LoG (Laplacian of Gaussian; see Section 3.3.8). In their approach, the Gaussian standard deviation represents the scale parameter. It is interesting to note that the second derivative of the Gaussian, used to calculate the LoG of an image, has become one of the most popular continuous mother wavelets, also known as *Marr wavelet* [Arnèodo et al., 1995; Grossmann, 1988].

The 2D continuous wavelet transform, which presents important applications in both image processing and physics, has been developed through the Ph.D. thesis of Romain Murenzi in Louvain-la-Neuve, Belgium, under the supervision of Jean-Pierre Antoine [Antoine et al., 1993; Murenzi, 1990].

7.2 Fourier-Based Multiscale Curvature

This section introduces several shape analysis tools based on the multiscale Fourier transform-based approach to curvature estimation introduced in [Cesar-Jr. & Costa, 1995, 1996, 1997]. Further information about Fourier contour representation can be found in the important context of Fourier descriptors (see Section 6.5). The multiscale approach to curvature estimation leads to the so-called *curvegram*, where the curvature values are presented in a scale-space representation (see Section 7.1.1). As already noted, digital curvature estimation presents two main problems: the lack of an analytical representation of the original contour (from which the curvature could be explicitly calculated) and the fact that numerical differentiation is required. The first problem can be tackled through the application of the Fourier theory, namely the derivative property, which allows the estimation of the *j*-th derivative $u^{(j)}(t)$ from the Fourier transform $U(s)$ of $u(n)$. The second problem is addressed by low-pass filtering the signal, which actually allows the introduction of the multiscale paradigm into the framework. Some simple solutions are discussed for the traditional problem of *contour shrinking*, which generally occurs when Gaussian contour filtering is applied. This section also presents some results on performance assessment of the methods with respect to precision.

7.2.1   Fourier-Based Curvature Estimation

Curvature estimation using the Fourier derivative property can be carried out both when the contour is represented by the $x(t)$ and $y(t)$ signals or by the complex signal $u(t)$, which are presented in the following:

Case I: $c(t) = (x(t), y(t))$

Let $c(t) = (x(t), y(t))$ be the parametric representation of the contour. As discussed in Chapter 2, the curvature $k(t)$ of $c(t)$ is defined as:

$$k(t) = \frac{\dot{x}(t)\,\ddot{y}(t) - \ddot{x}(t)\,\dot{y}(t)}{\left(\dot{x}(t)^2 + \dot{y}(t)^2\right)^{\frac{3}{2}}}. \tag{7.2}$$

In order to calculate the curvature, it is necessary to estimate the first and the second derivatives of the signals $x(t)$ and $y(t)$. Let $X(f)$ and $Y(f)$ be the Fourier transforms of $x(t)$ and $y(t)$. The Fourier derivative property asserts that [Papoulis, 1962]:

$$\dot{X}(f) = j2\pi f\, X(f),$$
$$\dot{Y}(f) = j2\pi f\, Y(f),$$
$$\ddot{X}(f) = -(2\pi f)^2\, X(f),$$
$$\ddot{Y}(f) = -(2\pi f)^2\, Y(f),$$

where j is the complex number and the abbreviations $\dot{X}(f)$, $\dot{Y}(f)$, $\ddot{X}(f)$ and $\ddot{Y}(f)$ denote the Fourier transforms of $\dot{x}(t)$, $\ddot{x}(t)$, $\dot{y}(t)$ and $\ddot{y}(t)$, respectively, and *not the derivative with respect to the frequency variable f*. Therefore, the application of the above Fourier property followed by the inverse Fourier transform allows the estimation of the curvature by applying equation (7.2) in terms of the Fourier transforms of the signals $x(t)$ and $y(t)$, i.e.,

$$\dot{x}(t) = F^{-1}\left\{\dot{X}(f)\right\},$$

$$\dot{y}(t) = F^{-1}\left\{\dot{Y}(f)\right\},$$

$$\ddot{x}(t) = F^{-1}\left\{\ddot{X}(f)\right\},$$

$$\ddot{y}(t) = F^{-1}\left\{\ddot{Y}(f)\right\}.$$

From equation (7.2), we have

$$
\begin{aligned}
\dot{x}(t)\,\ddot{y}(t) - \ddot{x}(t)\,\dot{y}(t) &= F^{-1}\left\{\dot{X}(f)\right\} F^{-1}\left\{\ddot{Y}(f)\right\} - F^{-1}\left\{\ddot{X}(f)\right\} F^{-1}\left\{\dot{Y}(f)\right\} \\
&= F^{-1}\{j2\pi f\, X(f)\}\, F^{-1}\left\{-(2\pi f)^2\, Y(f)\right\} \\
&\quad - F^{-1}\left\{-(2\pi f)^2\, \ddot{X}(f)\right\} F^{-1}\{j2\pi f\, Y(f)\} \\
&= -j(2\pi)^3\Big(F^{-1}\{f\, X(f)\}\, F^{-1}\left\{f^2\, Y(f)\right\} \\
&\quad - F^{-1}\left\{f^2\, X(f)\right\} F^{-1}\{f\, Y(f)\}\Big)
\end{aligned}
$$

and

$$
\begin{aligned}
\left(\dot{x}(t)^2 + \dot{y}(t)^2\right)^{\frac{3}{2}} &= \left(F^{-1}\left\{\dot{X}(f)\right\}^2 + F^{-1}\left\{\dot{Y}(f)\right\}^2\right)^{\frac{3}{2}} \\
&= \left(F^{-1}\{j2\pi f\, X(f)\}^2 + F^{-1}\{j2\pi f\, Y(f)\}^2\right)^{\frac{3}{2}} \\
&= \left((j2\pi)^2\left(F^{-1}\{f\, X(f)\}^2 + F^{-1}\{f\, Y(f)\}^2\right)\right)^{\frac{3}{2}} \\
&= (j2\pi)^3\left(F^{-1}\{f\, X(f)\}^2 + F^{-1}\{f\, Y(f)\}^2\right)^{\frac{3}{2}}.
\end{aligned}
$$

By substituting the above two expressions into equation (7.2), and considering the fact that $j^3 = -j$, we have the following curvature expression:

$$
k(t) = \frac{F^{-1}\{f\, X(f)\}(t)\, F^{-1}\left\{f^2\, Y(f)\right\}(t) - F^{-1}\left\{f^2\, X(f)\right\}(t)\, F^{-1}\{f\, Y(f)\}(t)}{\left[\left(F^{-1}\{f\, X(f)\}(t)\right)^2 + \left(F^{-1}\{f\, Y(f)\}(t)\right)^2\right]^{\frac{3}{2}}}.
$$

Case II: $u(t) = x(t) + jy(t)$

The above formulation for curvature estimation using the Fourier transform can be analogously derived for the complex contour representation $u(t)$. The derivatives of $u(t)$ with respect to the parameter t are defined as

$$\dot{u}(t) = \dot{x}(t) + j\,\dot{y}(t),$$
$$\ddot{u}(t) = \ddot{x}(t) + j\,\ddot{y}(t).$$

The Fourier transforms of $\dot{u}(t)$ and of $\ddot{u}(t)$ are defined from $U(f)$ as

$$\dot{U}(f) = j2\pi f\, U(f) \tag{7.3}$$

and

$$\ddot{U}(f) = -(2\pi f)^2\, U(f). \tag{7.4}$$

Therefore $\dot{u}(t) = F^{-1}\left\{\dot{U}(f)\right\}$ and $\ddot{u}(t) = F^{-1}\left\{\ddot{U}(f)\right\}$. It can be easily verified that

$$\dot{u}(t)\,\ddot{u}^*(t) = \dot{x}(t)\,\dot{y}(t) + \ddot{x}(t)\,\ddot{y}(t) - j(\dot{x}(t)\,\ddot{y}(t) - \ddot{x}(t)\,\dot{y}(t)). \tag{7.5}$$

The imaginary part of equation (7.5) is equivalent (except for the -1 multiplying factor) to the numerator of equation (7.2). Furthermore, it is easy to verify that

$$|\dot{u}(t)|^3 = \left(\sqrt{\dot{x}(t)^2 + \dot{y}(t)^2}\right)^3 = \left(\dot{x}(t)^2 + \dot{y}(t)^2\right)^{\frac{3}{2}} \tag{7.6}$$

is equivalent to the denominator of equation (7.2). Therefore, the curvature can now be expressed in terms of the complex signal derivatives by properly substituting equations (7.5) and (7.6) in equation (7.2), yielding

$$k(t) = \frac{-\,\mathrm{Im}\{\dot{u}(t)\ddot{u}^*(t)\}}{|\dot{u}(t)|^3}. \tag{7.7}$$

The Fourier derivative property can be again applied with respect to the signal $u(t)$, as done for the signals $x(t)$ and $y(t)$, in order to estimate $\dot{u}(t)$ and $\ddot{u}(t)$ from the Fourier transform $U(f)$ of $u(t)$. Nevertheless, for practical purposes, it is important to note that signals obtained from noisy digital images, as well as their spatially sampled nature, constitute a problem for the direct application of equation (7.7). In fact, the differentiation acts as a high-pass filter that accentuates the influence of high frequency noise, which can completely undermine the curvature estimation. This problem can be circumvented through the introduction of a low-pass filter in the derivatives estimation step. Before proceeding in this direction, it is important to discuss some issues regarding the computational implementation of the above expressions.

7.2.2 Numerical Differentiation Using the Fourier Property

The correct numerical implementation of the above-discussed Fourier-based signal differentiation requires the proper estimation of the continuous Fourier transform and its inverse which, given the discrete nature of the signals, has to be done in terms of the discrete Fourier transform (preferably in its fast version, i.e., the FFT). It is worth emphasizing that the topics covered in this section are fundamental for correctly implementing many of the shape analysis techniques discussed in this book, including Fourier-based curvature estimation.

Assume the following function is to be differentiated:

$$g(t) = \cos(2t) + \sin\left(t^2\right),$$

which is shown in Figure 7.6 (a). Let us also suppose that the signal $g(t)$ has been

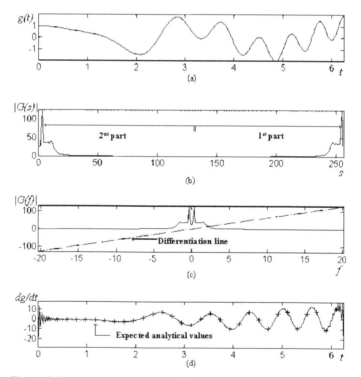

Figure 7.6: *Example of numerical differentiation using the Fourier trans-*
form: original signal **(a)***; resulting discrete Fourier transform*
obtained by FFT **(b)***; and estimated differentiated signal ob-*
tained by using the derivative property **(c)***.*

sampled with $N = 256$ points, with $t \in [0, 2\pi)$, i.e., the sampling interval is

$$T = \frac{2\pi}{N-1}.$$

Recall that the discrete Fourier transform of a signal is periodic with period N. The discrete Fourier transform (and, consequently, the FFT) usually produces the coefficients respective to the negative frequencies after those for positive frequencies (i.e., the frequency origin $f = 0$ coincides with the first element of the vector representing the discrete Fourier transform result), as illustrated in Figure 7.6 (b) (see Section 2.7 and [Brigham, 1988] for further detail). This figure presents the modulus $|G(s)|$ of the Fourier transform of $g(t)$, calculated using the following discrete Fourier transform equation:

$$U(s) = \sum_{n=0}^{N-1} u(n) e^{-j2\pi ns/N}, \quad s = 0, \dots, N-1,$$

which is defined for $s = 0, \dots, N-1$. Figure 7.6 (c) presents the signal $G(s)$ shifted so as to have the frequency origin in the middle of the vector. This operation for correctly assembling the Fourier coefficients is commonly known as *FFT shift* (see Section 2.7.7). In the 1D case, this is equivalent to defining the frequency variable s as

$$s = -\text{floor}\left(\frac{N}{2}\right), \dots, 0, \dots, N - \text{floor}\left(\frac{N}{2}\right) - 1.$$

The plot in Figure 7.6 (c) also illustrates the relationship between s and f, the discrete and the continuous frequency variables, respectively. This relationship is important because of the fact that the Fourier derivative property is defined with respect to the continuous Fourier transform of the signal. Recall that s is related to f as $f = \frac{s}{NT}$. Therefore, special care should be taken with the actual values of f when the derivative property is applied. The estimation of the first derivative of the signal involves multiplying $G(f)$ by $(j2\pi f)$, a complex straight line function also appearing in Figure 7.6 (c).

Once the multiplication corresponding to the Fourier property has taken place, the discrete Fourier coefficients $(j2\pi f) G(f)$ must be unshifted in order to obtain a suitable discrete Fourier representation (i.e., the second half of the period before the first half of the next period), to be input to FFT or DFT algorithms. After applying this second shift procedure, the inverse discrete Fourier transform can be applied in order to complete the estimation of the differentiated version of $g(t)$, which is shown in Figure 7.6 (d). By observing that

$$\frac{dg}{dt}(t) = -2 \sin(2t) + 2t \cos\left(t^2\right),$$

it is possible to plot some of the analytical values of $\frac{dg}{dt}$, which have been indicated by "+" in Figure 7.6 (d). It is worth emphasizing that the Fourier-based differentiation procedure can lead to good results, except for the oscillations near the signal extremities, where the original signal is discontinuous. Such effects almost vanish for periodic signals (e.g., closed contours).

It is important to note that the unshifting step is not necessarily equivalent to applying the same shift procedure, and that appropriate *shift* and *unshift* functions should be implemented. In order to better understand these procedures, suppose that the following vector is to be shifted and unshifted:

$$\begin{bmatrix} 0 & 1 & 2 & 3 \end{bmatrix}.$$

Shifting the above vector leads to

$$\begin{bmatrix} 2 & 3 & 0 & 1 \end{bmatrix}.$$

The application of the same shift procedure to this vector results in

$$\begin{bmatrix} 0 & 1 & 2 & 3 \end{bmatrix},$$

i.e., the original vector. Nevertheless, the correct results are not obtained if the same procedures are applied to a vector containing an odd number of elements, e.g.,

$$\begin{bmatrix} 0 & 1 & 2 & 3 & 4 \end{bmatrix}.$$

The resulting shifted vector is

$$\begin{bmatrix} 3 & 4 & 0 & 1 & 2 \end{bmatrix}.$$

However, the application of the same shift procedure to the above-shifted vector results in

$$\begin{bmatrix} 1 & 2 & 3 & 4 & 0 \end{bmatrix}.$$

That is, the 0 coefficient (the DC component) has been moved to the last position of the vector, instead of the first one (which would be correct). Therefore, two different procedures, *shift* and *unshift*, should be implemented in order to avoid the above problems. Section 2.7.7 also discusses some issues related to the application of the derivative property.

While the above discussion has covered the main procedure for obtaining the differentiated version of the signal, additional difficulties are implied by the Gaussian filtering, which induces an energy reduction of the contours (an effect known as *contour shrinking*). Methods for preventing this undesirable effect involve multiplying the signals by a normalizing factor, as explained in Section 7.2.4. Therefore, as a consequence of such a normalization, it is unimportant to use s or f in equations (7.3) and (7.4), since the linear relation between the two variables ($f = \frac{s}{NT}$) disappears if the signal energy is normalized. Nevertheless, due care should be taken regarding the other aspects discussed in this section (e.g., FFT shifting).

7.2.3 Gaussian Filtering and the Multiscale Approach

Although the discussion presented in this section concentrates on the complex contour representation $u(t)$, an analogous formulation can be easily developed for signals represented in terms of $x(t)$ and $y(t)$.

As discussed above (see also Section 3.3.2), signal differentiation can be viewed as high-pass filtering, which accentuates high frequency noise. This undesirable effect can be controlled through the introduction of a multiscale filtering mechanism in the curvature estimation process. In this context, the Fourier-based numerical differentiation process should be combined with Gaussian low-pass filtering in order to attenuate the high-frequency energy. The Gaussian filter bandwidth can be adjusted through the Gaussian standard deviation a, in such a way that the so-called multiscale approach is implemented by taking this value a as a continuous variable. Recall that such a Gaussian low-pass filtering can be understood as convolving the original signal and a Gaussian in the time domain, for the Fourier transform of a Gaussian is also a Gaussian (see Section 2.7.3):

$$g(t) = \frac{1}{a\sqrt{2\pi}} \exp\left(\frac{-t^2}{2a^2}\right) \Leftrightarrow G(f) = \exp\left(\frac{-(2\pi)^2 f^2}{2\tau^2}\right), \quad \tau = \frac{1}{a}. \tag{7.8}$$

The discussion about energy normalization at the end of Section 7.2.4 also applies to the different Gaussian normalizations. Refer again to equation (7.8). The standard deviation parameter a of the time-domain Gaussian is associated with the *analyzing scale*, while $\tau = \frac{1}{a}$ is the *filter frequency bandwidth*, being both inversely related to each other (which explains why large scales are generally associated with small frequencies and vice versa). Therefore, the application of a $\frac{1}{a}$ bandwidth filter is related to obtaining a smoothed contour observed for the scale a. A family of smoothed curves is defined by varying the scale parameter a, which produces a filter bank $G_{\frac{1}{a}}(f)$ indexed by this parameter, i.e.,

$$G(f) = G_{\frac{1}{a}}(f).$$

The family of smoothed curves $\hat{u}(t, a)$ is defined by filtering $U(f)$ with the filter bank $G_{\frac{1}{a}}(f)$ followed by the inverse Fourier transform, i.e.,

$$\hat{u}(t, a) = u * g_a(t) = F^{-1}\left\{U_{\frac{1}{a}}(f)\right\} = F^{-1}\left\{U(f)G_{\frac{1}{a}}(f)\right\}.$$

Therefore, $U_{\frac{1}{a}}(f)$ denotes $U(f)$ filtered by $G_{\frac{1}{a}}(f)$. Analogously, the multiscale differentiation expressions of $u(t)$ can be defined by filtering the respective Fourier transforms by $G_{\frac{1}{a}}(f)$, i.e.,

$$\hat{\dot{u}}(t, a) = \dot{u} * g_a(t) = F^{-1}\left\{\dot{U}_{\frac{1}{a}}(f)\right\} = F^{-1}\left\{\dot{U}(f)G_{\frac{1}{a}}(f)\right\} \quad \text{and}$$

$$\hat{\ddot{u}}(t, a) = \ddot{u} * g_a(t) = F^{-1}\left\{\ddot{U}_{\frac{1}{a}}(f)\right\} = F^{-1}\left\{\ddot{U}(f)G_{\frac{1}{a}}(f)\right\},$$

where $\dot{U}(f)$ and $\ddot{U}(f)$ are defined by equations (7.3) and (7.4), respectively.

7.2.4 Some Solutions for the Shrinking Problem

Gaussian filtering modifies the signal representation amplitude of contours, implying the so-called *shrinking effect* of the filtered contours. It is important to emphasize that contour shrinking directly affects the estimated curvature since the latter depends on the curve scale. Two possible solutions for this problem are introduced below.

Shrinking Prevention through Energy Conservation

Suppose that the contour has been normalized with respect to translation so that its center of mass is placed at the origin and, therefore, $U(0) = 0$. The energy of the contour $u(t)$ is defined as

$$E = \int |U(f)|^2 \, df.$$

As observed above, the energy E is modified by the filter function $G_{\frac{1}{a}}(f)$, and the energy of the filtered contour as a function of the scale parameter is defined as

$$E(a) = \int \left| U_{\frac{1}{a}}(f) \right|^2 \, df.$$

The total energy loss caused by the filtering process may be compensated by introducing a normalizing coefficient $\Omega(a)$, which depends on the scale a, i.e.,

$$\Omega(a) = \sqrt{\frac{E}{E(a)}}.$$

The normalization is achieved by multiplying the signals $\hat{u}(t, a)$, $\hat{\dot{u}}(t, a)$ and $\hat{\ddot{u}}(t, a)$, so that the energy of the reconstructed filtered contour is the same as the original contour:

$$u(t, a) = \hat{u}(t, a)\,\Omega(a),$$

$$\dot{u}(t, a) = \hat{\dot{u}}(t, a)\,\Omega(a),$$

$$\ddot{u}(t, a) = \hat{\ddot{u}}(t, a)\,\Omega(a).$$

Shrinking Prevention through Perimeter Conservation

An alternative to avoid the contour shrinking effect is to keep constant the perimeter of the original contour after each Gaussian filtering. In other words, all curves of the family of smoothed contours $u(t, a)$ should therefore have the same perimeter.

The perimeter conservation principle presents a particularly interesting physical interpretation: imagine a wire bent in the form of the initial contour and imagine further that someone starts unbending it so that the wire contour becomes an elliptical shape. Clearly, during this unbending process, which is analogous to the Gaussian contour filtering, the wire perimeter remains constant (the actual length of the wire). This physical interpretation is central to the bending energy concept, explored in Section 7.4.1. In order to be physically realistic, in the above context, the contour unbending process should preserve the original perimeter, for the actual wire length cannot shrink.

While the perimeter L of a continuous contour $u(t)$ is defined as

$$L = \int |\dot{u}(t)| \, dt,$$

the perimeter of its spatially sampled version can be estimated by using finite differences, i.e.,

$$L = \sum_{n=0}^{N-1} |u(n) - u(n-1)|.$$

As an alternative, the perimeter can be estimated from $\dot{u}(n)$, which can be calculated from the Fourier descriptors of $u(n)$ by applying the Fourier derivative property. This is exactly the same process that is used to estimate the signal derivatives required for the curvature. The perimeter can be estimated from $\dot{u}(n)$ as

$$L = \frac{2\pi}{N} \sum_{n=0}^{N-1} |\dot{u}(n)|.$$

Similarly, the perimeter $L(a)$ of the smoothed curve $u(t, a)$ is defined as

$$L(a) = \int |\dot{u}(t, a)| \, dt.$$

As in the previous case, it can also be estimated by using finite differences or $\dot{u}(n, a)$, i.e.,

$$L(a) = \sum_{n=0}^{N-1} |\hat{u}(n, a) - \hat{u}(n-1, a)|$$

or

$$L(a) = \frac{2\pi}{N} \sum_{n=0}^{N-1} |\dot{\hat{u}}(n, a)|.$$

The normalization factor $P(a)$, which depends on the analyzing scale a, is defined as

$$P(a) = \frac{L}{L(a)}.$$

The normalization factor $P(a)$ is therefore defined as the ratio between the original and the smoothed contour perimeters. Analogously to the previous case of energy-based normalization, the equations for perimeter-based shrinking prevention are defined as:

$$u(t, a) = \hat{u}(t, a)\, P(a),$$

$$\dot{u}(t, a) = \dot{\hat{u}}(t, a)\, P(a),$$

$$\ddot{u}(t, a) = \ddot{\hat{u}}(t, a)\, P(a).$$

Particularly, $u(t, a)$ represents a family of smoothed versions of $u(t)$, with $u(t, a_0)$ being a smoothed version of $u(t)$ obtained by convolving the original contour with a Gaussian of scale a_0. As the scale a varies, $u(t, a)$ defines the *evolution* of u, and the respectively generated curves can be shown in the so-called *morphograms* (see Figures 7.7 and 7.8 for examples).

Figure 7.7: *A morphogram: 4 curves for increasing scales.*

Note: *Different Approaches to Shrinking Prevention*

It is important to note that shrinking does not affect all the contour points in the same way, being more accentuated near high curvature points. In fact, there are

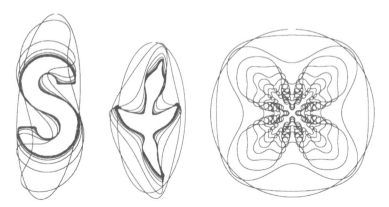

Figure 7.8: *A morphogram: Superposed views (all curves).*

some techniques that attempt to account for shrinking prevention (e.g., [Mokhtarian & Mackworth, 1992] with basis on the technique developed by Lowe [1989]). This section has discussed some simpler approaches that can be applied without substantial numerical errors and are suitable for many practical situations. Further references on this subject can be found in [Marshall, 1989; Oliensis, 1993; Sapiro & Tannenbaum, 1995]. In the case of analysis over small spatial scales, the shrinking effect can naturally be avoided by considering the factor (i.e., the sampling interval) implied by the numerical integration of the Fourier transform [Estrozi et al., 2003].

7.2.5 The Curvegram

The normalized multiscale expressions of $\dot{u}(t, a)$ and $\ddot{u}(t, a)$ defined in the previous sections, together with complex curvature expression of $k(t)$ given by equation (7.7), allow the definition of the multiscale curvature description, or *curvegram* of $u(t)$ as

$$k(t, a) = \frac{-\,\mathrm{Im}\{\dot{u}(t, a)\ddot{u}^*(t, a)\}}{\left|\dot{u}(t, a)\right|^3}. \tag{7.9}$$

The algorithm below summarizes the curvegram generation by using the aforementioned Fourier properties and perimeter normalization. In fact, this algorithm calculates the curvature for a given fixed scale a. In order to obtain the curvegram, it is enough to apply the algorithm below for a series of different scale values a.

Algorithm: *Fourier-Based Curvature Estimation*

1. Let u be the complex signal representing the contour;
2. $U \leftarrow$ FFT(u);
   ▷ Calculate the first and of the second derivatives of u
3. Apply the Fourier derivative property to calculate dU and ddU;
4. Apply the Gaussian filter of bandwidth $1/a$ to U, dU and to ddU;
   ▷ Reconstructed smooth contour
5. $ru \leftarrow$ IFFT(FFTUNSHIFT(U));
   ▷ Estimated derivatives
6. $du \leftarrow$ IFFT(FFTUNSHIFT(dU));
7. $ddu =$ IFFT(FFTUNSHIFT(ddU));
   ▷ Shrinking prevention normalization constant
8. $C \leftarrow$ PERIMETER(u)/PERIMETER(ru);
9. Multiply all elements of du and of ddu by the normalization constant C;
10. Calculate the curvature using Equation 7.9

Figure 7.9 (a) shows a digitized contour of the map of Brazil, while Figures 7.9 (b) through (h) show smoothed versions of the original contour, which is represented at seven different scales. The curvegram of the contour in Figure 7.9 is shown as a surface in Figure 7.10. The axes are labeled as t (the independent parameter of the signal), τ (the frequency bandwidth, with $\tau = \frac{1}{a}$) and k (the curvature). The curvegram is also shown as a gray-level image in Figure 7.11 (a). The curvegram $k(t, a)$ expresses the contour curvature in terms of a set of the analyzing scales. In this graphical representation of the curvegram $k(t, a)$, the horizontal axis is associated with the time parameter t (the parameter along the contour), while the vertical axis represents the scale parameter a, which increases downward. Higher curvature values have been indicated by brighter pixels. Note that the curvegram presents a strong response around the points associated with the contour corners. For smaller scales, the curvegram detects finer details. As the analyzing scale increases, the curvegram becomes less sensitive to fine details, and to high frequency noise, thus progressively reflecting the global shape structure. Therefore, in this type of image the signal parameter is the horizontal axis, while the scale parameter is the vertical axis. The curvature is shown in terms of gray-level intensities. The images are created by normalizing (in a logarithmic scale) the curvatures in the range from 0 to 255. In such images, darker regions correspond to lower (possibly negative) curvatures, while lighter regions correspond to higher curvatures. Figure 7.11 (b) presents the scale-space image of the maximum absolute curvature points, extracted from the curvegram of Figure 7.11 (a). This example illustrates the possibility of deriving scale-space images from the curvegram.

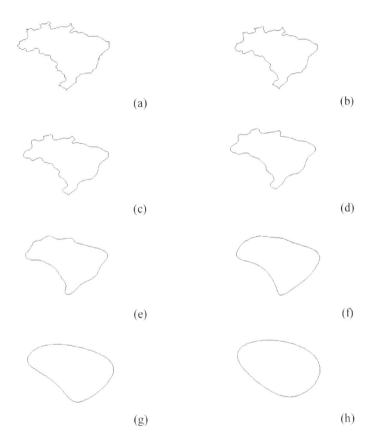

(a)

(b)

(c)

(d)

(e)

(f)

(g)

(h)

Figure 7.9: *Multiscale representation of a contour: original contour **(a)** and respective smoothed versions **(b)** through **(h)**.*

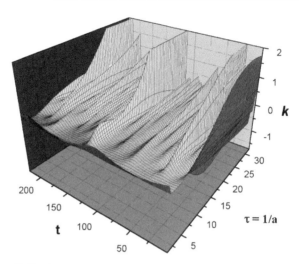

Figure 7.10: *Curvegram of the contour in Figure 7.9 (a) presented as a rendered surface.*

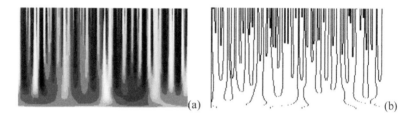

Figure 7.11: *Curvegram of the contour of Figure 7.9 (a) displayed as an image **(a)** and respective maximum absolute curvature scale-space **(b)**.*

Figure 7.12 (a) presents another example using the synthetic contour defined by the following equations:

$$x(t) = 4 \, \cos\left(\frac{t}{128}2\pi\right) + 2 \, \cos\left(\frac{t}{128}16\pi\right),$$

$$y(t) = 4 \, \sin\left(\frac{t}{128}2\pi\right) + 2 \, \sin\left(\frac{t}{128}16\pi\right).$$

Figures 7.12 (b) through (e) present the smoothed versions of the original contour. Note that the internal loops degenerate into singular points and, subsequently, to concave regions. This can be perceived in the curvegram by the abrupt change between the points of positive maximum curvature to negative minimum in Figure 7.13. The curvegram tends toward a region of constant positive curvature.

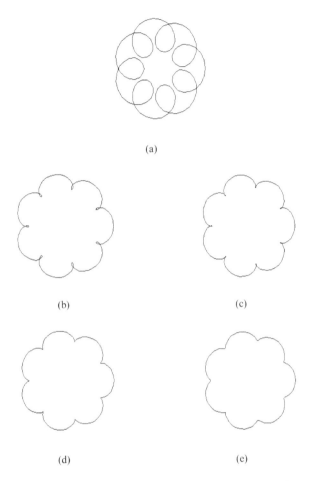

(a)

(b)   (c)

(d)   (e)

Figure 7.12: *Multiscale representation of a synthetic contour: original contour (a) and some of its smoothed versions (b) through (e).*

To probe further: *Fourier-Based Multiscale Curvature Estimation*

The curvegram has been described in [Cesar-Jr. & Costa, 1995, 1996, 1997], being a powerful multiscale contour analysis tool. As a multiscale representation, it can be visualized in a number of ways, and the reader is referred to [Grossmann et al., 1989] for further information on visualizing multiscale transforms. Fourier-based curvature estimation has been assessed extensively with respect to its precision, and results about its remarkable performance can be found in [Cesar-Jr. & Costa, 1996, 1997; Estrozi et al., 2003]. Corner detection and contour segmentation for polygonal approximation using the curvegram have been carried out in [Cesar-Jr.

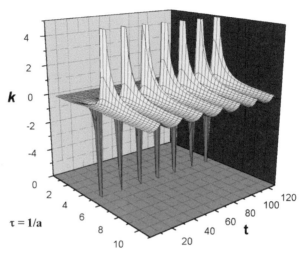

Figure 7.13: *Curvegram of the contour in Figure 7.12 (a) presented as a rendered surface.*

& Costa, 1995]. An approach for curvature estimation from the tangent orientation along the contour, which is also calculated using the Fourier transform, has been described in [Cesar-Jr. & Costa, 1995]. The experimental results presented in this section have been adapted from [Cesar-Jr. & Costa, 1996]. The reader is also referred to [Baroni & Barletta, 1992] for a related approach.

7.2.6 Curvature-Scale Space

One of the most popular multiscale curvature representations of 2D curves is the *curvature scale-space* [Mokhtarian & Mackworth, 1986], which has been later improved and applied in different works [Cesar-Jr. & Costa, 1995; Lowe, 1989; Mokhtarian & Mackworth, 1992; Pernus et al., 1994; Rattarangsi & Chin, 1992]. Basically, these methods obtain a description of the curvature as the curve is convolved with a series of Gaussians with the standard deviation varying continuously. The paper by [Mokhtarian & Mackworth, 1992] explores the fact that, if arc length parameterization is adopted, then the curvature equation (7.2) can be expressed as

$$k(t) = \dot{x}(t)\,\ddot{y}(t) - \ddot{x}(t)\,\dot{y}(t).$$

Furthermore, a series of Gaussian smoothed curves can be obtained from the parametric contour $c(t)$ as

$$c(t, a) = \{x(t, a), y(t, a)\},$$

where

$$x(t, a) = x(t) * g_a(t),$$
$$y(t, a) = y(t) * g_a(t),$$

and $g_a(t)$ denotes a Gaussian with standard deviation a. The multiscale derivatives of $x(t)$ and of $y(t)$ can be obtained by convolving the original contour signals with the derivatives of $g_a(t)$, i.e.,

$$\dot{x}(t, a) = x(t) * \dot{g}_a(t),$$
$$\dot{y}(t, a) = y(t) * \dot{g}_a(t),$$
$$\ddot{x}(t, a) = x(t) * \ddot{g}_a(t),$$
$$\ddot{y}(t, a) = y(t) * \ddot{g}_a(t).$$

The multiscale curvature is defined as:

$$k(t, a) = \dot{x}(t, a)\ddot{y}(t, a) - \ddot{x}(t, a)\dot{y}(t, a).$$

The work of [Mokhtarian & Mackworth, 1992] discusses the problems of shrinking and reparameterization. Such ideas have been extensively applied and generalized to 3D curves in subsequent works [Mokhtarian, 1995, 1997]. This approach is equivalent to analyzing the curve evolution guided by the heat (diffusion) equation. Let Γ_0 be the original contour. A family of curves Γ generated by Γ_0 is obtained by solving the heat equation assuming Γ_0 as the initial condition, i.e., [Widder, 1975] of

$$\frac{\partial \Gamma}{\partial \sigma} = \frac{\partial^2 \Gamma}{\partial t^2},$$

where t defines the parameterization along the curve and σ stands for the scale. Different representations can be obtained from the curve family Γ, generally defined in terms of the zeros, local maxima and minima of curvature of each curve of the family. Alternative models for curve flow have been more recently proposed in [Kimia et al., 1992, 1995; Sapiro & Tannenbaum, 1993].

7.3 | Wavelet-Based Multiscale Contour Analysis

7.3.1 | Preliminary Considerations

Section 7.2 described a shape analysis technique based on the multiscale curvature description of 2D contours, i.e., the *curvegram*. Alternative multiscale analysis techniques based on the wavelet transform are examined in this section, with special attention given to the *w-representation*, which is obtained through the application of the continuous wavelet transform of the contour complex signal $u(t)$. This approach was introduced in [Antoine et al., 1997]. The *w*-representation is especially useful for shape analysis in vision problems because wavelet theory provides many powerful properties generally required by shape analysis techniques. In this context, algorithms for dominant point detection and characterization and local periodic pattern analysis are available when wavelets are used. In fact, wavelets constitute a particularly efficient mathematical tool for detection and characterization of signals' singularities, as well as for the analysis of instantaneous frequencies [Torrésani, 1995].

Multiscale approaches to shape analysis should always consider the following two important points:

① Important shape structures are generally associated with transient events in the object (recall Attneave's psychophysical experiments discussed in Chapter 6).

② Different events may occur at different scales.

By an event it is usually meant a specific characteristic along the signal, such as a discontinuity, singularity or local frequency. It is important to note that these principles equally apply to several different problems in vision, signal, image and shape processing. Wavelet theory has become a natural mathematical theory to deal with such problems, unifying different, independently developed approaches. For example, wavelets have already been applied to fractal analysis, image compression and analysis, edge detection, texture segmentation and for modeling neural receptive fields. As far as shape analysis is concerned, the wavelet transform has shown to be useful for contour representation, corner detection from 2D images and from tangential representation of contours, and for the extraction of shape measures to be used in object recognition.

One of the main features of the wavelet transform is its ability to separate transient events from the signal, avoiding that local signal modifications affect the whole representation. Such a representation can be highly desirable for shape analysis problems where different situations, such as partial occlusion, may imply local interferences to the shape, therefore avoiding one of the main disadvantages of global shape descriptors, such as the Fourier descriptors and moments.

In this context, the w-representation allows a unified approach to different problems of contour-based shape analysis. Furthermore, many of the issues discussed in this section also apply to other multiscale transforms, in such a way that a good understanding of the wavelet transform and its properties provide a powerful background when analyzing alternative representations. A particularly important aspect of the (continuous) wavelet transform is the concept of the transform *skeleton* (which is different from *shape skeletons* discussed in Chapter 5), a compact representation that allows the extraction of useful information about the shape of interest.

To probe further: *Wavelets*

Wavelet theory has become as mathematically sound and vast as Fourier theory, constituting a research field in itself. Therefore, discussing it in detail is far beyond the scope of this book. The wavelet theory has a rich and interesting history, with roots in different works developed during the past two centuries, starting with the work of Jean-Baptiste Fourier, in 1807. Some interesting facts about the history of the wavelet theory can be found in the books by Barbara Hubbard [Hubbard, 1998] and by Yves Meyer [Meyer, 1993], as well as in a paper by Ingrid Daubechies [Daubechies, 1996]. The discussion in this book is not exhaustive, presenting only the basic concepts necessary to understand the shape analysis tools presented here. Castleman's book [Castleman, 1996] and Rioul and Vetterli's introductory paper [Rioul & Vetterli, 1991] are suitable for introducing the reader to this stimulating field. There are also many textbooks on wavelet theory, and the reader is referred to the list below for further detail.

The wavelet digest (http://www.wavelet.org/) is the most important Web resource related to wavelets. Some especially important references are listed below:

☞ Introduction to wavelets (papers): [Antoine, 1994; Daubechies, 1992; Grossmann et al., 1989; Mallat, 1996; Rioul & Vetterli, 1991];

☞ Introduction to wavelets (books): [Daubechies, 1992; Holschneider, 1995; Mallat, 1998; Meyer, 1993; Torrésani, 1995; Vetterli & Kovacevic, 1995];

☞ Wavelets (important and classical papers): [Daugman, 1985, 1988; Gaudart et al., 1993; Grossmann & Morlet, 1984; Mallat & Hwang, 1992; Mallat & Zhong, 1992; Mallat, 1989];

☞ Wavelets and image processing: [Antoine et al., 1993; Freeman & Adelson, 1991; Grossmann, 1988; Prasad et al., 1997; Simoncelli et al., 1992; Starck et al., 1998; Stollnitz et al., 1995a,b, 1996; Watson, 1987];

☞ Wavelets and shape analysis: [Antoine et al., 1997; Cesar-Jr. & Costa, 1998b; Chen et al., 1995; Khalil & Bayoumi, 2000; Lee et al., 1995, 1993; Nakamura & Yoshida, 1994; Tieng & Boles, 1997a,b];

☞ 2D wavelets and biomedical imaging: [Jelinek et al., 2003, 2007; Soares et al., 2006];

☞ Wavelets and fractals: [Arnèodo et al., 1995; Muzy et al., 1994];

☞ Related issues on multiscale transforms: [Koenderink, 1984; Lindeberg, 1994; Marr, 1982; Witkin, 1983].

7.3.2 The w-Representation

Let $u(t)$ be the complex signal representing the contour to be analyzed. The w-representation of $u(t)$ is defined as the continuous wavelet transform of this signal, i.e.:

$$W(b,a) = \frac{1}{a} \int_R \Psi\left(\frac{t-b}{a}\right) u(t)\, dt, \tag{7.10}$$

$$U_\psi(b,a) = \sqrt{a} \int_{-\infty}^{\infty} \Psi^*(af)\, U(f)\, e^{j2\pi fb}\, df, \tag{7.11}$$

where $b \in \mathbb{R}$ represents the shifting parameter, while $a > 0$ defines the dilation parameter of the wavelet transform. The parameter space is defined as the superior semi-plane of $\mathbb{R}^2$, $H = \{(b,a) \mid b \in \mathbb{R}, a > 0\}$. The analyzing wavelet $\psi(t)$ should have a null DC component, which is known as the simplified admissibility condition. This condition provides the wavelet transform with an exact reconstruction formula defined as:

$$u(t) = c_\psi \int_{-\infty}^{\infty} \int_{-\infty}^{\infty} a^{-\frac{1}{2}} U_\psi(b,a)\, \psi\left(\frac{t-b}{a}\right) \frac{db\, da}{a^2},$$

where the normalization constant c_ψ only depends on the analyzing wavelet.

The existence of the above-presented reconstruction means that the signal $u(t)$ is expanded, in an invertible way, in terms of the wavelets $\psi_{(b,a)}(t) = a^{\frac{1}{2}} \psi\left(a^1(t-b)\right)$, which are shifted and dilated copies of the analyzing wavelet (or "mother wavelet") ψ. Besides the admissibility condition, two further properties are usually required from the analyzing wavelet: that it is progressive (i.e., $\psi(f)$ should be real and $\psi(f) = 0, f \leqslant 0$) and that it has a certain number of vanishing moments, i.e.,

$$\int_{-\infty}^{\infty} t^r \psi(t)\, dt = 0, \quad r = 0, 1, \ldots, K.$$

The reason for requiring the vanishing moments property can be understood as follows. First, it is important to emphasize that both the wavelet $\psi(t)$ and its Fourier transform $\psi(f)$ should be well located in both domains, subjected to the uncertainty principle (see Section 7.1.4). The vanishing moments property implies that the analyzing wavelet is "blind" to polynomials of degree smaller or equal to K, particularly, the regular parts of the signal.

The w-representation presents many useful and desirable properties for a shape analysis scheme, including,

Unicity: because the wavelet transform has an inverse formula, the w-representation is unique, and two different shapes present different w-representations, which is important under the shape recognition point-of-view.

Invariance to geometric transforms: the linearity and the covariance properties of the wavelet transform (refer to [Holschneider, 1995] for an introduction to these and other wavelet properties) endow the w-representation with the following important properties associated with geometric transformations of the original contour.

Translation: translating an object in an image by a vector $\mathbf{v} = (x_0, y_0)$ is equivalent to adding a complex constant $z = x_0 + jy_0$ to the signal $u(t)$. This operation does not affect the w-representation, except for numerical errors due to the grid sampling and image pre-processing for contour extraction, because.

$$U\left[\psi, u(t) + z\right] = U\left[\psi, u(t)\right] + U\left[\psi, z\right] = U\left[\psi, u(t)\right], \quad z \in \mathbb{C}.$$

Scaling: homogeneous or isotropic contour scaling by a constant c affects the magnitude and the parameterization of $u(t)$. In this case, the w-representation is covariant to this transformation, because.

$$U\left[\psi, c_1 u\left(\frac{t}{c_2}\right)\right](b, a) = c_1 U\left[\psi, u(t)\right]\left(\frac{b}{c_2}, \frac{a}{c_2}\right), \quad c_1, c_2 \in \mathbb{R}^+.$$

If arc length representation is adopted, then $c = c_1 = c_2$ and the constant c can be estimated.

Rotation: rotating an object by an angle θ generally implies two modifications to the signal $u(t)$: the signal $u(t)$ is multiplied by $e^{j\theta}$, and the parameterization is shifted by a constant t_0 (because of the change of the starting point, which occurs for many contour extraction schemes). This geometrical transformation, therefore, modifies the w-representation in the following manner:

$$U\left[\psi, e^{j\theta} u(t - t_0)\right](b, a) = e^{j\theta} U\left[\psi, u(t)\right](b - t_0, a), \quad t_0 \in \mathbb{C}, \theta \in [0, 2\pi).$$

Local modifications of the shape: the w-representation is mainly affected locally by local modifications of the shape because of the fast decaying characteristic of most wavelets. This property is important in applications where, for instance, the objects can be partially occluded. This is one of the main advantages of the wavelet transform with respect to global transforms such as Fourier: local modifications on signals are taken into account by the whole transform.

Efficiency and easy implementation: calculating the w-representation involves 1D circular convolutions (closed contours are periodic), which can be easily implemented both in the time and frequency domains, and special fast algorithms are available. These algorithms generally explore the fact that the wavelet coefficients are highly correlated between adjacent scales because of the similarity of the dilated wavelets that are used in the transform. Alternatively, the w-representation can be obtained from the Fourier descriptors of the contour by using a discrete version of equation (7.11). In this sense, the Fourier-based curvature estimation algorithm of Section 7.2.5 can be easily adapted in order to generate the continuous wavelet transform of signals.

Additional resources: *Software for multiscale shape analysis*

The multiscale curvature and the w-representation are available in the software `http://code.google.com/p/imagerecognitionsystem/`. 2D wavelets have been used for vessel segmentation of fundus images (retina) and a corresponding software is available at `http://retina.incubadora.fapesp.br/portal`.

7.3.3 Choosing the Analyzing Wavelet

From the definition of the wavelet transform it is clear that the chosen analyzing wavelet ψ plays a central role. Two very common analyzing wavelets for signal analysis are discussed below: the family of wavelets defined by the Gaussian derivatives and the Morlet wavelet.

Gaussian Derivatives

Since its introduction by Marr and Hildreth as a resource for edge detection [Marr & Hildreth, 1980], the second derivative of the Gaussian has become one of the most popular analyzing wavelets, known as the *Marr wavelet* or *Mexican hat*, which is presented in the following:

$$\psi_g^{(2)}(t) = (2\pi)^{-\frac{1}{2}} \left(t^2 - 1\right) e^{-t^2/2} \Leftrightarrow \psi_g^{(2)}(f) = -(2\pi f)^2 e^{-f^2/2}.$$

This is a real and non-progressive wavelet, with two vanishing moments, being blind to constant and linear components of the signal, which explains why this wavelet is frequently used as a means to analyze singularities. A generalization often explored for signal analysis involves taking an arbitrary derivative of the Gaussian as a valid wavelet. Such derivatives can be defined in terms of the Fourier transform as

$$\psi_g^{(r)}(t) = \frac{d^r g(t)}{dt^r} \Leftrightarrow \psi_g^{(r)}(f) = (j2\pi f)^r g(f).$$

Figure 7.14 presents a Gaussian and its first and second derivatives for a series of different scales. The wavelet $\psi_g^{(r)}$ defined by the above equation has r vanishing moments (refer to the previous section).

This family of wavelets has been used for multifractal analysis of signals in turbulence and DNA sequences. In fact, these wavelets are also explored in this

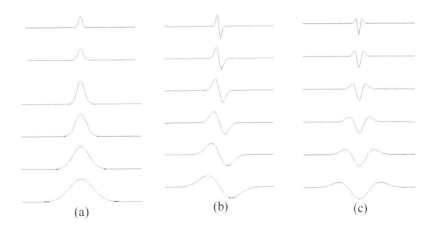

(a)   (b)   (c)

Figure 7.14: *Gaussians of different scales (a) and the respective first (b) and second derivatives (c).*

book for contour analysis within the context of corner detection. Let $u^{(r)}(t)$ denote the r-th derivative of $u(t)$. It can be shown that

$$U\left[\psi_g^{(r)}; u\right] = a^r U\left[\psi_g; u^{(r)}\right],$$

indicating that the analysis of $u(t)$ by the wavelet transform with $\psi_g^{(r)}$ as the mother wavelet is related to the analysis of the r-th derivative of $u(t)$.

The Morlet Wavelet

The first wavelet introduced within the context of wavelet theory was created by the French researcher J. Morlet. The Morlet wavelet is defined as a complex exponential modulated Gaussian, i.e.,

$$\psi_M(t) = e^{j2\pi f_0 t} e^{-t^2/2} + \eta(t).$$

The correction term η is necessary because the modulated Gaussian alone is not an admissible wavelet (this term is negligible whenever f_0 is large enough). This wavelet is especially suitable for the analysis of locally periodic patterns, i.e., local frequencies, as it will be illustrated, because it is well localized in the frequency domain. Its dilated versions can be seen as band-pass filters of bandwidth $\frac{1}{a}$. The Morlet wavelet is similar to the Gabor filters, with the important differences being that the filter bandwidth varies with the frequency (scale). In addition, observe that both the Morlet and Gabor functions are Gaussians multiplied by a complex exponential. The difference between them is that, while the Morlet wavelet varies the Gaussian standard deviation, keeping constant the complex exponential frequency, the Gabor transform varies the complex exponential frequency, keeping constant

the Gaussian standard deviation. Figure 7.15 presents the real and imaginary parts
of the Morlet wavelet for a series of different scales.

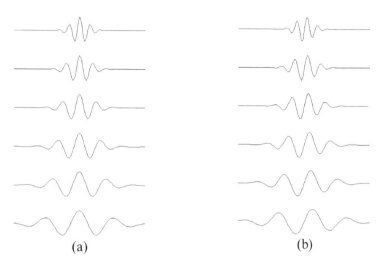

(a)   (b)

Figure 7.15: *Real (a) and imaginary (b) parts of Morlet wavelet at different scales.*

7.3.4 Shape Analysis from the w-Representation

Although there are many different manners of visually analyzing multiscale transforms [Grossmann et al., 1989], it is important to develop algorithms for automatic
shape analysis from the w-representation. Since this representation is generally
complex-valued, because of the complex signal $u(t)$, it is interesting to consider the
polar representation defined as

$$U(b, a) = M(b, a) \, e^{j \varphi(b,a)},$$

where:

$$M(b, a) = |U(b, a)| \quad \text{and} \quad \varphi(b, a) = \arg(U(b, a)).$$

Figure 7.16 illustrates the derivation of the w-representation of a P-shape. Only
the outer contour is considered, being extracted from the upper-left corner in clockwise direction.

The complex representation $u(t)$ in Figure 7.16 (b) has been obtained from the
tracked contour. The w-representation is shown in terms of the modulus $M(b, a)$ and
modulus maxima of $U(b, a)$ in Figure 7.16 (b), respectively (assuming a
Mexican hat wavelet). An alternative scheme for analyzing $U(b, a)$ is through

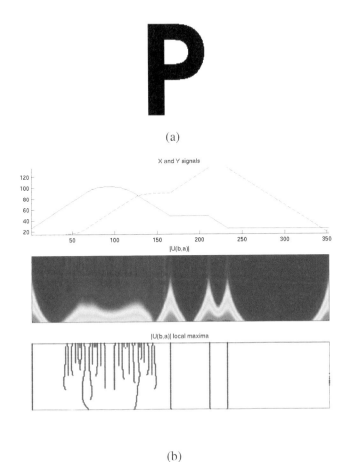

(a)

(b)

Figure 7.16: *General scheme of the w-representation: original shape (a);
parametric signals, wavelet modulus and wavelet local max-
ima, respectively (b).*

its *scalogram*, which is analogous to the *spectrogram* of the short-time Fourier
transform, representing the signal energy distribution on the time-scale plane, i.e.,
$M(b, a)^2$.

The information present in the continuous wavelet transform (and, hence, also
in the w-representation) is highly redundant, which can be explored by the analysis methods. In fact, a powerful approach for extracting information from the
w-representation considers the local maxima of $M(b, a)$ in order to eliminate redundancy and noise. Some methods for dominant point detection and natural
scales identification from the local maxima of the w-representation are presented
below.

7.3.5 Dominant Point Detection Using the w-Representation

If a signal $u(t)$ has a singularity at $t = t_0$, then the wavelet transform has a local vertical maxima crest that converges toward t_0 when $a \to 0$. Because a corner can be thought of as a singularity in the parametric representation $u(t)$, the vertical local maxima crests can be used for dominant point detection. Figure 7.17 illustrates how the modulus of the w-representation respond to such singularities by funneling toward one of the vertices of the L-shaped contour, denoted by t_0 in that figure.

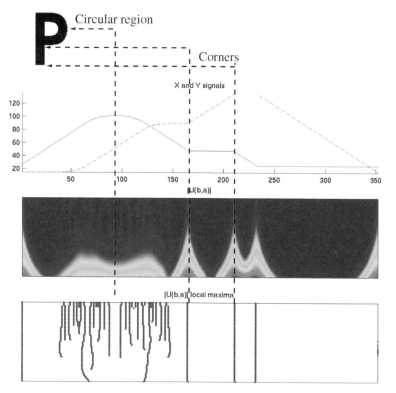

Figure 7.17: *Corner detection using wavelets: parametric signals, wavelet modulus and wavelet modulus maxima.*

Figure 7.17 presents the vertical maxima crests of the w-representation obtained from the modulus $M(b, a)$ for the above example. These maxima points correspond not only to the dominant points of the contour, but also to noise and secondary points. Therefore, it is important to identify the most relevant maxima crests, which are supposed to be associated with the prominent dominant points to be detected. First, some useful definitions are introduced, assuming that $H = \{(b, a) \mid b \in \mathbb{R}, a > 0\}$.

Vertical Local Maximum Point: A point $(b_0, a_0) \in H$ is defined as a vertical local maximum point if $\exists \varepsilon > 0 \mid M(b_0, a_0) > M(b, a_0), \forall b \in (b_0 - \varepsilon, b_0)$ and $M(b_0, a_0) \geqslant M(b, a_0), \forall b \in (b_0, b_0 + \varepsilon)$, or, alternatively, if $\exists \varepsilon > 0 \mid M(b_0, a_0) \geqslant M(b, a_0), \forall b \in (b_0 - \varepsilon, b_0)$ and $M(b_0, a_0) > M(b, a_0), \forall b \in (b_0, b_0 + \varepsilon)$.

Vertical Local Maxima Line: A line $L_v = \{(b, a) \in H\}$ is called a vertical local maxima line if $\forall (b_0, a_0) \in L_v$ is a vertical local maximum point.

Vertical Skeleton: The set of all vertical local maxima lines (or crests), denoted as $\{L_v\}$, is called the vertical skeleton of the w-representation.

The algorithm presented in Section 7.1.5 can be used in order to detect the vertical skeleton of a multiscale transform.

The detection of the maxima lines corresponding to the contour dominant points is done through the analysis of a relevance measure $f(L_v)$ that is calculated for each line of the vertical skeleton. The dominant points are detected by thresholding $f(L_v)$. Therefore, if $f(L_v) > T$, where T is the threshold, then the selected L_v is a valid vertical line and, if $L_v \to b_0$ (i.e., L_v tends to b_0) when $a \to 0$ (i.e., small scales), then $u(b_0)$ is taken as a dominant point. The adopted relevance measure is based on the scale-space lifetime [Lindeberg, 1993], being defined as the difference between the logarithms of the maximum and minimum line scale. Suppose that the line L_v starts at $(b_0, a_0) \in H$ and finishes at $(b_1, a_1) \in H, a_0 < a_1$. Therefore, the relevance measure $f(L_v)$ is defined as:

$$f(L_v) = \log a_1 - \log a_0.$$

Because in practice all maxima crests start at the minimum scale $a_0 = a_{min}$, the relevance measure can be defined as

$$f(L_v) = \log a_1.$$

It is important to note that the lifetime criterion is independent of the scale parameter discretization, being easy and fast to calculate. The algorithm has been tested on real images, and its ability to correctly identify the *perceptually important contour points* is extensively illustrated in [Antoine et al., 1997]. The wavelet-based contour segmentation algorithm can be summarized as follows:

Algorithm: *Dominant Point Detection Using the w-Representation*

---

1. Detect all the vertical maxima lines, i.e., the vertical skeleton;
2. Calculate the relevance measure for the vertical maxima lines;
3. Threshold the skeleton based on the relevance measure;
4. Determine the points b_k to which the thresholded vertical lines tend to;
5. Take $u(b_k)$ as the dominant points of $u(t)$;

The wavelet capabilities of detecting such dominant points (because of being a good tool to analyze signal singularities) may be explored to analyze the self-similar structure across the different scales of this fractal shape. Figures 7.18 and 7.19 show the *w*-representation of different wavelet shapes where the self-similar aspects are emphasized both on the modulus and the phase parts.

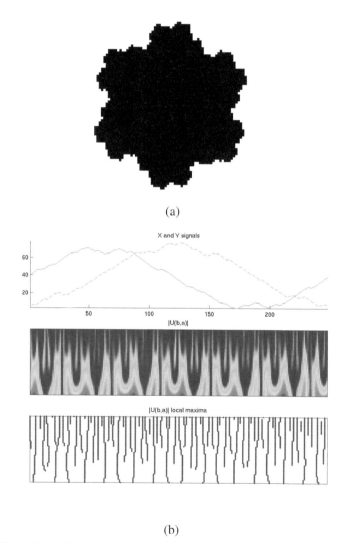

(a)

(b)

Figure 7.18: *W-representation of a fractal shape (1): original shape* **(a)***; parametric signals, wavelet modulus and wavelet local maxima, respectively* **(b)***.*

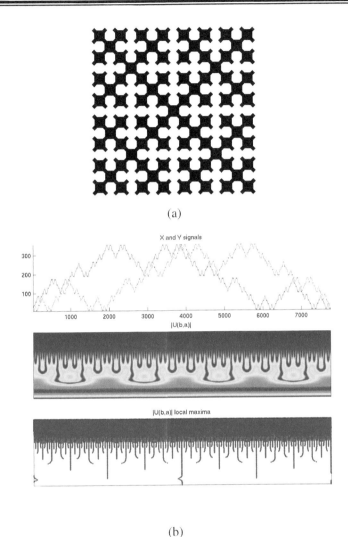

(a)

(b)

Figure 7.19: *W-representation of a fractal shape (2): original shape (a);*
parametric signals, wavelet modulus and wavelet local max-
ima, respectively (b).

7.3.6 Local Frequencies and Natural Scales

The concept of natural scales is related to the way in which a natural partition of
a contour can be defined. Natural partition refers to a segmentation process where
all the points of each segment share a set of common features, as illustrated in
Figure 7.20.

Figure 7.20 (a)-left shows a square where the four vertices may naturally be

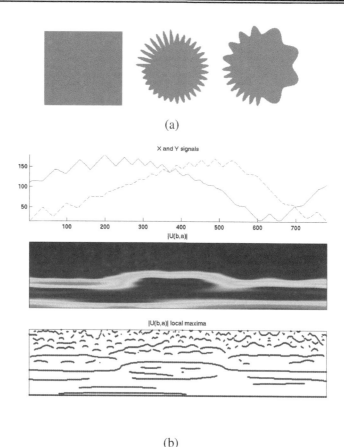

(a)

(b)

Figure 7.20: *Top row (a): three types of natural shape segmentation: based on corners, based on amplitude, and based on frequency. Bottom row (b): W-representation of the third shape above using the Morlet wavelet.*

taken as the segmentation points and, in this case, the points of each segment share the common property of having the same orientation. Figure 7.20 (a)-middle shows a radially textured circle with the sinusoidal textures of different amplitudes. The commonly shared property of each segment, in this case, is the amplitude. Finally, Figure 7.20 (a)-right shows a circle with three segments of sinusoidal texture radially added to the contour. In this case, the frequency of each sinusoidal texture defines the common property shared by the corresponding contour points. It is important to note that different contour segmentations could be equally defined for each shape of the example: for instance, the circle of Figure 7.20 (a)-right could be represented by a polygonal approximation with the high curvature points marked as dominant points. Nevertheless, it is argued that the criteria based on local frequency and amplitude of the sinusoidal texture are much more natural than curvature in this case. Furthermore, it is important to emphasize that the perception of periodic patterns—which are generally transient, i.e., extend for a limited curve parameter interval—is an important element in visual perception [Valois & Valois, 1990] that, as such, can be used to define a certain type of natural scale.

Let $u(t)$ be the complex representation of the shape shown in Figure 7.20 (a)-right, which presents three sinusoidally textured segments with frequencies f_i, $i = 1, 2, 3$. It is worth noting that $u(t)$ represents the shape outline. Figure 7.20 (b) shows the corresponding signals $x(t)$ and $y(t)$ and the respective w-representation. As discussed above, this shape has three natural portions or segments.

The w-representation of Figure 7.20 (b) has been generated assuming Morlet wavelets $\psi_M(t)$. Because the Fourier transform of the Morlet wavelet is a Gaussian in the Fourier domain, centered at $\frac{f_0}{a}$, the w-representation presents four lines of horizontal maxima points or crests. This set of horizontal lines is called the *horizontal skeleton*, describing the shape in terms of its locally periodic patterns. As it can be noted from the modulus of the w-representation in Figure 7.20 (b), the wavelet transform responds more strongly when the analyzing wavelet is properly tuned to the frequencies f_i, $i = 1, 2, 3$. Furthermore, the w-representation also indicates, through the vertical maxima lines and the constant phase lines (see Figure 7.21), the transition points between the different segments, illustrating the zoom property of the wavelet transform [Grossmann et al., 1989]. Therefore, the natural scales can be characterized by the horizontal maxima crests of $M(b, a)$, in the same way that vertical crests can be used in order to detect dominant points, as was shown in the last section. The contour partition is characterized by the horizontal skeleton, with each horizontal maxima line being the signature of a curve segment.

These concepts are formalized by the following definitions:

Horizontal Local Maximum Point: A point $(b_0, a_0) \in H$ is defined as a horizontal local maximum point if $\exists \varepsilon > 0 \mid M(b_0, a_0) > M(b_0, a), \forall a \in (a_0 - \varepsilon, a_0)$ and $M(b_0, a_0) \geqslant M(b_0, a), \forall a \in (a_0, a_0 + \varepsilon)$, or, alternatively, if $\exists \varepsilon > 0 \mid M(b_0, a_0) \geqslant M(b_0, a), \forall a \in (a_0 - \varepsilon, a_0)$ and $M(b_0, a_0) > M(b_0, a), \forall a \in (a_0, a_0 + \varepsilon)$.

Horizontal Local Maxima Line: A line $L_h = \{(b, a) \in H\}$ is called a horizontal local maxima line if $\forall (b_0, a_0) \in L_h$ is a horizontal local maximum point.

Horizontal Skeleton: The set of all horizontal local maxima lines, denoted as $\{L_h\}$, is called the horizontal skeleton of the w-representation.

The algorithm presented in Section 7.1.5 can be adapted to detect the horizontal skeleton of a multiscale transform. The concept of natural scales and the local frequency capabilities of the Morlet wavelet may also be explored to analyze fractal shapes. Figure 7.22 shows the w-representation (modulus, skeleton and phase) of a shape composed by two different Sierpinski curves (Figure 7.22(a)).

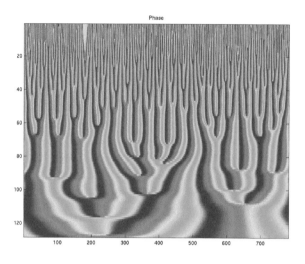

(a)

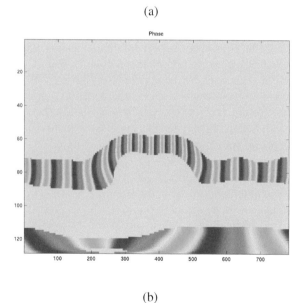

(b)

Figure 7.21: *(a) Phase information of Figure 7.20 (b); (a) Phase information of Figure 7.20 (b) thresholded by the modulus.*

7.3.7 Contour Analysis Using the Gabor Transform

The *short-time* or *windowed Fourier transform* is an alternative technique to the
wavelet transform for signal, image and shape analysis, presenting several features
similar to the wavelets, particularly the *Gabor transform*, i.e., the short-time Fourier

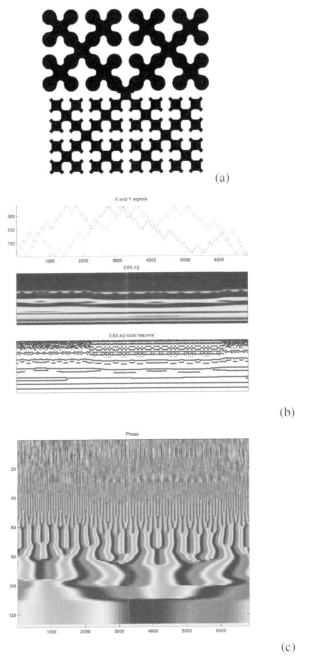

Figure 7.22: *(a) Fractal shape (the contour is extracted from the upper-left point); (b) w-representation; (c) phase information. The two different shape parts may be identified by modulus, skeleton and phase representations.*

transform with a Gaussian window (see Section 7.1.2). One of the main differences between the wavelets and the *gaborettes* (i.e., the kernel functions of the Gabor transform) is that, while the latter oscillate faster for the high frequencies, the former only change their scale, keeping the number of oscillations constant (see Figure 7.23).

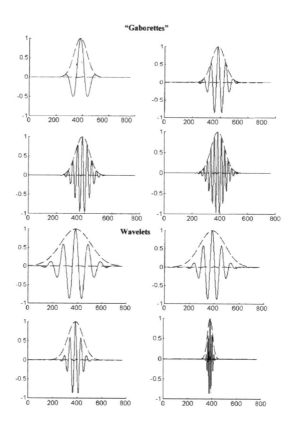

Figure 7.23: *Gaborettes and wavelets.*

An alternative to the wavelet transform that has been widely used in signal and image processing, including some problems of shape analysis, is the Gabor transform [Cesar-Jr. & Costa, 1996; Eichmann et al., 1990], which is discussed in this section. The Gabor transform $U(b, f)$ of a complex-represented contour $u(t)$ is defined as

$$U(b, f) = \int_{-\infty}^{\infty} g^*(t - b)\, u(t)\, e^{-j2\pi f t}\, dt.$$

The transform $U(b, f)$ can be thought of as a measure of the frequency energy distribution of the signal along the contour. The first and most classical work on time-frequency signal characterization was proposed by D. Gabor more than 50 years ago, when he addressed the problem of signal decomposition in terms of Gaussians and complex exponentials, a technique that presents several useful features such as the fact that it is the optimal decomposition with respect to the uncertainty principle. A brief analysis of the above shown Gabor transform expression suggests that, given a fixed analyzing frequency f_0, $U(b, f_0)$ is the response of a band-pass filter tuned to the frequency f_0. These are the so-called *Gabor filters*, which have motivated several computer vision applications such as modeling biological receptive fields, image compression, texture segmentation and perceptual grouping. Like the continuous wavelet transform, the Gabor transform is also highly redundant, for it is not an orthogonal transform.

To probe further: *Gabor Transform for Image Analysis*

The most classical works on the Gabor transform and its relation to vision are [Bastiaans, 1980; Daugman, 1980, 1985; Gabor, 1946; Marcelja, 1980]. Additional starting points on the field are [Allen & Rabiner, 1977; Bovik et al., 1990; Cohen, 1989; Daugman, 1988; Dunn & Higgins, 1995; Kruger & Sommer, 1999; Reed & Weschsler, 1990].

7.3.8 Comparing and Integrating Multiscale Representations

This section discusses some features of the multiscale contour representations (refer to [Cesar-Jr. & Costa, 1998a] for a detailed analysis). A comparison among the four representations discussed in this chapter leads to the following conclusions:

The Curvegram: the main advantage of the curvegram is the fact that it is based on the contour curvature, one of the most important features of 2D curves. It is worth noting that the curvegram generally detects the contour vertices, making a zoom as the analyzing scale decreases. Furthermore, the curvegram oscillates along textured regions for some small scales, but these oscillations vanish as the scale increases. The curvegram is more computationally expensive because its estimation is equivalent to two wavelet transforms, one for the first and one for the second signal derivative.

***w*-Representation Using the Marr Wavelet:** it can be easily noted that both the *w*-representation using the Marr wavelet and the curvegram are very similar to each other, a fact that can be explained because of the differential nature of this wavelet. This representation has two main advantages with respect to the curvegram: it is more efficient to compute and presents the properties of wavelet theory, which can be considered for the development of contour analysis algorithms. On the other hand, the superb curvature properties, that can be explored in the case of the curvegram, do not apply in the case of the *w*-representation. Corner detection can be performed both by this *w*-representation and by the curvegram, but the latter discriminates between concave and convex contour portions. Phase information may be explored in the case of the *w*-representation, which is composed of complex coefficients.

***w*-Representation Using the Morlet Wavelet:** Morlet wavelet represents an intermediate situation between the Marr wavelet-based *w*-representation and the Gabor transform of the contour. Therefore, it is still capable of zooming around the vertices (but not so explicitly as the Marr wavelet) and, at the same time, it has a stronger horizontal response when tuned to the local periodic texture.

The Gabor Transform of the Contour: comments on the previous representation also apply to the Gabor transform. Nevertheless, there is no zoom effect around the vertices, as already mentioned.

7.4 Multiscale Energies

While different multiscale shape representation techniques have been introduced in the previous sections, it is important to emphasize that the generation of a multiscale representation such as the curvegram or the *w*-representation is just the first step in many shape analysis practical problems. Another important related issue is to derive shape measures from the multiscale representations so that, for instance, statistical classifiers can be used for automatic shape classification. This section discusses how shape measures can be defined from the multiscale representations using the *multiscale energy* concept. Particularly, the *bending energy*, which is obtained from the *curvegram*, and the *multiscale wavelet energy*, which is obtained from the *w*-representation, are presented. Some results for the automatic classification of neural cells are also presented.

7.4.1 The Multiscale Bending Energy

The *bending energy*, also known as *boundary energy*, has been introduced in the context of imaging applications by I. T. Young and collaborators [Bowie & Young, 1977; Young et al., 1974] as a global shape measure for object analysis and classification. It has since been applied to several biomedical and object recognition problems, and generalizations to gray-level and 3D images have been proposed [van Vliet & Verbeeck, 1993]. Therefore, the bending energy has achieved a representative position as a classical and well-established shape descriptor. This measure has been motivated by a physical inspiration from elasticity theory, expressing the required amount of energy necessary to transform a given close contour into a circle with the same perimeter as the original contour. Consequently, this contour descriptor is invariant to translation, rotation and reflection, being easily normalized with respect to changes in scale (only 2D geometrical transformations are under consideration) and frequently used as a complexity measure. The *mean bending energy* is defined, in the discrete case, as the sum of squared curvature values along the contour, divided by the number of points, i.e.,

$$\hat{B} = \frac{1}{N} \sum_{n=0}^{N-1} k(n)^2.$$

The multiscale version of the bending energy is defined by applying the above equation to each scale of the rows in the curvegram, indexed by the scale parameter a, i.e.,

$$\hat{B}(a) = \frac{1}{N} \sum_{n=0}^{N-1} k(a, n)^2.$$

It is important to note that this definition of the multiscale bending energy is scale dependent, because the curvature actually inversely depends on the scale. This means that the above-defined bending energy is related to a circle with the same perimeter as the original contour. If scale invariance is desirable, the bending energy should be normalized, leading to the so-called *normalized multiscale bending energy* (*NMBE*):

$$B(a) = \frac{L^2}{N} \sum_{n=0}^{N-1} k(a, n)^2.$$

The energy defined by the above equation is invariant to the scale of the original curve, which is achieved by the incorporation of the L^2 factor, where L represents the contour perimeter. It is worth emphasizing that the choice of the bending energy expression depends on each specific application, which actually defines whether scale invariance is required. This is because, in some applications, the object size itself may be an important discriminating feature. In this case, the scale dependent bending energy would be preferable.

The multiscale bending energy presents a particularly suitable feature for expressing shape complexity measures, generally allowing performance superior to traditional complexity descriptors such as the squared perimeter divided by the area and the mean absolute curvature (see Section 6.2.18). Multiscale bending energy diagrams can be obtained by considering the bending energy for different values of the scale a, leading to a complexity measure associated with *each analyzing scale*. In the case of small scales, i.e., small values of a, the curvature is calculated for contours that have undergone little smoothing, consequently retaining most of their original detail. As the scale increases, such detail information vanishes, and the bending energy tends to reflect the more global structure of the contour. A detailed description of the application of the multiscale bending energy for neuromorphometry may be found in [Cesar-Jr. & Costa, 1997]. In general, the analyzing scale is associated with the following shape attributes:

☞ small scales are associated with the presence of local branching and contour roughness;

☞ intermediary scales are associated with the presence of long protrusions and branches;

☞ large scales are associated with the overall shape elongation.

These facts play a central role in shape analysis: if the shapes to be classified differ with respect to very small structures along the contour, then the bending energy should be calculated considering small scales. On the other hand, if the shapes differ because of the presence of a different number of protrusions for each shape class, then intermediate scale bending energies are good feature candidates. Finally, differences between global shape structure can be quantized by large scale bending energies.

7.4.2 | The Multiscale Wavelet Energy

This section discusses an alternative approach analogous to the bending energy, but now adopts the w-representation instead of the curvegram as the multiscale contour representation. The *normalized multiscale wavelet energy* (NMWE) describes the w-representation normalized mean energy as a function of the analyzing scale a. Therefore, the NMWE is defined as:

$$\text{NMWE}_\Psi(a) = \frac{1}{K^2} \frac{1}{L} \int_R |U_\Psi(b, a)| \, db,$$

where K is a scale normalization factor [Cesar-Jr. & Costa, 1998b]. While different scale normalization schemes can be adopted, the length of the shape's major axis (see Section 6.2.9) is henceforth adopted in this section. As with the NMBE, the NMWE is also invariant to scaling, rotation, translation, and reflection of the original shape. The discrete version of the NMWE is defined as:

$$\text{NMWE}_\Psi(a_j) = \frac{1}{K^2} \frac{1}{N} \sum_{m=1}^{N} |U_\Psi(b_m, a_j)|$$

In order to gain some insight about the way that a multiscale transform reflects and emphasizes the subjective meaning of shape complexity, refer to Figures 7.24 (a) and (b), which show two neural cells created by using artificial stochastic grammars [Costa et al., 1999].

The neuron in Figure 7.24 (b) is more complex, exhibiting more intricate, small-scale branching patterns than the neuron in Figure 7.24 (a).

The modulus of the respective w-representations can be visualized in Figures 7.25 (a) and (b), indicating that the branching points can be perceived in the w-representation as singularities of the signals.

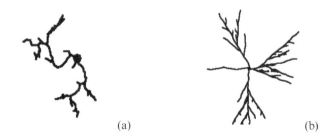

(a)   (b)

Figure 7.24: *Two neural cells with simpler (a) and more complex (b) morphology.*

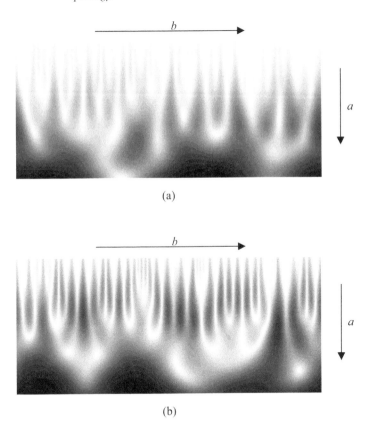

Figure 7.25: *The respective magnitudes of the w-representations of the cells in Figures 7.24 (a) and (b).*

Therefore, the fact that the cell of Figure 7.24 (b) is more complex than that
of Figure 7.24 (a) with respect to small-scale branching can be observed in the w-
representation because of the presence of new structures as the scale decreases, as
it can be noted from Figure 7.25 (b). The NMWE attempts to capture these shape
features through a global analysis, and successful application of it to neural cells
classification is discussed in Section 8.5.1.

To probe further: *Multiscale Energies*

The bending energy has been introduced as a shape descriptor by Young and collaborators in [Young et al., 1974], and has since been widely used as a complexity
measure [Bowie & Young, 1977; Castleman, 1996; Duncan et al., 1991; Marshall,
1989; Pavlidis, 1980; van Vliet, 1993; van Vliet & Verbeeck, 1993]. An early multiscale version has been applied by Pernus et al. [1994] and been generalized and
extensively assessed in [Cesar-Jr. & Costa, 1997, 1998a]. The wavelet energy has
been introduced in the context of shape analysis in [Cesar-Jr. & Costa, 1998a]. The
issue of feature extraction from the wavelet transform is discussed in [Pittner &
Kamarthi, 1999].

... classification is, at base, the task of recovering the
model that generated the patterns...

Duda et al. [2000]

8

Shape Recognition

Chapter Overview

THIS CHAPTER ADDRESSES the particularly relevant issues of pattern recognition, concentrating on the perspective of shape classification. The common element characterizing each of these approaches is the task of assigning a specific class to some observed individual, with basis on a selected set of measures. One of the most important steps in pattern recognition is the selection of an appropriate set of features with good discriminative power. Such measures can then be normalized by using statistical transformations, and fed to a classification algorithm. Despite the decades of intense research in pattern recognition, there are no definitive and general solutions to choosing the optimal features and obtaining an optimal classification algorithm. Two main types of classification approaches are usually identified in the literature: supervised and unsupervised, which are characterized by the availability of prototype objects. In this chapter we present, discuss and illustrate two techniques representative of each of these two main types of classification approaches.

8.1 Introduction to Shape Classification

This section presents basic notions in classification that are essential for the proper understanding of the rest of this chapter, as well as for practical applications of shape classification. The first important fact to be noted is that classification is a general, broad and not completely developed area, in such a way that shape classification is but a specific case where the objects to be classified are limited to shapes. It is important to realize that many important contributions to the theory of pattern classification have been made by the most diverse areas, from biology to human sciences, in such a way that the related literature has become particularly vast and eclectic. Since the concepts and results obtained by classification approaches with respect to a specific area can often be immediately translated to other areas, including shape analysis, any reader interested in the fascinating issue of pattern classification should be prepared to consider a broad variety of perspectives.

8.1.1 | The Importance of Classification

In its most general context, to classify means to *assign* classes or categories to items according to their properties. As a brief visit to any supermarket or shop will immediately show, humans like to keep together items which, in some sense, belong together: trousers are stored close to shirts, tomatoes close to lettuces, and oranges close to watermelons. But our passion for classifying goes even further, extending to personal objects, hobbies, behavior, people and, of course, science and technology. As a matter of fact, by now humans have classified almost all known living species and materials on earth. Remarkably, our very brain and thoughts are themselves inherently related to classification and association, since not only are the neurons and memories performing similar operations packed together in the brain (proximity is an important element in classification), but our own flow of thoughts strongly relies on associations and categorizations. Interestingly, the compulsion for collecting and organizing things (e.g., postal stamps, banners, cars, etc.) exhibited by so many humans is very likely a consequence of the inherent role classification plays in our lives. For all that has been said above, the first important conclusion about classification therefore is:

> 1: **To classify** *is human.*

But what are the reasons that make classification so ubiquitous and important? One of the most immediate benefits of classifying things is that each obtained class usually subsumes and emphasizes some of the main general properties shared by its elements. For instance, all items in the clothes section of any store will have some value for dressing, even if they are completely different as far as other properties, such as color, are concerned. Such a property can be more effectively explored through *hierarchical classification schemes*, such as those commonly considered for species taxonomy: we humans, for instance, are first living beings, then animals, mammals and primates, as illustrated in Figure 8.1.

The beauty of such hierarchical classifications is that, by subsuming and unifying the description of common characteristics into the superior hierarchies, they allow substantial savings in the description and representation of the involved objects, while clearly identifying relationships between such entities. In other words, every subclass inherits the properties of the respective superclass. For instance, to characterize humans, it is enough to say they are primates, then mention only those human characteristics that are not already shared by primates. In this specific case, we could roughly say humans are like apes in the sense of having two legs but are

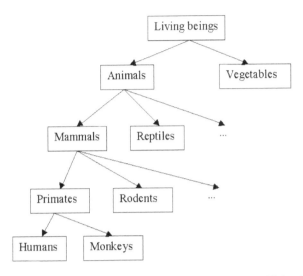

Figure 8.1: *A (rather simplified) hierarchical classification of living be-ings.*

less hairy and (allegedly) more intelligent beings. Such a redundancy reduction ac-counts for one of the main reasons why our own brain and thoughts are so closely related to categorizations. Whenever we need to make sense of some new concept (let us say, a pink elephant), all we need to do is to find the most similar concept (i.e., elephant), and to add the new uncommon features (i.e., pink). And hence we have verified another important fact about classification:

> 2: **To classify** *removes redundancy.*

Observe that humans have always tended to classify in greater accuracy those entities that present higher survival value. For instance, in pre-historic days humans had to develop a complete and detailed classification of those fruits and animals that were edible or not. Nowadays, human interest is being focused on other tasks, such as trying to classify businesses and stocks according to their potential for profits. It should also be observed that, since classification is so dear to humans, its study cannot only lead to the substitution of humans in repetitive and/or dangerous tasks, but also provide a better understanding about an important portion of our own essence.

8.1.2 Some Basic Concepts in Classification

One of the most generic definitions of classification, adopted henceforth in this book, is:

> 3: **To classify** *is the act of assigning objects to classes.*

As already observed, this is one of the most inherently human activities, which is often performed in a subjective fashion. For instance, although humans have a good agreement on classifying facial emotions (e.g., to be glad, sad, angry, pensive, etc.), it is virtually impossible to clearly state what are the criteria subjectively adopted for such classifications. Indeed, while there is no doubt that our brains are endowed with sophisticated and effective algorithms for classification, little is known about what they are or what features they take into account. Since most categories in our universe have been defined by humans using subjective criteria, one of the greatest challenges in automated classification resides precisely in trying to select features and devise classification algorithms compatible with those implemented by humans. Indeed, powerful as they certainly are, such human classification algorithms are prone to biasing and errors (e.g., the many human prejudices), in such a way that automated classification can often supply new perspectives and corrections to human attitudes. In this sense, the study of automated classification can produce important lessons regarding not only how humans conceptualize the world, but also for revising some of our misconceptions.

However, it should be borne in mind that classification does not always need to suit the human perspective. Indeed, especially in science, situations arise where completely objective specific criteria can be defined. For instance, a mathematician will be interested in classifying matrices as being invertible or not, which is a completely objective criterion leading to a fully precise respective classification of matrices. More generally speaking, the following three main situations are usually found in general pattern classification:

Imposed criteria: the criteria are dictated by the specific practical problem. For instance, one might be interested in classifying as mature all those chickens whose weight exceeds a specific threshold. Since the criteria are clearly stated from the outset, all that remains to be done is to implement suitable and effective means for measuring the features. Consequently, this is the easiest situation in classification.

By example (or supervised classification): one or more examples, known as *training set*, of each previously known class of objects are provided as prototypes for classifying additional objects. For instance, one can be asked to develop a strategy for classifying people as likely movie stars (or not) by taking into account their similarity with a set of specific prototypes, such as Clark Gable, Clint Eastwood and Mel Gibson. Such a problem is usually more difficult than classification by imposed criteria, since the features to be considered are not evident and typically are not specified (if not, what makes somebody similar to Gable?). However, the discriminative power of each possible feature can be immediately verified by applying them to the supplied prototypes. Such a type of classification usually involves two stages: (i) *learning*, corresponding to the stage where the criteria and methods are tried on the prototypes; and (ii) *recognition*, when the trained system is used to classify new entities.

Open criteria (or unsupervised classification): you are given a set of objects and asked to find adequate classes, but no specific prototypes or suggested features and criteria are available. This is the situation met by taxonomists while trying to make sense of the large variety of living beings, and by babies while starting to make sense of the surrounding world. Indeed, the search for classification criteria and suitable features in such problems characterize a process of discovery through which new concepts are created and relationships between these objects are identified. When the adopted classification scheme consists of trying to obtain classes in such a way as to maximize the similarity between the objects in each class and minimize the similarity between objects in different classes, unsupervised classification is normally called *clustering*, and each obtained group of objects a *cluster*. As the reader may well have anticipated, unsupervised classification is usually much more difficult than supervised classification.

It is clear from the above situations that classification is always performed with respect to some *properties* (also called *features, characteristics, measurements, attributes* and *scores*) of the *objects* (also called *subjects, cases, samples, data units, observations, events, individuals, entities* and *OTUs*—operational taxonomic units). Indeed, the fact that objects present the same property defines an equivalence relation partitioning the object space. In this sense, a sensible classification operates in such a way as to group together into the same class things that share some properties, while distinct classes are assigned to things with distinct properties. Since the features of each object can often be quantified, yielding a feature vector in the respective feature space, the process of classification can be understood as organizing the feature space into a series of classes, i.e.,

> 4: **To classify** *is to organize a feature space into regions corresponding to the several classes.*

In addition to producing relevant classes, it is often expected that a good classification approach can be applied to treat additional objects without requiring a new training stage. This property is known as *generalization*. It is also important to notice that the organization of the feature space into regions corresponding to classes can eventually produce disconnected or overlapping regions. Let us consider the above concepts and possibilities in terms of the example discussed in the following section.

8.1.3 A Simple Case Study in Classification

Suppose that a bakery packs vanilla cookies into square boxes, and chocolate cookies into circular boxes, and that an automated scheme has to be devised to allow the identification of the two types of box, in order to speed up the storage and distribution. The squares are known to have side a, $0.7 < a \leqslant 1.5$ (arbitrary units),

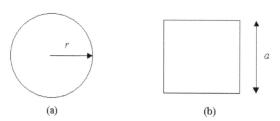

(a)   (b)

Figure 8.2: *Top view of the two types of cookie boxes: circular (a) and square (b).*

and the circles radius r, $0.5 < r \leqslant 1$ (arbitrary units), as illustrated in Figure 8.2. A straightforward possibility for classification of the boxes consists of using as features the perimeter (P) and area (A) of the shapes, indicated in Table 8.1 defining an (Area × Perimeter) feature space.

	Circle	**Square**
Perimeter(P)	$P(r) = 2\pi r$	$P(a) = 4a$
Area(A)	$A(r) = \pi r^2$	$A(a) = a^2$

Table 8.1: *The perimeter and area of square and circular regions.*

It is clear that each circular region with radius r will be mapped into the feature vector $\vec{F}(r) = (P(r), A(r))$, and each square with side a will be mapped into $\vec{F}(a) = (P(a), A(a))$. Observe that these points define the respective parametric curves $\vec{F}(r) = \left(2\pi r, \pi r^2\right)$ and $\vec{F}(a) = \left(4a, a^2\right)$. In addition, since $P = 2\pi r \Rightarrow r = \frac{P}{2\pi}$, we have $A = \pi r^2 = \pi \left(\frac{P}{2\pi}\right)^2 = \frac{P^2}{4\pi}$, indicating that the feature vectors corresponding to the circular regions are continuously distributed along a parabola. Similarly, since $P = 4a \Rightarrow a = \frac{P}{4}$, we have $A = a^2 = \left(\frac{P}{4}\right)^2 = \frac{P^2}{16}$, implying that the feature points corresponding to squares are also distributed along a continuous parabola. Figure 8.3 illustrates both of these parabolas.

Clearly the points defined by squares and circles in the (Area × Perimeter) feature space correspond to two segments of parabolas that will never intercept each other, once r and a are always different from 0. As a matter of fact, an intersection would theoretically only take place at the point $(0, 0)$. This geometric distribution of the feature points along the feature space suggests that a straightforward classification procedure is to verify whether the feature vector falls over any of these two parabolas. However, since there are no perfect circles or squares in the real world, but only approximated and distorted versions, the feature points are not guaranteed to fall exactly over one of the two parabolas, a situation that is illustrated in Figure 8.4.

A possible means for addressing such a problem is to use a third parabola $A = \frac{P^2}{k}$, where $4\pi < k < 16$ and $P > 0$, to separate the feature space into two

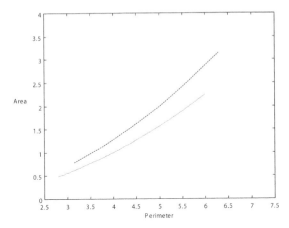

Figure 8.3: *Position of the feature points defined by circles (dashed) and squares (dotted).*

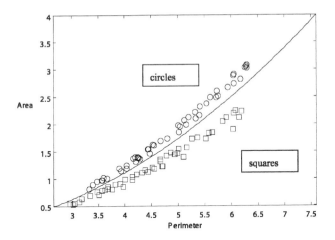

Figure 8.4: *Possible mapping of real circles (represented by circles) and squares (represented by squares) in the* (Area × Perimeter) *feature space. One of the possible decision boundaries, namely* $A = \frac{P^2}{3.8^2}$, *is also shown as the intermediate solid curve.*

regions, as shown in Figure 8.4 with respect to $A = \frac{P^2}{3.8^2}$. Such separating curves are traditionally called *decision boundaries*. Now, points falling to the left (right) of the separating parabola are classified as circles (squares). Observe, however, that these two semi-planes are not limited to squares and circles, in the sense that other shapes will produce feature vectors falling away from the two main parabolas, but still contained in one of the two semi-planes. Although this binary partition of the feature space is fine in a situation where only circles and squares are considered, additional partitions may become necessary whenever additional shapes are also presented as input.

In case the dispersion is too large, as illustrated in Figure 8.5, it could become impossible to find a parabola that properly partitions the space. This by no means

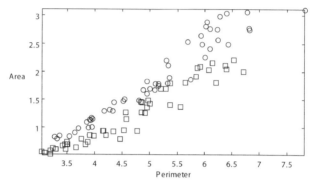

Figure 8.5: *Mapping of substantially distorted circles and squares in the (Area×Perimeter) feature space. No parabola can be found that will perfectly separate this feature space.*

implies that no other curve exists which allows such a clear-cut separation. Indeed, it can be proven (see [Duda et al., 2000], for example), that it is always possible to find a decision region, however intricate, perfectly separating the two classes.

But let us now go back to the ideal situation illustrated in Figure 8.3. Since the feature points corresponding to each of the two classes fall along parabolas, it is clear that a logarithmic transformation of both features, i.e., $\vec{F}(r) = (\log(P(r)), \log(A(r))$ and $\vec{F}(a) = (\log(P(a)), \log(A(a)))$, will produce straight line segments in such a transformed parameter space, as shown in Figure 8.6. Now, such a loglog feature space allows us to define a separating straight line instead of a parabola, as illustrated in Figure 8.6.

While proper classification is possible by using two features, namely area and perimeter, it is always interesting to consider if a smaller number of features, in this case a single one, could produce similar results. A particularly promising possibility would be to use the relation $C = \frac{\text{Area}}{\text{Perimeter}^2}$, a dimensionless measure commonly called *thinness ratio* in the literature (see Section 6.2.18 for additional information about this interesting feature). Circles and squares have the following thinness ratios:

$$C\,(\text{circle}) = \frac{A}{P^2} = \frac{\pi r^2}{4\pi^2 r^2} = \frac{1}{4\pi} \quad \text{and}$$

$$C\,(\text{square}) = \frac{A}{P^2} = \frac{a^2}{16a^2} = \frac{1}{16}.$$

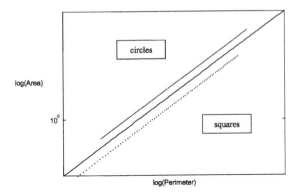

Figure 8.6: *The loglog version of the feature space in Figure 8.3, and one of the possible decision boundaries (solid straight line).*

Observe that, by being dimensionless, this feature does not vary with r or a, in such a way that any perfect circle will be mapped into exactly the same feature value $F = \frac{1}{4\pi}$, while squares are mapped into $F = \frac{1}{16}$. In such a reduced feature space, it is enough to compare the value of the measured thinness ratio with a predefined threshold $\frac{1}{16} < T < \frac{1}{4\pi}$ in order to assign the class to each respective shape. However, since some dispersion is expected in practice because of imperfections in the objects and measuring process, the mapping into the thinness ratio feature space will not be limited to the points $F = \frac{1}{4\pi}$ and $F = \frac{1}{16}$, but to clouds around these points. Figure 8.7 presents the one-dimensional feature space obtained for the same situation illustrated in Figure 8.4.

Let us now assume that a special type of cookie was produced during the holiday season and packed into both square with $1.3 < a \leqslant 1.5$ and circular boxes with $0.8 < r \leqslant 1$. The first important thing to note regarding this new situation is that the single feature approach involving only the thinness ratio measure is no longer suitable because the special and traditional cookie box sizes overlap each other. Figure 8.8 presents the two segments of parabolas corresponding to such boxes superimposed onto the previous two parabola segments corresponding to the circular and square boxes. It is clear that a disconnected region of the feature space has been defined by the boxes containing special cookies. In addition, this new class also presents overlapping with substantial portions of both parabola segments defined by the previous classes (square and circular boxes), in such a way that we can no longer identify for certain if boxes falling over these regions contain vanilla (i.e., square boxes), chocolate (i.e., circular boxes) or special cookies (both square and circular boxes, but at specific range sizes).

The above two problems, namely the *disconnected regions* in the feature space and the *overlapping regions* related to different classes, have distinct causes. In

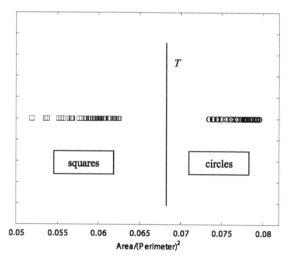

Figure 8.7: *Distribution of feature vectors in the one-dimensional feature space corresponding to the thinness ratio measure, and a possible threshold T allowing perfect classification of the shapes. Square boxes are represented as squares and circular boxes as circles.*

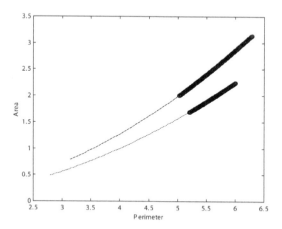

Figure 8.8: *The set of feature points corresponding to the special cookie packs (thin lines) are parabola segments overlapping both the circular (dashed) and square (dotted) parabolas. Compare with Figure 8.3.*

the first case, the problem was the arbitrary decision of using such different boxes for the same type of cookie. The second problem, namely the *overlap between distinct classes*, is a direct consequence of the fact that the considered features (i.e., area and perimeter) are not enough for distinguishing among the three classes of cookie boxes. In other words, the classes cannot be bijectively represented in the Area × Perimeter feature space, since distinct objects will be mapped into the same feature points. Although it would still be possible to classify a great deal of the cookie boxes correctly, there will be situations (i.e., larger sizes) where two classes would be possible. For instance, the upper emphasized region in Figure 8.8 could correspond to both chocolate and special cookie boxes, while the lower emphasized region can be understood as securely indicating both vanilla and special cookie boxes. This problem can be addressed by incorporating additional discriminative information into the feature vector. For instance, in case the boxes used for special cookies are known to have width 0.2 (arbitrary units), while the traditional boxes have width 0.1 (arbitrary units), a third feature indicating the width could be used, thus producing a feature space similar to that shown in Figure 8.9. Observe that, provided the dispersion of the width measures is not too high, the three classes of cookie boxes can now be clearly distinguished in this enhanced feature space. On the other hand, i.e., in case there are no additional features distinguishing between the special and traditional cookie boxes, it will not be possible to remove the overlap. Indeed, such situations are sometimes verified in the real world as a consequence of arbitrary and subjective definitions of classes and incomplete information about the analyzed objects.

As a final possibility regarding the cookie box example, consider that, for some odd reason, the bakery packs chocolate cookies in circular boxes and vanilla cookies in square boxes from June to December but, during the rest of the year, uses square boxes for chocolate cookies and circular boxes for vanilla cookies. In such case, the only way for properly identifying the product (i.e., type of cookies) is to take into account, as an additional feature, the production time. Such situations make it clear, as indicated in quotation at the beginning of this chapter, that to classify means to understand and take into account as much information as possible about the processes generating the objects.

8.1.4 Some Additional Concepts in Classification

Although extremely simple, the above example allowed us not only to illustrate the general approach to classification, involving the selection of features and partitioning of the feature space, but also characterize some important situations often found in practice. The above case study shows that even simple situations demand special care in defining suitable features and interpreting the dispersions of feature vectors in the feature space. At the same time, it has shown that arbitrary or subjective definition of classes, as well as incomplete information about the objects, can make the classification much more difficult or even impossible. In brief, it should be clear by now that:

> 5: **To classify** *is not easy.*

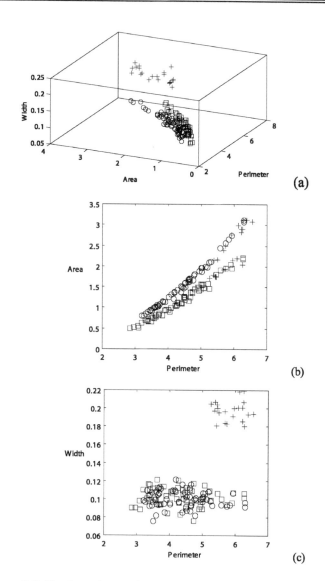

Figure 8.9: *The three classes of cookie boxes become perfectly distinguishable in the augmented feature space* (Width × Area × Perimeter), *shown in* **(a)**. *The projections in* **(b)** *and* **(c)** *help illustrate the separation of the classes.*

Yet, there are several important issues that have not been covered in the previous example and should normally be considered in practice. Although it is not practical to consider all the possible properties of objects while performing classifications, care has to be invested in identifying particularly useful specific features. For instance, in the previous example in Section 8.1.1 trousers and shirts were understood as presenting the property of being *clothes*, tomatoes and lettuces as being *vegetables*, and oranges and watermelons as being *fruits*.

Since we can have several alternative criteria for classifying the same set of objects, *several distinct classifications can be obtained*. If we considered as criterion "to be green," watermelons would be classified together with green shirts. In brief, the adopted criterion determines the obtained classes. However, it should be observed that the application of some criteria might not always be as clear-cut as we would wish. The criterion "to be green," for instance, may be difficult to implement in a completely objective way by humans—a watermelon looking green to one person may be perceived as yellowish by another. Even if computer-based color analysis were used, such a decision would still depend on one or more *thresholds* (see Section 3.4.1). It is also clear that the application of any such criteria operates over some set of object *features* (e.g., color, size, shape, taste, etc.), which have to be somehow accurately *measured* or *estimated*, a process that often involves parameters, such as the illumination while acquiring the images, which in turn can substantially affect the color of the objects. The difficulties in classification are aggravated by the fact that there is no definitive procedure exactly prescribing what features should be used in each specific case. Indeed, except for a few general basic guidelines presented in Section 8.1.5, *classification is not a completely objective procedure*. This is an extremely important fact that should be borne in mind at all times, being emphasized as:

> 6: **To classify** *involves choosing amongst several features, distance, classification criteria, and parameters, each choice leading to possibly different classifications. There are no exact rules indicating how to make the best choices.*

One additional reason why classification is so inherently difficult is the number of implied possibilities in which p objects can be classified into q classes. Table 8.2 illustrates this number for $p = 30$ objects and $q = 1, 2, 3, 4,$ and 5 classes [Anderberg, 1973]. Observe that the magnitude of such values is so surprisingly large that

Number of classes	Number of possibilities
1	1
2	536870911
3	3.4315e13
4	4.8004e16
5	7.7130e18

Table 8.2: *The number of possibilities for arranging $p = 30$ objects into $q = 1, 2, 3, 4,$ and 5 classes.*

the complete exploration of all possible classifications of such objects is completely beyond human, or even computer, capabilities.

Figure 8.10 summarizes the three basic steps (in italics) involved in the traditional pattern classification approach.

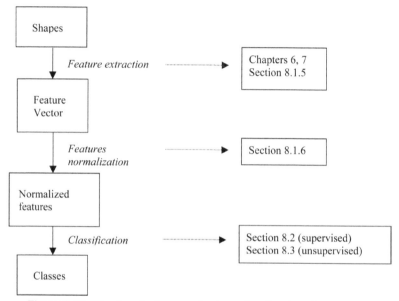

Figure 8.10: *The three basic stages in shape classification: feature extraction, feature normalization and classification.*

The process initiates with the *extraction* of some features from the shape, and follows by possibly *normalizing* such features, which can be done by transforming the features in such a way as to have zero mean and unit variance (see Section 2.6.2), or by using principal component analysis (see Section 8.1.6). Finally, the normalized features are used as input to some suitable classification algorithm. These fundamental stages in shape classification are discussed in more detail in the following sections.

Observe that a fourth important step should often be considered in shape classification, namely the *validation* of the obtained results. Since there are no closed solutions to classification, the obtained solutions may not correspond to an adequate solution of the problem, or present some specific unwanted behavior. Therefore, it is important to invest some efforts in order to verify the quality and generality of the obtained classification scheme. More detail about these important issues can be found in Sections 8.3.4, 8.4 and 8.5.

Before addressing the issues in more detail, Table 8.3 presents the classification related abbreviation conventions henceforth adopted in this book, and the accompanying box provides an example of their usage.

Total number of objects	N
Number of features	M
Number of classes	K
Number of objects in class C_p	N_p
Data (or feature) matrix containing all the features (represented along columns) of all the objects (represented along rows)	F
Feature vector representing an object p, corresponding to the transposed version of the p-th row of matrix F	$\vec{f}_p$
Data (or feature) matrix containing all objects in class C_p. The features are represented along columns and the objects along rows.	F_p
Mean feature vector for objects in class C_p	$\vec{\mu}_p = \frac{1}{N_p} \sum_{i \in C_k} \vec{f}_i$
Global mean feature vector (considering all objects)	$\vec{M} = \frac{1}{N} \sum_{i=1}^{N} \vec{f}_i$

Table 8.3: *Main adopted abbreviations related to classification issues.*

Example: *Classification Representation Conventions*

The following table includes seven objects and their specific classes and features. Represent this information in terms of the abbreviations in Table 8.3.

Object #	Class	Feature 1	Feature 2
1	C_3	9.2	33.2
2	C_2	5.3	21.4
3	C_3	8.8	31.9
4	C_1	2.9	12.7
5	C_3	9.0	32.4
6	C_1	1.5	12.0
7	C_1	1.2	11.5

Solution:

We clearly have $N = 7$ objects, organized into $K = 3$ classes, characterized in terms of $M = 2$ features. Class C_1 contains $N_1 = 3$ objects, C_2 contains $N_2 = 1$

object, and C_3 contains $N_3 = 3$ objects. The data (or feature) matrix is

$$F = \begin{bmatrix} 9.2 & 33.2 \\ 5.3 & 21.4 \\ 8.8 & 31.9 \\ 2.9 & 12.7 \\ 9.0 & 32.4 \\ 1.5 & 12.0 \\ 1.2 & 11.5 \end{bmatrix}$$

The involved feature vectors, corresponding to each row of F, are

$$\vec{f_1} = \begin{bmatrix} 9.2 \\ 33.2 \end{bmatrix}; \qquad \vec{f_2} = \begin{bmatrix} 5.3 \\ 21.4 \end{bmatrix}; \qquad \vec{f_3} = \begin{bmatrix} 8.8 \\ 31.9 \end{bmatrix}; \qquad \vec{f_4} = \begin{bmatrix} 2.9 \\ 12.7 \end{bmatrix};$$

$$\vec{f_5} = \begin{bmatrix} 9.0 \\ 32.4 \end{bmatrix}; \qquad \vec{f_6} = \begin{bmatrix} 1.5 \\ 12.0 \end{bmatrix}; \qquad \vec{f_7} = \begin{bmatrix} 1.2 \\ 11.5 \end{bmatrix}$$

The global mean vector is

$$\vec{M} = \begin{bmatrix} 5.4143 \\ 22.1571 \end{bmatrix}$$

The matrices representing each class follow (observe that these matrices are not unique, in the sense that any matrix containing the same rows will also represent each class):

$$F_1 = \begin{bmatrix} 2.9 & 12.7 \\ 1.5 & 12.0 \\ 1.2 & 11.5 \end{bmatrix}; \qquad F_2 = \begin{bmatrix} 5.3 & 21.4 \end{bmatrix}; \qquad F_3 = \begin{bmatrix} 9.2 & 33.2 \\ 8.8 & 31.9 \\ 9.0 & 32.4 \end{bmatrix}.$$

The respective mean feature vectors are

$$\vec{\mu_1} = \begin{bmatrix} 1.8667 \\ 12.0667 \end{bmatrix}; \qquad \vec{\mu_2} = \begin{bmatrix} 5.3 \\ 21.4 \end{bmatrix}; \qquad \vec{\mu_3} = \begin{bmatrix} 9.0 \\ 32.5 \end{bmatrix}$$

8.1.5 Feature Extraction

The feature extraction problem involves the following three important issues: (a) how to *organize and visualize the features*; (b) *what features* to extract; and (c) *how to measure* the selected features from the objects. Issues (a) and (b) are discussed in this section, while the equally important problem of feature measurement is addressed in Chapters 6 and 7. It is observed that, although several types of features are often defined in the related literature [Anderberg, 1973; Romesburg, 1990], the present book is mostly constrained to real features, i.e., features whose values extend along a real interval.

Feature Organization and Visualization

Since typical data analysis problems involve many observations, as well as a good number of respective features, it is important to organize such data in a sensible way before it can be presented and analyzed by humans and machines. One of the most traditional approaches, and the one adopted henceforth, consists of using a table where the objects are represented in rows, and the respective features in columns. Table 8.4 illustrates this kind of organization for a hypothetical situation.

Object	Feature 1 (area in cm^2)	Feature 2 (volume in cm^3)
1	32.67	68.48
2	28.30	63.91
3	24.99	71.95
4	26.07	59.36
5	31.92	70.33
6	31.32	68.40
7	25.14	81.00

Table 8.4: *Tabular organization of objects and respective features.*

Although this tabular representation can provide a reasonable general view of data inter-relationship for a small number of objects and features, it becomes virtually impossible to make sense of larger data sets. This is precisely the point where computational assistance comes into play. However, if data is to be analyzed by computers, it must be stored in some adequate format. This can be achieved naturally by representing the data tables as matrices. For instance, the data in Table 8.4 can be represented in terms of the following matrix F, whose rows and columns correspond to the objects and features, respectively:

$$
F = \begin{bmatrix}
32.67 & 68.48 \\
28.30 & 63.91 \\
24.99 & 71.95 \\
26.07 & 59.36 \\
31.92 & 70.33 \\
31.32 & 68.40 \\
25.14 & 81.00
\end{bmatrix} =
\begin{bmatrix}
\leftarrow \vec{f}_1^T \rightarrow \\
\leftarrow \vec{f}_2^T \rightarrow \\
\leftarrow \vec{f}_3^T \rightarrow \\
\leftarrow \vec{f}_4^T \rightarrow \\
\leftarrow \vec{f}_5^T \rightarrow \\
\leftarrow \vec{f}_6^T \rightarrow \\
\leftarrow \vec{f}_7^T \rightarrow
\end{bmatrix}
$$

Observe that the vectors obtained by transposing each row in such matrices correspond to the respective *feature vectors*. Thus, the seven feature vectors corresponding to the seven objects in the above example are as presented below:

$$\vec{f_1} = \begin{bmatrix} 32.67 \\ 68.48 \end{bmatrix}; \qquad \vec{f_2} = \begin{bmatrix} 28.30 \\ 63.91 \end{bmatrix}; \qquad \vec{f_3} = \begin{bmatrix} 24.99 \\ 71.95 \end{bmatrix}; \qquad \vec{f_4} = \begin{bmatrix} 26.07 \\ 59.36 \end{bmatrix};$$

$$\vec{f_5} = \begin{bmatrix} 31.92 \\ 70.33 \end{bmatrix}; \qquad \vec{f_6} = \begin{bmatrix} 31.32 \\ 68.40 \end{bmatrix}; \qquad \vec{f_7} = \begin{bmatrix} 25.14 \\ 81.00 \end{bmatrix}$$

Another important possibility to be considered as a means of providing a first contact with the measured features requires their proper visualization, in such a way that the proximity and distribution of the objects throughout the feature space become as explicit as possible. An example of such a visualization is provided in Figure 8.11 (a), where each of the objects in Table 8.4 has been represented by a small circle. Note that the axes are in the same scale, thus avoiding distance dis-

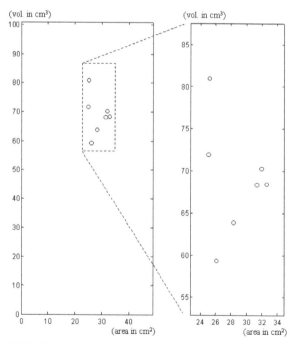

Figure 8.11: *Two possible visualizations of the data in Table 8.4: **(a)** by including the origin of coordinates (absolute visualization) and **(b)** by zooming at the region containing the objects (relative visualization). The axes are presented at the same scale in both situations.*

tortions. This type of visual presentation including the coordinate system origin provides a clear characterization of the absolute value of the considered features, and is henceforth called *absolute visualization*. Figure 8.11 (b) illustrates the possibility of windowing (or zooming) the region of interest in the feature space, in order to allow a more detailed representation of the relative position of the objects represented in the feature space; this possibility is henceforth called *relative visualization*.

While the utility of visualization becomes evident from the above examples, it is unfortunately restricted to situations involving a small number of features, generally up to a maximum number of three, since humans cannot see more than three dimensions. However, research efforts are being directed at trying to achieve suitable visualizations of higher dimensional spaces, such as by projecting the points into 1, 2- or 3-dimensional spaces.

Feature Selection

As we have verified in Section 8.1.3, the choice of features is particularly critical, since it can greatly impact the final classification result. Indeed, the process of selecting suitable features has often been identified [Ripley, 1996] as being even more critical than the classification algorithms. Although no definitive rules are available for defining what features to use in each specific situation, there are a few general guidelines that can help such a process, including

① Look for *highly discriminative* features regarding the objects under consideration. For instance, in case we want to classify cats and lions, size or weight are good features, but color or presence of whiskers are not. Observe that previous knowledge about the objects can be highly valuable.

② *Avoid highly correlated features.* Correlated measures tend to be redundant, implying additional computing resources (time and storage). However, there are cases, such as those illustrated in Section 8.1.6, in which even correlated features can prove to be decisive for effective classification.

③ *Keep the number of features as small as possible.* In addition to implying higher computational costs, a large number of features make visual and automated explorations of the feature space more difficult, and also demands more effective and powerful classification schemes. Indeed, if an excessive number of features is used, the similarity among the objects calculated in terms of such features tend to become smaller, implying less discriminative power.

④ Frequently, but not always, it is interesting to consider features that are *invariant* to specific geometric transformations such as rotation and scaling. More specifically, in case shape variations caused by specific transformations are to be understood as similar, it is important to identify the involved transformations and to consider features that are invariant to them (see Section 4.9).

⑤ Use features that can be *measured objectively by methods not involving too many parameters*. We have already seen in the previous chapters that most of the algorithms for image and shape analysis involve several parameters, many of which are relatively difficult to be tuned to each specific case. The consideration of features involving such parameters will complicate the classification process. In case such sensible parameters cannot be avoided, extreme care must be taken in trying to find suitable parameter configurations leading to appropriate classifications, a task that can be supported by using data mining concepts.

⑥ The choice of adequate features becomes more natural and simple as the user gets progressively more *acquainted and experienced* with the classification area and specific problems. Before you start programming, search for previous related approaches in the literature, and learn from them. In addition, get as familiar as possible with the objects to be classified and their more representative and inherent features. Particularly in the case of shape analysis, it is important to carefully visualize and inspect the shapes to be classified. Try to identify what are the features you naturally and subjectively would use to separate the objects into classes—do not forget humans are expert classifiers.

⑦ Dedicate special attention to those objects that do not seem to be typical of their respective classes, henceforth called *outliers*, since they often cause damaging effects during classification, including overlapping in the feature space. Try to identify which of their properties agree with those from other objects in their class, and which make them atypical. It is also important to take special care with outliers that are particularly similar to objects in other classes. In case such objects are supplied as prototypes, consider the possibility of them having been originally misclassified.

⑧ Get acquainted with the largest number of possible features, their discriminative power and respective computational cost. Chapters 6 and 7 of this book are dedicated to presenting and discussing a broad variety of shape features.

Let us illustrate the above concepts with respect to a real example pertaining to the classification of four species of plants by taking into account images of their leaves, which are illustrated in Figure 8.12.

Observe that each class corresponds to a row in this figure. We start by visually inspecting the leaves in each class trying to identify specific features with higher discriminative power. We immediately notice that the leaves in class 1 tend to be more elongated than the others, and that the leaves in class 4 tend to exhibit two sharp vertices at their middle height. These two features seem to be unique to the respective two classes, exhibiting good potential for their recognition. In other words, there is a chance that most entities from class 1 can be immediately set apart from the others based only on the elongation (and similarly for class 4 regarding the two sharp vertices). On the other hand, the leaves in classes 2 and 3 exhibit rather similar shapes, except for the fact that the leaves in class 3 tend to present a more acute angle at both extremities. However, the leaves in class

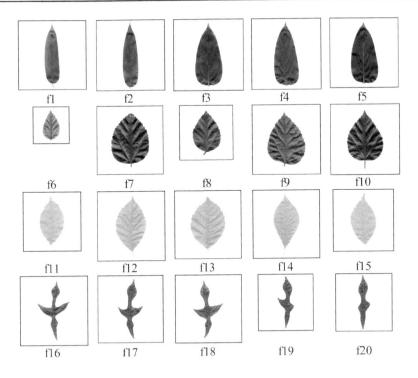

Figure 8.12: *Five examples of each of the four classes of leaves. The classes are shown as rows.*

2 are substantially darker than those in class 3, a feature that is very likely to be decisive for the separation of these two classes. In brief, a first visual inspection suggests the classification criteria and choice of features (in bold italics) illustrated in Figure 8.13.

Such a structure is normally called a *decision tree* (see, for example, [Duda et al., 2000]). Observe that such a simple initial inspection has allowed a relatively small number of features to be considered. In addition, the three selected features are easily verified not to be at all correlated, since they have completely different natures (one is related to the object's gray levels, the other to the presence of local vertices and the third to the overall distribution of the shape).

Although such simple preliminary criteria will very likely allow us to correctly classify most leaves, they will almost certainly fail to correctly classify some outliers. For instance, leaves f_3 and f_5 are not particularly elongated, and may be confused with leaves f_6, f_{11}, f_{14} and f_{15}. On the other hand, leaf f_6 has a particularly fair interior and consequently can be confused as belonging to class 3. In class 4, leaf f_{19} has only one sharp vertex at its middle height, and leaf f_{20} does not have any middle sharp vertices. In addition, leaf f_{19}, and particularly f_{20}, are more elongated than the others in their class, which may lead to a subpartition of this

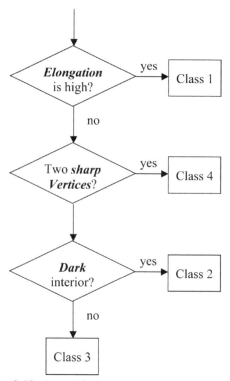

Figure 8.13: *A possible decision tree for the leaves example.*

class in case clustering approaches are used (see Section 8.3 for a more detailed discussion of this problem). The way out of such problems is to consider additional features, such as local curvature to characterize the sophisticated contour of the shapes in class 4, and texture in order to identify the corrugated surface of the leaves in class 2. Such a refining process usually involves interactively selecting features, performing validation classifications, revising the features and repeating the process over and over.

Dimensionality reduction

The previous section discussed the important topic of selecting good features to design successful pattern recognition systems. In fact, this is a central topic in most pattern recognition studies that has been receiving intense attention over the years. Also known as *dimensionality reduction*, this problem has an interesting statistical structure that may be explored in the search for good solutions. The first important fact is that the performance of the classifier may deteriorate as the number of features increases if the training set size is kept constant. In this context, the *dimensionality* is associated to the number of features (i.e., the feature space dimension). Figure 8.14 helps to understand this situation, which is often observed in experimental conditions. This figure illustrates the so-called U-curve because the classifier error often presents a U-shaped curve as a function of the dimensionality if the training set size is kept constant. This fact arises because of two other phenomena also illustrated in Figure 8.14: as the dimensionality increases, the **mixture** among the different classes tends to decrease, i.e. the different classes tend to be further from each other. This is a good thing that helps to decrease the classifier error. Nevertheless, as the dimensionality increases, because the number of samples used to train the classifier is kept constant, the **estimation error** also increases (because more samples would be needed to estimate more and more classifier parameters). The composition of these two curves lead to the U-curve of classifier error, as illustrated in Figure 8.14.

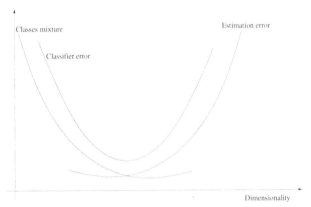

Figure 8.14: *The U-curve: classifier error often presents a U-shaped curve as a function of the dimensionality if the training set size is kept constant.*

This problem motivated the development of different dimensionality reduction methods. There are two basic approaches that may be taken: *feature fusion* and *feature selection*. Feature fusion methods explore mathematical techniques to build smaller feature spaces from larger ones. The most popular method for this is PCA (Section 2.6.6). Additional approaches include Fourier, wavelets and linear discriminant analysis (LDA).

On the other hand, feature selection methods search for a smaller subset of an initial set of features. Because this is a combinatorial search, it may be formulated as an optimization problem with two important components:

- Criterion function

- Optimization algorithm

  Different criterion functions may be used, such as

- Classifier error: although seemingle a natural choice, it may be difficult to be used in practice for analytical expressions are seldom available and error estimation procedures may be required (e.g. leave-one-out, cross-validation, bolstered error, etc.);

- Distances between classes, including fuzzy distances;

- Entropy and mutual information;

- Coefficient of determination;

    On the other hand, different optimization algorithms have been also described:

- Exhaustive search (though optimal, it may only be adopted in small problems because of the combinatorial computational explosion);

- Branch-and-bound;

- Sequential searchers (Sequential Forward Search - SFS and Sequential Backward Search - SBS);

- Floating searchers (Sequential Floating Forward Search - SFFS and Sequential Forward Backward Search - SFBS);

- Genetic algorithms;

To probe further: *Dimensionality reduction*

The reader is referred to [Barrera et al., 2007; Braga-Neto & Dougherty, 2004; Campos et al., 2001; Jain & Zongker, 1997; Jain et al., 2000; Martins-Jr et al., 2006; Pudil et al., 1994; and Somol et al., 1999] for further references on dimensionality reduction.

Additional resources: *Dimensionality reduction software*

Dimensionality reduction software is available on the Internet. In particular, an open-source software is available at http://dimreduction.incubadora.fapesp.br/portal, including a video demo at http://dimreduction.incubadora.fapesp.br/portal/downloads/video-dimreduction-fs-640x480.mov.

8.1.6 Feature Normalization

This section discusses two of the most popular alternatives for feature normalization: normal transformation and principal component analysis.

Normal Transformation of Features

One important fact to be noted about features is that they are, usually, dimensional entities. For instance, the size of a leaf can be measured in inches or centimeters. As a graphical example, consider a simple two-dimensional feature space (*Width* (cm) × *Weight* (g)) as shown in Figure 8.15 (a), including the three feature points *a*, *b* and *c*. Observe that, in this space, point *a* is closer to point *b* than to *c*. Figure 8.15 (b) presents the same three objects mapped into a new feature space (*Width* (in) × *Weight* (g)), where the abscissa unit has been converted from centimeters to inches. Remarkably, as a consequence of this simple unit change, point *a* is now closer to point *c* than to point *b*. Not surprisingly, such similarity changes can imply great changes in the resulting classification. Since the choice of units affects the distance in the feature space, it is often interesting to consider some standardization procedure for obtaining dimensionless versions of the features, which can be done by adopting a reference and taking the measures relative to it. For instance, in case you are interested in expressing the heights of a soccer team in dimensionless units, you can take the height of any player (let us say, the tallest) as a standard, and redefine the new heights as (dimensionless height) = (dimensional height) / (highest height).

A possibility to further reduce the arbitrariness implied by the choice of the feature units is to *normalize* the original data. The most frequently used normalization strategy consists in applying equation (8.1), where μ and σ stand for the mean and standard deviation of the feature j, respectively, which can be estimated as described in Section 2.6.4. This operation corresponds to a transformation of the original data in the sense that the new feature set is now guaranteed to have zero mean and unit standard deviation. It can be easily verified that this transformation also yields dimensionless features, most of which falling within the interval $[-2, 2]$.

$$\hat{f}(i, j) = \frac{f(i, j) - \mu_j}{\sigma_j} \tag{8.1}$$

In the box entitled *Normal Transformation of Features* we present the application of the above normalization procedure in order to obtain a dimensionless version of the features for the example in Section 8.1.5, and Figure 8.16 presents the visualization of those data before and after such a normalization.

Example: *Normal Transformation of Features*

Apply the normal transformation in order to obtain a dimensionless version of the features in the example in Section 8.1.5.

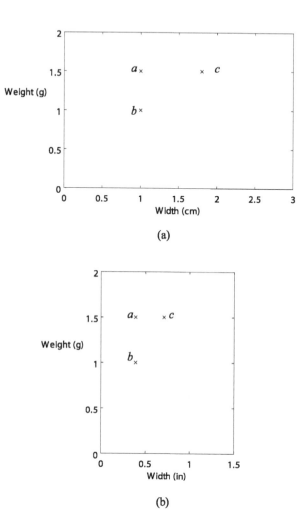

Figure 8.15: *Three objects represented in the* (Width *(cm)* × Weight *(g))* *(a)*
and (Width *(in)* × Weight *(g))* *(b) feature spaces. By changing*
the relative distances between feature vectors, a simple unit
conversion implies different similarities between the objects.

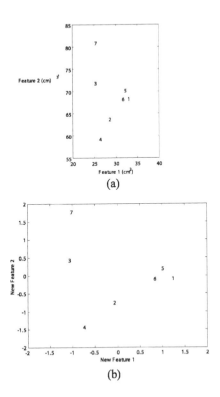

Figure 8.16: *Data from the example in Section 8.1.5 before **(a)** and after **(b)** normal transformation.*

Solution:

We start with the original feature matrix:

$$F = \begin{bmatrix} 32.67\,\text{cm}^2 & 68.48\,\text{cm}^3 \\ 28.30\,\text{cm}^2 & 63.91\,\text{cm}^3 \\ 24.99\,\text{cm}^2 & 71.95\,\text{cm}^3 \\ 26.07\,\text{cm}^2 & 59.36\,\text{cm}^3 \\ 31.92\,\text{cm}^2 & 70.33\,\text{cm}^3 \\ 31.32\,\text{cm}^2 & 68.40\,\text{cm}^3 \\ 25.14\,\text{cm}^2 & 81.00\,\text{cm}^3 \end{bmatrix}$$

Now, the mean and standard deviation of the respective features are obtained as

$$\mu = \begin{bmatrix} 29.63\,\text{cm}^2 & 69.06\,\text{cm}^3 \end{bmatrix} \quad \text{and} \quad \sigma = \begin{bmatrix} 3.3285\,\text{cm}^2 & 6.7566\,\text{cm}^3 \end{bmatrix}$$

by applying equation (8.1), we obtain the normalized features:

$$\tilde{F} = \begin{bmatrix} 1.2138 & -0.0861 \\ -0.0991 & -0.7624 \\ -1.0936 & 0.4275 \\ -0.7691 & -1.4358 \\ 0.9884 & 0.1878 \\ 0.8082 & -0.0979 \\ -1.0485 & 1.7669 \end{bmatrix}$$

The original and transformed (dimensionless) feature spaces are shown in Figures 8.16 (a) and (b), respectively.

Observe that other transformations of the feature vectors are also possible, including nonlinear ones such as the logarithmic transformation of the perimeter and area values used in Section 8.1.3. However, it should be borne in mind that the inherent properties allowed by such transformations might, in practice, correspond to either benefits or shortcomings, depending on the specific case. For example, the above-mentioned logarithmic transformation allowed us to use a straight line as a decision boundary. Other cases are where the classes are already linearly separable, and logarithmic or other nonlinear transformations could complicate the separation of the classification regions. By changing the relative distances, even the normal transformation can have adverse effects. As a matter of fact, no consensus has been reached in the literature regarding the use of the normal transformation to normalize the features, especially in the sense that the distance alteration implied by this procedure tends to reduce the separation between classes in some cases. A recommended pragmatic approach is to consider both situations, i.e., features with and without normalization, in classification problems, choosing the situation that provides the best results.

Principal Component Analysis

Another main approach to feature normalization is to use the principal component analysis approach, which is directly related to the Karhunen-Loève transform presented in Section 2.6.6. In other words, the covariance matrix of the feature values is estimated (see Section 2.6.4), and its respective eigenvectors are then used to define a linear transformation whose main property is to minimize the covariance between the new transformed features along the main coordinate axes, thus maximizing the variance along each new axis, in such a way that the new features become perfectly uncorrelated. Consequently, the principal component methodology is suitable for removing redundancies between features (by uncorrelating them). Let us illustrate this strategy in terms of the following example.

Assume somebody experimentally (and hence with some error) measured the length of two types of objects in centimeters and inches, yielding the following data matrix F:

$$F = \begin{bmatrix} 5.3075 & 2.1619 \\ 2.8247 & 1.1941 \\ 3.0940 & 1.2318 \\ 2.3937 & 0.9853 \\ 5.2765 & 2.0626 \\ 4.8883 & 1.9310 \\ 4.6749 & 1.8478 \\ 3.5381 & 1.4832 \\ 4.9991 & 1.9016 \\ 3.4613 & 1.3083 \\ 2.8163 & 1.0815 \\ 4.6577 & 1.7847 \end{bmatrix}.$$

The respective distribution of the feature vectors in the feature space is graphically illustrated in Figure 8.17 (a). It is clear that the cloud of feature points concentrates

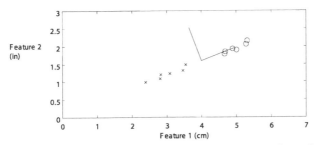

Figure 8.17: *Representation in the feature space of the data obtained by measuring objects in centimeters and inches. The principal orientations are shown as solid lines.*

along a single straight line, thus indicating that the two features are strongly correlated. Indeed, the small dispersion is the sole consequence of the above mentioned experimental error.

As to the mathematical detail, we have that the respective covariance (K) and correlation coefficient (*CorrCoef*) matrices are

$$K = \begin{bmatrix} 1.1547 & 0.4392 \\ 0.4392 & 0.1697 \end{bmatrix} \quad \text{and}$$

$$CorrCoef = \begin{bmatrix} 1.0 & 0.992 \\ 0.992 & 1.0 \end{bmatrix}$$

As expected from the elongated cloud of points in Figure 8.17, the above correlation coefficient matrix confirms that the two features are highly correlated. This indicates that a single feature may be enough to represent the observed measures.

Actually, because the correlation matrix is necessarily symmetric (Hermitian in the case of complex data sets) and positive semidefinite (in practice, it is often positive definite), its eigenvalues are real and positive and can be ordered as $\lambda_1 \geqslant \lambda_2 \geqslant 0$. Let $\vec{v}_1$ and $\vec{v}_2$ be the respective orthogonal eigenvectors (the so-called *principal components*). Let us organize the eigenvectors into the following 2×2 orthogonal matrix L:

$$L = \begin{bmatrix} \uparrow & \uparrow \\ \vec{v}_1 & \vec{v}_2 \\ \downarrow & \downarrow \end{bmatrix}.$$

It should be observed that this matrix is organized differently from the Ω matrix in Section 2.6.6 in order to obtain a simpler "parallel" version of this transformation. Let us start with the following linear transform:

$$\tilde{\vec{F}}_i = (L)^T \vec{F}_i, \tag{8.2}$$

which corresponds to the Karhunen-Loève transform. Observe that all the new feature vectors can be obtained in "parallel" (see Section 2.2.5) from the data matrix F by making:

$$\tilde{F} = \left((L)^T F^T \right)^T = FL.$$

The new features are shown in Figure 8.18, from which it is clear that the maximum dispersion occurs along the abscissae axis.

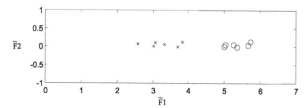

Figure 8.18: *The new feature space obtained after the Karhunen-Loève transformation.*

Because the features are predominantly distributed along the abscissae axis, it is possible to consider using only the new feature associated with this axis. This can be done immediately by defining the following truncated version of the matrix L:

$$L_{(1)} = \begin{bmatrix} \uparrow \\ \vec{v}_1 \\ \downarrow \end{bmatrix} \quad \text{and making} \quad \tilde{F} = FL_{(1)}.$$

Remember that it is also possible to use the following equivalent approach:

$$L_{(1)}^T = \begin{bmatrix} \leftarrow & \vec{v}_1 & \rightarrow \end{bmatrix} \quad \text{and making} \quad \tilde{F} = L_{(1)}^T F.$$

Observe that the two classes of objects can now be readily separated by using just a single threshold T, as illustrated in Figure 8.19.

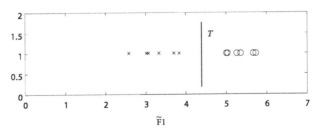

Figure 8.19: *The two classes in the above example can be perfectly distinguished by using a single threshold, T, along the new feature space.*

No matter how useful the principal component approach may initially seem, especially considering examples such as those above, there are situations where this strategy will lead to poor classification. One such case is illustrated in Figure 8.20. Although both involved features exhibit a high correlation coefficient, calculated as

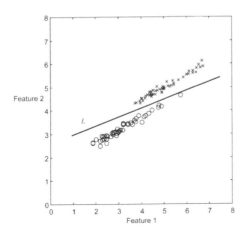

Figure 8.20: *A situation where applying the principal components lead to substantially overlapping classes.*

0.9668, the use of principal components on this data will imply substantial overlapping between these two classes.

An even more compelling example of a situation where the use of principal components will lead to worse results is the separation of square and circular boxes discussed in Section 8.1.3, since whole regions of the parabola segments are highly correlated, yielding complete overlap when a principal component is applied. While the decision of applying this technique can be made easily in classi-

fication problems involving just a few features by performing a visual exploration
of the distribution of the feature vectors in the feature space, it becomes very difficult to conduct such visualizations in higher dimensional spaces. In such situations,
the remaining alternative is again to consider both situations, i.e., using principal
component analysis or not, and then adopting the approach leading to better classification results.

Note: *Artificial Neural Networks*

Another popular approach, which can be applied to both supervised and unsupervised classification, is based on artificial neural networks (ANNs). Initially
inspired by biological neural systems, ANNs usually provide a black-box approach
to classification that can nevertheless be of interest in some situations. The interested reader should refer to [Allen et al., 1999; Anderson, 1995; Fausett, 1994;
Hertz et al., 1991; Ripley, 1996; Schalkoff, 1992; Schürmann, 1996].

8.2 | Supervised Pattern Classification

Having discussed how to perform an initial exploration of the feature space, select
adequate features and eventually normalize them, it is time to consider methods for
implementing the classification itself. As introduced in Section 1.3.3, there are two
main approaches to automated classification: *supervised* and *unsupervised* (i.e.,
clustering), characterized respectively by considering or not considering samples
or prototypes of the involved classes. The current section presents, illustrates and
discusses in an introductory fashion the powerful approach to supervised classification known as *Bayesian* classification. Provided a good statistical model (i.e.,
the conditional probability density functions) of the studied situation is available,
Bayesian classification can be shown to supply the statistically optimum solution
to the problem of supervised classification, thus defining a standard against which
many alternative approaches are usually compared.

8.2.1 | Bayes Decision Theory Principles

Assume you need to predict the sex of the inhabitants of a given town. Let the
female and male classes be abbreviated as C_1 and C_2, respectively. In addition,
let $P(C_1)$ and $P(C_2)$ denote the probabilities of an individual belonging to either
class C_1 or C_2, respectively. It follows from the definition of probability (see Section 2.6.1) that:

$$P(C_1) = \frac{\text{Number of females}}{\text{total population}} \quad \text{and} \quad P(C_2) = \frac{\text{Number of males}}{\text{total population}}.$$

In case such total figures are not known, they can always be estimated by randomly sampling N individuals from the population and making:

$$P(C_1) = \frac{\text{Number of females in the sample}}{N}$$

and

$$P(C_2) = \frac{\text{Number of males in the sample}}{N}.$$

The larger the value of N, the better the estimation. Given the probabilities $P(C_1)$ and $P(C_2)$, the first criterion for deciding whether an observed individual is male or female would simply be to take the class with larger probability. For example, if the only information we have is that $P(C_1) > P(C_2)$, indicating that there is an excess of women in the population, it can be shown that the best strategy is to classify the individual as being of class C_1. Such an extremely simplified, but statistically optimal, strategy can be summarized as

$$P(C_1) \underset{C_2}{\overset{C_1}{\gtrless}} P(C_2). \tag{8.3}$$

The box titled *Bayes Decision Theory I* gives an illustrative example of the application of this criterion.

Example: *Bayes Decision Theory I*

You are required to identify the class of a single leaf in an image. All you know is that this image comes from a database containing 200 laurel leaves and 120 olive leaves.

Solution:

Let us understand C_1 as *laurel* and C_2 as *olive*. The probabilities $P(C_1)$ and $P(C_2)$ can be estimated as

$$P(C_1) = \frac{\text{Number of laurel leaves}}{\text{total population}} = \frac{200}{320} = 0.625$$

and

$$P(C_2) = \frac{\text{Number of olive leaves}}{\text{total population}} = \frac{120}{320} = 0.375.$$

By using equation (8.3):

$$P(C_1) > P(C_2) \quad \Rightarrow \quad \text{Select } C_1.$$

Thus, the best bet is to classify the leaf in the image as being a laurel leaf.

Going back to the male/female classification problem, it is intuitive (and correct) that better guesses can generally be obtained by considering additional information about the sampled individuals, such as their measured height, defining a new random variable henceforth identified by h. In such a situation, the natural approach would be to extend the previous criterion in terms of conditional probabilities $P(C_1 \mid h)$, indicating the probability of an individual with height h being of class C_1 (according to the above convention, a woman), and $P(C_2 \mid h)$, representing the probability of an individual with height h being of class C_2. Provided such information is available or can be reasonably estimated, the optimal classification criterion would be, if $P(C_1 \mid h) > P(C_2 \mid h)$, decide for C_1, and if $P(C_2 \mid h) > P(C_1 \mid h)$, decide for C_2. In practice, the problem with this approach is that the required conditional probabilities are rarely available. Fortunately, Bayes law (see Section 2.6.1) can be applied in order to redefine the above criterion in terms of the density functions $f(h \mid C_1)$ and $f(h \mid C_2)$, i.e., the conditional density functions of h given that an individual is either female or male, respectively. The advantage of doing so is that the conditional density functions $f(h \mid C_1)$ and $f(h \mid C_2)$, if not available, can often be estimated. For instance, the conditional density function for $f(h \mid C_1)$ can be estimated by separating a good number of individuals of class C_1 (i.e., women) and applying some estimation approach (such as Parzen's windows [Duda & Hart, 1973; Duda et al., 2000]). In case the nature of such functions is known (let us say, we know they are Gaussians), it would be enough to estimate the respective parameters by applying parametric estimation techniques.

The derivation of the new classification criterion, now in terms of functions $f(h \mid C_1)$ and $f(h \mid C_2)$, is straightforward and starts by considering Bayes law (see Section 2.6.1), which states that

$$P(C_i \mid h) = \frac{f(h \mid C_i)\, P(C_i)}{\sum_{k=1}^{2} f(h \mid C_k)\, P(C_k)}.$$

Now, the criterion in equation (8.3) can be rewritten as

$$\frac{P(C_1)\, f(h \mid C_1)}{\sum_{k=1}^{2} f(h \mid C_k)\, P(C_k)} \overset{C_1}{\underset{C_2}{\gtrless}} \frac{P(C_2)\, f(h \mid C_2)}{\sum_{k=1}^{2} f(h \mid C_k)\, P(C_k)}.$$

This new criterion, called *Bayes decision rule*, is obtained by eliminating the denominators, as presented in equation (8.4). The box titled *Bayes Decision Theory*

II exemplifies an application of this criterion.

$$P(C_1) \, f(h \mid C_1) \quad \overset{C_1}{\underset{C_2}{\gtrless}} \quad P(C_2) \, f(h \mid C_2). \tag{8.4}$$

Example: *Bayes Decision Theory II*

You are required to identify the class of an isolated leaf in an image. As in the previous example, you know that this image comes from a database containing 200 laurel leaves (class C_1) and 120 olive leaves (class C_2), but now you also know that the conditional density functions characterizing the length (h in cm) distribution of a leaf, given that it is of a specific species, are

$$f(h \mid C_1) = \frac{1}{\Gamma(2)} h e^{-h} \quad \text{and} \quad f(h \mid C_2) = \frac{4}{\Gamma(2)} h e^{-2h},$$

where Γ is the gamma function (which can be obtained from tables or mathematical software).

Solution:

From the previous example, we know that $P(C_1) = 0.625$ and $P(C_2) = 0.375$. Now we should measure the length of the leaf in the image. Let us say that this measure yields 3 cm. By using the Bayes decision rule in equation (8.4):

$$P(C_1) \, f(h = 3 \mid C_1) \quad \overset{C_1}{\underset{C_2}{\gtrless}} \quad P(C_2) \, f(h = 3 \mid C_2) \Rightarrow$$

$$\Rightarrow 0.625 \, f(3 \mid C_1) \quad \overset{C_1}{\underset{C_2}{\gtrless}} \quad 0.375 \, f(3 \mid C_2) \Rightarrow$$

$$\Rightarrow (0.625)(0.1494) \quad \overset{C_1}{\underset{C_2}{\gtrless}} \quad (0.375)(0.0297) \Rightarrow$$

$$\Rightarrow 0.0934 \quad \overset{C_1}{\underset{C_2}{\gtrless}} \quad 0.0112 \Rightarrow C_1$$

Thus, the best bet is to predict C_1. The above situation is illustrated in Figure 8.21. Observe that the criterion in equation (8.4) defines two regions along the graph domain, indicated as R_1 and R_2, in such a way that whenever h falls within R_1, we

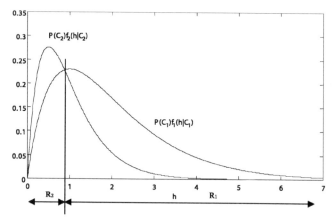

Figure 8.21: *The conditional weighted probability density functions*
$P(C_1)f(h \mid C_1)$ *and* $P(C_2) f(h \mid C_2)$ *and the decision regions*
R_1 *and* R_2.

had better choose C_1; and whenever h is within R_2, we had better choose C_2.

By defining the function $L(h)$ as in equation (8.5), which is known as the *likeli-hood ratio*, and the threshold T in equation (8.6), the above criterion can be rewrit-ten as equation (8.7):

$$L(h) = \frac{f(h \mid C_2)}{f(h \mid C_1)}, \tag{8.5}$$

$$T = \frac{P(C_1)}{P(C_2)}, \tag{8.6}$$

and

$$T \begin{array}{c} C_1 \\ \gtrless \\ C_2 \end{array} L(h) \tag{8.7}$$

This totem pole arrangement of inequalities should be read as

$$\begin{cases} \text{if } T \geqslant L(h) \text{ then } C_1 \\ \text{else } C_2. \end{cases}$$

The Bayes decision criterion can be further generalized by considering the costs implied by, respectively, taking hypothesis H_2 (i.e., the observation is of class C_2) when the correct is H_1 (i.e., the observation is of class C_1), and taking hypothesis H_1 when the correct is H_2, which are henceforth identified as k_1 and k_2, respectively. In this case, the new criterion is as follows (see [Duda & Hart, 1973; Duda et al.,

2000]):

$$k_2 \, P(C_1) \, f(h \mid C_1) \quad \overset{C_1}{\underset{C_2}{\gtrless}} \quad k_1 \, P(C_2) \, f(h \mid C_2). \tag{8.8}$$

The above simple results, which underlie the area known as *Bayes decision theory* (alternatively *Bayes classification*), are particularly important in pattern supervised classification because they can be proved to be statistically optimal in the sense that they minimize the chance of misclassification [Duda & Hart, 1973]. The main involved concepts and respective abbreviations are summarized in Table 8.5. A

H_1	The object is of class C_1.
H_2	The object is of class C_2.
H	Random variable (e.g., measured height of an individual).
C_1	One of the classes (e.g., female).
C_2	The other class (e.g., male).
$f(h \mid C_1)$	The conditional density function of the random variable h given that the individual is of class C_1.
$f(h \mid C_2)$	The conditional density function of h given that the individual is of class C_2.
$P(C_1)$	The probability (mass) of an individual being of class C_1.
$P(C_2)$	The probability (mass) of an individual being of class C_2.
k_1	The cost of concluding H_1 when the correct is H_2.
k_2	The cost of concluding H_2 when the correct is H_1.
$P(C_1 \mid h)$	The probability of choosing class C_1 after measuring h.
$P(C_2 \mid h)$	The probability of choosing class C_2 after measuring h.

Table 8.5: *The required elements in Bayes classification and the adopted conventions for their identification.*

practical drawback with the Bayes classification approach is that the conditional density functions $f(h \mid C_i)$ are frequently not available. Although they often can be estimated (see Section 2.6.4), there are practical situations, such as in cases where just a few observations are available, in which these functions cannot be accurately estimated, and alternative approaches have to be considered.

8.2.2 | Bayesian Classification: Multiple Classes and Dimensions

The concepts and criteria presented in the previous section can be immediately generalized to situations involving more than two classes and multiple dimensional feature spaces. First, let us suppose that we have K *classes* and the respective conditional density functions $f(h \mid C_i)$. The criteria in Equations (8.3) and (8.4) can now be respectively rewritten as:

$$\text{If } P(C_i) = \max_{k=1,K} \{P(C_k)\} \text{ then select } C_i \tag{8.9}$$

and

$$\text{If } f_i(h \mid C_i) P(C_k) = \max_{k=1,K} \{f_k(h \mid C_k) P(C_k)\} \text{ then select } C_i. \tag{8.10}$$

Figure 8.22 illustrates Bayesian classification involving three classes C_1, C_2 and C_3 and their respective classification regions R_1, R_2 and R_3 as defined by equation (8.10).

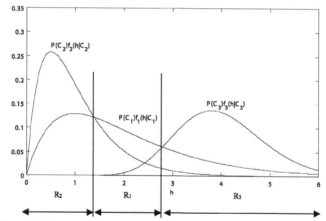

Figure 8.22: *Decision regions defined by a problem involving three classes and respective probabilities.*

Another natural extension of the results presented in the previous section addresses situations where there are *more than a single measured feature*. For instance, taking the female/male classification example, we could consider not only height h as a feature, but also *weight, age*, etc. In other words, it is interesting to have a version of the Bayes decision rule considering *feature vectors*, henceforth represented by $\vec{x}$. Equation 8.11 presents the respective generalizations of the original criteria in equation (8.4) to multivariate features:

$$P(C_1) f(\vec{x} \mid C_1) \underset{C_2}{\overset{C_1}{\gtrless}} P(C_2) f(\vec{x} \mid C_2). \qquad (8.11)$$

Figure 8.23 illustrates Bayesian classification involving two classes (C_1 and C_2) and two measures or random variables (x and y). The two bivariate weighted Gaussian functions $P(C_1) f(x, y \mid C_1)$ and $P(C_2) f(x, y \mid C_2)$, are shown in (a), and their level curves and the respectively defined decision boundary are depicted in (b). In

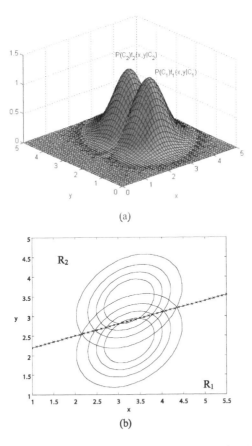

(a)

(b)

Figure 8.23: *The weighted multivariate probability density functions $P(C_1) f(x, y \mid C_1)$ and $P(C_2) f(x, y \mid C_2)$ **(a)** and some of the level curves and respectively defined decision regions R_1 and R_2 corresponding to the upper and lower semi-planes defined by the straight decision boundary marked as a barbed wire **(b)**.*

this case, the intersection between the bivariate Gaussian curves defines a straight *decision boundary* in the (x, y) feature space, dividing it into two semi-planes corresponding to the decision regions R_1 and R_2. Observe that such intersections are not always straight lines (see [Duda et al., 2000] for additional information). In case an object is found to produce a specific measure (x, y) within region R_1, the optimal decision is to classify it as being of class C_1. Once the decision boundary is defined, it provides enough subsidy to implement the classification.

In practice, many applications require the use of *multiple features and classes*. The extension of Bayes decision rule for such a general situation is given by equation (8.12):

$$\text{If } f(\vec{x} \mid C_i) \, P(C_k) = \max_{k=1,K} \{ f(\vec{x} \mid C_k) \, P(C_k) \} \text{ then select } C_i. \tag{8.12}$$

8.2.3 | Bayesian Classification of Leaves

In this section we illustrate the Bayesian approach with respect to the classification of plant leaves. We consider 50 observations of each of three types of leaves, which are illustrated in Figure 8.24. Out of these 50 observations, 25 were randomly

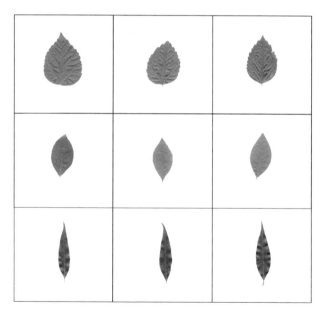

Figure 8.24: *Three examples of each of the considered leaf classes.*

selected for defining the decision regions (the "training" stage), and 25 were left for assessing the classification. As discussed in Section 8.1.5, the first step toward achieving a good classification is to select a suitable set of features.

A preliminary subjective analysis of the three types of leaves (see Figure 8.24) indicates that one of the most discriminating measures is elongation, in the sense that leaves in class 1 are less elongated than those in class 2, which in turn are less elongated than those in class 3. In addition, leaves in class 1 tend to be more circular than those in class 2, which in turn are less circular than leaves in class 3.

Figure 8.25 presents the two-dimensional feature space defined by the circularity and elongation with respect to the two sets of 25 observations, respective to the training and evaluating sets. As is evident from this illustration, in the sense that a

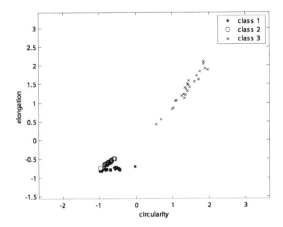

Figure 8.25: *The two-dimensional feature space defined by the circularity and elongation measures, after normal transformation of the feature values. Each class is represented in terms of the 25 observations.*

straight cloud of points is obtained for objects in each class, the features elongation and circularity, as could be expected, are positively correlated. However, in spite of this fact, this combination of features provides a particularly suitable choice in the case of the considered leaf species. Indeed, in the present example, it led to no classification errors.

Figure 8.26 presents the bivariate Gaussian density functions defined by the mean and covariance matrices obtained for each of the 25 observations representing the three classes.

8.2.4 Nearest Neighbors

The non-parametric supervised technique known as the *nearest neighbor* approach constitutes one of the simplest approaches to classification. Assuming that we have a set S of N samples already classified into M classes C_i and that we now want to classify a new object $\vec{x}$, all that is needed is

> Identify the sample in S that is closest to $\vec{x}$ and take its class.

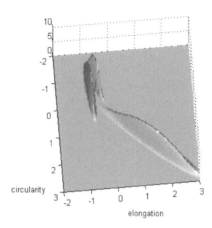

Figure 8.26: *The bivariate Gaussian density functions defined by the mean and covariance matrices of each of three classes in the training set of observations.*

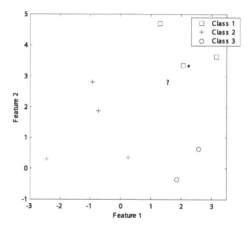

Figure 8.27: *What is the class of the object identified by the question mark? According to the nearest neighbor approach, its class is taken as being equal to the class of the nearest object in the feature space. In this case, the nearest neighbor, identified by the asterisk, is of class 1.*

Consider, as an example, the situation illustrated in Figure 8.27. Here we have three classes of objects represented in a two-dimensional feature space. In case we want to assign a class to the object represented as a question mark in this figure, the nearest neighbor approach consists in taking the class of its nearest neighbor, which is marked by an asterisk. Therefore, the new object is of class 1. It should be observed that the performance of the nearest neighbor approach is generally inferior to the Bayes decision criterion (see, for instance, [Duda & Hart, 1973]).

The nearest neighbor approach can be immediately extended to the *k-nearest neighbors* method. In this case, instead of taking the class of the nearest neighbor, k (where k is an integer positive number) nearest neighbors are determined, and the class is taken as that exhibited by the majority of these neighbors (in case of tie, one of the classes can be selected arbitrarily). Theoretically, it can be shown that for a very large number of samples, there are advantages in using large values of k. More specifically, if k tends to infinity, the performance of the k-neighbors method approaches the Bayes rate [Duda & Hart, 1973]. However, it is rather difficult to predict the performance in general situations.

8.3 | Unsupervised Classification and Clustering

This section addresses the important and challenging issue of unsupervised classi-fication, also called clustering. We start by presenting and discussing the involved basic concepts and related issues and then proceed by introducing how scatter ma-trices can be calculated and used to define the popular similarity-clustering crite-rion. The two main types of clustering algorithms, namely partitional and hierar-chical, are introduced next. Finally, a complete comparative example regarding leaf classification is presented and discussed.

8.3.1 | Basic Concepts and Issues

We have already seen that unsupervised classification, or clustering, differs from su-pervised classification in the sense that neither prototypes of the classes nor knowl-edge about pattern generation is available. All that is usually supplied is a set of observations represented in terms of its respective feature vectors. Figure 8.28 illus-trates a few of the possible feature vector distributions in two-dimensional feature spaces. A brief inspection of many of the depicted situations immediately sug-gests "natural" partition possibilities. For instance, the situation in (a) very likely corresponds to two main classes of objects that can be linearly separated (i.e., a straight decision boundary is enough for proper classification). While situations (b) through (d) also suggest two clusters, these are no longer linearly separable. How-ever, the reasonably uniform feature vector distribution represented in (e) does not suggest any evident cluster. Situation (f) provides an amazing example where a small displacement of just one of the points in (a) implied an ambiguity, in the sense that the feature vector distribution now seems to flip between one or two clusters. Such points between well-defined clusters are often called *noise points*. Another particular type of point, known as *outlier*, is illustrated in (g). In this case, in addition to the well-defined cluster, we have two isolated points, i.e., the outliers.

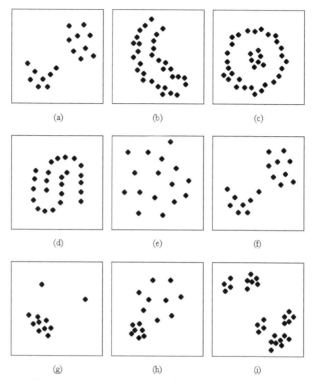

Figure 8.28: *A few illustrations of possible feature vector distributions in a two-dimensional feature space. See text for discussion.*

It is usually difficult to determine whether these points belong to some of the other more defined clusters, or if they correspond to poorly sampled additional clusters. Another important issue in clustering, namely the coexistence of spatial scales, is illustrated in (h), where a tension has been induced by separating the two clusters, each characterized by different relative distances between its elements, and a single cluster including all objects. Finally, situation (i) illustrates the possibility of having a hierarchy of clusters, or even a fractal organization (in the sense of having clusters of clusters of clusters...). Observe that even more sophisticated situations can be defined in higher dimensional spaces.

While the above discussion illustrates the variability and complexity of the possible situations found in clustering, it was completely biased by the use of Euclidean distances and, more noticeably, by our own subjective grouping mechanisms (such as those studied by Gestalt [Rock & Palmer, 1990], which reflect our natural tendencies to clustering). Indeed, it should be stressed at the outset that there is no general or unique clustering criterion. For instance, in the above example we were biased by proximity between the elements and our own subjective perceptual mechanisms. However, there are infinite choices, involving several combinations of proximity and dispersion measures, and even more subtle and nonlinear possibilities. In practice, clustering is by no means an easy task, since the selected features, typically involving higher dimensional spaces, are often incomplete (in the sense of providing a degenerate description of the objects) and/or not particularly suitable. Since no general criterion exists, any particular choice will define how the data is ultimately clustered. In other words, the clustering criterion imposes a structure onto the feature vectors that may or may not correspond to that actually underlying the original observations. Since this is a most important fact to be kept in mind at all times while applying and interpreting clustering, it is emphasized in the following:

> The adopted clustering criterion in great part defines the obtained clusters, therefore imposing structure over the observations.

In practice, every piece of knowledge or reasonable hypothesis about the data or the processes through which it is produced is highly valuable and should be carefully considered while deciding about some specific clustering criterion. Observe that the existence of such clues about the clustering structures corresponds to some sort of supervision. However, no information is supplied about the nature of the clusters, all that we can do is to consider several criteria and judge the outcome in a subjective way. In practice, one of the most frequently adopted clustering criteria can be summarized as

> **The Similarity Criterion:** Group things so that objects in the same class are as similar as possible and objects from any two distinct clusters are as different as possible.

Observe that the above definition depends on the adopted type of similarity (or distance). In addition, it can be shown that, as illustrated in the following section, the above definition is actually redundant, in the sense that to maximize similarity with the clusters automatically implies minimizing dissimilarity between objects from distinct clusters. Another important and difficult problem in clustering regards how to define the correct number of clusters, which can have substantial effects on the results achieved. Two situations arise: (1) this number is provided and (2) the number of clusters has to be inferred from the data. Naturally, the latter situation is usually more difficult than the former.

8.3.2 Scatter Matrices and Dispersion Measures

One approach to formalize the similarity-clustering criterion can be achieved in terms of *scatter matrices*, which are introduced in the following by using the abbreviations introduced in Section 8.1.4. Conceptually, in order to qualify as a cluster candidate, each set of points should include elements that are narrowly dispersed. At the same time, a relatively high dispersion should be expected between points belonging to different clusters. The *total scatter matrix*, S, indicating the overall dispersion of the feature vectors, is defined as

$$S = \sum_{i=1}^{N} \left(\vec{f}_i - \vec{M}\right)\left(\vec{f}_i - \vec{M}\right)^{T},$$ (8.13)

the *scatter matrix for class* C_i, hence S_i, expressing the dispersion of the feature vectors within each class, is defined as

$$S_i = \sum_{i \in C_i} \left(\vec{f}_i - \vec{\mu}_i\right)\left(\vec{f}_i - \mu_i\right)^{T},$$ (8.14)

the *intraclass scatter matrix*, hence S_{intra}, indicates the combined dispersion in each class and is defined as

$$S_{\text{intra}} = \sum_{i=1}^{K} S_i,$$ (8.15)

and the *interclass scatter matrix*, hence S_{inter}, expresses the dispersion of the classes (in terms of their centroids) and is defined as

$$S_{\text{inter}} = \sum_{i=1}^{K} N_i \left(\vec{\mu}_i - \vec{M}\right)\left(\vec{\mu}_i - \vec{M}\right)^{T}.$$ (8.16)

It can be demonstrated [Jain & Dubes, 1988] that, whatever the class assignments, we necessarily have

$$S = S_{\text{intra}} + S_{\text{inter}},$$ (8.17)

i.e., the sum of the interclass and intraclass scatter matrices is always preserved. The box entitled *Scatter Matrices* presents a numeric example illustrating the calculation of the scatter matrix and this property. Scatter matrices are important because it is possible to quantify the intra- and interclass dispersion of the feature vectors in terms of functionals, such as the trace and determinant, defined over them (see [Fukunaga, 1990] for additional detail). It can be shown [Jain & Dubes, 1988] that the scattering conservation is also verified for the trace measure, i.e.,

$$\text{trace}(S) = \text{trace}\left(S_{\text{intra}}\right) + \text{trace}\left(S_{\text{inter}}\right).$$

Example: *Scatter Matrices*

Calculate the scatter matrices for the data in Example Box in Section 8.1.4 and verify the scattering conservation property.

Solution:

Recalling that $\vec{M} = \begin{bmatrix} 5.4143 \\ 22.1571 \end{bmatrix}$, we have from equation (8.13):

$$S = \sum_{i=1}^{N} \left(\vec{f}_i - \vec{M}\right)\left(\vec{f}_i - \vec{M}\right)^T$$

$$= \begin{bmatrix} 9.2 - 5.4143 \\ 33.20 - 22.1571 \end{bmatrix}\begin{bmatrix} 9.2 - 5.4143 & 33.20 - 22.1571 \end{bmatrix} + \cdots +$$

$$+ \begin{bmatrix} 1.2 - 5.4143 \\ 11.5 - 22.1571 \end{bmatrix}\begin{bmatrix} 1.2 - 5.4143 & 11.5 - 22.1571 \end{bmatrix}$$

$$= \begin{bmatrix} 78.0686 & 220.0543 \\ 220.0543 & 628.5371 \end{bmatrix}.$$

Applying equation (8.14) for class C_1:

$$S_1 = \sum_{i\in C_1} \left(\vec{f}_i - \vec{\mu}_1\right)\left(\vec{f}_i - \mu_1\right)^T$$

$$= \begin{bmatrix} 2.9 - 1.8667 \\ 12.7 - 12.0667 \end{bmatrix}\begin{bmatrix} 2.9 - 1.8667 & 12.7 - 12.0667 \end{bmatrix} +$$

$$+ \begin{bmatrix} 1.5 - 1.8667 \\ 12 - 12.0667 \end{bmatrix}\begin{bmatrix} 1.5 - 1.8667 & 12 - 12.0667 \end{bmatrix} +$$

$$+ \begin{bmatrix} 1.2 - 1.8667 \\ 11.5 - 12.0667 \end{bmatrix}\begin{bmatrix} 1.2 - 1.8667 & 11.5 - 12.0667 \end{bmatrix}$$

$$= \begin{bmatrix} 1.6467 & 1.0567 \\ 1.0567 & 0.7267 \end{bmatrix}.$$

Applying equation (8.14) for class C_2:

$$S_2 = \sum_{i\in C_2} \left(\vec{f}_i - \vec{\mu}_2\right)\left(\vec{f}_i - \mu_2\right)^T = \begin{bmatrix} 5.3 - 5.3 \\ 21.4 - 21.4 \end{bmatrix}\begin{bmatrix} 5.3 - 5.3 & 21.4 - 21.4 \end{bmatrix}$$

$$= \begin{bmatrix} 0 & 0 \\ 0 & 0 \end{bmatrix}.$$

Applying equation (8.14) for class C_3:

$$S_3 = \sum_{i \in C_3} \left(\vec{f}_i - \vec{\mu}_3 \right) \left(\vec{f}_i - \mu_3 \right)^T$$

$$= \begin{bmatrix} 9.2 - 9 \\ 33.2 - 32.5 \end{bmatrix} \begin{bmatrix} 9.2 - 9 & 33.2 - 32.5 \end{bmatrix} + \begin{bmatrix} 8.8 - 9 \\ 31.9 - 32.5 \end{bmatrix} \begin{bmatrix} 8.8 - 9 & 31.9 - 32.5 \end{bmatrix} +$$

$$+ \begin{bmatrix} 9 - 9 \\ 32.4 - 32.5 \end{bmatrix} \begin{bmatrix} 9 - 9 & 32.4 - 32.5 \end{bmatrix}$$

$$= \begin{bmatrix} 0.08 & 0.26 \\ 0.26 & 0.86 \end{bmatrix}.$$

Therefore, from equation (8.15), we have that the intraclass scatter matrix is

$$S_{\text{intra}} = \sum_{i=1}^{K} S_i = S_1 + S_2 + S_3 = \begin{bmatrix} 1.7267 & 1.3167 \\ 1.3167 & 1.5867 \end{bmatrix}$$

and, from equation (8.16), we have

$$S_{\text{inter}} = \sum_{i=1}^{K} N_i \left(\vec{\mu}_i - \vec{\mathbf{M}} \right) \left(\vec{\mu}_i - \vec{\mathbf{M}} \right)^T$$

$$= (3) \begin{bmatrix} 1.8667 - 5.4143 \\ 12.0667 - 22.1571 \end{bmatrix} \begin{bmatrix} 1.8667 - 5.4143 & 12.0667 - 22.1571 \end{bmatrix} +$$

$$+ (1) \begin{bmatrix} 5.3 - 5.4143 \\ 21.4 - 22.1571 \end{bmatrix} \begin{bmatrix} 5.3 - 5.4143 & 21.4 - 22.1571 \end{bmatrix} +$$

$$+ (3) \begin{bmatrix} 9 - 5.4143 \\ 32.5 - 22.1571 \end{bmatrix} \begin{bmatrix} 9 - 5.4143 & 32.5 - 22.1571 \end{bmatrix}$$

$$= \begin{bmatrix} 72.3419 & 218.7376 \\ 218.7376 & 626.9505 \end{bmatrix}.$$

Now, observe that

$$S_{\text{intra}} + S_{\text{inter}} = \begin{bmatrix} 1.7267 & 1.3167 \\ 1.3167 & 1.5867 \end{bmatrix} + \begin{bmatrix} 76.3419 & 218.7376 \\ 218.7376 & 626.9505 \end{bmatrix}$$

$$\cong \begin{bmatrix} 78.0686 & 220.0543 \\ 220.0543 & 628.5371 \end{bmatrix}$$

$$= S,$$

In addition, we also have

$$\text{trace}\,(S_{\text{intra}}) + \text{trace}\,(S_{\text{inter}}) = 3.3133 + 703.2924 \cong 706.6057 = \text{trace}(S),$$

where the approximation symbols are used because of numerical round-off errors.

8.3.3 Partitional Clustering

By partitional clustering (also called non-hierarchical clustering), it is usually meant that the clusters are obtained as a definite partition of the feature space with respect to a fixed number of clusters. A simple partitional clustering algorithm can be immediately obtained in terms of the trace-based dispersion measures introduced in the previous section, which can be used to implement the similarity clustering criterion, in the sense that a good clustering should exhibit low intraclass dispersion and high interclass dispersion. However, as the overall dispersion is preserved, these two possibilities become equivalent. A possible clustering algorithm based on such criteria is

---

Algorithm: *Clustering*

1. Assign random classes to each object;
2. **while** unstable
3.   **do**
4.   Randomly select an object and randomly change its class,
           avoiding to leave any class empty;
5.   If the intraclass dispersion, measured for instance in terms of the
           trace of the intraclass scatter matrix, increased, reassign the
           original class.

The termination condition involves identifying when the clusters have stabilized, which is achieved, for instance, when the number of unchanged successive classifications exceeds a pre-specified threshold (typically two). An important point concerning this algorithm is that the number of clusters usually is pre-specified. This is a consequence of the fact that the intraclass dispersion tends to decrease with larger numbers of clusters (indeed, in the extreme situation where each object becomes a cluster, the scattering becomes null), which tends to decrease the number of clusters if the latter is allowed to vary.

Figure 8.29 presents the progression of decreasing intraclass configurations (the intermediate situations leading to increased intraclass dispersion are not shown) obtained by the above algorithm, together with the respective total, inter and intraclass dispersions. Although the convergence is usually fast, as just a few interactions are

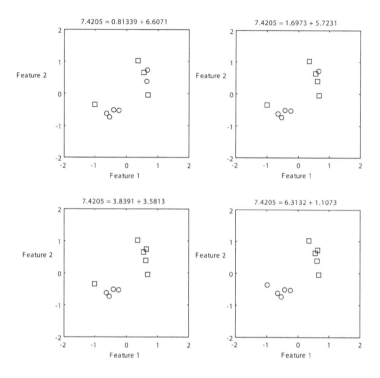

Figure 8.29: *The traces of the scatter matrices ("trace(S) = trace(S_{inter}) +*
trace(S_{intra})") for a sequence of cluster configurations. The
last clustering allows the smallest intracluster scattering.

usually required, this methodology unfortunately is not guaranteed to converge to
the absolute minimal intraclass dispersion (the local minimum problem), a problem
that can be minimized by using simulated annealing (see, for instance, [Press et al.,
1989; Rose et al., 1993]). In addition, if the trace of the scatter matrices is used,
different clusters can be obtained in case the coordinate axes of the feature space
are scaled [Jain & Dubes, 1988]. It can also be shown [Jain & Dubes, 1988] that
the quantification of the intraclass dispersion in terms of the trace of the respective
scatter matrix corresponds to a popular partitional clustering technique known as
square-error method, which tries to minimize the sum of the squared Euclidean
distances between the feature vectors representing the objects in each cluster and
the respective mean feature vectors. This can be easily perceived by observing
that the trace of the intraclass scatter matrix corresponds to the sum of the squared
distances.

An alternative clustering technique based on the minimal intraclass dispersion
criterion is commonly known as *k-means*, which can be implemented in increasing

degrees of sophistication. Here we present one of its simplest, but useful, versions. Figure 8.30 presents the overall steps typically involved in this approach, which are also characteristic of the hierarchical classification methods to be discussed in the next section. This scheme is similar to that generally used in classification

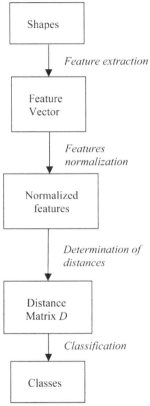

Figure 8.30: *Stages usually involved in distance-based partitional (and hierarchical) clustering.*

(see Figure 8.10 in Section 8.1.4), except for the additional stage corresponding to the *determination of the distances* between the feature vectors, yielding a *distance matrix D*. Basically, each entry at a specific row i and column j in this matrix, which is symmetric, corresponds to the distance between the feature vectors i and column j. Although it is also possible to consider a similarity matrix instead of a distance matrix, which can be straightforwardly done, this situation is not pursued further in this book.

The k-means technique starts with N objects, characterized in terms of their respective feature vectors, and tries to classify them into K classes. Therefore, the number of classes has to be known *a priori*. In addition, this method requires K initial prototype points P_i (or *seeds*), which may be supplied (characterizing some supervision over the classification) or somehow automatically estimated (for instance, by randomly selecting K points, if possible uniformly distributed along the considered region of the feature space). It is important to note that each centroid defines an *area of influence* in the feature space corresponding to the points that are closer to that centroid than to any other centroid. In other words, the set of centroids defines a Voronoi tessellation (see Section 5.10) of the feature space. The k-means method proceeds as follows:

Algorithm: *k-means*

1. Obtain the K initial prototype points and store them into the list W;
2. **while** unstable
3.   **do**
4.   Calculate all distances between each prototype point (or mean) P_i and each feature vector, yielding a $K \times N$ distance matrix D;
5.   Use the matrix D to identify the feature points that are closest to each prototype P_i (this can be done by finding the minimum values along each column of D). Store these points into a respective list L_i;
6.   Obtain as new prototype points the centroids of the feature points stored into each respective L_i;

A possible stability criterion corresponds to the situation when the maximum displacement of each centroid is smaller than a previously defined threshold. Observe in the above algorithm that one of the classes can become empty, which is caused by the fact that one (or more) of the centroids defines an empty area of influence (i.e., no feature vector is closer to that centroid than to the other centroids). To avoid such an effect, in case the number of classes has to be kept constant, one can change the prototype point of the empty class to some other value and recalculate the whole configuration. After termination, the objects closer to each specific resulting centroid are understood as belonging to the respectively defined class. The box titled *K-Means Classification* presents a numeric example of an application of this procedure.

Since the convergence to the smallest dispersion is not guaranteed by such interactive algorithms, it is particularly interesting and effective to consider several initial prototype points and to take as the best solution that configuration leading to the smallest dispersion, such as that measured in terms of the trace of the intraclass scattering matrix. Several additional variations and enhancements of this basic technique have been reported in the literature, including the possibility of merging the clusters corresponding to centroids that are too close (with respect to some supplied threshold) and splitting in two a cluster exhibiting too high a dispersion (this parameter has also to be determined *a priori*). Both strategies are used in the well-known ISODATA clustering algorithm [Gose et al., 1996].

Example: *k-means classification*

Apply the k-means algorithm in order to cluster into two classes the points characterized in terms of the following features:

Object	Feature 1	Feature 2
X_1	1	1
X_2	3	4
X_3	5	4

Consider as initial prototype points the vectors $P_1 = (0,0)$ and $P_2 = (3,3)$ and use 0.25 as minimum value for the termination criterion.

Solution:

① The initial distance matrix is

$$D = \begin{bmatrix} \sqrt{2} & 5 & \sqrt{41} \\ 2\sqrt{2} & 1 & \sqrt{5} \end{bmatrix} \quad \text{and} \quad L_1 = (X_1) \quad \text{and} \quad L_2 = (X_2, X_3).$$

Hence:

$$P_1 = mean\{X_1\} = X_1 \quad \text{and} \quad P_2 = mean\{X_2, X_3\} = (1,4) \quad \text{and}$$

$$m = \max\left\{\left\|\tilde{P}_1 - \vec{P}_1\right\|, \left\|\tilde{P}_2 - \vec{P}_2\right\|\right\} = \max\left\{5, \sqrt{5}\right\} = 5.$$

② As $m > 0.25$, we have a new interaction:

$$D = \begin{bmatrix} 0 & \sqrt{13} & 5 \\ \sqrt{18} & 1 & 1 \end{bmatrix} \quad \text{and} \quad L_1 = (X_1) \quad \text{and} \quad L_2 = (X_2, X_3).$$

Hence:

$$P_1 = mean\{X_1\} = X_1 \quad \text{and} \quad P_2 = mean\{X_2, X_3\} = (1,4) \quad \text{and}$$

$$m = \max\left\{\left\|\tilde{P}_1 - \vec{P}_1\right\|, \left\|\tilde{P}_2 - \vec{P}_2\right\|\right\} = 0.$$

Since $m < 0.25$, the procedure terminates, yielding as classes $C_1 = \{X_1\}$ and $C_2 = \{X_2, X_3\}$. The above two stages are illustrated in Figure 8.31, where the feature points are represented by crosses and the prototype points by squares.

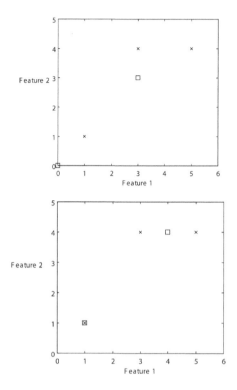

Figure 8.31: *The two stages in the above execution of the k-means algorithm.*

In the above classical k-means algorithm, at any stage each object is understood as having the class of the nearest mean. By allowing the same object to have probabilities of belonging to several classes, it is possible to obtain a variation of the k-means algorithm, which is sometimes known as *"fuzzy" k-means* (see, for instance, [Duda et al., 2000]. Although this method presents some problems, especially the fact that the probabilities depend on the number of clusters, it provides a clustering alternative worth trying in practice. The basic idea of the fuzzy k-means algorithm is described in the following.

Let the probability that an object p_j (recall that $j = 1, 2, \ldots, N$) belongs to the class C_i; $i = 1, 2, \ldots, K$; be represented as $P\left(C_i \mid p_j\right)$. At each step of the algorithm, the probabilities are normalized in such a way that for each object p_j we have:

$$\sum_{i=1}^{K} P\left(C_i \mid p_j\right) = 1.$$

The mean for each class at any stage of the algorithm is calculated as:

$$P_i = \frac{\sum_{j=1}^{N} \left[P\left(C_i \mid p_j \right) \right]^a p_j}{\left[\sum_{j=1}^{N} P\left(C_i \mid p_j \right) \right]^a},$$

where a is a real parameter controlling the interaction between each observation and the respective mean value. After all the new means P_i have been obtained by using the above equation, the new probabilities are calculated as follows:

$$P\left(C_i \mid p_j \right) = \frac{\left\| p_j - P_i \right\|^{\frac{2}{1-a}}}{\sum_{q=1}^{K} \left\| p_j - P_q \right\|^{\frac{2}{1-a}}}.$$

As in the classical k-means, this algorithm stops once the mean values stabilize.

8.3.4 Hierarchical Clustering

By hierarchical clustering it is usually meant that the grouping of M objects into K classes is performed *progressively* according to some parameter, typically the distance or similarity between the feature vectors representing the objects. In other words, the objects that are more similar to one another (e.g., the distance between them is smaller) are grouped into subclasses before objects that are less similar, and the process ends once all the objects have been joined into a single cluster. Observe that, unlike the partitional clustering methodology, which produces a single partition of the objects, hierarchical clustering provides several possible partitions, which can be selected in terms of a distance (or similarity) parameter. Although it is also possible to start with a single cluster and proceed by splitting it into subclusters (called divisive approach), the present book is limited to the more popular *agglomerative* approach, which starts with single element clusters and proceeds by merging them.

The basic stages in typical hierarchical clustering approaches are illustrated in Figure 8.30. In addition to feature extraction, feature normalization, and classification, we have an additional intermediate step consisting of the determination of a distance (or similarity) matrix indicating the distances (similarities) between each pair of feature vectors. As observed above, the progressive merging is performed by taking into account such a distance matrix, which is updated along the process. The obtained hierarchical classification can be represented as a (usually binary) tree, commonly called a *dendrogram*, which is illustrated in Figure 8.32. In this case, 20 objects have been grouped according to the distance values indicated at the ordinate axis (in this example, the minimal distance between two sets). The first pair of objects to be grouped are those identified by the numbers 6 and 8 along the abscissa axis, and the process continues by joining objects 11 and 12, 1 and 2, and so on, until all the subgroups are ultimately joined as a single cluster. For

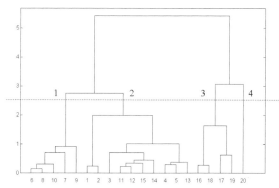

Figure 8.32: *A typical dendrogram clearly indicating the subgroups re-*
sulting from a hierarchical classification, in this case the
single linkage approach defined by the minimal distance be-
tween two sets.

generality's sake, it is convenient to consider each of the original objects as a subclass containing a single element.

A particularly interesting feature of hierarchical clustering schemes is the fact
that they make the clustering structure, defined by the similarity between objects,
very clear, emphasizing the relationship between the several clusters. However, observe that the dendrogram by itself does not indicate the correct number of classes.
Indeed, it is usually possible to define any number of classes, from 1 (the top of the
dendrogram) to the number of objects, M, except when more than two objects are
joined simultaneously into a single subgroup. The different number of classes are
obtained by horizontally cutting the dendrogram at different distance values. Provided the number of classes K has been accordingly chosen, the respective objects
can be obtained by tracing down the subtrees defined while cutting the dendrogram
at a distance point corresponding to the K branches. This process is illustrated in
Figure 8.32, where the horizontal line at the distance 2.5 (indicated by the dashed
line) has defined four clusters, which are arbitrarily numbered from left to right
as 1, 2, 3 and 4. The respective objects can be obtained by following down the
subtrees, yielding:

Cluster #	Objects
1	$\{6, 7, 8, 9, 10\}$
2	$\{1, 2, 3, 4, 5, 11, 12, 13, 14, 15\}$
3	$\{16, 17, 18, 19\}$
4	$\{20\}$

Observe that dendrograms are inherently similar to the hierarchical taxonomies
normally defined in biological sciences. However, the two approaches generally
differ in that the classification criterion (e.g., the adopted features and distance

values) typically remains the same during the whole determination of dendrograms, while it can vary in biological taxonomies.

The remainder of this section presents several possible distances between sets, which define the respective hierarchical clustering methods, including single and complete linkage, average, centroid and Ward's.

Distances between Sets

Although usually understood with respect to two points, the concept of distance can be extended to include distances between two sets A and B. There are several alternative possibilities for doing so, four of the most popular are given in Table 8.6, together with the respectively defined hierarchical cluster techniques. Figure 8.33 illustrates the minimal, maximal and centroid distances between two

Distance between two sets A and B	Comments	Hierarchical clustering
$\text{dist}\{A, B\} = \min_{\substack{x \in A \\ y \in B}} (\text{dist}\{x, y\})$	Minimal distance between any of the points of A and any of the points of B.	Single linkage
$\text{dist}\{A, B\} = \max_{\substack{x \in A \\ y \in B}} (\text{dist}\{x, y\})$	Maximum distance between any of the points of A and any of the points of B.	Complete linkage
$\text{dist}\{A, B\} = \frac{1}{N_A N_B} \sum_{\substack{x \in A \\ y \in B}} \text{dist}(x, y)$	Average of the distances between each of the N_A points of A and each of the N_B points of B.	Group average
$\text{dist}\{A, B\} = \text{dist}\{C_A, C_B\}$	Distance between the centers of mass (centroids) of the points in set A (i.e., C_A) and B (i.e., C_B).	Centroid

Table 8.6: *Four definitions of possible distances between two sets A and B.*

clusters. For instance, the minimal distance, which corresponds to the minimal distance between any two points respectively taken from each of the two sets, defines the single linkage clustering algorithm. It is interesting to observe that the average group distance represents an intermediate solution between the maximal and minimal distances. Observe also that each of the presented distances between two sets can comprise several valid distances for $\text{dist}(x, y)$, such as Euclidean, city-block and chessboard. The choice of such distances, together with the adopted metrics

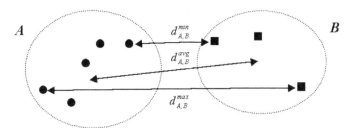

Figure 8.33: *Minimal ($d_{A,B}^{min}$), maximal ($d_{A,B}^{max}$), and average ($d_{A,B}^{avg}$) distances between the sets A and B.*

(usually Euclidean), completely define the specific properties of the hierarchical clustering technique based on the typical algorithm to be described in the next section.

Distance-Based Hierarchical Clustering

Once the distance between two sets and the respective metrics have been chosen (e.g., complete linkage with Euclidean metrics), the following linkage procedure is performed in order to obtain the hierarchical clustering:

① Construct a distance matrix D including each of the distances between the initial N objects, which are understood as the initial single element clusters C_i, $i = 1, 2, \ldots, N$;

② $n = 1$;

③ While $n < N$:

    (a) Determine the minimal distance in the distance matrix, dmin, and the respective clusters C_j and C_k, $j < k$, defining that distance;

    (b) Join these two clusters into a new single cluster C_{N+n}, which is henceforth represented by the index j;

    (c) $n = n + 1$;

    (d) Update the distance matrix, which becomes reduced by the row and column corresponding to the index k.

Since the dendrograms obtained by hierarchical clusters are usually considered to be a binary tree (i.e., only branches involving two subclusters are allowed), matches between distances must usually be resolved by using some pre-specified criterion (such as randomly selecting one of the cases). The above simple algorithm is illustrated in terms of a real example and with respect to single linkage with Euclidean distance in the box entitled *Single Linkage Hierarchical Clustering*.

Example: *Single Linkage Hierarchical Clustering*

Group the objects into the following data matrix by using single linkage with Euclidean distance:

$$F = \begin{bmatrix} 1.2 & 2.0 \\ 3.0 & 3.7 \\ 1.5 & 2.7 \\ 2.3 & 2.0 \\ 3.1 & 3.3 \end{bmatrix}.$$

Solution:

① We have $N = 5$ objects, with respective distance matrix:

$$D^{(1)} = \begin{bmatrix} 0 & & & & \\ 2.4759 & 0 & & & \\ 0.7616 & 1.8028 & 0 & & \\ 1.1000 & 1.8385 & 1.0630 & 0 & \\ 2.3022 & 0.4123 & 1.7088 & 1.5264 & 0 \end{bmatrix} \begin{matrix} C_1 \\ C_2 \\ C_3 \\ C_4 \\ C_5 \end{matrix}.$$

② $n = 1$, and the minimal distance is dmin $= 0.4123$, with $j = 2$ and $k = 5$, which leads to the new cluster $C_{1+5} = C_6 = \{C_2 C_5\}$. The new distance matrix is

$$D^{(2)} = \begin{bmatrix} 0 & & & \\ 2.3022 & 0 & & \\ 0.7616 & 1.7088 & 0 & \\ 1.1000 & 1.5264 & 1.0630 & 0 \end{bmatrix} \begin{matrix} C_1 \\ C_2 C_5 = C_6 \\ C_3 \\ C_4 \end{matrix}.$$

③ $n = 2$, and the minimal distance is dmin $= 0.7616$, with $j = 1$ and $k = 3$, which leads to the new cluster $C_{2+5} = C_7 = \{C_1 C_3\}$. The new distance matrix is

$$D^{(2)} = \begin{bmatrix} 0 & & \\ 1.7088 & 0 & \\ 1.0630 & 1.5264 & 0 \end{bmatrix} \begin{matrix} C_1 C_3 = C_7 \\ C_2 C_5 = C_6 \\ C_4 \end{matrix}.$$

④ $n = 3$, and the minimal distance is dmin $= 1.0630$, with $j = 1$ and $k = 3$, which leads to the new cluster $C_{3+5} = C_8 = \{C_1 C_3 C_4\}$. The new distance matrix is

$$D^{(2)} = \begin{bmatrix} 0 & \\ 1.5264 & 0 \end{bmatrix} \begin{matrix} C_1 C_3 C_4 = C_8 \\ C_2 C_5 = C_6 \end{matrix}.$$

⑤ $n = 4$, and the minimal distance is dmin $= 1.5264$, with $j = 1$ and $k = 2$, which leads to the last cluster $C_{3+5} = C_9 = \{C_1 C_3 C_4 C_2 C_5\}$. The obtained dendrogram is illustrated in Figure 8.34, and the nested structure of the clusters is shown in Figure 8.35.

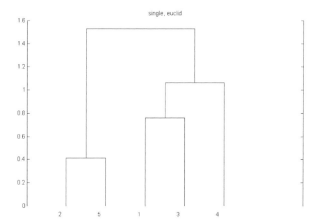

Figure 8.34: *The obtained dendrogram.*

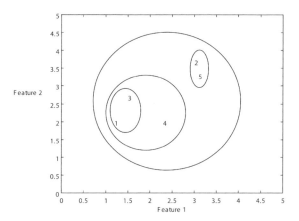

Figure 8.35: *The nested structure of the obtained clusters.*

Dispersion Based Hierarchical Clustering—Ward's Method

Instead of considering distance matrices, it is also possible to use the intraclass (or interclass) dispersion as a clustering criterion. This works as follows: at first, each feature vector is understood as a cluster, and the intraclass dispersion (such as that measured by the trace) is therefore null. The pairs of points to be merged into a cluster are chosen in such a way as to ensure the smallest increase in the intraclass dispersion as the merges are successively performed. Although such a property is guaranteed in practice, it should be borne in mind that the partition obtained for a specific number of clusters is not necessarily optimal as far as the overall resulting intraclass dispersion is concerned. One of the most popular dispersion-based hierarchical cluster algorithms is known as Ward's [Anderberg, 1973; Jain, 1989; Romesburg, 1990]. Other variations of this method are described in [Jain, 1989].

Hierarchical Clustering Validation

Having obtained a dendrogram, how much confidence can we have that it indeed reflects the original structure of the data? Despite its practical importance, this is a particularly difficult problem in clustering analysis that has not yet been settled definitely, despite the several approaches described in the literature (e.g., [Aldenderfer & Blashfield, 1984; Anderberg, 1973; Jain, 1989]. In this section we present two of the simplest possibilities for hierarchical clustering validation, namely the cophenetic correlation coefficient and replication.

Cophenetic Correlation Coefficient

The *cophenetic correlation coefficient* is defined in terms of the cross-correlation coefficient between the elements in the lower diagonal (remember that this excludes the main diagonal) of the original distance matrix D and the *cophenetic matrix*. The latter is defined as having as entries the distance at which two objects first appeared together in the same cluster. The determination of the cophenetic correlation coefficient is illustrated in the box entitled *Cophenetic Correlation Coefficient*. The higher the value of this coefficient, the more representative the result should be. However, it should be noted that the value of such a coefficient as an indication of the hierarchical cluster validity has been strongly criticized. An in-depth treatment of the cophenetic coefficient and other validation measures can be found in [Jain & Dubes, 1988].

Example: *Cophenetic Correlation Coefficient*

Determine the cophenetic matrix and the cophenetic correlation coefficient for the hierarchical single linkage clustering in the example in the box entitled *Single Linkage Hierarchical Clustering*.

Solution:

The original distance matrix is:

$$D = \begin{bmatrix} 0 & & & & \\ 2.4759 & 0 & & & \\ 0.7616 & 1.8028 & 0 & & \\ 1.1000 & 1.8385 & 1.0630 & 0 & \\ 2.3022 & 0.4123 & 1.7088 & 1.5264 & 0 \end{bmatrix}.$$

The objects 2 and 5 appear together for the first time at distance 0.4123, hence

$$CP = \begin{bmatrix} 0 & & & & \\ - & 0 & & & \\ - & - & 0 & & \\ - & - & - & 0 & \\ - & 0.4123 & - & - & 0 \end{bmatrix}.$$

The next merge, which occurred at distance 0.7616, brought together for the first time objects 1 and 3, hence

$$CP = \begin{bmatrix} 0 & & & & \\ - & 0 & & & \\ 0.7616 & - & 0 & & \\ - & - & - & 0 & \\ - & 0.4123 & - & - & 0 \end{bmatrix}.$$

The cluster $\{C_1 C_3 C_4\}$ defined at distance 1.0630 brings object 4 together with objects 1 and 3, hence

$$CP = \begin{bmatrix} 0 & & & & \\ - & 0 & & & \\ 0.7616 & - & 0 & & \\ 1.0630 & - & 1.0630 & 0 & \\ - & 0.4123 & - & - & 0 \end{bmatrix}.$$

Finally, at distance 1.5264, all the objects were joined, implying

$$CP = \begin{bmatrix} 0 & & & & \\ 1.5264 & 0 & & & \\ 0.7616 & 1.5264 & 0 & & \\ 1.0630 & 1.5264 & 1.0630 & 0 & \\ 1.5264 & 0.4123 & 1.5264 & 1.5264 & 0 \end{bmatrix}.$$

The cophenetic correlation coefficient can now be obtained as the correlation coefficient between the elements in the low diagonal portions of matrices D and CP (excluding the main diagonal), yielding 0.90, which suggests a good clustering quality.

Replication

Replication is a validation mechanism motivated by the fact that a sound clustering approach should be stable with respect to different sets of sampled objects and to different clustering methodologies. In the former case, several sets of objects are clustered separately and the results compared. In case the obtained clusters are practically the same, this is understood as some limited support for the validation of the results. In the second situation, the same set of objects are clustered by several hierarchical clustering methods and the results compared. A reasonable agreement between the obtained results can be taken as an indication of the validity of the obtained classes. Indeed, in case the clusters are well separated, such as in Figure 8.28 (a), most clustering approaches will lead to the same result. However, in practice the different natures of the clustering algorithms will tend to amplify specific characteristics exhibited by the feature vector dispersion, almost always leading to different clustering structures.

Determining the Relevance and Number of Clusters

Since hierarchical clustering approaches provide a way for organizing the N original objects into an arbitrary number of clusters $1 \leqslant K \leqslant N$, the important issue of selecting a suitable number of clusters is inherently implied by this kind of clustering algorithm. Not surprisingly, there is no definitive criterion governing such a choice, but only tentative guidelines, a few of which are briefly presented and discussed in the following.

One of the most natural indications about the *relevance* of a specific cluster is its *lifetime*, namely the extent of the distance interval defined from the moment of its creation up to its merging with some other subgroup. Therefore, a possible criterion for selecting the clusters (and hence their number) is to take into account the clusters with the highest lifetime. For instance, the cluster $\{C_2 C_5\}$ in Figure 8.34 exhibits the longest lifetime in that situation and should consequently be taken as one of the resulting clusters. A related approach to determining the number of clusters consists of identifying the largest jumps along the clustering distances (e.g., [Aldenderfer & Blashfield, 1984]). For instance, in the case of Figure 8.32, we have:

Number of clusters	Distance	Distance jump
4	0.4123	0.3493
3	0.7616	0.3014
2	1.0630	0.4643
1	1.5264	—

Thus, in this case we verify that the maximum jump for two clusters indicates that this is a reasonable choice for the number of clusters.

A Brief Comparison of Methods

The above-presented hierarchical methods exhibit some interesting specific properties worth discussing. In addition, several investigations (usually considering simulated data, e.g., by using Monte Carlo) have been made intending to theoretically and experimentally quantify the advantages and disadvantages of each method. Some of the more relevant properties and tendencies of the considered hierarchical cluster algorithms are briefly reviewed in the following:

Single Linkage: Although this method presents some distinctive mathematical properties, such as *ultrametrics* or *monotonicity* (see, for instance, [Jain & Dubes, 1988]), which guarantee that the clusters always occur at increasing distance values, in practice its performance has often been identified as among the poorest. It has been found to be particularly unsuitable for Gaussian data [Bayne et al., 1980]; but it is less affected by outliers [Milligan, 1980] and is one of few methods that work well for nonellipsoidal clusters such as those shown in Figure 8.28 (b), (c) and (f) [Anderberg, 1973; Everitt, 1993]. The obtained clusters, however, have a tendency to present *chaining*, i.e., the tendency to form long strings [Anderberg, 1973; Everitt, 1993] which, while not being properly a problem, tends to merge well-separated clusters linked by a few points [Everitt, 1993].

Complete Linkage: This alternative also exhibits the ultrametric property [Anderberg, 1973; Jain & Dubes, 1988], but seeks ellipsoidal, compact clusters. It has been identified as being particularly poor for finding high density clusters [Hartigan, 1985].

Group Average Linkage: Tends to produce clustering results similar to those obtained by the complete linkage method [Anderberg, 1973] method, but performs poorly in the presence of outliers [Milligan, 1980].

Centroid Linkage: Suggested for use only with Euclidean distance [Jain & Dubes, 1988], this technique presents as shortcoming the fact that the merging distances at successive mergings are not monotonic [Anderberg, 1973; Jain & Dubes, 1988]. It has been identified as being particularly suitable for treating clusters of different sizes [Hands & Everitt, 1987].

Ward's Linkage: This dispersion-based clustering approach has often been identified as a particularly superior, or even the best, hierarchical method (e.g., [Anderberg, 1973; Blashfield, 1976; Gross, 1972; Kuiper & Fisher, 1975; Mojena, 1975]). It seeks ellipsoidal and compact clusters, and is more effective when the clusters have the same size [Hands & Everitt, 1987; Milligan & Schilling, 1985], tending to absorb smaller groups into larger ones [Aldenderfer & Blashfield, 1984]. It is monotonic regarding the successive merges [Anderberg, 1973], but performs poorly in the presence of outliers [Milligan, 1980].

From the above-mentioned works, we conclude that the issue of identifying the best clustering approach, even considering specific situations, is still far from being settled. In practice, the selection of a method should often involve applying several alternative methods and then adopting the one most compatible with what is expected.

8.4 A Case Study: Leaves Classification

This section presents and discusses several situations and problems of clustering with respect to the classification of leaves. More specifically, four species of leaves (illustrated in Figure 8.12), totaling 79 examples, have been scanned with the same resolution. In order to carry out the classification, we applied the single, complete and average linkage, centroid and Ward's hierarchical clustering algorithms. The enumeration of these algorithms is given in Table 8.7.

Method #	Method
1	Single linkage
2	Complete linkage
3	Average group linkage
4	Centroid
5	Ward's

Table 8.7: *The considered five hierarchical clustering methods.*

Since the original leaf classes are known, they can be used as a standard for comparing misclassifications, which allows us to discuss and illustrate several important issues on hierarchical clustering, including

☞ how the choice of clustering method affects the performance;

☞ the effects of the metrics choice;

☞ how the adopted features affect the performance of the clustering algorithms;

☞ the influence of normal transformation;

☞ validation of the obtained clusters in terms of misclassifications and cophenetic correlation coefficient.

Table 8.8 presents the eight considered features, which include an eclectic selection of different types of simple measures.

Feature #	Feature
1	Area
2	Perimeter
3	Circularity
4	Elongation
5	Symmetry
6	Gray-level histogram average
7	Gray-level histogram entropy
8	Gray-level variation coefficient

Table 8.8: *The eight features considered in the leaves example.*

The henceforth presented results have been obtained by comparing the obtained clusters with the known original classes, thus determining the number of misclassifications. In every case, the number of clusters was pre-defined as four, i.e., the number of considered plant species. Determining the misclassification figures is not trivial and deserves some special attention. Having obtained the clusters, which are enumerated in an arbitrary fashion, the problem consists of making these arbitrary labels correspond with the original classes. In order to do so, a matrix is constructed whose rows and columns represent, respectively, the new (arbitrary) and the original class numbers. Then, for each row, the number of elements in the new class corresponding to each original class is determined and stored into the respective columns, so that each row defines a histogram of the number of original objects included into the respective new cluster. For instance, the fact that the cell at row 3 and column 2 of this matrix contains the number 5 indicates that the new cluster number 3 contains 5 elements of the original class number 2. Having defined such a matrix, it is repeatedly scanned for its maximum value, which defines the association between the classes corresponding to its row and column indexes, and lastly the respective data corresponding to these classes is removed from the table. The process continues until all original and arbitrary class numbers have been placed in correspondence. Then, all that remains is to compare how many objects in the obtained clusters have been wrongly classified.

It should be emphasized that it would be highly tendentious and misleading if we generalized the results obtained in terms of simplified evaluations such as that presented in this section. However, the obtained results clearly illustrate some of the most representative problems and issues encountered while applying hierarchical clustering algorithms.

8.4.1 Choice of Method

Figure 8.36 presents the average and standard deviation of the misclassifications produced by each of the five considered hierarchical methods by taking into account all possible combinations of features in both a 2-by-2 (a) and a 3-by-3 (b) fashion (after unit variance normalization and adopting Euclidean metrics). Both situations

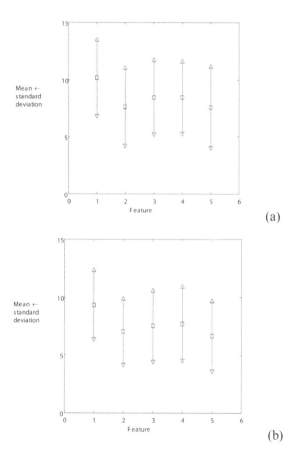

(a)

(b)

Figure 8.36: *The average and standard deviations of the misclassifications by each of the five considered hierarchical methods considering all possible combinations of 2 (a) and 3 (b) features (unit variance normalized).*

led to substantially similar results, with clear advantage to the complete linkage and Ward's methods. The single linkage represented the poorest overall performance. While such results tend to corroborate some of the general tendencies discussed in this chapter, they should not be immediately generalized to other situations. Given the obtained results, the following sections are limited to Ward's approach.

8.4.2 Choice of Metrics

The choice of metrics remains one of the unsettled issues in clustering. However, given the isotropy (implying invariance to rotations of the feature space) of Euclidean metrics, it is particularly suitable for most practical situations, and has by far been the most frequently adopted option in the literature. Figure 8.37 presents the dendrograms obtained by Ward's hierarchical clustering approach adopting Euclidean (a) and city-block (b) metrics. The considered features included circularity, histogram average and entropy (after unit variance normalization). Although

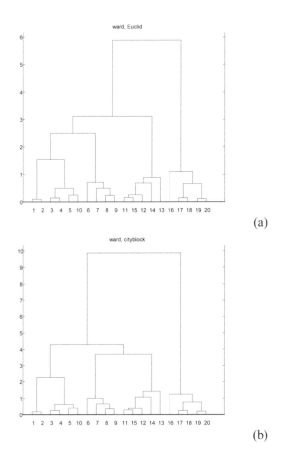

(a)

(b)

Figure 8.37: *Hierarchical clustering of leaf data through the Ward's method; considering circularity, histogram average and entropy as features (after unit variance normalization); adopting Euclidean (a) and city-block (b) metrics.*

different clustering structures have been obtained for large distances, no variations have been observed for a total of four clusters, a tendency also observed for the other considered clustering algorithms. The Euclidean metrics is adopted henceforth in this section.

8.4.3 Choice of Features

As already observed in this chapter, the feature choice is usually much more critical than the choice of methods or normalization. We have characterized the performance of both the Ward hierarchical clustering approach and the k-means partitional clustering in terms of the number of misclassifications considering all possible 2-by-2 and 3-by-3 combinations of the features in Table 8.8, which is graphically depicted in Figures 8.38 and 8.39.

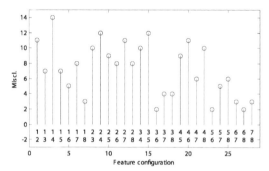

Figure 8.38: *Average misclassifications by Ward's and k-means methods considering all possible combinations of 2 and 3 features, after unit variance normalization: Ward for 2 features.*

The features have been normalized to unit variance and zero mean. The first interesting result is that the number of misclassifications varies widely in terms of the selected features. In other words, *the choice of features is confirmed as being crucial for proper clustering*. In addition, a careful comparative analysis of the results obtained for Ward's and k-means techniques does not indicate an evident advantage for either of these methods, except for a slight advantage of Ward's approach, especially for the 3-feature combinations. Moreover, the feature configurations tend to imply similar clustering quality in each method. For instance, the combinations involving features 6 (histogram average), 7 (histogram entropy) and 8 (histogram variation coefficient) tended to consistently provide less misclassifications despite the adopted clustering method. Such results confirm that the proper selection of feature configurations is decisive for obtaining good results. It has been experimentally verified that the incorporation of a larger number of features did not improve the clustering quality for this specific example.

Figure 8.40 and Figure 8.41 present the misclassification figures corresponding to those in Figure 8.38 and Figure 8.39, obtained without such a normalization strategy.

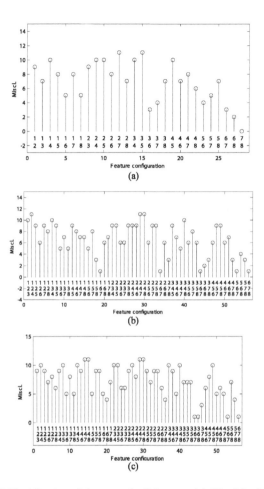

Figure 8.39: *(Continued) k-means for 2 features (a); Ward for 3 features (b); and k-means for 3 features (c). Each feature configuration is identified by the list of number at the bottom of each graph. For instance, the leftmost feature configuration in Figure 8.38 corresponds to features 1 and 2 from Table 8.8, identified respectively as area and perimeter.*

8.4.4 Validation Considering the Cophenetic Correlation Coefficient

Figure 8.42 presents the cophenetic correlation coefficient in terms of the misclassification figures obtained from Ward's method considering three normalized features.

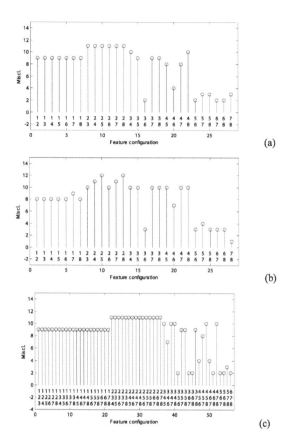

Figure 8.40: *Misclassification figures corresponding to those in Figure 8.38, but without unit variance normalization or principal component analysis. Refer to the caption of Figure 8.38.*

Although it could be expected that the misclassification figure would be negatively correlated with the cophenetic coefficient, it is clear from this graph that these two parameters, at least in the case of the present data, were not correlated. Although not conclusive, such a result confirms the criticisms of the use of the cophenetic correlation coefficient as a measure of clustering structure quality.

8.5 Evaluating Classification Methods

Since there is no consensus about the choice of classification and clustering methods considering general problems, and given the large number of alternative approaches described in the literature, it is important to devise means for comparing

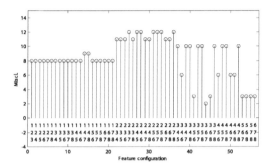

Figure 8.41: *(Continued) Misclassification figures corresponding to those in Figure 8.38, but without unit variance normalization or principal component analysis. Refer to the caption of Figure 8.38.*

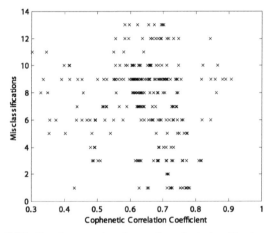

Figure 8.42: *Graph expressing the number of misclassifications in terms of the cophenetic correlation coefficient. Although such measures could be expected to be negatively correlated, the obtained results indicate that they are substantially uncorrelated.*

the available alternatives. Unfortunately, this has proven to be a difficult problem. To begin with, several representative problems should be defined, involving several distributions of feature vectors. A particularly interesting approach consists of using not only real data, but also simulated sets of objects, which allows a more complete control over the properties and characteristics of the feature vector organization in the feature space. In addition, a large number of suitable features should be considered, yielding a large number of combinations. The situation becomes more manageable when the methods are compared with respect to a single classification problem. A simple example has been presented in Section 8.4 regarding the comparison of five hierarchical approaches applied to leaf classification. Usually, methods can be compared with respect to performance parameters such as the overall execution time, complexity of the methods, sensitivity to the choice of features and, more importantly, the overall number of misclassifications. The weight of each of these parameters will depend on the specific applications. Real-time problems, for instance, will pay greater attention to the execution time. In the case of misclassifications, in addition to the overall number of mistakes, it is often interesting to consider what is sometimes called the *confusion matrix*. Basically, this is a square matrix whose rows and columns are associated with each of the original classes. Each element (i, j) of this matrix represents the number of objects that were of class i but were classified as belonging to class j. Therefore, the more such a matrix approximates to the diagonal matrix, the better the classification performance. In addition, it is possible to characterize bias in the classification. Consider, as an example, the following confusion matrix, which considers 5 classes:

$$\text{Confusion\_Matrix} = \begin{bmatrix} 23 & 5 & 2 & 0 & 0 \\ 10 & 30 & 5 & 1 & 1 \\ 9 & 2 & 12 & 1 & 3 \\ 25 & 2 & 3 & 5 & 9 \\ 0 & 0 & 0 & 0 & 49 \end{bmatrix}.$$

It is clear from this matrix that no error has been obtained while classifying objects of class 5, but the majority of objects of classes 3 and 4 have been incorrectly classified. A strong tendency to misclassify objects originally in class 4 as class 1 is also evident. Observe that the sum along each row i corresponds to the total number of objects originally in class i.

Another particularly promising alternative for comparing and evaluating classification methods is to use *data mining* approaches. More specifically, this involves considering a substantially large number of cases representing several choices of features, classification methods and parameters, and using statistical and artificial intelligence methods. For instance, the genetic algorithm [Bäck, 1996; Holland, 1975] could be used to search for suitable feature configurations while considering the correct classification ratios as the fitness parameter.

8.5.1 | Case Study: Classification of Ganglion Cells

Some illustrative results of using the NMWE and the NMBE (refer to Chapter 7) for automatic neural cell classification are presented in the following experiment (adapted from [Cesar-Jr. & Costa, 1998b]), which is based on the classification of cat retinal ganglion cells (α-cells and β-cells). This type of cell has interested neuroscientists during the last decades, being an excellent example of the interplay between form and function. Indeed, a good consistency has been found between the above morphological types and the two physiological classes known as X- and the Y-cells. The former cells, that present a morphology characteristic of β-class, normally respond to small-scale stimuli, while the latter, related to the α-class, are associated with the detection of rapid movements. Boycott and Wässle have proposed the morphological classes for α-cells and β-cells (as well as γ-cells, which are not considered here) based on the neural dendritic branching pattern [Boycott & Wässle, 1974]. Generally, the α-cells dendritic branching spreads around a larger area, while the β-cells are more densely concentrated with respect to their dendrites, with less small-scale detail. Examples of some of these cells are presented in Figure 8.43 with respect to prototypical synthetic cells.

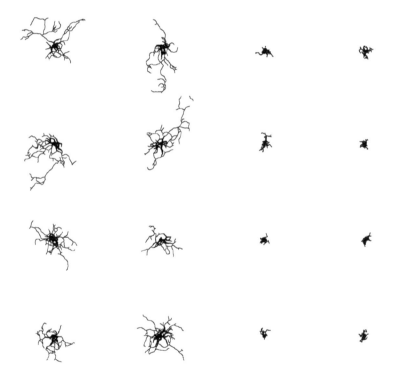

Figure 8.43: *Two morphological classes of cat ganglion cells: α-cells **(a)** and β-cells **(b)**. The cells have been artificially generated by using stochastic formal grammars [Costa et al., 1999].*

The considered 53 cells followed previous classifications by experts [Boycott & Wässle, 1974; Fukuda et al., 1984; Kolb et al., 1981; Leventhal & Schall, 1983;

Saito, 1983]. Each cell image was pre-processed by median filtering and morphological dilation in order to reduce spurious noise and false contour singularities.

All cells were edited in order to remove their self-intersections, which was followed by contour extraction. The original contours of the database have, in general, between $1,000$ and $10,000$ points, which implies two additional difficulties that must be circumvented. First, it is more difficult to establish fair criteria to make comparisons among contours of different lengths. Furthermore, the more efficient implementations of FFT algorithms require input signals of length equal to an integer power of 2. In order to address these problems, all contours have been interpolated and resampled (sub-pixel resolution), in order to have the same number of points (in the case of the present experiment, $8192 = 2^{13}$).

8.5.2 The Feature Space

Some neuromorphometric experiments have been devised in order to explore the aforementioned energy-based shape analysis techniques, illustrating their capabilities (Chapter 7). The experiments involved two main aspects of pattern recognition problems, namely feature selection for dimensionality reduction and pattern classification. Five cells from each class were used as the training set for two simple statistical classifiers, a k-nearest neighbors and a maximum-likelihood classifier. The maximum-likelihood classifier adopted a multivariate normal density distribution with equal *a priori* probabilities for both classes. A total of 100 features have been calculated for each cell, and the features are stored in two arrays, one for α-cells and the other for β-cells, which are explained below:

Fractal Dimension (FD): The fractal dimension is denoted as $M_{\alpha,1}(j)$ for the j-th α-cell and as $M_{\beta,1}(j)$ for the j-th β-cell.

Normalized Multiscale Bending Energy (NMBE): The NMBE has been calculated for 32 different scales, being denoted in the current experiment as $M_{\alpha,m}(j)$ for the j-th α-cell and as $M_{\beta,m}(j)$ for the j-th β-cell, with $m = 2, 3, \ldots, 33$. The NMBEs are in coarse-to-fine order, i.e., decreasing in scale, with the larger scale corresponding to $m = 2$ and the smallest to $m = 33$.

Dendritic Arborization Diameter (DAD): This feature is denoted as $M_{\alpha,34}(j)$ for the j-th α-cell and as $M_{\beta,34}(j)$ for the j-th β-cell.

Soma Diameter (SD): This feature is represented as $M_{\alpha,35}(j)$ for the j-th α-cell and as $M_{\beta,35}(j)$ for the j-th β-cell.

Normalized Multiscale Wavelet Energy (NMWE): The NMWE was also calculated for 32 different scales, being denoted as $M_{\alpha,m}(j)$ for the j-th α-cell and as $M_{\beta,m}(j)$ for the j-th β-cell, with $m = 36, 37, \ldots, 65$. The NMWEs similarly are in coarse-to-fine order, i.e., decreasing in scale, with the larger scale corresponding to $m = 36$ and the smallest to $m = 65$.

Fourier Descriptors (FDs): A set of 30 FDs NFD(s), as defined in Chapter 6, were calculated for each cell, being denoted as $M_{\alpha,m}(j)$ for the j-th α-cell and as $M_{\beta,m}(j)$ for the j-th β-cell, with $m = 66, 67, \ldots, 98$.

FD of Shen, Rangayyan and Desautels (FF): As explained in Chapter 6, the descriptors NFD(s) are used in the definition of the following FD [Shen et al., 1994]:

$$FF = \frac{\left[\sum_{s=-(N/2)+1}^{N/2} \frac{|\text{NFD}(s)|}{|s|} \right]}{\sum_{s=-(N/2)+1}^{N/2} |\text{NFD}(s)|}.$$

Therefore, the FF descriptor was also considered in the current experiment, being represented as $M_{\alpha,99}(j)$ for the j-th α-cell and as $M_{\beta,99}(j)$ for the j-th β-cell.

Fourier Energy (FE): The last shape descriptor to be included is the energy of NFD(s) defined as follows (see Chapter 6):

$$EF = \sum_{s=-(N/2)+1}^{N/2} |\text{NFD}(s)|^2.$$

This measure is denoted as $M_{\alpha,100}(j)$ for the j-th α-cell and as $M_{\beta,100}(j)$ for the j-th β-cell.

The logarithm of all measured features was taken in order to attenuate the effects of large variation in their magnitude (see Section 3.2.1). Furthermore, all features have been normalized in order to fit within a similar dynamic range.

8.5.3 Feature Selection and Dimensionality Reduction

The design of a pattern classifier includes an attempt to select, among a set of possible features, a minimum subset of weakly correlated features that better discriminate the pattern classes. This is usually a difficult task in practice, normally requiring the application of heuristic knowledge about the specific problem domain. Nevertheless, some useful clues can be provided by feature ordering techniques and trial-and-error design experiments. An example of such feature ordering techniques is the so-called *class separation distance* [Castleman, 1996], which is considered here in order to assess the above-defined set of 100 possible features. Let $\mu_{\alpha,m}$ and $\mu_{\beta,m}$ be the estimated mean of the m−th feature for the α and β classes, respectively; and let $\sigma_{\alpha,m}^2$ and $\sigma_{\beta,m}^2$ be the estimated variance of the m−th feature for the α and β classes, respectively, for $m = 1, 2, \ldots, 100$. These values can be easily estimated from the features database as follows:

$$\mu_{\alpha,m} = \frac{1}{N_\alpha} \sum_{j=1}^{N_\alpha} M_{\alpha,m}(j),$$

$$\mu_{\beta,m} = \frac{1}{N_\beta} \sum_{j=1}^{N_\beta} M_{\beta,m}(j),$$

$$\sigma_{\alpha,m}^2 = \frac{1}{N_\alpha} \sum_{j=1}^{N_\alpha} \left(M_{\alpha,m}(j) - \mu_{\alpha,m}\right)^2,$$

$$\sigma_{\beta,m}^2 = \frac{1}{N_\beta} \sum_{j=1}^{N_\beta} \left(M_{\beta,m}(j) - \mu_{\beta,m}\right)^2,$$

where N_α and N_β is the total number of α and β cells of the image database, respectively. The class separation distance between the α and β classes with respect to the m-th feature is defined as:

$$D_{\alpha,\beta,m} = \frac{\left|\mu_{\alpha,m} - \mu_{\beta,m}\right|}{\sqrt{\sigma_{\alpha,m}^2 + \sigma_{\beta,m}^2}}.$$

The potential for discrimination capabilities of each feature (alone) increases with $D_{\alpha,\beta,m}$. In the performed experiments, the features with larger $D_{\alpha,\beta,m}$ correspond to the small-scale bending energies, followed by the wavelet energies and the dendritic diameter. The remaining features led to poorer performance. The performance of the small-scale bending energies can be explained because the α-cell dendrites spread more sparsely than the β-cell dendrites, especially with respect to the soma diameter. Furthermore, the α-cells have, in general, a larger number of terminations and more ragged segments. These shape characteristics are observed both in the bending and the wavelet energy for small scales, thus emphasizing the complexity differences between the shape classes. Furthermore, adjacent scale energies tend to have similar class separation distance values between the classes. On the other hand, while small-scale energies show a good discrimination potential in the case of this pattern classification problem, the performance of intermediary and large-scale energies considerably decreases. In fact, the bending energy presents some of the smaller class separation distances, illustrates an important and very common problem in multiscale shape analysis: although shapes may be composed of several structures of different scales, it is important to attempt to identify the *good* analyzing scales.

As already observed, adjacent scale energies present similar $D_{\alpha,\beta,m}$ values. In fact, the information between neighboring scales is highly correlated and redundant, which should be taken into account by the feature selection process that defines the feature vectors used by statistical classifiers. The reason for this correlation and redundancy among neighbor energies is that the signal is analyzed by similar kernels. Therefore, suppose that we want to define a 2D feature vector, i.e., a feature vector composed of two features, using the wavelet energies extracted from 33 ganglion cells, as explained before. An important related question is whether is better to choose a large and a small scale, or two different small scales. If only the class separation distance is taken into account, the latter option seems to be more appropriate, since the small scale energies show larger class separation distances than do larger scales. Nevertheless, after a deeper analysis, it turns out that this is not necessarily true. In fact, the features extracted from similar scales are highly correlated, indicating that one of the two features can be eliminated, for high correlations between features of a feature vector can be undesirable for statistical pattern classification. The paper [Cesar-Jr. & Costa, 1998b] discusses several automatic classification results of the aforementioned cells considering these features.

To probe further: *Morphological Analysis of Neurons*

Many of the techniques discussed in this book have been successfully applied to many different problems in neuromorphology. For instance, the terminations and branch points of neural dendrites can be properly identified by using contour representation and curvature-based corner detection (see Figure 8.44) [Cesar-Jr. & Costa, 1999].

A series of interesting works in neural cell shape analysis are listed by subject in Table 8.9 (see also [Rocchi et al., 2007] for a recent review).

Approach	Papers
Sholl diagrams	[Sholl, 1953]
Ramifications density	[Caserta et al., 1995; Dacey, 1993; Dann et al., 1988; Troilo et al., 1996]
Fractal dimension	[Caserta et al., 1990; Jelinek & Fernandez, 1998; Jr. et al., 1996, 1989; Montague & Friedlander, 1991; Morigiwa et al., 1989; Panico & Sterling, 1995; Porter et al., 1991]
Curvature, wavelets and multiscale energies	[Cesar-Jr. & Costa, 1997, 1998b; Costa et al., 1999; Costa & Velte, 1999]
Dendrograms	[Cesar-Jr. & Costa, 1997, 1999; Costa et al., 2000; Poznanski, 1992; Schutter & Bower, 1994; Sholl, 1953; Turner et al., 1995; Velte & Miller, 1995]

Table 8.9: *Shape analysis approaches for neural morphology.*

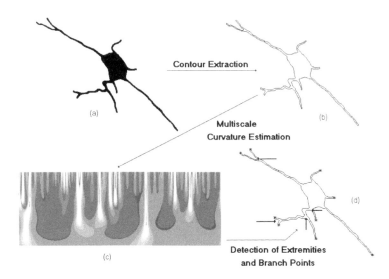

Contour Extraction

(a)

(b)

Multiscale
Curvature Estimation

(d)

(c)

Detection of Extremities
and Branch Points

Figure 8.44: *Multiscale curvature-based detection of dendritic terminations and branch points for neuron morphology: binary
cell **(a)**; respective contour **(b)**; curvogram **(c)**, and detected
terminations and branch points **(d)**.*

To probe further: *Classification*

Classification is covered in a vast and varied literature. The classical related literature, covering both supervised and unsupervised approaches, includes [Duda &
Hart, 1973; Duda et al., 2000; Fukunaga, 1990; Schalkoff, 1992; Theodoridis &
Koutroumbas, 1999]. An introductory overview of some of the most important
topics in clustering, including the main measures and methods, validation techniques and a review of the software and literature in the area can be found in the
short but interesting book [Aldenderfer & Blashfield, 1984]. Two other very read-
able introductory texts, including the description of several algorithms and comments on their applications, are [Everitt, 1993] and [Everitt & Dunn, 1991], which
deliberately keep the mathematical level accessible while managing not to be su-
perficial. The book by [Romesburg, 1990] also provides a very accessible introduction to clustering and its applications, concentrating on hierarchical clustering
approaches and presenting several detailed examples. A classical reference in this
area, covering partitional and hierarchical clustering in detail, as well as several
important related issues such as cluster results interpretation and comparative evaluation of cluster methods, is [Anderberg, 1973]. A more mathematical and comprehensive classic textbook on clustering algorithms is [Jain & Dubes, 1988], which
includes in-depth treatments of data representation, clustering methods, validation,

and applications. Several classification books also dedicate at least a section to clustering, including [Chatfield & Collins, 1980; Gnanadesikan, 1977; Young & Calvert, 1974]. The latter two require a relatively higher mathematic skill, but are particularly rewarding. The book by [Sokal & Sneath, 1963], in great part responsible for the initial interest in hierarchical clustering, has become a historical reference still worth reading.

Structural Shape Recognition

Chapter Overview

STRUCTURAL PATTERN RECOGNITION takes a different path to characterize and recognize shapes: the image objects are divided into parts, which are then organized in terms of relationships among such parts. The present chapter reviews some of the basic concepts behind this approach, discussing different possibilities for shape representation, characterization and recognition. An application of this approach to image segmentation and parts recognition is provided.

9.1 Introduction

The shape analysis paradigm discussed so far is based on extracting shape features (Chapters 6 and 7) possibly followed by statistical classification of the obtained feature vectors (Chapter 8). An important alternative approach that has been developed since the early 1960s is based on decomposing the shape into a set of parts before other tasks are processed. The shape is then seen as a set of parts that are spatially arranged through spatial relations among them. For instance, the human body shape may be decomposed as head, torso, arms and legs. The head is *above* the torso, the left arm is *to the left of* the torso while the right arm is *to the right* of the torso. The legs are *below* the torso (Figure 9.1).

There are hence two main aspects regarding structural shape analysis: the *parts* that compose the shape and the *relations* between such parts, which can be resumed as:

$$\boxed{\textbf{shape} = parts + relations.}$$

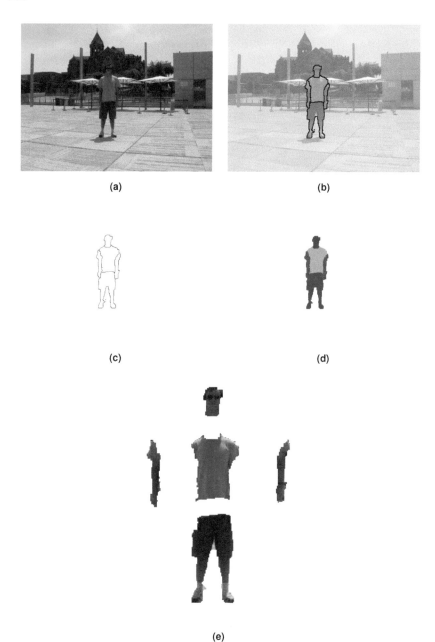

(a)   (b)

(c)   (d)

(e)

Figure 9.1: *Shape = parts + relations:* *(a) original image;* *(b) body parts with superposed contours;* *(c) parts contours;* *(d) body parts represented as 2D shapes (connected components)—the perceived shape is a function of the different parts with a suitable spatial arrangement, i.e., suitable spatial relations among the different parts;* *(e) the different separated body parts.*

Therefore, the shape is seen as a collection of parts which, in turn, may be composed of subparts. From a shape perception perspective, this is a subjective matter. As far as 2D digital shapes are concerned, this decomposition may be carried out up to the pixel level (or even subpixel). Such scheme can be understood as a multiscale decomposition: shape, parts, parts of parts, and so forth. There is no *a priori* rules to define how such decomposition is carried out and when it should stop. It is necessary to identify the different parts that compose the shape and there are different ways to carry out this task:

Top-down: Starts with the whole shape and decomposes it. An example of such approach is contour segmentation using curvature and other salient points. The shape is represented by its contour which can be segmented in terms of its salient points. Each portion delimited by a pair of salient points is taken as a part.

Bottom-up: The shape is divided into a set of small subparts (eventually up to the pixel level). The subparts are merged in order to form the parts. An example of such approach is image oversegmentation through watershed. Each shape part is composed by a set of subparts defined by the watershed basins. In order to identify the shape parts, it is necessary to group the correct basins of each part.

The structural approach addresses many important problems in computer vision such as

Object segmentation: There are methods that perform image segmentation by exploring a structural model of the expected object to guide the segmentation process.

Object and parts recognition: Structural models are useful to recognize the different parts that compose an object, allowing the recognition of the object itself.

Shape decomposition: As stated in the first chapter, shape decomposition is an important problem in shape analysis. A representative approach to perform shape decomposition explores a structural model of the shape and to recognize its parts. Therefore, the shape is decomposed into the recognized parts.

Shape matching: A natural way of matching two or more shapes is to decompose and to recognize the respective parts. Shape matching is then carried out by creating correspondences among the respective parts.

A key-issue in structural shape recognition is shape representation, which should allow storage of the shape parts as well as the relationships among them. Two important types of representations have been explored in the literature: *formal grammars* and *graphs*. In the case of grammars, the shape parts are represented by terminal symbols of a formal grammar. The shape is then represented as a string of such symbols. The relations are implicitly represented as juxtapositions of the symbols. On the other hand, graphs provide a powerful and natural way of representing the shape in a structural way: the shape parts are associated with the graph vertices while the relations among such parts are represented as edges. The characterization of each shape part can be performed using standard shape features, since each part may be taken as a shape itself. Moreover, it is important to characterize the spatial relations among the different parts, which consists of an important interdisciplinary research topic. These topics are addressed in the subsequent sections.

9.2 Syntactic Pattern Recognition

An alternative approach to pattern classification is known as syntactic or structural pattern recognition [Fu, 1982], which is briefly discussed in the following. In order to gain some insight about this approach, suppose that we want to classify some geometrical figure, such as a square with side 3. Consider the following unit norm vectors $\vec{a} = (1,0)$, $\vec{b} = (0,1)$, $\vec{c} = (-1,0)$ and $\vec{d} = (0,-1)$, which point to the right, upward, to the left and downward, respectively. The square can now be represented in terms of the string indicating the sequence of vector additions necessary to follow the shape outline, i.e.,

$$\vec{a} + \vec{a} + \vec{a} + \vec{b} + \vec{b} + \vec{b} + \vec{c} + \vec{c} + \vec{c} + \vec{d} + \vec{d} + \vec{d}.$$

This representation means that any perfect square of size n, where n is an integer value, can be represented by the string $(\vec{a} +)^n (\vec{b} +)^n (\vec{c} +)^n (\vec{d} +)^{n-1} \vec{d}$, where the vectors $\vec{a}$, $\vec{b}$, $\vec{c}$ and $\vec{d}$ are said to be the geometrical *primitives* composing the square pattern. Observe that the above exponent should not be taken as an indication of arithmetic exponentiation, but rather as the repetition of the respective strings, i.e., $(\vec{a} +)^4 = \vec{a} + \vec{a} + \vec{a} + \vec{a} + $.

The choice of primitives to be used depends on each specific application, as well as on the patterns' inherent nature. In other words, there is no universal set of primitives that can be considered for general use. In addition, observe that the primitives represent only one of the aspects related to the syntactical approach. The second important concept is the *relation* between primitives, which defines the way in which primitives are associated with each other in order to represent the patterns. As an example, the above situation involving a square adopted concatenation (i.e., by appending the vectors) as a means of composing the patterns. In the traditional syntactic approach, the primitives and the relations between them are formalized by an area known as *formal grammars theory* (including stochastic grammars) and, in this sense, syntactic analysis of the input strings provides the means for pattern recognition. Particularly, the process known as *parsing* involves investigating whether a specific sentence (representing an object) belongs to some given grammar.

$\boxed{9.3}$ Region Decomposition

An important aspect of structural shape analysis is that the shape should be decomposed into a number of parts or subparts. For instance, the syntactic approach assumes that the shape is decomposed into a number of primitives which are associated to terminal symbols of a string. Shape decomposition directly depends on the shape representation adopted. If the primitives are straight lines or vectors, then the shape may be initially represented by its contour, which undergoes a subsequent polygonal approximation (Section 5.4.1). Each edge of the polygonal approximation would be associated to a symbol to compose the shape string. The list below indicates some shape decomposition possibilities commonly found in the literature:

Pixels: The most straightforward decomposition is by using the shape pixels themselves, which is often the case for Markov random fields methods [Samet, 1980]. A useful generalization is by sampling the shape, i.e., through sets of points sampled from the shape. This is often explored in geometric modelling [Medeiros et al., 2007];

Quadtrees: Quadtrees are data structures widely used in image processing and computer graphics in order to represent shapes (more generally, images) [Samet, 1980]. The shape image is initially associated to a tree root. Because such an image normally has pixels that belong to the shape and others that do not belong (henceforth referred to as shape and non-shape pixels), the image is divided into four quadrants associated to four children nodes of the root node. The same rule is applied recursively until all subregions become homogeneous (i.e., only contain either shape or non-shape pixels). Eventually, some of these subregions may be composed of a single pixel.

Meshes: Quadtrees may be seen as a mesh approximation of the shape composed of squared tiles of different sizes. There are some alternative methods that produce more regular meshes, or meshes with some specific desirable properties. Among such possibilities, triangular meshes play a special role and have been widely studied in the literature [Cuadros-Vargas et al., 2005]. Mesh approximation leads to a decomposition by polygons which are taken as the shape parts (or subparts). See also [Faustino & de Figueiredo, 2005] for centroidal Voronoi diagrams which may also be used in this context.

Image oversegmentation: If a shape composed of multiple parts should be obtained by image segmentation, it may be difficult to segment each part in an accurate way because of typical image segmentation difficulties. However, it may be possible to oversegment the shape in a way that each part is actually present in the form of a collection of subregions, e.g., by using the watershed algorithm, as shown in Figure 9.2. In this case, each region produced by the watershed oversegmentation is taken as an image (or object) subpart.

Contour decomposition: As discussed above, a shape can also be decomposed in
terms of its boundary properties. Recall, for instance, the polygonal approximation algorithms described in Section 5.4.1: each portion of the contour is
subdivided into a number of portions, which can then be approximated by a
straight line. The resulting representation may be seen as a shape decomposition in terms of its straight parts (see Figure 5.25). The paper of [Felzen-
szwalb & Schwartz, 2007] describes a shape detection approach based on
this strategy.

9.4 | Graph Models

A shape can be represented by a graph model in a natural way. The shape is decomposed into a number of parts which are associated to the graph vertices. The
spatial relations among the different parts are then represented as edges linking the
respective vertices. There are three important aspects that characterize the graph
representation of the shape:

Graph structure: It is related to the adopted structure (or topology) of the graph to
represent the shape and mainly reflects how the vertices are linked by edges.
For instance, in case of an adjacency graph, two vertices are linked by an
edge if the corresponding parts are adjacent in the shape. On the other hand,
if a complete graph is assumed, all vertices are linked to each other.

Part features: It is related to the features adopted to describe each part. Examples
of valid part features include color, texture and shape features (e.g., area,
perimeter, etc.).

Relational features: These are related to the features adopted to describe the relations between shape parts. Examples of important relational features include
distance between parts, orientation and semantical relations such as symmetry, betweenness, alongness, etc.

As discussed above, the shape parts are associated to the vertices of the graph
while the relationships between parts are associated to the edges linking the corresponding vertices. Hence, part features are stored into the graph vertices while
relational features are stored into the edges. A graph that stores part and relational
information in this way is called an *attributed relational graph* . Let $\tilde{G} = (V, E)$
denote a directed graph where V represents the set of vertices of $\tilde{G}$ and $E \subseteq V \times V$
the set of edges. Two vertices a, b of V are adjacent if $(a, b) \in E$. We define an attributed relational graph as $G = (V, E, \mu, \nu)$, where $\mu : V \rightarrow L_V$ assigns an attribute
vector to each vertex of V. Similarly, $\nu : E \rightarrow L_E$ assigns an attribute vector to
each edge of E. We typically have $L_V = \mathbb{R}^m$ and $L_E = \mathbb{R}^n$, where m and n are the
numbers of vertex and edge attributes, respectively.

(a)

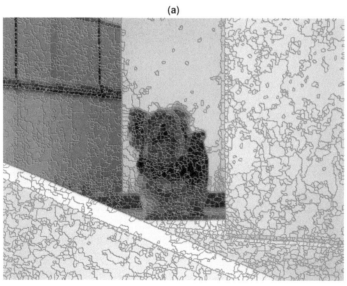

(b)

Figure 9.2: *Watershed decomposition of an image: **(a)** Input image; **(b)** watershed oversegmentation.*

Figure 9.3: *Spatial relations: the book* BETWEEN *the laptop and the wood-craft.*

In order to represent a shape by a graph, the three aspects above should be defined: graph structure, part and relational features. Such a representation starts by decomposing the shape into its parts followed by the creation of the associated graph, which defines its structure. Then, part and relational features are measured from the image and stored in the graph. There are different ways to decompose a shape, each being more suitable to specific problems at hand, as seen in the previous section. It is possible to use the original image (color or grey-level, for instance) in order to produce a structural representation even if the shape has not been segmented. Indeed, this approach may be useful when shape segmentation is a difficult task and shape information may be used to guide the segmentation process. For instance, the image may be oversegmented using the watershed algorithm [Beucher & Meyer, 1993; Vincent & Soille, 1991] and each segmented region is taken as a shape part. Actually, in most cases, such regions would correspond to subparts to be merged in order to form the expected shape parts. Section 9.7 describes an example of such an application.

9.5 Spatial Relations

Suppose you are somewhere in your house, say the kitchen, and you want to ask someone to go get the *Shape Analysis* book in you bedroom (see Figure 9.3). A precise way to indicate the position of the book is to define a coordinate system for the bedroom (e.g., a canonical system defined on the bottom-left corner of the bedroom) and to ask,

"Please, go get the Shape Analysis book on the position $(31.7, 43.3, 10.4)$ *of the canonical coordinate system of the bedroom."*

Although very precise, such a strategy is not only unnatural, but very difficult to implement in practice and highly unnecessary. It would be much easier just to ask,

"Please, go get the Shape Analysis book ɪɴ *the bedroom, which is* ᴏɴ *the table,* ʙᴇᴛᴡᴇᴇɴ *the laptop and the woodcraft."*

The latter approach is less precise than the former, but much more natural, powerful and robust. In fact, it is more powerful and robust *because* it is imprecise and it relies on *spatial relations* between objects: *in* the bedroom, *on* the table, **between** the laptop and the woodcraft. This example illustrates the importance of the spatial relations and why it has been studied in multidisciplinary fields such as linguistics, cognitive science, geosciences, natural and computational vision [Bloch, 1999, 2005; Egenhofer & Shariff, 1998; Freeman, 1975; Gasós & Saffiotti, 1999; Miyajima & Ralescu, 1994]. A growing literature about this topic is available in different fields, though related research efforts have been continuously increasing in recent years. In fact, much of the difficulty of deriving computational methods to analyze spatial relations in images arises because often such relations are linguistic/cognitive imprecise concepts that may depend on the shape of the objects in hand.

The spatial relations can be classified in different ways (see [Bloch, 2005; Hudelot et al., 2008] for reviews). A possibility is to divide them into the three following classes [Takemura, 2008]:

Topological relations: Topological relations can be identified by being invariant to some standard geometrical transformations such as rotation and scaling. They are often related to concepts such as adjacency, inclusion and connectivity [Schneider & Behr, 2006]. Examples of this type of spatial relation include *inside, outside* and *adjacent* [Bloch et al., 1997; Klette & Rosenfeld, 2004; Krishnapuram et al., 1993; Rosenfeld & Klette, 1985].

Distance relations: In contrast to topological relations, distance relations are often linked to quantitative measures, being generally modeled with the help of some metrics. Although the use of mathematical metrics helps a lot, the shape of the involved objects plays a central role in the perceived relation, thus requiring sophisticated mathematical models and algorithms to suitably extract them from images. For instance, we can say that the two shapes in Figure 9.4 (a) are far from each other. This perceived distance relation may be modeled and measured by calculating the distance between the centroids of both shapes. Now, consider the two shapes in Figure 9.4 (b). Should we consider that the shapes are far from each other? Observe that the distance between the centroids of the shapes in Figure 9.4 (b) may be very similar to that in Figure 9.4 (a). Fuzzy modeling plays a central role in order to deal with such difficulties [Bloch, 2006; Chaudhuri & Rosenfeld, 1996].

Directional relations: Directional relations are characterized by orientation or angle-based aspects regarding some reference. For instance, if we have two objects A and B, we may characterize whether A is *to the left of, to the right of, above* or *below* B depending on the angle between the segment $\overline{AB}$ and

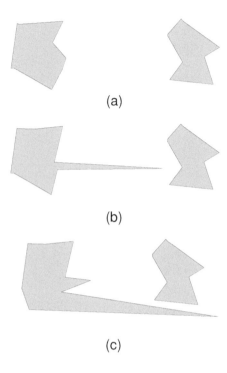

(a)

(b)

(c)

Figure 9.4: *Spatial relations: The shapes in **(a)** can be considered to be far from each other, but not necessarily in **(b)**. In case of **(c)**, the leftmost shape could be considered to the left, below or both, regarding the rightmost shape.*

x-axis. The comments about the object shapes regarding distance relations apply to directional relations: is the first shape in Figure 9.4 (c) *to the left* of the second shape? Or is it *below*? Or both? Different approaches to address such ambiguous situations have been proposed in the literature, such as non-linear functions [Krishnapuram et al., 1993; Miyajima & Ralescu, 1994], positioning based on characteristic points [Bloch & Ralescu, 2003], mathematical morphology operators [Bloch, 1999] and histograms [Bloch & Ralescu, 2003; Matsakis & Andrefouet, 2002].

In general, there are four important levels of the research about spatial relationships:

Linguistic/cognitive characterization: This level regards the analysis of the conceptual meaning behind the relation of interest. What it means to be *above*, *to the right* or *along* a given object? The studies in this level are more abstract, being the most multidisciplinary.

Mathematical modeling: Once the spatial relation of interest is characterized in the previous level, it is important to derive mathematical models that capture the main spatial relation concepts. For instance, spatial relations such as *far* or *near* could be modeled with the help of Euclidean distance.

Algorithmic level: In order to explore spatial relations in image analysis situations, it is important to design algorithms to extract (measure) the spatial relation from digital images based on the created mathematical model.

Applications: Finally, specific applications may be addressed to use spatial relations measured from digital images. For instance, the graph matching algorithms explained below use spatial relations to characterize the relative position among objects, which help to recognize them.

Let's consider the specific case of the spatial relation *between* [Bloch et al., 2006; Takemura et al., 2005], as in *the book is* BETWEEN *the laptop and the woodcraft*:

Linguistic/cognitive characterization: It is very difficult to completely define this relation because of the different meanings that it may assume depending on the context and the object shapes. Also, it is a ternary relation, in the sense that it involves three objects (in our example, the book, the laptop and the woodcraft). An analysis based on intuitive meaning or dictionary definitions leads to the concept that an object A is *between* objects B and C if it intersects the region between B and C. Therefore, we can divide the characterization of the degree to which A is *between* B and C into two steps: (1) calculate the region *between* B and C and (2) calculate the degree to which A intersects this region.

Mathematical modeling: There are different ways to model the region *between* B and C (see [Bloch et al., 2006] for a review with discussion). An intuitive approach is to define the region *between* B and C as the convex hull of the union of these objects but excluding the objects themselves:

$$\beta_{CH}(B, C) = CH(B \cup C) \cap B^C \cap C^C \qquad (9.1)$$

where $CH(X)$ denotes the convex hull of X and X^C denotes the complement of X. Although this definition is suitable if B and C are convex shapes, counter-examples for more complex shapes can be easily presented, implying the need for more robust models [Bloch et al., 2006]. Different models to measure the degree of intersection between A and the *between* region can be defined.

Algorithmic level: The previous definition of the *between* region based on convex hulls may be implemented using the convex hull and set operations methods already discussed in the previous chapters. Figure 9.5 shows a result respective to the example in Figure 9.3.

Applications: Different applications may be derived to explore this relation. For instance, a medical image segmentation method which uses this relation to guide the segmentation process is described in [Moreno et al., 2008].

9.6 Graph Matching

Many important shape analysis problems may be solved by matching two or more graphs that represent the analyzed shapes. In order to match two graphs G_1 and G_2 it is necessary to find a correspondence between those graphs, i.e., to find a mapping between the respective vertices V_1 and V_2. There are different possibilities to define a mapping $f : V_1 \rightarrow V_2$ between two graphs G_1 and G_2 that play important roles in structural shape analysis:

Graph isomorphism: f is an isomorphism if any $a, b \in V_1$ are adjacent if and only if $f(a), f(b) \in V_2$ are adjacent. In other words, a graph isomorphism is a one-to-one mapping of V_1 onto V_2 that preserves edges on both graphs.

Subgraph isomorphism: This type of mapping is a generalization of the graph isomorphism. The problem is to find an isomorphism f between G_1 and a subgraph of G_2. This problem is also known as subgraph matching.

Graph homomorphism: f is a homomorphism if $f(a), f(b) \in V_2$ are adjacent whenever $a, b \in V_1$ are. In other words, a homomorphism preserves the edges of G_1. It is important to note a different aspect regarding isomorphism: the homomorphism allows many-to-one mappings, which is very important for different shape analysis problems.

Inexact match: In some situations, the homomorphism constraint of mapping adjacent vertices to adjacent ones should be relaxed. Such mappings of V_1 to V_2 which not necessarily preserve connectedness are called inexact matchings of G_1 and G_2.

Figure 9.6 illustrates these different types of possible graph matchings.

In general, there are many possible matches between two given graphs and it is necessary to evaluate which are most suitable for a given application. The natural approach is to define an objective function that reflects the suitability of each match. The problem of finding a suitable match between two graphs can then be expressed as an optimization problem where the objective function should be optimized. For instance, in order to evaluate the similarity between two shapes, a similarity measure between the respective graphs can be defined as an objective function to be optimized by matching the graphs. The similarity between the shapes is then evaluated by the optimized value of the objective function. In this context, the comparison between two shapes is expressed as an optimization problem: in order to compare two shapes, the respective graphs should be matched, which is

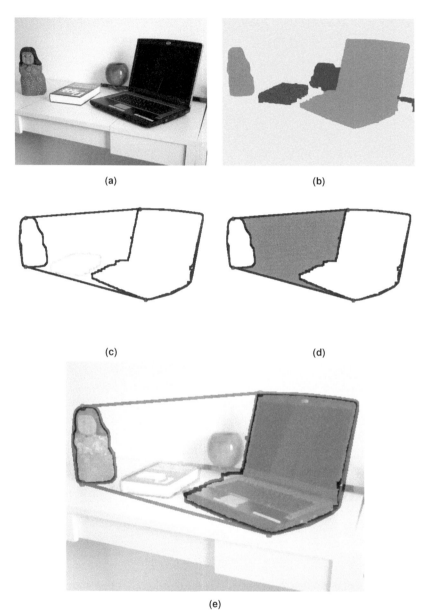

(a) (b)

(c) (d)

(e)

Figure 9.5: *Spatial relations: the book is* BETWEEN *the laptop and the woodcraft. (a) original image; (b) segmented image; (c) contour of the different objects, including the convex hull; (d) region* BETWEEN *the laptop and the woodcraft calculated using the convex hull approach; (e) results superposed to the original image using transparency.*

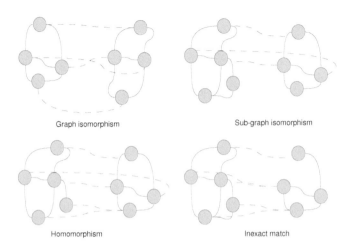

Figure 9.6: *Illustrative examples of possible graph matchings.*

done by optimizing an objective function. In most situations, there are some known shapes used as references. An a priori known shape is represented by a *model graph* in the sense that the graph is a model for the given class of shapes. Actually, a shape class can be represented by a set of model graphs. This is analogous to a supervised training set of samples. There is an important difference though. In the case of a model graph, besides the shape class, we also have its parts labeled, i.e., the shape parts are also known and represented in the model. On the other hand, the unknown shape to be analyzed (e.g., recognized or decomposed) is represented by an *input graph*. In general, shape analysis problems are solved in this structural approach by matching an input graph to one or more model graphs. The list below indicates how different shape analysis problems can be solved based on this approach:

Shape recognition: Shape recognition is performed by classifying the unknown shape. Each shape class should be represented by a set of prototypical graphs (possibly a single one). In order to recognize a given shape, it should be matched to each graph of each class and the objective function used to decide which is the best match, thus recognizing the shape.

Recognition of shape parts: An important shape analysis task that can be solved by structural approaches is the recognition of shape parts. In order to recognize the parts of a given shape, it should be represented as a graph where each part is associated to a vertex. By matching this graph to a model graph, each unknown part is mapped into a model vertex, thus recognizing it.

Recognition of shape subparts: The above approach to recognize the shape parts implicitly assumes that each shape part is correctly represented by a graph vertex, i.e., that the shape has been suitably decomposed into the correct parts. In many situations, this is not possible and the expected shape parts are split into different pieces or subparts. In order to identify the shape parts, such subparts should be recognized and merged. This problem may also be solved by shape matching. The input shape is represented by a graph where each subpart is associated to a vertex. Similarly to the previous approach, the graph is matched to a model graph. However, in this case, multiple vertices of the input graph may be mapped onto a single model vertex (many-to-one mapping). All vertices of the input graph mapped onto a single model vertex are recognized as subparts that compose a single shape part.

Shape detection: Shape detection is the problem of finding a given shape in an input image. In order to solve this problem in a structural framework, the shape to be detected should be represented by a model graph. The input image should be represented by an input graph. The object to be detected is hence a subgraph of such input graph. Shape detection may then be carried out by subgraph matching, i.e., by looking for suitable matches of the model graph to subgraphs of the input graph. Each suitable match defines a detected shape in the input image.

Shape tracking: Shape tracking in a video sequence may be performed as a series of shape matching steps. A model graph of the shape to be tracked should be initially created and used to detect the shape in the video stream (see shape detection above). Once the shape is detected, it may be tracked by updating the model graph information followed by detecting the shape in the subsequent frame. Temporal coherence may be explored to delimit the image region of the subsequent frame (alternatively, the input graph vertices) where shape detection is to be performed. It is worth noticing that shape tracking through this structural approach allows not only tracking the whole shape, but also its parts.

9.7 | Case Study: Interactive Image Segmentation

This section describes an approach to interactive image segmentation using graph matching based on the works described in [Consularo et al., 2007; Noma et al., 2008a]. Interactive image segmentation is an important approach to segment images by allowing the user to provide some type of input in order to guide the segmentation process. The reader may be familiar with the Magic Wand tool for smart region selection available in most interactive image processing software such as GIMP® or Photoshop®. This is a special application of such algorithm where the selected region is actually a segmented image region defined from seeds defined by the user. The region growing algorithm described in Section 3.4.2 is another example of such interactive image segmentation methods. Powerful interactive segmentation methods have been developed such as the watershed with markers [Vincent &

Soille, 1991], IFT [Falcão et al., 2004], graph cuts [Boykov & Jolly, 2001; Rother et al., 2004] and the random walker [Grady, 2006], for example. Most of these methods are based on finding segmentation regions connected to the user-provided seeds. On the other hand, the structural image segmentation method described here does not require that the user traces lie perfectly on each region to be segmented. Also, the graph model created by the user can be employed to segment a series of similar objects, as long as the overall image structure is suitably captured by the model and preserved on the images to be segmented.

The basic idea behind the method is to generate an oversegmentation of the input image and to represent it by an ARG, the input ARG denoted as G_I. Then, the user should make some labeled traces over the input image in order to identify the main regions to be segmented. The intersection between the user traces and each oversemented region is calculated and used to generate a model-graph, denoted as G_M, with labeled vertices. The input and model graphs are then matched, thus providing the final image segmentation with the correct parts recognized (i.e., labeled). Figure 9.7 summarizes these steps.

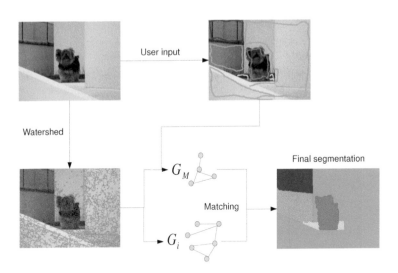

Figure 9.7: *Framework for structural image segmentation and parts recognition.*

The proposed method detailed in [Consularo et al., 2007; Noma et al., 2008a] uses gray-scale information as object attributes and vectors between region centroids as relational attributes. Such relational attribute codes two types of spatial relations: orientation and distance (i.e., the orientation and the modulus of the vector). An efficient graph-matching algorithm that explores an auxiliary data structure, the *deformation graph*, to solve topological problems is used.

Figures 9.8 and 9.9 present some illustrative results obtained by using the proposed approach.

A particularly interesting feature of this approach is that the generated model may be reused to segment other images with the same structure. This is because the image segmentation task is seen as a parts recognition problem which is solved by matching input images to the generated model G_M. The basic idea is to produce a model for one image with the help of the operator and then to match other similar images using the same model. Figure 9.11 explains this method. In order to illustrate a possible application, refer to Figure 9.12, which shows a series of subsequent photos taken using a high-speed burst mode of a simple digital camera. The first image has been segmented using the structural segmentation method explained above. The resulting graph model was then applied to the other images of the series with no need for new models. The results are shown in Figure 9.13, thus illustrating the method capabilities. This approach can be used for shape tracking with multiple parts recognition, as explored in [Paixão et al., 2008].

Additional resources: *Structural image segmentation software*

An open-source software that implements the structural image segmentation method is available at `http://segmentacao.incubadora.fapesp.br/portal`. The software implements both the simple approach and the model reuse. A video demo is available at `http://segmentacao.incubadora.fapesp.br/portal/documentos/video-short-4min.swf`.

9.8 | Complex Networks for Image and Shape Analysis

Another recently introduced possibility to explore graphs for shape analysis rely on complex networks (Section 2.8). As a consequence of their flexibility for representing and modeling virtually any discrete system, complex networks are also potentially useful in shape analysis and vision [Costa, 2004, 2006a]. In this section we illustrate such possibilities in terms of three specific problems: image segmentation, texture characterization and gaze movement. Other applications of complex networks to imaging and vision have been reported in [Bruno et al., 2008b].

9.8.1 | Image Segmentation

The basic principle underlying the application of complex networks to image segmentation is that the pixels in the regions to be separated present locally similar features (e.g., gray level, texture, color, etc.). So, by representing each pixel as a node and defining edges with weights which are inversely proportional to the distance between the respectively attached nodes and directly proportional to the similarity between the respective pixels, it is possible to represent the given image

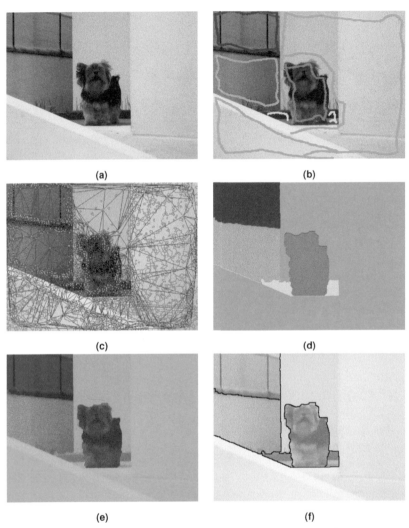

(a)

(b)

(c)

(d)

(e)

(f)

Figure 9.8: *Structural image segmentation results I: (a) Original image; (b) traces defined by the user of each image part of interest. The different trace colors represent different labels; (c) over-segmented region with the model ARG superposed; (d) recognized parts; (e) segmented image with the recognized parts superposed with transparency; (f) segmented image with object boundaries superposed.*

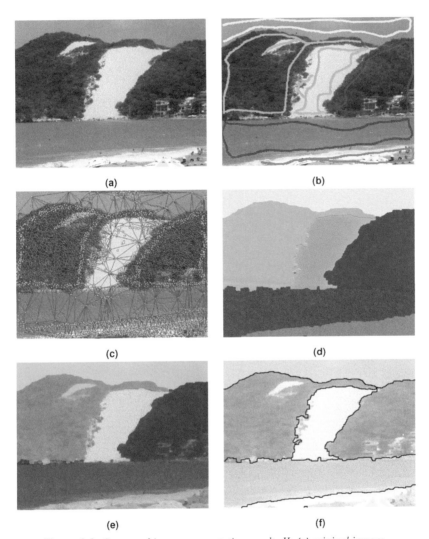

(a)

(b)

(c)

(d)

(e)

(f)

Figure 9.9: *Structural image segmentation results II: (a) original image; (b) traces defined by the user of each image part of interest. The different trace colors represent different labels; (c) over-segmented region with the model ARG superposed; (d) recognized parts; (e) segmented image with the recognized parts superposed with transparency; (f) segmented image with object boundaries superposed.*

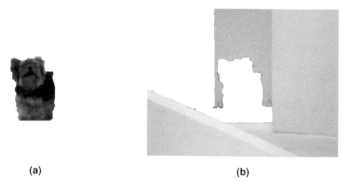

(a)   (b)

Figure 9.10: *Examples of two objects segmented in Figure 9.8: (a) the dog and (b) the walls.*

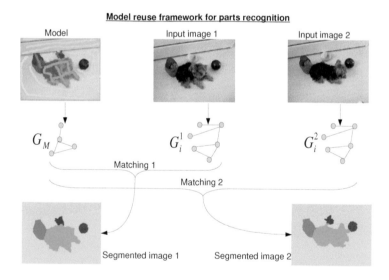

Figure 9.11: *Framework for structural image segmentation and parts recognition in multiple images with similar structure. The model graph G_M is generated from one image and applied to different images with similar structure, thus generating different input graphs (G_i^1 and G_i^2 in the figure). The model and input graphs are matched to recognize the corresponding parts in the input images without requiring the production of new traces by the user for each different input image.*

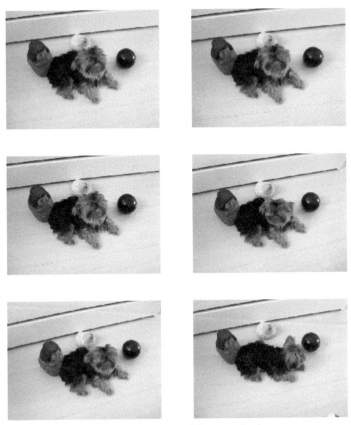

Figure 9.12: *High-speed burst mode photos.*

as a complex network, and then obtain the segmented regions by community finding [Rodrigues et al., 2006]. For instance, it is possible to assign a feature vector
to each node (pixel) containing local properties around the respective pixel (e.g.,
by using a circular window centered at that pixel), such as average and standard
deviation of gray levels, texture features, etc.

9.8.2 Texture Characterization

Several types of textures are characterized by well-defined statistical distribution
of specific properties, such as orientation, frequency, etc. Once an image has been
mapped into a complex network, e.g., by using the procedure described in the previous section, it is possible to obtain statistical distributions of the respective connectivity in terms of several measurements available for complex network characterization [Costa et al., 2007b]. As a simple illustration, let us assume that a texture
image has been mapped into a complex network by the above procedure. We can
now measure the degree and clustering coefficient of each node, and obtain the

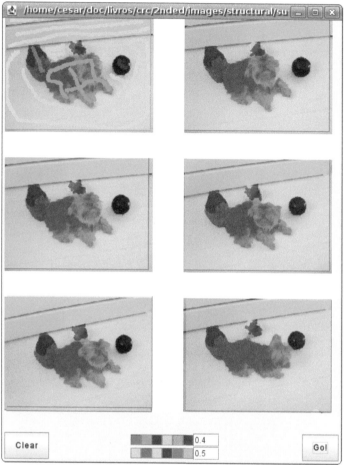

Figure 9.13: *Segmentation results of the images shown in Figure 9.12. The
graph model was generated for the first photo and directly
applied to the others with no need for additional model generation by the user.*

respective statistical density probabilities, from which global measurements of the
connectivity of the original texture image (e.g., average and standard deviation) can
be obtained. Therefore, the original texture image is mapped into a feature vector
containing information about the connectivity of the respective complex network.
Now, classification methods can be immediately applied to such feature vectors in
order to classify them. Promising results of the application of complex networks to
characterize and classify textures have been reported in the literature [Chalumeau
et al., 2006; Costa, 2004].

9.8.3 Gaze Movement

The high-resolution ability of our eyes is limited to a relatively small window focused at the fovea, corresponding approximately to the center of fixation or *gaze*. The rather limited extent of such a high-resolution window can be easily appreciated by looking at one of the characters in a dense piece of text and then trying to identify how neighboring characters can be easily discerned. Consequently, in order to achieve high vision performance, the human eyes are allowed to move along space, changing gaze. Only by integrating visual information during scanning gaze movements we are able to obtain a comprehensive representation of presented scenes. Usually, gaze movements take place among points of particularly high *saliency* in the presented scenes, such as vertices and high contrast borders. However, as several sequence of gaze movements can be performed along the salient points for any given image, an interesting optimization problem arises concerning the speed in which important information can be extracted from the presented scene through gaze movements.

By representing salient points in terms of nodes and establishing edges between such nodes with weights proportional to the respective visual properties and saliencies of the nodes, it is possible to model gaze movements in terms of random walks [Costa, 2006b]. Basically, a random walk is a dynamical process in which one or more agents move along the network nodes, choosing randomly among the possible outgoing edges. The dynamics of gaze movement among salient points in an image can therefore be modeled in terms of random walks in complex networks [Costa, 2006b], where the high accuracy region of the retina is understood as the moving agent. There are several ways in which to assign weights to pairs of salient points, each leading to a different model of the gaze movement dynamics. For instance, it is reasonable to make the weights inversely proportional to the distance between the salient points, as it is less likely that our gaze will change abruptly to a distant point of the image. At the same time, it is possible to avoid long-term gaze fixation at salient points with similar properties. This can be enforced by making the weights directly proportional to the difference between the visual features of the salient points.

A preliminary investigation of gaze movement using random walks in complex networks derived from images has been reported [Costa, 2006b]. In this work, several modeling hypotheses were considered. First, the border of the shapes in the images were identified and the vector field tangent to them determined by using the Fourier method described in Section 3.3.6 of this book. Each border pixel was represented as a node. For each border point i, the other points falling along the straight line determined by the border tangent at that point were identified and connected to the point i. Such a weighting scheme implies in which gaze movements would follow the border tangent. After the original images were transformed into the respective weighted complex network, according to the above methodology, the frequency of visits to each node was determined by using a spectral methodology (eigenvalues of the transition stochastic matrix). Higher frequencies of visits were found for the vertices as well as for points corresponding to the convergence of several border tangents. Another modeling approach investigated in [Costa, 2006b] involved the determination of weights considering the proximity and size of the different objects in the image. As a result, larger objects at the denser portions of the images tended to receive more visits by the moving gaze.

To probe further: *Structural pattern recognition*

The reader is referred to the classical books [Fu, 1974, 1982; Pavlidis, 1977; Tou & Gonzalez, 1974], as well as to the works of [Allen et al., 1999; Bloch, 1999; Casacuberta & de la Higuera, 1999; Hopcroft & Ullman, 1979; Lafarge et al., 2006; Ledley, 1964; Prabhu & Pande, 1999; Ristad & Yianilos, 1998; Yang, 1991; Zhu et al., 2006] for additional information about syntactic pattern recognition related concepts and methods. A syntactic approach for neural shape analysis and dendrogram generations has been proposed in [Cesar-Jr. & Costa, 1999].

Graph models have been extensively explored in vision and pattern recognition in recent years. Some important early works have been proposed in the 1960s and 1970s, presenting much intersection with AI and Knowledge Engineering [Barrow & Popplestone, 1971; Fischler & Elschlager, 1973; Minsky, 1974]. Two important scientific meetings have been regularly held, the *Joint IAPR International Workshops on Structural and Syntactic Pattern Recognition* and the *Workshop on Graph-Based Representations in Pattern Recognition*. Traditional journals such as *Pattern Recognition* and *PAMI* organized Special Issues on the subject: *Pattern Recognition, Volume 38, Issue 9, 2005* and *IEEE PAMI, Volume 27, Issue 7, 2005*. Important review information may be found at [Bunke, 2000a,b; Conte et al., 2004; Graciano, 2007; Kandel et al., 2007]. Additional references on the subject, including the important topic of graph learning, include [Bunke & Allermann, 1983; Caetano et al., 2007; Cesar-Jr. et al., 2005; Colliot et al., 2004; Conte et al., 2003; Dickinson et al., 2005; Felzenszwalb & Schwartz, 2007; Felzenszwalb & Huttenlocher, 2005; Graciano et al., 2007; Haxhimusa et al., 2006; Ion et al., 2006; Luo et al., 2003; Luo & Hancock, 2001; Noma et al., 2008b; Paixão et al., 2008].

Please check Section 9.5 for good starting points in the literature for the study of spatial relations.

Graph structures have been playing a central role in many approaches for image processing and analysis. The reader interested in mathematical morphology, watershed and graphs may find a good start in the literature [Beucher & Meyer, 1993; Roerdink & Meijster, 2000; Trémeau & Colantoni, 2000; Vincent, 1989; Vincent et al., 1992; Vincent & Soille, 1991]. It is also important to check the IFT proposed by Falcão and collaborators [Falcão et al., 2004; Falcão et al., 2002]. The vision community have also recently paid a lot of attention to graph models and algorithms, mainly regarding Markov random fields, graph-cuts, Bayesian networks and belief propagation [Boykov & Jolly, 2001; Dick et al., 2000; Felzenszwalb & Huttenlocher, 2004, 2006; Grady, 2006; Rother et al., 2004].

Epilogue

Chapter Overview

THIS CONCLUDING chapter highlights some of the main features of this book and indicates some interesting research topics for the coming years.

A.1 Future Trends in Shape Analysis and Classification

From all that has been presented and discussed in this book, the following important characteristics and problems in shape analysis and classification should have become clear:

① the inherent *inter-* and *multidisciplinary nature* of research in these areas, involving computer science, mathematics, electronic engineering, physics, artificial intelligence and psychology, among many others;

② the often incomplete representations of scenes as images and objects in terms of features, implying several distinct objects to be mapped into the same representation;

③ the importance of pattern recognition not only for identifying specific objects, but also even in the particularly challenging task of segmenting an image into its constituent parts;

④ the *lack of precise guidelines* for choosing not only the shape measures (features)—be it for shape analysis or classification—but also the classification methods to be used in general problems;

609

⑤ the importance of the *specific aspects of each problem and our knowledge about them* for defining suitable measures and selecting effective classification methods;

⑥ the importance of 2D shapes;

⑦ the importance of *properly validating* not only the algorithms for image processing and shape analysis, but also the obtained classification results;

⑧ the *broad application potential* of shape analysis and classification in a variety of areas, ranging from biology to spatial research;

⑨ the relationship between *continuous and discrete* approaches.

Any prognostics about the future of shape analysis and classification should carefully take into account the above listed facts, which shall define the developments in these areas for at least the following one or two decades. Although some perspectives on such advancements are presented in the following, it should be borne in mind that these reflect the authors' personal points-of-view and are certainly incomplete and biased.

① **Inter- and Multidisciplinarity:** The integration of the continuous advances in all the areas encompassed by shape analysis and classification, from the advent of faster hardware and larger memories to the better understanding of biological vision, will likely provide the principal means through which advances will be accomplished in these research areas. This will possibly lead to automated shape analysis systems with capabilities similar to those exhibited by the human visual system. Of particular interest will be the advances in artificial intelligence (here understood at its broadest scope, including data-mining, syntactic methods, neural networks, multi-agents, knowledge databases, statistics, dynamic systems, etc.), since these fields are likely to provide the means for replicating "human intelligence," which is necessary for providing suitable constraints that can be used to solve several ill-posed and ill-conditioned problems in vision and shape research. At the same time, developments in computer hardware, especially parallel systems, are likely to provide the computational power required for the extraction of more sophisticated features, as well as for performing more powerful classification methods. Results from neuroscience and psychology are also particularly promising, as they will unveil the solutions adopted in the most powerful shape analyzer (i.e., the human visual system). Another important trend should be the progressive incorporation of more sophisticated mathematics into shape analysis, such as multilinear algebra and singularities theory.

② **Guidelines:** It is expected that further results from statistics and mathematics, as well as from more formal and careful experimental performance assessments of the existing features and classification methods, will allow a more complete analytical understanding of the properties of shape measures and classification algorithms, allowing the selection of effective solutions for each class of problems.

③ **Specific Problem Aspects:** A promising research line in shape analysis and classification is a more in-depth and complete study of how the specific nature of each problem or class of problems, as well as available knowledge about them, can be used to help the user define suitable features and classification methods. A particularly interesting possibility for obtaining such knowledge is through the application of data-mining methodology considering several typical applications.

④ **3D Shapes:** Although the current book concentrates on 2D shapes, there is already a reasonably well-developed and established literature for 3D shapes (and video sequences), which should become more vigorous as the computational power and theoretical basis of shape analysis evolves. In this respect, it is expected that most of the knowledge in 2D shape analysis will be used, in a suitably adapted and extended fashion. Related areas which are likely to play an important role in such developments include 3D vision, mathematics (e.g., singularities theory, differential geometry and nonlinear systems), computer graphics and parallel systems.

⑤ **Validation and Comparative Assessment:** Before improving an algorithm, it is essential to ensure not only that it has been correctly implemented, but also to compare it with similar previous approaches. In this way, the positive and negative aspects of each method/algorithm can be identified, providing valuable information for redesign or code correction. Although the developments in shape analysis and computer vision were initially rather empirical, often lacking a more solid mathematics foundation and comprehensive validations, a shift has occurred toward more careful analysis of the results typically obtained by each method and algorithm. Yet, the sheer complexity usually implied by shape analysis and classification, often involving dozens of measures defining multiple dimensional feature spaces, has constrained such approaches. It is expected that more effective data-mining techniques, used jointly with more powerful (and possibly parallel) computer systems should allow more comprehensive approaches to the validation and comparison of methods and implementations.

⑥ **Application Potential:** The several examples and applications considered in the present book correspond to but a small sample of the whole universe of problems that can be solved by using shape analysis and classification. Undoubtedly, the above-identified trends and advances should be paving the way to the progressive and inexorable incorporation of such solutions to the most varied areas, especially the Internet, virtual reality, medicine, art and graphical design, to name just a few. A particularly interesting possibility is the use of shape analysis in investigations relating neural shape and function, as well as helping to identify the influence of specific genes over the morphology of the microscopic and macroscopic properties of living beings.

⑦ **Continuous and Discrete Approaches:** The current approaches to shape processing and analysis can be divided into two major classes respectively to the nature of the adopted concepts and tools, namely continuous and discrete. In the *continuous approach*, the discrete shapes are treated in terms of concepts and tools derived from continuous formulations (typically differential operators) and typically the shapes are processed in terms of numerical approximations of the original operators. Examples of continuous approaches to shape analysis include the Laplacian of the Gaussian, the curvature-based methodology described in Chapter 7, and level sets [Sethian, 1999]. The so-called *discrete approaches* involve treating discrete shapes in terms of concepts and tools closely related to their own discrete domain. The prototypical example of such a class of shape analysis approach is mathematical morphology, where Minkowsky's algebra and set theory are used to formalize every operation and result. While the latter type of methods provides a sound basis for treating low-level shape operations, it is felt that these methods are somewhat dissociated from important families of algorithms and concepts dealing at a higher levels of abstraction, such as the differential formulation underlying the large mass of geometrical knowledge in the continuum. A particularly promising perspective to bridge the gap between these two types of approaches is through a series of increasing levels of hierarchical abstractions and representations. In addition, it would be interesting to relate and integrate mathematical morphology with more general concepts and tools from discrete geometry. For instance, it is interesting to express middle-level shape concepts such as the SAT and distance transforms in terms of basic morphological operators, and then use them as scaffolding upon which even higher concepts and operators (such as discrete convolution, fractal dimension and shape decomposition in terms of sausages) could be defined and characterized, paving the way for integration with a large number of discrete geometry algorithms, including those with syntactic nature.

⑧ **Integration with Graph Theory and Complex Networks:** Because graphs and networks provide an effective means for representing virtually any discrete structure or system including digital images, as discussed and illustrated in this book, they stand out as being particularly promising for addressing several critical tasks in image and shape analysis. We believe the respective literature will be growing steadily in the next years.

We conclude the present book by observing that the perspectives in shape analysis and classification are not only particularly promising as candidates for outstanding theoretical works, but also represent an impressive potential for a series of important applications to a myriad of important areas. We hope to have contributed a little toward presenting the basic concepts and techniques that could be involved in re*shaping* our world in the forthcoming decades.

Bibliography

AGUADO, A. S., NIXON, M. S. & MONTIEL, M. E. (1998). Parameterizing arbitrary shapes via Fourier descriptors for evidence-gathering extraction. *Computer Vision and Image Understanding* **69**(2), 202–221.

AHNERT, S. & COSTA, L. D. F. (2008). Connectivity and Dynamics of Neuronal Networks as Defined by the Shape of Individual Neurons. *Arxiv preprint arXiv:0805.1640*.

ALBERT, R. & BARABÁSI, A. (2002). Statistical mechanics of complex networks. *Reviews of Modern Physics* **74**(1), 47–97.

ALDENDERFER, M. S. & BLASHFIELD, R. K. (1984). *Cluster Analysis*. Sage Publications.

ALFEREZ, R. & WANG, Y.-F. (1999). Geometric and illumination invariants for object recognition. *IEEE Transactions on Pattern Analysis and Machine Intelligence* **21**(6), 505–536.

ALLEN, F. T., KINSER, J. M. & CAULFIELD, H. J. (1999). A neural bridge from syntactic to statistical pattern recognition. *Neural Networks* **12**(3), 519–526.

ALLEN, J. B. & RABINER, L. R. (1977). A unified approach to short-time Fourier analysis and synthesis. *Proceedings IEEE* **65**(11), 1558–1564.

ALOIMONOS, Y. (1993). *Active Perception*. Lawrence Erlbaum Associates Publishers.

ANDERBERG, M. R. (1973). *Cluster Analysis for Applications*. Academic Press.

ANDERSON, J. A. (1995). *An Introduction to Neural Networks*. Cambridge, Mass.: The MIT Press.

ANGULO, J. M. & MADRIGAL, R. I. (1986). *Vision Artificial por Computador*. Madrid: Paraninfo.

ANTOINE, J.-P. (1994). Wavelet analysis: A new tool in signal processing. *Physicalia Magazine* **16**, 17–42.

ANTOINE, J.-P., BARACHE, D., CESAR-JR., R. & COSTA, L. D. F. (1997). Shape characterization with the wavelet transform. *Signal Processing* **62**(3), 265–290.

ANTOINE, J.-P., CARRETTE, P., MURENZI, R. & PIETTE, B. (1993). Image analysis with two-dimensional continuous wavelet transform. *Signal Processing* **31**, 241–272.

APOSTOL, T. M. (1969). *Calculus*. Wiley International Edition.

ARAÚJO, H. & DIAS, J. M. (1996). An introduction to the log-polar mapping. In: *Proceedings of II Workshop on Cybernetic Vision* (PRESS, I. C. S., ed.). São Carlos, SP, Brasil.

ARBTER, K., SNYDER, W. E., BURKHARDT, H. & HIRZINGER, G. (1990). Application of affine-invariant Fourier descriptors to recognition of 3-D objects. *IEEE Transactions on Pattern Analysis and Machine Intelligence* **12**(7), 640–647.

ARNÈODO, A., ARGOUL, F., BACRY, E., ELEZGARAY, J. & MUZY, J.-F. (1995). *Ondelettes, Multifractales et Turbulences: de l'ADN aux Croissances Cristallines.* Paris: Diderot Editeur, Arts et Sciences. In French.

ASADA, H. & BRADY, M. (1986). The curvature primal sketch. *IEEE Transactions on Pattern Analysis and Machine Intelligence* **8**, 2–14.

ATTNEAVE, F. (1954). Some informational aspects of visual perception. *Psychological Review* **61**, 183–193.

AURENHAMMER, F. (1991). Voronoi diagrams - a survey of a fundamental geometric data structure. *ACM Computing Surveys* **23**(3), 345–405.

BALLARD, D. H. & BROWN, C. M. (1982). *Computer Vision.* Englewood Cliffs, NJ: Prentice-Hall.

BARABÁSI, A. & ALBERT, R. (1999). Emergence of Scaling in Random Networks. *Science* **286**(5439), 509.

BARNETT, S. (1992). *Matrices: Methods and Applications.* Clarendon Press.

BARONI, M. & BARLETTA, G. (1992). Digital curvature estimation for left ventricular shape analysis. *Image and Vision Computing* **10**(7), 485–494.

BARRERA, J., CESAR-JR, R. M., HUMES-JR, C., MARTINS-JR, D. C., PATRÃO, D. F. C., SILVA, P. J. S. & BRENTANI, H. (2007). A feature selection approach for identification of signature genes from SAGE data. *BMC Bioinformatics* **8**(169).

BARRERA, J., DOUGHERTY, E. R. & TOMITA, N. S. (1997). Automatic programming of binary morphological machines by design of statistically optimal operators in the context of computational learning theory. *Journal of Electronic Imaging,* 54–67.

BARROW, H. & POPPLESTONE, R. (1971). Relational descriptions in picture processing. *Machine Intelligence* **6**(377-396), 3–2.

BASTIAANS, M. J. (1980). Gabor's expansion of a signal into gaussian elementary signals. *Proceedings of the IEEE* **68**(4), 538–539.

BAXES, G. A. (1994). *Digital Image Processing: Principles and Applications.* New York: John Wiley and Sons.

Bayne, C. K., Beauchamp, J. J., Begovich, C. L. & Kane, V. E. (1980). Monte carlo comparisons of selected clustering procedures. *Pattern Recognition* **12**, 51–62.

Bell, D. J. (1990). *Mathematics of Linear and Nonlinear Systems.* Clarendon Press.

Belongie, S., Malik, J. & Puzicha, J. (2001). Matching shapes. *Proc. of ICCV.*

Bennett, J. R. & Donald, J. S. M. (1975). On the measurement of curvature in a quantized environment. *IEEE Transactions on Computers* **C-24**(8), 803–820.

Bertero, M., Poggio, T. A. & Torre, V. (1988). Ill-posed problems in early vision. *Proceedings of the IEEE* **76**, 869–889.

Beucher, S. & Meyer, F. (1993). The morphological approach to segmentation: The watershed transform. In: *Mathematical Morphology in Image Processing* (Dougherty, E. R., ed.), chap. 12. Marcel Dekker Inc., 433–481.

Beus, H. L. & Tiu, S. S. H. (1987). An improved corner detection algorithm based on chain-coded plane curves. *Pattern Recognition* **20**, 291–296.

Biederman, I. (1985). Human image understanding: Recent research and a theory. *Computer Vision, Graphics and Image Processing* **32**, 29–73.

Blake, A. & Isard, M. (1998). *Active Contours.* Springer.

Blakemore, C. & Over, R. (1974). Curvature detectors in human vision? *Perception* **3**, 3–7.

Blashfield, R. (1976). Mixture model tests of cluster analysis: Accuracy of four agglomerative hierarchical methods. *Psychological Bulletin* **83**(3), 377–388.

Bloch, I. (1999). Fuzzy relative position between objects in image processing: A morphological approach. *IEEE Transactions on Pattern Analysis and Machine Intelligence* **21**(7), 657–664.

Bloch, I. (2005). Fuzzy spatial relationships for image processing and interpretation: a review. *Image and Vision Computing* **23**(2), 89–110.

Bloch, I. (2006). Spatial reasoning under imprecision using fuzzy set theory, formal logics and mathematical morphology. *International Journal of Approximate Reasoning* **41**(2), 77–95.

Bloch, I., Colliot, O. & Cesar-Jr., R. M. (2006). On the ternary spatial relation "between". *IEEE Transactions on Systems, Man and Cybernetics, Part B.* **36**(2), 312–327.

Bloch, I. & Maître, H. (1995). Fuzzy mathematical morphologies: A comparative study. *Pattern Recognition* **28**(9), 1341–1387.

BLOCH, I., MAÎTRE, H. & ANVARI, M. (1997). Fuzzy adjacency between image objects. *Int. J. Uncertain. Fuzziness Knowl.-Based Syst.* **5**(6), 615–653.

BLOCH, I. & MAÎTRE, H. (1997). Data fusion in 2D and 3D image processing: An overview. In: *Proceedings of SIBGRAPI 97.* http://www.visgraf.impa.br/ sibgrapi97/invited/bloch.html.

BLOCH, I. & RALESCU, A. (2003). Directional relative position between objects in image processing: a comparison between fuzzy approaches. *Pattern Recognition* **36**(7), 1563–1582.

BLUM, H. (1967). *A transformation for Extracting New Descriptors of Shape.* Models for the Perception of Speech and Visual Form. Cambridge, MA: MIT Press. W. Wathen-Dunn, Ed., 362–380.

BOAS, M. L. (1996). *Mathematical Methods in the Physical Sciences.* New York: John Wiley and Sons.

BOLLOBÁS, B. (1979). *Graph Theory: An Introductory Course.* Springer Verlag.

BONDY, J. A. & MURTY, U. S. R. (1976). *Graph Theory with Applications.* London: Macmillan.

BOOKSTEIN, F. L. (1991). *Morphometric Tools for Landmark Data: Geometry and Biology.* Cambridge: Cambridge University Press. (Also 1997).

BORGEFORS, G. (1984). Distance transforms in arbitrary dimensions. *Computer Vision Graphics and Image Processing* **27**, 321–345.

BOVIK, A. C., CLARK, M. & GEISLER, W. S. (1990). Multichannel texture analysis using localized spatial features. *IEEE Transactions on Pattern Analysis and Machine Intelligence* **12**(1), 55–73.

BOWIE, J. E. & YOUNG, I. T. (1977). An analysis technique for biological shape - ii. *Acta Cytologica* **21**(3), 455–464.

BOYCOTT, B. B. & WÄSSLE, H. (1974). The morphological types of ganglion cells of the domestic cat's retina. *Journal of Physiology* **240**, 297–319.

BOYKOV, Y. & JOLLY, M. (2001). Interactive graph cuts for optimal boundary and region segmentation of objects in N-D images. *International Conference on Computer Vision* **1**, 105–112.

BRACEWELL, R. N. (1986). *The Fourier Transform and its Applications.* McGraw-Hill, 2nd ed.

BRACEWELL, R. N. (1995). *Two-Dimensional Imaging.* Englewood Cliffs, NJ: Prentice-Hall.

BRAGA-NETO, U. & DOUGHERTY, E. (2004). Bolstered error estimation. *Pattern Recognition* **37**(6), 1267–1281.

BRIGHAM, E. O. (1988). *The Fast Fourier Transform and its Applications*. Englewood Cliffs, NJ: Prentice Hall.

BRONSON, R. (1988). *Schaum's Outline of Matrix Operations*. New York: McGraw-Hill.

BRONSON, R. (1991). *Matrix Methods: An Introduction*. Academic Press.

BRUNO, O., DE OLIVEIRA PLOTZE, R., FALVO, M. & DE CASTRO, M. (2008a). Fractal dimension applied to plant identification. *Information Sciences*.

BRUNO, O., NONATO, L., PAZOTI, M. & NETO, J. (2008b). Topological multi-contour decomposition for image analysis and image retrieval. *Pattern Recognition Letters*.

BRUNO, O. M., CESAR-JR., R. M., CONSULARO, L. A. & COSTA, L. D. F. (1998). Automatic feature selection for biological shape classification in σ ynergos. In: *Proceedings of Brazilian Conference on Computer Graphics, Image Processing and Vision* (PRESS, I. C. S., ed.). Rio de Janeiro, RJ.

BUCK, R. C. (1978). *Advanced Calculus*. New York: McGraw-Hill.

BUHAN, C. L., JORDAN, T. & EBRAHIMI, T. (1997). Scalable vertex-based shape coding - S4h results. Tech. rep., ISO/IEC JTC1/SC29/WG11 MPEG-4 meeting, Bristol, UK. M2034.

BUNKE, H. (2000a). Graph matching: Theoretical foundations, algorithms, and applications. In: *Proceedings of Vision Interface 2000, Montreal*.

BUNKE, H. (2000b). Recent developments in graph matching. In: *ICPR*.

BUNKE, H. & ALLERMANN, G. (1983). Inexact graph matching for structural pattern recognition. *Pattern Recognition Letters* **1**(4), 245–253.

BURR, D. J. (1981). Elastic matching of line drawings. *IEEE Transactions on Pattern Analysis and Machine Intelligence* **3**(6), 708–713.

BURRUS, C. S., McCLELLAN, J. H., OPPENHEIM, A. V., PARKS, T. W., SCHAFER, R. W. & SCHUESSLER, H. W. (1994). *Computer-Based Exercises for Signal Processing using MATLAB®*. Englewood Cliffs, NJ: Prentice-Hall.

BURT, P. J. & ADELSON, E. H. (1983). The laplacian pyramid as a compact image code. *IEEE Transactions on Communications* **31**(4), 532–540.

BÄCK, T. (1996). *Evolutionary Algorithms in Theory and Practice*. Oxford University Press.

CAETANO, T., CHENG, L., LE, Q. & SMOLA, A. (2007). Learning Graph Matching. *IEEE 11th International Conference on Computer Vision – ICCV 2007*, 1–8.

CAMPOS, T. E., BLOCH, I. & CESAR-JR., R. M. (2001). Feature selection based on fuzzy distances between clusters: First results on simulated data. In: *Proc. ICAPR'2001 - International Conference on Advances in Pattern Recognition* (SINGH, S., MURSHED, N. & KROPATSCH, W., eds.), vol. 2013 of *Lecture Notes in Computer Science*, Rio de Janeiro, Brasil: Springer-Verlag.

CANNY, J. (1986). A computational approach to edge detection. *IEEE Transactions on Pattern Analysis and Machine Intelligence* **8**(6), 679–698.

CARMONA, R., HWANG, W.-L. & TORRÉSANI, B. (1998). *Practical Time-Frequency Analysis: Wavelet and Gabor Transforms with an Implementation in S.* Academic Press.

CASACUBERTA, F. & DE LA HIGUERA, C. (1999). Optimal linguistic decoding is a difficult computational problem. *Pattern Recognition Letters* **20**(8), 813–821.

CASERTA, F., ELDRED, E. D., FERNANDEZ, E., HAUSMAN, R. E., STANFORD, L. R., BULDEREV, S. V., SCHWARZER, S. & STANLEY, H. E. (1995). Determination of fractal dimension of physiologically characterized neurons in two and three dimensions. *Journal of Neuroscience Methods* **56**, 133–144.

CASERTA, F., STANLEY, H. E., ELDRED, W. D., DACCORD, G., HAUSMAN, R. E. & NITTMANN, J. (1990). Physical mechanisms underlying neurite outgrowth: A quantitative analysis of neuronal shape. *Physical Review Letters* **64**, 95–98.

CASTLEMAN, K. R. (1996). *Digital Image Processing.* Englewood Cliffs, NJ: Prentice Hall.

CESAR-JR., R. M., BENGOETXEA, E., BLOCH, I. & LARRAÑAGA, P. (2005). Inexact graph matching for model-based recognition: Evaluation and comparison of optimization algorithms. *Pattern Recognition* **38**(11), 2099–2113.

CESAR-JR., R. M. & COSTA, L. D. F. (1995). Piecewise linear segmentation of digital contours in O(n log(n)) through a technique based on effective digital curvature estimation. *Real-Time Imaging* **1**, 409–417.

CESAR-JR., R. M. & COSTA, L. D. F. (1996). Shape characterization by using the Gabor transform. In: *Proceedings of 7th IEEE Digital Signal Processing Workshop.* Loen, Norway.

CESAR-JR., R. M. & COSTA, L. D. F. (1997). Application and assessment of multiscale bending energy for morphometric characterization of neural cells. *Review of Scientific Instruments* **68**(5), 2177–2186.

CESAR-JR., R. M. & COSTA, L. D. F. (1998a). Análise multi-escala de formas bidimensionais. In: *Proc. 6th Congresso Iberoamericano de Inteligência Artificial*. Lisbon, Portugal.

CESAR-JR., R. M. & COSTA, L. D. F. (1998b). Neural cell classification by wavelets and multiscale curvature. *Biological Cybernetics* **79**(4), 347–360.

CESAR-JR., R. M. & COSTA, L. D. F. (1999). Dendrogram generation for neural shape analysis. *The Journal of Neuroscience Methods* **93**, 121–131.

CHALUMEAU, T., COSTA, L. D. F., LALIGANT, O. & MERIAUDEAU, F. (2006). Optimized texture classification by using hierarchical complex network measurements. *Proceedings of SPIE* **6070**, 60700Q.

CHARTRAND, G. (1985). *Introductory Graph Theory*. Courier Dover Publications.

CHASSERY, J. M. & MONTANVERT, A. (1991). *Géométrie Discrète*. Paris: Hermes. (In French).

CHATFIELD, C. & COLLINS, A. J. (1980). *Introduction to Multivariate Analysis*. Chapman and Hall.

CHAUDHURI, B. B. & ROSENFELD, A. (1996). On a metric distance between fuzzy sets. *Pattern Recognition Letters* **17**(11), 1157–1160.

CHEEVER, E. A., OVERTON, G. C. & SEARLS, D. B. (1991). Fast Fourier transform-based correlation of DNA sequences using complex plane encoding. *Computer Application in the Biosciences* **7**(2), 143–154.

CHELLAPPA, R. & BAGDAZIAN, R. (1984). Fourier coding of image boundaries. *IEEE Transactions on Pattern Analysis and Machine Intelligence* **6**(1), 102–105.

CHEN, C.-H., LEE, J.-S. & SUN, Y.-N. (1995). Wavelet transformation for gray-level corner detection. *Pattern Recognition* **28**(6), 853–861.

CHEN, H., SCHUFFELS, C. & ORWIG, R. (1996). Internet categorization and search: A self-organizing approach. *Journal of Visual Communication and Image Representation* **7**(1), 88–102.

CHEN, M.-H. & YAN, P.-F. (1989). A multiscaling approach based on morphological filtering. *IEEE Transactions on Pattern Analysis and Machine Intelligence* **11**(7), 694–700.

CHINN, W. G. & STEENROD, N. E. (1966). *First Concepts of Topology*. The Mathematical Association of America.

CHO, K., MEER, P. & CABRERA, J. (1997). Performance assessment through bootstrap. *IEEE Transactions on Pattern Analysis and Machine Intelligence* **19**(11), 1185–1198.

CHONG, M. M. S., GAY, R. K. L., TAN, H. N. & LIU, J. (1992). Automatic representation of fingerprints for data compression by B-spline functions. *Pattern Recognition* **25**(10), 1199–1210.

CHUANG, C.-H. & KUO, C.-C. J. (1996). Wavelet descriptor of planar curves: Theory and applications. *IEEE Transactions on Image Processing* **5**(1), 56–70.

CHURCHILL, R. V. & BROWN, J. W. (1989). *Complex Variables and Applications.* New York: McGraw-Hill.

COELHO, R. C. & COSTA, L. D. F. (1996). On the application of the Bouligand-Minkowisk fractal dimension for shape characterisation. *Applied Signal Processing* **3**, 163–176.

COHEN, L. (1989). Time-frequency distributions - a review. *Proceedings of the IEEE* **77**(7), 941–981.

COHEN, L. (1991). On active contour models and balloons. *CVGIP - Image Understanding* **53**(2), 211–218.

COLLIOT, O., TUZIKOV, A., CESAR-JR., R. & BLOCH, I. (2004). Approximate reflectional symmetries of fuzzy objects with an application in model-based object recognition. *Fuzzy Sets and Systems* **147**(1), 141–163.

CONSULARO, L. A., CESAR-JR., R. M. & BLOCH, I. (2007). Structural image segmentation with interactive model generation. In: *Proc. IEEE International Conference on Image Processing (ICIP-07).* Piscataway, NJ: IEEE.

CONTE, D., FOGGIA, P., SANSONE, C. & VENTO, M. (2003). Graph matching applications in pattern recognition and image processing. In: *ICIP (2).*

CONTE, D., FOGGIA, P., SANSONE, C. & VENTO, M. (2004). Thirty years of graph matching in pattern recognition. *IJPRAI* **18**(3), 265–298.

COOK, T. (1979). *The Curves of Life.* NY: Dover.

COOPER, J., VENKATESH, S. & KITCHEN, L. (1993). Early jump-out corner detectors. *IEEE Transactions on Pattern Analysis and Machine Intelligence* **15**(8), 823–828.

COOTES, T. F., TAYLOR, C. J., COOPER, D. H. & GRAHAM, J. (1995). Active shape models - their training and application. *Computer Vision and Image Understanding* **61**(1), 38–59.

CORMEN, T., LEISERSON, C. E. & RIVEST, R. L. (1990). *Introduction to Algorithms.* Cambridge, MA: MIT Press.

CORNIC, P. (1997). Another look at the dominant point detection of digital curves. *Pattern Recognition Letters* **18**, 13–25.

Costa, L. d. F. (1992). *Effective Detection of Line Segments with Hough Transform.* Ph.D. thesis, King's College, University of London.

Costa, L. d. F. (1995). Towards real-time detection of discrete straight features with a hybrid technique based on preliminary curve segmentation and zoomed-adaptive parametric mapping. *Real-Time Imaging* **1**(3), 203–214.

Costa, L. d. F. (2003). Enhanced multiscale skeletons. *J Real-Time Imaging* **9**, 315–319.

Costa, L. d. F. (2004). Complex Networks, Simple Vision. *Arxiv preprint cond-mat/0403346.*

Costa, L. d. F. (2005). Morphological complex networks: Can individual morphology determine the general connectivity and dynamics of networks? *Arxiv preprint q-bio.MN/0503041.*

Costa, L. d. F. (2006a). Complex networks: New concepts and tools for real-time imaging and vision. URL http://www.citebase.org/abstract?id=oai:arXiv.org:cs/0606060.

Costa, L. d. F. (2006b). Visual Saliency and Attention as Random Walks on Complex Networks. URL http://arxiv.org/abs/0711.2736.

Costa, L. d. F. (2007a). Knitted complex networks. URL http://arxiv.org/abs/0711.2736.

Costa, L. d. F. (2007b). The Path-Star Transformation and its Effects on Complex Networks. *Arxiv preprint arXiv:0711.1271.*

Costa, L. d. F., Campos, A. G., Estrozi, L. F., Rios, L. G. & Bosco, A. (2000). A biologically-motivated approach to image representation and its applications to neuromorphometry. In: *Lectures Notes in Computer Science*, vol. 1811.

Costa, L. d. F., Campos, A. G. & Manoel, E. T. M. (2001). An integrated approach to shape analysis: Results and perspectives. In: *Proceedings of International Conference on Quality Control by Artificial Vision*, vol. 1.

Costa, L. d. F., Cesar-Jr., R. M., Coelho, R. C. & Tanaka, J. S. (1999). Perspective on the analysis and synthesis of morphologically realistic neural networks. In: *Modeling in the Neurosciences: From Ionic Channels to Neural Networks* (Poznanski, R., ed.). Gordon and Breach Science Publishers.

Costa, L. d. F. & Consularo, L. A. (1999). The dynamics of biological evolution and the importance of spatial relations and shapes. In: *Proceedings of the 3rd International Workshop on Human and Machine Perception: Emergence, Attention and Creativity.* Cantoni, V., di Gesu, V., Setti, A. & Tegolo, D., eds. Pavia, Itália: Kluwer Academic.

COSTA, L. D. F., LENG, X., SANDLER, M. B. & SMART, P. (1991). A system for semi-automated analysis of clay samples. *Review of Scientific Instruments* **62**, 2163–2166.

COSTA, L. D. F., OLIVEIRA JR, O. N., TRAVIESO, G., RODRIGUES, F. A., BOAS, P. R. V., ANTIQUEIRA, L., VIANA, M. P. & DA ROCHA, L. E. C. (2007a). Analyzing and modeling real-world phenomena with complex networks: A survey of applications. URL http://arxiv.org/abs/0711.3199.

COSTA, L. D. F., RODRIGUES, F., TRAVIESO, G. & BOAS, P. (2007b). Characterization of complex networks: A survey of measurements. *Advances in Physics* **56**, 167–242.

COSTA, L. D. F., RODRIGUES TOGNETTI, M. & SILVA, F. (2008). Concentric characterization and classification of complex network nodes: Application to an institutional collaboration network. *Physica A: Statistical Mechanics and its Applications.*

COSTA, L. D. F. & SANDLER, M. (1993). Effective detection of bar segments with Hough transform. *Computer Vision Graphics and Image Processing* **55**, 180–191.

COSTA, L. D. F. & VELTE, T. J. (1999). Automatic characterization and classification of ganglion cells from the salamander retina. *Journal of Comparative Neurology* **404**(1), 35–51.

COSTA, L. D. F. & WECHSLER, H. (2000). Guest editorial: Special issue on the Hough transform - has the Hough transform come of age? *Real-Time Imaging* **6**, 77–78.

CRANE, R. (1996). *A Simplified Approach to Image Processing: Classical and Modern Techniques in C.* New York: Prentice-Hall.

CROWLEY, J. L. & PARKER, A. C. (1984). A representation for shape based on peaks and ridges in the difference of low-pass transform. *IEEE Transactions on Pattern Analysis and Machine Intelligence* **1**(2), 156–170.

CROWLEY, J. L. & SANDERSON, A. C. (1987). Multiple resolution representation and probabilistic matching of 2-D gray-scale shape. *IEEE Transactions on Pattern Analysis and Machine Intelligence* **9**(1), 113–120.

CUADROS-VARGAS, A., NONATO, L., MINGHIM, R. & ETIENE, T. (2005). Imesh: An Image Based Quality Mesh Generation Technique. *Proceedings of the XVIII Brazilian Symposium on Computer Graphics and Image Processing (SIBGRAPI'05).*

CUISENAIRE, O. (1999). *Distance Transformations: Fast Algorithms and Applications to Medical Image Processing.* Ph.D. thesis, Laboratoire de Telecommunications et Teledetection, Louvain-la-Neuve, Belgium.

CYGANSKY, D., NOEL, W. F. & ORR, J. A. (1990). The analytic Hough transform. In: *SPIE/SPSE Symposium on Electronic Imaging Science and Technology.* Sta. Clara, USA. Paper 1260-18.

DACEY, M. D. (1993). The mosaic of midget ganglion cells in the human retina. *The Journal of Neuroscience* **13**(12), 5334–5355.

DANN, J. F., BUHL, E. & PEICHL, L. (1988). Posnatal dendritic maturation of alpha and beta ganglion cells in cat retina. *The Journal of Neuroscience* **8**(5), 1485–1499.

DAUBECHIES, I. (1992). Ten lectures on wavelets. *CBMS-NSF Regional Conference Series in Applied Mathematics.*

DAUBECHIES, I. (1996). Where do wavelets come from? - a personal point of view. *Proceedings of the IEEE* **84**(4), 510–513.

DAUGMAN, J. G. (1980). Two-dimensional spectral-analysis of cortical receptive-field profiles. *Vision Research* **20**(10), 847–856.

DAUGMAN, J. G. (1985). Uncertainty relation for resolution in space, spatial-frequency and orientation optimized by two-dimensional visual cortical filters. *J. Optical Society of America* **2**(7), 1160–1169.

DAUGMAN, J. G. (1988). Complete discrete 2-D Gabor transforms by neural networks for image analysis and compression. *IEEE Transactions on Acoustics, Speech and Signal Processing* **36**(7), 1169–1179.

DAVENPORT-JR, W. B. (1970). *Probability and Random Processes - An Introduction for Applied Scientists and Engineers.* New York: McGraw-Hill.

DAVIES, A. & SAMUELS, P. (1996). *An Introduction to Computational Geometry for Curves and Surfaces.* Clarendon Press.

DAVIS, L. S. (1977). Understanding shapes: Angles and sides. *IEEE Transactions on Computers* **26**(3), 236–242.

DAVIS, L. S. (1979). Shape matching using relaxation techniques. *IEEE Transactions on Pattern Analysis and Machine Intelligence* **1**(1), 60–72.

DAVIS, P. J. & HERSH, R. (1999). *The Mathematical Experience.* Houghton Mifflin Co.

DEL BIMBO, A. (1999). *Visual Information Retrieval.* Academic Press.

DELOGNE, P. (1996). Internet, special section - video/image coding and transmission for the internet. In: *Proceedings of the IEEE International Conference on Image Processing (ICIP)* (DELOGNE, P., ed.). Lausanne, Switzerland: IEEE, Piscataway, NJ.

DEMIGNY, D. & KAMLÉ, T. (1997). A discrete expression of Canny's criteria for step edge detector performances evaluation. *IEEE Transactions on Pattern Analysis and Machine Intelligence* **19**(11), 1199–1211.

DETTMAN, J. W. (1988). *Mathematical Methods in Physics and Engineering*. Dover.

DICK, A., TORR, P. & CIPOLLA, R. (2000). Automatic 3D modelling of architecture. *British Machine Vision Conference, Bristol*, 372–381.

DICKINSON, S. J., SHOKOUFANDEH, A., KESELMAN, Y., DEMIRCI, M. F. & MACRINI, D. (2005). Object categorization and the need for many-to-many matching. In: *DAGM-Symposium*.

DO CARMO, M. P. (1976). *Differential Geometry of Curves and Surfaces*. Prentice-Hall.

DOBBINS, A., ZUCKER, S. W. & CYNADER, M. (1987). Endstopped neurons in the visual cortex as a substrate for calculating curvature. *Nature* **329**, 438–441.

DOROGOVTSEV, S. & MENDES, J. (2002). Evolution of networks. *Advances In Physics* **51**(4), 1079–1187.

DOUGHERTY, E. & LOTUFO, R. (2003). *Hands-on Morphological Image Processing*. SPIE Publications.

DOUGHERTY, E. R. & ASTOLA, J. T. (1994). *An Introduction to Nonlinear Image Processing*. Society of Photo-optical Instrumentation Engineers. Tutorial Texts in Optical Engineering, Vol Tt 16.

DOUGHERTY, E. R. & ASTOLA, J. T. (1999). *Nonlinear Filters for Image Processing*. SPIE-IEEE.

DRYDEN, I. L. & MARDIA, K. V. (1998). *Statistical Shape Analysis*. New York: John Wiley and Sons.

DUBERY, F. & WILLATS, J. (1972). *Drawing Systems*. Van Nostrand Reinhold.

DUDA, R. O. & HART, P. (1973). *Pattern Classification and Scene Analysis*. New York: John Wiley and Sons.

DUDA, R. O., HART, P. E. & STORK, D. G. (2000). *Pattern Classification*. New York: Wiley-Interscience.

DUNCAN, J. S., LEE, F. A., SMEULDERS, A. W. M. & ZARET, B. L. (1991). A bending energy model for measurement of cardiac shape deformity. *IEEE Transactions on Medical Imaging* **10**(3), 307–320.

DUNN, D. & HIGGINS, W. E. (1995). Optimal Gabor filters for texture segmentation. *IEEE Transactions on Image Processing* **4**(7), 947–964.

DuPain, Y., Kamae, T. & France, M. M. (1986). Can one measure the temperature of a curve? *Arch. Rational. Mech. Anal.* **94**, 155–163.

Edelman, S. (1999). *Representation and Recognition in Vision*. The MIT Press.

Edwards, C. H. & Penney, D. E. (1998). *Calculus with Analytic Geometry*. Prentice Hall.

Egenhofer, M. J. & Shariff, A. R. B. M. (1998). Metric details for natural-language spatial relations. *ACM Trans. Inf. Syst.* **16**(4), 295–321.

Eichmann, G., Lu, C., Jankowski, M. & Tolimieri, R. (1990). Shape representation by Gabor expansion. In: *Proceedings of SPIE Hybrid Image and Signal Processing II*, vol. 1297.

Erdos, P. & Renyi, A. (1960). On the evolution of random graphs. *Publ. Math. Inst. Hungar. Acad. Sci* **5**, 17–61.

Erdős, P. & Rényi, A. (1961). On the strength of connectedness of a random graph. *Acta Mathematica Hungarica* **12**(1), 261–267.

Esakov, J. & Weiss, T. (1989). *Data Structures: An Advanced Approach Using C.* Englewood Cliffs, NJ: Prentice Hall.

Estrozi, L., Rios-Filho, L., Campos Bianchi, A., Cesar, R. & Costa, L. (2003). 1D and 2D Fourier-based approaches to numeric curvature estimation and their comparative performance assessment. *Digital Signal Processing* **13**(1), 172–197.

Everitt, B. S. (1993). *Cluster Analysis*. Arnold, 3rd ed.

Everitt, B. S. & Dunn, G. (1991). *Applied Multivariate Data Analysis*. Edward Arnhold.

Fabbri, R., da F. Costa, L., Torelli, J. C. & Bruno, O. M. (2008). 2d euclidean distance transform algorithms: A comparative survey. *ACM Comput. Surv.* **40**(1), 1–44.

Faber, R. L. (1983). *Differential Geometry and Relativity Theory—An Introduction*. Marcel Dekker.

Fairney, D. P. & Fairney, P. T. (1994). On the accuracy of point curvature estimators in a discrete environment. *Image and Vision Computing* **12**(5), 259–265.

Falcão, A., Stolfi, J. & de Alencar Lotufo, R. (2004). The Image Foresting Transform: Theory, Algorithms, and Applications. *IEEE Transactions on Pattern Analysis and Machine Intelligence*, 19–29.

Falconer, K. (1990). *Fractal Geometry—Mathematical Foundations and Applications*. England: John Wiley and Sons Ltd.

FALCÃO, A., COSTA, L. D. F. & DA CUNHA, B. (2002). Multiscale skeletons by image foresting transform and its application to neuromorphometry. *Pattern Recognition* **35**(7), 1571–1582.

FALOUTSOS, M., FALOUTSOS, P. & FALOUTSOS, C. (1999). On power-law relationships of the Internet topology. *Proceedings of the conference on Applications, technologies, architectures, and protocols for computer communication*, 251–262.

FAUSETT, L. (1994). *Fundamentals of Neural Networks*. Prentice-Hall.

FAUSTINO, G. & DE FIGUEIREDO, L. (2005). Simple Adaptive Mosaic Effects. *Proc. 18th SIBGRAPI 2005*, 315–322.

FELSENSTEIN, J., SAWYER, S. & KOCHIN, R. (1982). An efficient method for matching nucleic acid sequences. *Nucleic Acid Research* **10**, 133–139.

FELZENSZWALB, P. & HUTTENLOCHER, D. (2004). Efficient Graph-Based Image Segmentation. *International Journal of Computer Vision* **59**(2), 167–181.

FELZENSZWALB, P. & HUTTENLOCHER, D. (2006). Efficient Belief Propagation for Early Vision. *International Journal of Computer Vision* **70**(1), 41–54.

FELZENSZWALB, P. & SCHWARTZ, J. (2007). Hierarchical matching of deformable shapes. *Proceedings of CVPR*.

FELZENSZWALB, P. F. & HUTTENLOCHER, D. P. (2005). Pictorial structures for object recognition. *International Journal of Computer Vision* **61**(1), 55–79.

FISCHLER, M. & ELSCHLAGER, R. (1973). The Representation and Matching of Pictorial Structures. *IEEE Transactions on Computers* **100**(22), 67–92.

FISCHLER, M. A. & WOLF, H. C. (1994). Locating perceptually salient points on planar curves. *IEEE Transactions on Pattern Analysis and Machine Intelligence* **16**(2), 113–129.

FISHER, R., PERKINS, S., WALKER, A. & WOLFART, E. (1996). *Hypermedia Image Processing Reference*. John Wiley and Sons.

FLANDRIN, P. (1999). *Time-Frequency/Time-Scale Analysis*. Academic Press.

FLORY, P. (1941). Molecular Size Distribution in Three Dimensional Polymers. II. Trifunctional Branching Units. *Journal of the American Chemical Society* **63**(11), 3091–3096.

FOLEY, J. D., VAN DAM, A., FEINER, S., HUGHES, J. & PHILLIPS, R. L. (1994). *Introduction to Computer Graphics*. Reading: Addison-Wesley.

FONG, C. F. C. C. M., KEE, D. D. & KALONI, P. N. (1997). *Advanced Mathematics for Applied and Pure Sciences*. Gordon and Breach Science Publishers.

FORTUNE, S. (1987). A sweepline algorithm for Voronoi diagrams. *Algorithmica* **2**, 153–174.

FREDERICK, D. K. & CARLSON, A. B. (1971). *Linear Systems in Communication and Control.* John Wiley and Sons.

FREEMAN, H. (1961). On the encoding of arbitrary geometric configurations. *IRE Trans. Elec. Comp.*, 260–268EC-10.

FREEMAN, H. (1970). *Boundary Encoding and Processing.* New York: Academic Press. In: B. S. Lipkin and A. Rosenfeld, Eds., Picture Processing and Psychopictorics, pp. 241–266.

FREEMAN, H. (1974). Computer processing of line-drawings images. *Computing Surveys* **6**(1), 57–95.

FREEMAN, H. & DAVIS, L. S. (1977). A corner finding algorithm for chain-coded curves. *IEEE Transactions on Computers* **26**, 297–303.

FREEMAN, J. (1975). The modelling of spatial relations. *Computer Graphics and Image Processing* **4**, 156–171.

FREEMAN, W. T. & ADELSON, E. (1991). The design and use of steerable filters. *IEEE Transactions on Pattern Analysis and Machine Intelligence* **13**(9), 891–906.

FU, K. S. (1974). *Syntactic methods in pattern recognition.* Academic Press.

FU, K. S. (1982). *Syntactic Pattern Recognition and Applications.* Englewood Cliffs, NJ: Prentice-Hall.

FUKUDA, Y., WATANABE, C.-F. & ITO, H. (1984). Physiologically identified y-, x- and w-cells in cat retina. *Journal of Neurophysiology* **52**(6), 999–1013.

FUKUNAGA, K. (1990). *Introduction to Statistical Pattern Recognition.* Academic Press.

GABOR, D. (1946). Theory of communication. *Proceedings of IEE* **93**(26), 429–457.

GARCIA, J. A., FDEZ-VALDIVIA, J. & MOLINA, R. (1995). A method for invariant pattern recognition using the scale-vector representation of planar curves. *Signal Processing* **43**, 39–53.

GASÓS, J. & SAFFIOTTI, A. (1999). Using fuzzy sets to represent uncertain spatial knowledge in autonomous robots. *Spatial Cognition and Computation* **1**(3), 205–226.

GAUDART, L., CREBASSA, J. & PETRAKIAN, J. P. (1993). Wavelet transform in human visual channels. *Applied Optics* **32**(22), 4119–4127.

GEIGER, D. & YUILLE, A. (1991). A common framework for image segmentation. *International Journal of Computer Vision* **6**(3), 227–243.

GERIG, G. & KLEIN, F. (1986). Fast contour identification through efficient Hough transform and simplified interpretation strategy. In: *8th International Joint Conference on Pattern Recognition*. Paris, France.

GHYKA, M. (1977). *The Geometry of Art and Life*. New York: Dover.

GIBLIN, P. J. & BRUCE, J. W. (1992). *Curves and Singularities*. Cambridge University Press.

GIBSON, J. J. (1933). Adaptation, after-effect and contrast in the perception of curved lines. *Journal of Experimental Psychology* **16**(1), 1–31.

GIRVAN, M. & NEWMAN, M. (2002). Community structure in social and biological networks. *Proceedings of the National Academy of Sciences* **99**(12), 7821.

GLASBEY, C. A. & HORGAN, G. W. (1995). *Image Analysis for the Biological Sciences*. Chichester: John Wiley and Sons.

GLEICK, J. (1987). *Chaos: Making a New Science*. New York: Viking.

GNANADESIKAN, R. (1977). *Methods for Statistical Data Analysis of Multivariate Observations*. John Wiley and Sons.

GOLDSTEIN, H. (1980). *Classical Mechanics*. Addison-Wesley.

GOLUB, G. H. & LOAN, C. F. V. (1989). *Matrix Computations*. The John Hopkins University Press.

GONZALEZ, R. C. & WINTZ, P. (1987). *Digital Image Processing*. Addison-Wesley.

GONZALEZ, R. C. & WOODS, R. E. (1993). *Digital Image Processing*. Addison-Wesley.

GOSE, E., JOHNSONBAUGH, R. & JOST, S. (1996). *Pattern Recognition and Image Analysis*. NJ: Prentice Hall.

GOSHTASBY, A. (1985). Description and discrimination of planar shapes using shape matrices. *IEEE Transactions on Pattern Analysis and Machine Intelligence* **7**(6), 738–743.

GOSHTASBY, A. (1992). Gaussian decomposition of two-dimensional shapes: A unified representation for CAD and vision applications. *Pattern Recognition* **25**(5), 463–472.

GOSHTASBY, A. (1993). Design and recovery of 2-D and 3-D shapes using rational gaussian curves and surfaces. *International Journal of Computer Vision* **10**(3), 233–256.

GOUTSIAS, J. & BATMAN, S. (2000). *Handbook of Medical Imaging: Volume 3 - Progress in Medical Image Processing and Analysis, chapter Morphological Methods for Biomedical Image Analysis.* SPIE Optical Engineering Press.

GRACIANO, A., CESAR-JR, R. & BLOCH, I. (2007). Graph-based Object Tracking Using Structural Pattern Recognition. *Proceedings of the XX Brazilian Symposium on Computer Graphics and Image Processing (SIBGRAPI 2007)*, 179–186.

GRACIANO, A. B. V. (2007). *Rastreamento de objetos baseado em reconhecimento estrutural de padrões.* Master's thesis, Institute of Mathematics and Statistics–University of São Paulo.

GRADY, L. (2006). Random Walks for Image Segmentation. *IEEE Transactions on Pattern Analysis and Machine Intelligence*, 1768–1783.

GRANLUND, G. H. (1972). Fourier preprocessing for hand print character recognition. *IEEE Transactions on Computers* **21**(2), 195–201.

GRAUSTEIN, W. C. (1966). *Differential Geometry.* Dover.

GROEN, F. C. A. & VERBEEK, P. W. (1978). Freeman-code probabilities of object boundary quantized contours. *Computer Vision Graphics and Image Processing* **7**, 391–402.

GROSS, A. L. (1972). A monte carlo study of the accuracy of a hierarchical grouping procedure. *Multivariate Behavioral Research* **7**, 379–389.

GROSSMANN, A. (1988). *Stochastic Processes in Physics and Engineering.* Wavelet Transforms and Edge Detection. Dordrecht: Reidel Publishing Company. In S. Albeverio, Ph. Blanchard, M. Hazewinkel, and L. Streit (Eds.).

GROSSMANN, A., KRONLAND-MARTINET, R. & MORLET, J. (1989). Reading and understanding continuous wavelet transform. In: *Proc. Marseille 1987* (J.-M. COMBES, A. G. & TCHAMITCHIAN, P., eds.), Wavelets, Time-Frequency Methods and Phase-Space. Berlin: Springer-Verlag.

GROSSMANN, A. & MORLET, J. (1984). Decomposition of hardy functions into squared integrable wavelets of constant shape. *SIAM J. Math. Analysis* **15**, 723–736.

GUEZ, J.-E., MARCHAL, P., GARGASSON, J.-F. L., GRALL, Y. & O'REGAN, J. K. (1994). Eye fixations near corners: Evidence for a centre of gravity calculation based on contrast, rather than luminance or curvature. *Vision Research* **34**, 1625–1635.

GUGGENHEIMER, H. W. (1977). *Differential Geometry.* Dover.

GULLBERG, J. (1997). *Mathematics: From the Birth of Numbers.* W. W. Norton and Co.

GÖCKELER, M. & SCHÜCKER, T. (1987). *Differential Geometry, Gauge Theories and Gravity*. Cambridge University Press.

HALMOS, P. R. (1958). *Finite-Dimensional Vector Spaces*. Van Nostrand Company.

HANDS, S. & EVERITT, B. S. (1987). A monte carlo study of the recovery of cluster structure in binary data by hierarchical clustering techniques. *Multivariate Behavioral Research* **22**, 235–243.

HARALICK, R. M. & SHAPIRO, L. G. (1993). *Computer and Robot Vision*, vol. 2. Reading, MA: Addison-Wesley Pub. Co.

HARALICK, R. M., STERNBERG, S. R. & ZHUANG, X. (1987). Image analysis using mathematical morphology. *IEEE Transactions on Pattern Analysis and Machine Intelligence* **9**(4), 532–550.

HARTIGAN, J. (1985). Statistical theory in clustering. *Journal of Classification* **2**(1), 63–76.

HAXHIMUSA, Y., ION, A. & KROPATSCH, W. G. (2006). Evaluating hierarchical graph-based segmentation. In: *ICPR (2)*.

HEARN, D. & BAKER, M. (1986). *Computer Graphics*. Englewood Cliffs, NJ: Prentice Hall.

HEARST, M. A. (1997). Interfaces for searching the web. *Scientific American* **276**(3), 68–72.

HEIJMANS, H. & GOUTSIAS, J. (1999). Multiresolution signal decomposition schemes, part 2: Morphological wavelets. Tech. Rep. PNA-R9905, CWI - Centrum voor Wiskunde en Informatica. 30.

HEIJMANS, H. J. A. M. (1994). *Morphological Image Operators*. Boston: Academic Press.

HELD, A., ABE, K. & ARCELLI, C. (1994). Towards a hierarchical contour description via dominant point detection. *IEEE Transactions on Systems Man and Cybernetics* **24**(6), 942–949.

HERTZ, J., PALMER, R. G. & KROGH, A. (1991). *Introduction to the Theory of Neural Computation*. Perseus.

HILDITCH, C. J. (1969). *Linear skeletons from square cupboards*. Machine Intelligence. Edinburgh Univ. Press. Meltzer, B.; Mitchie, D. (Eds), pp. 403–420.

HILDRETH, E. C. (1983). *The Measurement of Visual Motion*. MIT Press.

HLWATSCH, F. & BORDREAUX-BARTELS, G. F. (1992). Linear and quadratic time-frequency signal representations. *IEEE Signal Processing Magazine*, 21–67.

HOFFMAN, K. & KUNZE, R. (1971). *Linear Algebra*. Prentice-Hall.

HOLLAND, J. (1975). *Adaption in Natural and Artificial Systems*. Ann Arbon: The University of Michigan Press.

HOLSCHNEIDER, M. (1995). *Wavelets: An Analysis Tool*. Oxford: Clarendon Press.

HOPCROFT, J. E. & ULLMAN, J. D. (1979). *Introduction to Automata Theory, Languages and Computation*. MA: Addison-Wesley.

HORN, B. (1986). *Robot Vision*. Mass.: Cambridge, MIT Press.

HOUGH, P. V. C. (1962). Method and means for recognizing complex patterns. United States Patent Office. Patent 3,069654.

HUBBARD, B. B. (1998). *The World According to Wavelets*. A. K. Peters, second edition ed.

HUBEL, D. & WIESEL, T. (2005). *Brain and visual perception: the story of a 25-year collaboration*. Oxford University Press, Oxford; New York, NY.

HUBEL, D. H. (1995). *Eye, Brain and Vision*. NY: Scientific American Library.

HUDELOT, C., ATIF, J. & BLOCH, I. (2008). Fuzzy spatial relation ontology for image interpretation. *Fuzzy Sets Syst.* **159**(15), 1929–1951.

ILLINGWORTH, J. & KITTLER, J. (1988). A survey of the Hough transform. *Computer Graphics and Image Processing* **44**, 87–116.

INGLE, V. K. & PROAKIS, J. G. (1997). *Digital Signal Processing using MATLAB V.4*. PWS Publishing Company.

ION, A., KROPATSCH, W. G. & HAXHIMUSA, Y. (2006). Considerations regarding the minimum spanning tree pyramid segmentation method. In: *SSPR/SPR*.

IVERSON, L. A. & ZUCKER, S. W. (1995). Logical/linear operators for image curves. *IEEE Transactions on Pattern Analysis and Machine Intelligence* **17**(10), 982–996.

JACKWAY, P. T. & DERICHE, M. (1996). Scale-space properties of the multiscale morphological dilation erosion. *IEEE Transactions on Pattern Analysis and Machine Intelligence* **18**(1), 38–51.

JAIN, A. (1989). *Fundamentals of Digital Image Processing*. Prentice Hall.

JAIN, A. & ZONGKER, D. (1997). Feature selection - evaluation, application, and small sample performance. *IEEE Transactions on Pattern Analysis and Machine Intelligence* **19**(2), 153–158.

JAIN, A. K. & DUBES, R. (1988). *Algorithms for Clustering Data*. Prentice Hall.

JAIN, A. K., DUIN, R. P. W. & MAO, J. (2000). Statistical pattern recognition: A review. *IEEE Transactions on Pattern Analysis and Machine Intelligence* **22**(1), 4–37.

JAIN, R., KASTURI, R. & SCHUNCK, B. G. (1995). *Machine Vision*. McGraw-Hill.

JAMES, G. (1996). *Modern Engineering Mathematics*. Addison-Wesley.

JAMES, J. F. (1995). *A Student's Guide to Fourier Transforms with Applications in Physics and Engineering*. Cambridge University Press.

JANSING, E. D., ALBERT, T. A. & CHENOWETH, D. L. (1999). Two-dimensional entropic segmentation. *Pattern Recognition Letters* **20**(3), 329–336.

JELINEK, H., CESAR-JR., R. M. & LEANDRO, J. J. G. (2003). Exploring the structure-function relationship of cat retinal ganglion cells using wavelets. *Brain and Mind* **4**(1), 67–90.

JELINEK, H. F., CREE, M. J., LEANDRO, J. J. G., SOARES, J. V. B., CESAR-JR., R. M. & LUCKIE, A. (2007). Automated segmentation of retinal blood vessels and identification of proliferative diabetic retinopathy. *Journal of the Optical Society of America A* **24**(5), 1448–1456.

JELINEK, H. F. & FERNANDEZ, E. (1998). Neurons and fractals: How reliable and useful are calculations of fractal dimensions? *Journal of Neuroscience Methods* **81**(1–2), 9–18.

JENNINGS, A. (1977). *Matrix Computation for Engineers and Scientists*. John Wiley and Sons.

JIANG, X. Y. & BUNKE, H. (1991). Simple and fast computation of moments. *Pattern Recognition* **24**(8), 801–806.

JOHNSON, R. & WICHERN, D. (1998). *Applied multivariate statistical analysis*. Prentice Hall, Upper Saddle River, NJ.

JOHNSTON, E. & ROSENFELD, A. (1973). Angle detection on digital curves. *IEEE Trans. Comp.* **22**(7), 875–878.

JOLION, J.-M. & ROSENFELD, A. (1994). *A Pyramid Framework for Early Vision: Multiresolutional Computer Vision*. Dordrecht, The Netherlands: Kluwer Academic Publishers.

JR., T. G. S., LANGE, G. D. & MARKS, W. B. (1996). Fractal methods and results in cellular morphology—dimensions, lacunarity and multifractals. *Journal of Neuroscience Methods* **69**, 123–136.

JR., T. G. S., MARKS, W. B., LANGE, G. D., JR., W. H. S. & NEALE, E. A. (1989). A fractal analysis of cell images. *Journal of Neuroscience Methods* **27**, 173–180.

JULESZ, B. (1971). *Foundations of Cyclopean Perception*. Chicago: University of Chicago Press.

JÄHNE, B. (1997). *Digital Image Processing: Concepts, Algorithms and Scientific Applications*. Berlin: Springer, 4th ed.

KAHN, D. W. (1995). *Topology: An Introduction to the Point-Set and Algebraic Areas*. Dover.

KAMEN, E. W. & HECK, B. S. (1997). *Fundamentals of Signals and Systems Using MATLAB*. Prentice-Hall.

KANDEL, A., BUNKE, H. & LAST, M. (2007). *Applied Graph Theory in Computer Vision and Pattern Recognition (Studies in Computational Intelligence)*. Springer-Verlag New York, Inc. Secaucus, NJ, USA.

KANEKO, T. & OKUDAIRA, M. (1985). Encoding of arbitrary curves based on the chain code representation. *IEEE Transactions on Communications* **33**(7), 697–707.

KASHI, R. S., BHOI-KAVDE, P., NOWAKOWSKI, R. S. & PAPATHOMAS, T. V. (1996). 2-D shape representation and averaging using normalized wavelet descriptors. *Simulation* **66**(3), 164–178.

KASS, M., WITKIN, A. & TERZOPOULOS, D. (1988). Snakes: Active contour models. *International Journal of Computer Vision* **1**, 321–331.

KENDALL, D. G., BARDEN, D., CARNE, T. K. & LE, H. (1999). *Shape and Shape Theory*. John Wiley and Sons.

KHALIL, M. I. & BAYOUMI, M. M. (2000). Invariant 2D object recognition using the wavelet modulus maxima. *Pattern Recognition Letters* **21**(9), 863–872.

KIM, S. D., LEE, J. H. & KIM, J. K. (1988). A new chain-coding algorithm for binary images using run-length codes. *Computer Vision, Graphics and Image Processing* **41**, 114–128.

KIMIA, B. B., TANNENBAUM, A. & ZUCKER, S. W. (1992). On the evolution of curves via a function of curvature. i. the classical case. *Journal of Mathematical Analysis and Applications* **163**, 438–458.

KIMIA, B. B., TANNENBAUM, A. & ZUCKER, S. W. (1995). Shapes, shocks and deformations: The components of two-dimensional shape and the reaction-diffusion space. *International Journal of Computer Vision* **15**, 189–224.

KIRSCH, R. A. (1971). Computer determination of the constituent structure of biological images. *Computers and Biomedical Research* **4**, 315–328.

KIRYATI, N. & MAYDAN, D. (1989). Calculating geometric properties from Fourier representation. *Pattern Recognition* **22**(5), 469–475.

KITCHEN, L. & ROSENFELD, A. (1982). Gray-level corner detection. *Pattern Recognition Letters* **1**(2), 95–102.

KLETTE, R. & ROSENFELD, A. (2004). *Digital Geometry: Geometric Methods for Digital Picture Analysis*. Morgan Kaufmann.

KLINKER, G. K. (1993). *A Physical Approach to Color Image Understanding*. Wellesley, MA: A. K. Peters, Wellesley.

KOENDERINK, J. J. (1984). The structure of images. *Biological Cybernetics* **50**, 363–370.

KOENDERINK, J. J. & RICHARDS, W. (1988). Two-dimensional curvature operators. *Journal Optical Society of America A* **8**(7), 1136–1141.

KOENDERINK, J. J. & VAN DOORN, A. J. (1992). Generic neighborhood operators. *IEEE Transactions on Pattern Analysis and Machine Intelligence* **14**(6), 597–605.

KOLB, H., NELSON, R. & MARIANI, A. (1981). Amacrine cells, bipolar cells and ganglion cells of the cat retina: A golgi study. *Vision Research* **21**, 1081–1114.

KREYSZIG, E. (1991). *Differential Geometry*. Dover.

KREYSZIG, E. (1993). *Advanced Engineering Mathematics*. John Wiley and Sons.

KRISHNAPURAM, R., KELLER, J. & MA, Y. (1993). Quantitative analysis of properties and spatial relations of fuzzy image regions. *IEEE Transactions on Fuzzy Systems* **1**(3), 222–233.

KRUGER, V. & SOMMER, G. (1999). Affine real-time face tracking using a wavelet network. In: *Proc. ICCV'99 Workshop Recognition, Analysis and Tracking of Faces and Gestures in Real-time Systems*. Corfu, Greece.

KUBOTA, T. & ALFORD, C. O. (1995). Computation of orientational filters for real-time computer vision problems I: Implementation and methodology. *Real-Time Imaging* **1**(4), 261–281.

KUBOTA, T. & ALFORD, C. O. (1996). Computation of orientational filters for real-time computer vision problems II: Multi-resolution image decomposition. *Real-Time Imaging* **2**(2), 91–116.

KUBOTA, T. & ALFORD, C. O. (1997). Computation of orientational filters for real-time computer vision problems III: Steerable system and VLSI architecture. *Real-Time Imaging* **3**(1), 37–58.

KUIPER, F. K. & FISHER, L. A. (1975). Monte carlo comparison of six clustering procedures. *Biometrics* **31**, 777–783.

KÖRNER, T. W. (1996). *Fourier Analysis*. Cambridge University Press.

LAFARGE, F., DESCOMBES, X., ZERUBIA, J. & PIERROT-DESEILLIGNY, M. (2006). An automatic building reconstruction method: A structural approach using high resolution images. *Proc. IEEE International Conference on Image Processing (ICIP), Atlanta, USA, October.*

LANGSAM, Y., AUGENSTEIN, M. J. & TENENBAUM, A. M. (1996). *Data Structures using C and C++.* Prentice-Hall.

LANTUÈJOUL, C. (1980). *Skeletonization in Quantitative Metallography.* Issues of digital image processing. Sijthoff and Noordhoff. R. M. Haralick and J.-C. Simon (Eds).

LARA, A. & HIRATA JR, R. (2006). Motion Segmentation using Mathematical Morphology. *SIBGRAPI'06. 19th Brazilian Symposium on Computer Graphics and Image Processing, 2006.* 315–322.

LARSON, H. J. & SHUBERT, B. O. (1979). *Probabilistic Models in Engineering Sciences.* John Wiley and Sons.

LAWRENCE, J. D. (1972). *A Catalog of Special Plane Curves.* New York: Dover Publications.

LAWSON, T. (1996). *Linear Algebra.* John Wiley and Sons.

LEANDRO, J., CESAR JR, R. & COSTA, L. D. F. (2008). Automatic Contour Extraction from 2D Neuron Images. *eprint arXiv: 0804.3234.*

LEAVERS, V. (1992). *Shape Detection in Computer Vision Using the Hough transform.* Berlin: Springer-Verlag.

LEDLEY, R. (1964). High-Speed Automatic Analysis of Biomedical Pictures. *Science* **146**(3641), 216–223.

LEE, J.-S., SUN, Y.-N. & CHEN, C.-H. (1995). Multiscale corner detection by using the wavelet transform. *IEEE Transactions on Image Processing* **4**(1), 100–104.

LEE, J.-S., SUN, Y.-N., CHEN, C.-H. & TSAI, C.-T. (1993). Wavelet based corner detection. *Pattern Recognition* **26**(6), 853–865.

LEE, Y. H. & HORNG, S.-J. (1999). Optimal computing the chessboard distance transform on parallel processing systems. *Computer Vision and Image Understanding* **73**(3), 374–390.

LEITHOLD, L. (1990). *The Calculus of a Single Variable with Analytic Geometry.* New York: Harper-Collins.

LENGYEL, J. (1998). The convergence of graphics and vision. *Computer* **31**(7), 46–53.

LESTREL, P. E. (1997). *Fourier Descriptors and Their Applications in Biology*. Cambridge University Press.

LEVENTHAL, A. G. & SCHALL, J. D. (1983). Structural basis of orientation sensitivity of cat retinal ganglion cells. *The Journal of Comparative Neurology* **220**, 465–475.

LEVINE, M. D. (1985). *Vision in Man and Machine*. New York: McGraw-Hill.

LEYTON, M. (1988). A process-grammar for shape. *Artificial Intelligence* **34**, 213–247.

LEYTON, M. (1992). *Symmetry, Causality, Mind*. Cambridge: The MIT Press.

LI, S. Z. (1995). *Markov Random Field Modelling in Computer Vision*. Tokyo: Springer Verlag.

LI, Y. (1992). Reforming the theory of invariant moments for pattern recognition. *Pattern Recognition* **25**(7), 723–730.

LIM, J. S. (1990). *Two-Dimensional Signal and Image Processing*. Englewood Cliffs, NJ: Prentice Hall.

LIN, Y., DOU, J. & WANG, H. (1992). Contour shape description based on an arch height function. *Pattern Recognition* **25**(1), 17–23.

LINDEBERG, T. (1993). Effective scale: A natural unit for measuring scale-space lifetime. *IEEE Transactions on Pattern Analysis and Machine Intelligence* **15**(10), 1068–1074.

LINDEBERG, T. (1994). *Scale-Space Theory in Computer Vision*. Dordrecht, the Netherlands: Kluwer Academic Publishers.

LIU, H.-C. & SRINATH, M. D. (1990). Corner detection from chain-code. *Pattern Recognition* **23**, 51–68.

LONCARIC, S. (1998). A survey of shape analysis techniques. *Pattern Recognition* **31**(8), 983–1001.

LOWE, D. G. (1989). Organization of smooth image curves at multiple scales. *International Journal of Computer Vision* **3**, 119–130.

LUO, B., C. WILSON, R. & HANCOCK, E. (2003). Spectral embedding of graphs. *Pattern Recognition* **36**(10), 2213–2230.

LUO, B. & HANCOCK, E. (2001). Structural Graph Matching Using the EM Algorithm and Singular Value Decomposition. *IEEE Transactions on Pattern Analysis and Machine Intelligence* **23**(10), 1120–1136.

LYNCH, C. (1997). Searching the internet. *Scientific American* **276**(3), 52–56.

LYNN, P. A. (1984). *An Introduction to the Analysis and Processing of Signals*. MacMillan.

MA, J., WU, C. K. & LU, X. R. (1986). A fast shape descriptor. *Computer Vision, Graphics and Image Processing* **34**, 282–291.

MAHMOUD, S. A. (1994). Arabic character recognition using Fourier descriptors and character contour encoding. *Pattern Recognition* **27**(6), 815–824.

MALLAT, S. (1996). Wavelets for a vision. *Proceedings of the IEEE* **84**(4), 604–614.

MALLAT, S. (1998). *A Wavelet Tour of Signal Processing*. Academic Press.

MALLAT, S. & HWANG, W. L. (1992). Singularity detection and processing with wavelets. *IEEE Transactions on Information Theory* **38**(2), 617–643.

MALLAT, S. & ZHONG, S. (1992). Characterization of signals from multiscale edges. *IEEE Transactions on Pattern Analysis and Machine Intelligence* **14**(7), 710–732.

MALLAT, S. G. (1989). A theory for multiresolution signal decomposition: The wavelet representation. *IEEE Transactions on Pattern Analysis and Machine Intelligence* **11**(7), 674–693.

MANDELBROT, B. (1982). *The Fractal Geometry of Nature*. New York: W.H. Freeman and Company.

MARCELJA, S. (1980). Mathematical description of the responses of simple cortical cells. *Journal of the Optical Society of America* **70**(11), 1297–1300.

MARCHETTE, D. (2004). *Random Graphs for Statistical Pattern Recognition*. Wiley-Interscience.

MARION, J. B. & THORNTON, S. T. (1995). *Classical Dynamics of Particles and Systems*. Saunders College Publishing.

MARR, D. (1982). *Vision*. San Francisco, CA: W. H. Freeman and Company.

MARR, D. & HILDRETH, E. (1980). Theory of edge detection. In: *Proceedings of Royal Society of London B*, vol. 207.

MARROQUIN, J. L., MITTER, S. K. & POGGIO, T. (1986). Probabilistic solution of ill-posed problems in computational vision. *Journal of the American Statistical Association* **82**(397), 76–89.

MARSHALL, S. (1989). Review of shape coding techniques. *Image and Vision Computing* **7**(3), 281–294.

MARTINS-JR, D. C., CESAR-JR, R. M. & BARRERA, J. (2006). W-operator window design by minimization of mean conditional entropy. *Pattern Analysis & Applications*.

MATHERON, G. (1975). *Random Sets and Integral Geometry*. John Wiley.

MATSAKIS, P. & ANDREFOUET, S. (2002). The fuzzy line between among and surround. In: *FUZZ-IEEE'02. Proceedings of the 2002 IEEE International Conference on Fuzzy Systems*, vol. 2.

MEDEIROS, E., LEWINER, T., LOPES, H. & VELHO, L. (2007). Reconstructing Poisson disc samplings of solid objects with topological guarantees. Tech. rep., IMPA. URL http://www.visgraf.impa.br/cgi-bin/refQuery.cgi?author=Velho&year=2007&submit=Search&output=html.

MEDIONI, G. & YASUMOTO, Y. (1987). Corner detection and curve representation using cubic B-splines. *Computer Vision Graphics and Image Process.* **39**, 267–278.

MELTER, R. A., STOJMENOVIC, I. & ZUNIC, J. (1993). A new characterization of digital lines by least square fits. *Pattern Recognition Letters* **14**(2), 83–88.

MENDELSON, B. (1990). *Introduction to Topology*. Dover.

MEYER, F. & BEUCHER, S. (1990). Morphological segmentation. *Journal of Visual Communication and Image Representation* **1**(1), 21–46.

MEYER, Y. (1993). *Wavelets: Algorithms and Applications*. Philadelphia: Society for Industrial and Applied Mathematics.

MICHEL, A. N. & HERGET, C. J. (1981). *Applied Algebra and Functional Analysis*. Dover.

MILGRAM, S. (1967). The small world problem. *Psychology Today* **2**(1), 60–67.

MILIOS, E. (1989). Shape matching using curvature processes. *Computer Vision, Graphics and Image Processing* **47**(2), 203–226.

MILLIGAN, G. W. (1980). An examination of the effect of six types of error perturbation on fifteen clustering algorithms. *Psychometrica* **45**, 325–342.

MILLIGAN, G. W. & SCHILLING, D. A. (1985). Assymptotic and finite-sample characteristics of four external criterion measures. *Multivariate Behavioral Research* **22**, 235–243.

MINSKY, M. (1974). A Framework for Representing Knowledge. Tech. rep., Massachusetts Institute of Technology, Cambridge, MA, USA.

MIYAJIMA, K. & RALESCU, A. (1994). Spatial organization in 2D segmented images: representation and recognition of primitive spatial relations. *Fuzzy Sets Syst.* **65**(2–3), 225–236.

MOJENA, R. (1975). Hierarchical grouping methods and stopping rules: An evaluation. *J. Computer* **20**, 359–363.

MOJSILOVIC, A., KOVACEVIC, J., HU, J., SAFRANEK, R. J. & GANAPATHY, S. (2000). Matching and retrieval based on the vocabulary and grammar of color patterns. *IEEE Transactions on Image Processing* **9**(1), 38–54.

MOKHTARIAN, F. (1995). Silhouette-based isolated object recognition through curvature scale space. *IEEE Transactions on Pattern Analysis and Machine Intelligence* **17**(5), 539–544.

MOKHTARIAN, F. (1997). A theory of multi-scale, torsion-based shape representation for space curves. *Computer Vision and Image Understanding* **68**(1), 1–17.

MOKHTARIAN, F. & MACKWORTH, A. (1986). Scale-based description and recognition of planar curves and two-dimensional shapes. *IEEE Transactions on Pattern Analysis and Machine Intelligence* **8**, 34–43.

MOKHTARIAN, F. & MACKWORTH, A. K. (1992). A theory of multiscale, curvature-based shape representation for planar curves. *IEEE Transactions on Pattern Analysis and Machine Intelligence* **14**, 789–805.

MONTAGUE, P. R. & FRIEDLANDER, M. J. (1991). Morphogenesis and territorial coverage by isolated mammalian retinal ganglion cells. *The Journal of Neuroscience* **11**(5), 1440–1457.

MOREL, J.-M. & SOLIMINI, S. (1995). *Variational Methods in Image Segmentation*. Boston: Birkhauser.

MORENO, A., TAKEMURA, C. M., COLLIOT, O., CAMARA, O. & BLOCH, I. (2008). Using anatomical knowledge expressed as fuzzy constraints to segment the heart in CT images. *Pattern Recognition*.

MORIGIWA, K., TAUCHI, M. & FUKUDA, Y. (1989). Fractal analysis of ganglion cell dendritic branching patterns of the rat and cat retinae. *Neurocience Research* Suppl. 10, S131–S140.

MORRISON, N. (1994). *Introduction to Fourier Analysis*. Wiley Interscience - John Wiley and Sons.

MUMFORD, D. & SHAH, J. (1985). Boundary detection by minimizing functionals. In: *Proceedings of IEEE Conference on CVPR*. S. Francisco, CA: IEEE Computer Society.

MURENZI, R. (1990). *Ondelettes Multidimensionelles et Applications à l'Analyse d'Images*. Ph.D. thesis, FYMA, UCL, Belgium. (In French.)

MURRAY, J. D. (1995). Use and abuse of fractal theory in neuroscience. *The Journal of Comparative Neurology* **361**, 369–371.

MUZY, J. F., BACRY, E. & ARNÈODO, A. (1994). The multifractal formalism revisited with wavelets. *International Journal of Bifurcation and Chaos* **4**(2), 245–302.

MYLER, H. R. & WEEKS, A. R. (1993). *Computer Imaging Recipes in C.* Englewood Cliffs, NJ: Prentice Hall.

NAKAMURA, Y. & YOSHIDA, T. (1994). Learning two-dimensional shapes using wavelet local extrema. In: *Proceedings 12th International Conference on Pattern Recognition, Conference C.* Jerusalem, Israel: IEEE Computer Society Press, Los Alamitos, CA. 9–13.

NEEDHAM, T. (1997). *Visual Complex Analysis.* Clarendon Press.

NEWMAN, M. (2003). The structure and function of complex networks. *SIAM Review* **45**, 167–256.

NIEMANN, H., SAGERER, G., SCHRODER, S. & KUMMERT, F. (1990). ERNEST: a semantic network system for pattern understanding. *Pattern Analysis and Machine Intelligence, IEEE Transactions on* **12**(9), 883–905.

NOMA, A., GRACIANO, A. B. V., CONSULARO, L. A., CESAR-JR, R. M. & BLOCH, I. (2008a). A new algorithm for interactive structural image segmentation. URL http://www.citebase.org/abstract?id=oai:arXiv.org:0805.1854.

NOMA, A., PARDO, A. & CESAR-JR., R. M. (2008b). Structural matching of 2D electrophoresis gels using graph models. In: *Proc. SIBGRAPI 2008.* IEEE Computer Society Press.

ODEN, J. T. (1979). *Applied Functional Analysis.* Prentice-Hall.

OGNIEWICZ, R. L. (1992). *Discrete Voronoi Skeletons.* Ph.D. thesis, Swiss Federal Institute of Technology, Zurich, Switzerland.

O'HIGGINS, P. (1997). *Methodological Issues in the Description of Forms.* Cambridge University Press. In P. E. Lestrel (Ed.), Fourier Descriptors and Their Applications in Biology.

OLIENSIS, J. (1993). Local reproducible smoothing without shrinkage. *IEEE Transactions on Pattern Analysis and Machine Intelligence* **15**(3), 307–312.

OPPENHEIM, A. V. & SCHAFER, R. W. (1975). *Digital Signal Processing.* Prentice-Hall.

OPPENHEIM, A. V. & SCHAFER, R. W. (1989). *Discrete-Time Signal Processing.* Prentice Hall.

PAIXÃO, T., GRACIANO, A. B. V., CESAR-JR, R. M. & HIRATA-JR., R. (2008). Backmapping approach for graph-based object tracking. In: *Proc. SIBGRAPI 2008.* IEEE Computer Society Press.

PANICO, J. & STERLING, P. (1995). Retinal neurons and vessels are not fractal but space-fulfilling. *The Journal of Comparative Neurology* **361**, 479–490.

PAPOULIS, A. (1962). *The Fourier Integral and its Applications.* New York: McGraw-Hill.

PAPOULIS, A. (1984). *Signal Analysis.* New York: McGraw-Hill.

PARKER, J. R. (1997). *Algorithms for Image Processing and Computer Vision.* John Wiley and Sons.

PAVLIDIS, T. (1973). Waveform segmentation through functional approximation. *IEEE Transactions Computers* **C22**(7), 689–697.

PAVLIDIS, T. (1977). *Structural Pattern Recognition.* New York: Springer-Verlag.

PAVLIDIS, T. (1980). Algorithms for shape analysis of contours and waveforms. *IEEE Transactions on Pattern Analysis and Machine Intelligence* **2**(4), 301–312.

PAVLIDIS, T. (1986). Editorial. *IEEE Transactions on Pattern Analysis and Machine Intelligence* **8**(1).

PAVLIDIS, T. & HOROWITZ, S. L. (1974). Segmentation of plane curves. *IEEE Transactions on Computers* **C23**(8), 860–870.

PEITGEN, H. & SAUPE, D. (1988). *The Science of Fractal Images.* New York: Springer-Verlag.

PEITGEN, H.-O., JÜRGENS, H. & SAUPE, D. (1992). *Chaos and Fractals: New Frontiers of Science.* New York: Springer-Verlag.

PEITGEN, H.-O. & SAUPE, D. (1998). *The Science of Fractal Images.* New York: Springer-Verlag.

PERNUS, F., LEONARDIS, A. & KOVACIC, S. (1994). Two-dimensional object recognition using multiresolution non-information-preserving shape features. *Pattern Recognition Letters* **15**, 1071–1079.

PERRET, D. I. & ORAM, M. W. (1993). Neurophysiology of shape processing. *Image and Vision Computing* **11**(6), 317–333.

PERSOON, E. & FU, K.-S. (1977). Shape discrimination using Fourier descriptors. *IEEE Transactions on Systems Man and Cybernetics* **7**, 170–179.

PHILLIPS, D. (1997). *Image Processing in C: Analyzing and Enhancing Digital Images.* RandD Press Books.

PITAS, I. (2000). *Digital Image Processing Algorithms and Applications.* John Wiley and Sons.

PITAS, I. & VENETSANOPOULOS, A. N. (1990). Morphological shape decomposition. *IEEE Transactions on Pattern Analysis and Machine Intelligence* **12**(1), 38–45.

PITAS, I. & VENETSANOPOULOS, A. N. (1992). Morphological shape representation. *Pattern Recognition* **25**(6), 555–565.

PITTNER, S. & KAMARTHI, S. V. (1999). Feature extraction from wavelet coefficients for pattern recognition tasks. *IEEE Transactions on Pattern Analysis and Machine Intelligence* **21**(1), 83–88.

POGGIO, T. & EDELMAN, S. (1990). A network that learns to recognize three-dimensional objects. *Nature* **343**(18), 263–266.

POGGIO, T., TORRE, V. & KOCH, C. (1985). Computational vision and regularization. *Nature* **317**(26), 314–319.

PORTER, R., GHOSH, S., LANGE, G. D. & JR., T. G. S. (1991). A fractal analysis of pyramidal neurons in mammalian motor cortex. *Neuroscience Letters* **130**, 112–116.

POULARIKAS, A. D. (2000). *The Transforms and Applications Handbook*. Boca Raton: CRC Press, 2nd ed.

POZNANSKI, R. R. (1992). Modelling the electrotonic structure of starbust amacrine cells in the rabbit retina: Functional interpretation of dendritic morphology. *Bulletin of Mathematical Biology* **54**(6), 905–928.

PRABHU, B. S. & PANDE, S. S. (1999). Intelligent interpretation of CAD drawings. *Computers and Graphics* **23**(1), 25–44.

PRASAD, L., IYENGAR, S. S. & AYENGAR, S. S. (1997). *Wavelet Analysis with Applications to Image Processing*. Boca Raton, FL: CRC Press.

PRATT, W. (1991). *Digital Image Processing*. John Wiley and Sons, 2nd ed.

PRATT, W. K., FAUGERAS, O. D. & GAGALOWICZ, A. (1981). Applications of stochastic texture field models to image processing. *Proceedings of the IEEE* **69**(5), 542–551.

PREPARATA, F. & SHAMOS, M. (1985). *Computational Geometry: An Introduction*. Springer.

PRESS, W. H., FLANNERY, B. P. & TEUKOLSKY, S. (1989). *Numerical recipes in C: The Art of Scientific Computing*. Cambridge: Cambridge University.

PREWITT, J. M. S. & MENDELSOHN, M. L. (1966). The analysis of cell images. *The New York Academy of Science Annals* **128**, 1035–1053.

PUDIL, P., NOVOVICOVÁ, J. & KITTLER, J. (1994). Floating search methods in feature selection. *Pattern Recognition Letters* **15**, 1119–1125.

RAGNEMALM, I. (1993). *The Euclidean Distance Transform*. Linköping Studies in Science and Technology.

RAMER, U. (1972). An iterative procedure for the polygonal approximation of plane curves. *Computer Graphics and Image Processing* **1**, 244–256.

RAMON Y CAJAL, S. R. (1989). *Recollections of my Life*. Cambridge: MIT Press.

RANDEN, T. & HUSOY, J. (1999). Filtering for texture classification: A comparative study. *IEEE Transactions on Pattern Analysis and Machine Intelligence* **21**(4), 291–310.

RATTARANGSI, A. & CHIN, R. T. (1992). Scale-based detection of corners of planar curves. *IEEE Transactions on Pattern Analysis and Machine Intelligence* **14**(4), 430–449.

REED, T. R. & WESCHSLER, H. (1990). Segmentation of textured images and gestalt organization using spatial/spatial-frequency representations. *IEEE Transactions on Pattern Analysis and Machine Intelligence* **12**(1), 1–12.

RICHARD, C. W. & HEMANI, H. (1974). Identification of 3-dimensional objects using Fourier descriptors of boundary curves. *IEEE Transactions on Systems Man and Cybernetics* **4**, 371–378.

RICHARDS, W. R., DAWSON, B. & WHITTINGTON, D. (1986). Encoding contour shape by curvature extrema. *Journal of the Optical Society of America* **3**(9), 1483–1491.

RIOUL, O. & VETTERLI, M. (1991). Wavelets and signal processing. *IEEE Signal Processing Magazine*, 14–89.

RIPLEY, B. D. (1996). *Pattern Recognition and Neural Networks*. Cambridge University Press.

RISTAD, E. S. & YIANILOS, P. N. (1998). Learning string-edit distance. *IEEE Transactions on Pattern Analysis and Machine Intelligence* **20**(5), 522–532.

RITTER, G. X. & WILSON, J. N. (1996). *Handbook of Computer Vision Algorithms in Image Algebra*. Boca Raton, FL: CRC Press.

ROCCHI, M., SISTI, D., ALBERTINI, M. & TEODORI, L. (2007). Current trends in shape and texture analysis in neurology: Aspects of the morphological substrate of volume and wiring transmission. *Brain Research Reviews* **55**(1), 97–107.

ROCK, I. & PALMER, S. (1990). The legacy of gestalt psychology. *Scientific American*, 84–90.

RODRIGUES, F., TRAVIESO, G. & COSTA, L. D. F. (2006). Fast Community Identification by Hierarchical Growth. *Arxiv preprint physics/0602144*.

ROERDINK, J. B. T. M. & MEIJSTER, A. (2000). The watershed transform: Definitions, algorithms and parallelization strategies. *Fundamenta Informaticae* **41**, 187–228.

ROMESBURG, H. C. (1990). *Cluster Analysis for Researchers*. Robert E. Krieger.

ROSE, K., GUREWITZ, E. & FOX, G. C. (1993). Deterministic annealing approach to constrained clustering. *IEEE Transactions on Pattern Analysis and Machine Intelligence* **15**, 785–794.

ROSENFELD, A. & KAK, A. C. (1982). *Digital Picture Processing*. New York: Academic Press, 2nd ed.

ROSENFELD, A. & KLETTE, R. (1985). Degree of adjacency or surroundedness. *Pattern Recognition* **18**(2), 169–177.

ROSENFELD, A. & WESZKA, J. S. (1975). An improved method of angle detection on digital curves. *IEEE Transactions on Computers* **24**, 940–941.

ROSIN, P. L. (1993). Multiscale representation and matching of curves using codons. *CVGIP - Graphical Models and Image Processing* **55**(4), 286–310.

ROSIN, P. L. & VENKATESH, S. (1993). Extracting natural scales using Fourier descriptors. *Pattern Recognition* **26**(9), 1383–1393.

ROSIN, P. L. & WEST, G. A. W. (1995). Curve segmentation and representation by superellipses. *IEE Proceedings of Vision, Image and Signal Processing* **142**(5), 280–288.

ROTHER, C., KOLMOGOROV, V. & BLAKE, A. (2004). "GrabCut": interactive foreground extraction using iterated graph cuts. *ACM Transactions on Graphics (TOG)* **23**(3), 309–314.

ROWLEY, H., BALUJA, S. & KANADE, T. (1998). Neural network-based face detection. *IEEE Transactions on Pattern Analysis and Machine Intelligence* **20**(1), 23–38.

RUSS, J. C. (1995). *Image Processing Handbook*. Boca Raton, FL: CRC Press, 2nd ed.

SAFAEE-RAD, R., SMITH, K. C., BENHABIB, B. & TCHOUKANOV, I. (1992). Application of moment and Fourier descriptors to the accurate estimation of elliptical shape parameters. *Pattern Recognition Letters* **13**(7), 497–508.

SAITO, H.-A. (1983). Morphology of physiologically identified x-, y- and w-type retinal ganglion cells of the cat. *The Journal of Comparative Neurology* **221**, 279–288.

SALARI, E. & BALAJI, S. (1991). Recognition of partially occluded objects using B-spline representation. *Pattern Recognition* **24**(7), 653–660.

SAMET, H. (1980). Region representation: quadtrees from boundary codes. *Communications of the ACM* **23**(3), 163–170.

SANCHIZ, J. M., IÑESTA, J. M. & PLA, F. (1996). A neural network-based algorithm to detect dominant points from chain-code of a contour. In: *Proceedings of 13th International Conference on Pattern Recognition*, vol. 4. Technical University of Vienna, Austria. Track D, 325–329.

SAPIRO, G. & TANNENBAUM, A. (1993). Affine invariant scale-space. *International Journal of Computer Vision* **11**(1), 25–44.

SAPIRO, G. & TANNENBAUM, A. (1995). Area and length preserving geometric invariant scale-spaces. *IEEE Transactions on Pattern Analysis and Machine Intelligence* **17**(1), 67–72.

SARKAR, D. (1993). A simple algorithm for detection of significant vertices for polygonal-approximation of chain-coded curves. *Pattern Recognition Letters* **14**(12), 959–964.

SCHALKOFF, R. (1989). *Digital Image Processing and Computer Vision*. Singapore: John Wiley and Sons.

SCHALKOFF, R. J. (1992). *Pattern Recognition: Statistical, Structural and Neural Approaches*. John Wiley and Sons.

SCHEY, H. M. (1997). *Div Grad Curl and All That—An Informal Text on Vector Calculus*. W. W. Norton and Co.

SCHNEIDER, M. & BEHR, T. (2006). Topological relationships between complex spatial objects. *ACM Trans. Database Syst.* **31**(1), 39–81.

SCHUTTER, E. D. & BOWER, J. M. (1994). An active membrane model of the cerebellar purkinjie cell i. simulation of current clamps in slice. *Journal of Neurophysiology* **71**, 375–400.

SCHÜRMANN, J. (1996). *Pattern Classification—A Unified View of Statistical and Neural Approaches*. John Wiley and Sons.

SCIENTIFIC AMERICAN (1997). Special issue: The internet—filfilling the promise. *Scientific American* **276**(3).

SERRA, J. (1982). *Image Analysis and Mathematical Morphology*. Academic Press.

SERRA, J. (1986). Introduction to mathematical morphology. *Computer Vision, Graphics and Image Processing* **35**, 283–305.

SERRA, J. (1989). Special issue on advances in mathematical morphology. *Signal Processing* **16**(4), 297–431. Guest ed.

SETHIAN, J. (1999). *Level Set Methods and Fast Marching Methods*. Cambridge University Press.

SHAHRARAY, B. & ANDERSON, D. J. (1985). Uniform resampling of digitized contours. *IEEE Transactions on Pattern Analysis and Machine Intelligence* **7**(6), 674–682.

SHAHRARAY, B. & ANDERSON, D. J. (1989). Optimal estimation of contour properties by cross-validated regularization. *IEEE Transactions on Pattern Analysis and Machine Intelligence* **11**(6), 600–610.

SHANMUGAN, K. S. & BREIPOHL, A. M. (1988). *Random Signals - Detection, Estimation and Data Analysis*. John Wiley and Sons.

SHEN, L., RANGAYYAN, R. M. & DESAUTELS, J. E. L. (1994). Application of shape analysis to mammographic calcifications. *IEEE Transactions on Medical Imaging* **13**(2), 263–274.

SHERBROOKE, E. C., PATRIKALAKIS, N. M. & WOLTER, N. M. (1996). Differential and topological properties of medial axis transform. *Graphical Models and Image Processing* **58**(6), 574–592.

SHOLL, D. A. (1953). Dendritic organization in the neurons of the visual and motor cortices of the cat. *Journal of Anatomy* **87**, 387–406.

SIDDIQI, K. & KIMIA, B. B. (1995). Parts of visual form: Computational aspects. *IEEE Transactions on Pattern Analysis and Machine Intelligence* **17**(3), 239–251.

SIMON, H. (1955). On a Class of Skewed Distribution Functions. *Biometrika* **42**(3/4), 425–440.

SIMONCELLI, E. P., FREEMAN, W. T., ADELSON, E. H. & HEEGER, D. J. (1992). Shiftable multiscale transforms. *IEEE Transactions on Information Theory* **38**(2), 587–607.

SMALL, C. G. (1996). *The Statistical Theory of Shape*. New York: Springer Verlag.

SNEDDON, I. N. (1995). *Fourier Transforms*. Dover.

SOARES, J. V. B., LEANDRO, J. J. G., CESAR-JR., R. M., JELINEK, H. F. & CREE, M. J. (2006). Retinal vessel segmentation using the 2-D Gabor wavelet and supervised classification. *IEEE Transactions on Medical Imaging* **25**, 1214–1222.

SOILLE, P. (1999). *Morphological Image Analysis: Principles and Applications*. Springer Verlag.

SOKAL, R. & SNEATH, P. (1963). *Principles of Numerical Taxonomy*. W. H. Freemand.

SOMOL, P., PUDIL, P., NOVOVICOVÁ, J. & PACLÍK, P. (1999). Adaptive floating search methods in feature selection. *Pattern Recognition Letters* **20**, 1157–1163.

SPIEGEL, M. R. (1995). *Fourier Analysis with Applications to Boundary Value Problems*. Schaum's Theory and Problems.

STAIB, L. H. & DUNCAN, J. S. (1992). Boundary finding with parametrically deformable models. *IEEE Transactions on Pattern Analysis and Machine Intelligence* **14**(11), 1061–1075.

STARCK, J. L., BIJAOUI, A. & MURTAGH, F. (1998). *Image Processing and Data Analysis: The Multiscale Approach*. Cambridge Univ. Press.

STEGER, C. (1998). An unbiased detector of curvilinear structures. *IEEE Transactions on Pattern Analysis and Machine Intelligence* **20**(2), 113–125.

STEIN, F. & MEDIONI, G. (1992). Structural indexing efficient 2D object recognition. *IEEE Transactions on Pattern Analysis and Machine Intelligence* **14**(12), 1198–1204.

STIX, G. (1997). Finding pictures in the web. *Scientific American* **276**(3), 54–55.

STOLLNITZ, E. J., DeROSE, T. D. & SALESIN, D. H. (1995a). Wavelets for computer graphics: A primer, part 1. *IEEE Computer Graphics and Applications* **15**(3), 76–84.

STOLLNITZ, E. J., DeROSE, T. D. & SALESIN, D. H. (1995b). Wavelets for computer graphics: A primer, part 2. *IEEE Computer Graphics and Applications* **15**(4), 75–85.

STOLLNITZ, E. J., DEROSE, T. D., SALESIN, D. H. & DeROSE, A. D. (1996). *Wavelets for Computer Graphics: Theory and Applications*. Morgan Kaufmann Publishers.

STOLYAR, A. (1984). *Introduction to Elementary Mathematical Logic*. Courier Dover Publications.

TAKEMURA, C. M. (2008). *Modelagem de posições relativas de formas complexas para análise de configuração espacial:* ENTRE *e* AO LONGO DE. Ph.D. thesis, IME-USP.

TAKEMURA, C. M., CESAR-JR., R. M. & BLOCH, I. (2005). Fuzzy modeling and evaluation of the spatial relation along. In: *Progress in Pattern Recognition, Image Analysis and Applications*, vol. 3773 of *X Iberoamerican Congress on Pattern Recognition*. Havana.

TANIMOTO, S. L. & PAVLIDIS, T. (1975). A hierarchical data structure for picture processing. *Computer Graphics and Image Processing* **4**(2), 104–119.

THEODORIDIS, S. & KOUTROUMBAS, K. (1999). *Pattern Recognition*. Academic Press.

THERRIEN, C. W. (1992). *Discrete Random Signals and Statistical Signal Processing*. Prentice Hall.

THOMPSON, D. W. (1992). *On Growth and Form—The Complete Revised Edition.* Dover.

TIENG, Q. M. & BOLES, W. W. (1997a). Recognition of 2D object contours using the wavelet transform zero-crossing representation. *IEEE Transactions on Pattern Analysis and Machine Intelligence* **19**(8), 910–916.

TIENG, Q. M. & BOLES, W. W. (1997b). Wavelet-based affine invariant representation - a tool for recognizing planar objects in 3D space. *IEEE Transactions on Pattern Analysis and Machine Intelligence* **19**(8), 846–857.

TIMNEY, B. N. & MACDONALD, C. (1978). Are curves detected by 'curvature detectors'? *Perception* **7**, 51–64.

TOLSTOV, G. P. (1976). *Fourier Series.* Dover.

TORRÉSANI, B. (1995). *Analyse Continue par Ondelettes.* Paris: InterEditions et CNRS Editions. (In French.0

TOU, J. & GONZALEZ, R. (1974). *Pattern recognition principles.* Reading, MA: Addison-Wesley.

TRAVERS, J. & MILGRAM, S. (1969). An experimental study of the small world problem. *Sociometry* **32**(4), 425–443.

TRAVIS, D. (1991). *Effective Color Displays: Theory and Practice.* London: Academic Press.

TRICOT, C. (1995). *Curves and Fractal Dimension.* New York: Springer.

TRIER, O. D., JAIN, A. K. & TAXT, T. (1996). Feature extraction methods for character recognition - a survey. *Pattern Recognition* **29**(4), 641–662.

TROILO, D., XIONG, M., CROWLEY, J. C. & FINLAY, B. L. (1996). Factors controlling the dendritic arborization of retinal ganglion cells. *Visual Neuroscience* **13**, 721–733.

TRUCCO, E. & VERRI, A. (1998). *Introductory Techniques for 3-D Computer Vision.* Prentice-Hall.

TRÉMEAU, A. & COLANTONI, P. (2000). Regions adjacency graph applied to color image segmentation. *IEEE Transactions on Image Processing* **9**(4), 735–744.

TSANG, W. M., YUEN, P. C. & LAM, F. K. (1994). Detection of dominant points on an object boundary: a discontinuity approach. *Image and Vision Computing* **12**(9), 547–557.

TURNER, D. A., LI, X.-G., PYAPALI, G. K., YLINEN, A. & BUZSAKI, G. (1995). Morphometric and electrical properties of reconstructed hippocampal ca3 neurons recorded in vivo. *The Journal of Comparative Neurology* **356**, 556–580.

VALOIS, R. L. D. & VALOIS, K. K. D. (1990). *Spatial Vision*. Oxford Sciences Publications.

VAN DER HEIJDEN, F. (1994). *Image Based Measurement Systems*. New York: John Wiley and Sons.

VAN OTTERLOO, P. J. (1991). *A Contour-Oriented Approach to Shape Analysis*. Englewood Cliffs, NJ: Prentice-Hall,

VAN VLIET, L. J. (1993). *Grey-Scale Measurements in Multi-Dimensional Digitized Images*. Technische Universiteit Delft, Delft University Press.

VAN VLIET, L. J. & VERBEECK, P. W. (1993). Curvature and bending energy in digitized 2d and 3d images. In: *Proceedings of the 8th Scandinavian Conference on Image Analysis* (K. A. HOGDA, B. B. & HEIA, K., eds.), vol. 2. Norway: NONIM-Norwegian Soc. Image Process. and Pattern Recognition.

VANEY, D. (1994). Patterns of neuronal coupling in the retina. *Progress in Retinal and Eye Research* **13**, 301–355.

VAQUERO, D., BARRERA, J. & HIRATA JR, R. (2005). A Maximum-Likelihood Approach for Multiresolution W-Operator Design. *Computer Graphics and Image Processing, 2005. SIBGRAPI 2005. 18th Brazilian Symposium on*, 71–78.

VELHO, L., FRERY, A. & GOMES, J. (2008). *Image Processing for Computer Graphics and Vision*. Springer.

VELTE, T. J. & MILLER, R. F. (1995). Dendritic integration in ganglion cells of the mudpuppy retina. *Visual Neuroscience* **12**, 165–175.

VETTERLI, M. & KOVACEVIC, J. (1995). *Wavelets and Subband Coding*. Englewood Cliffs, NJ: Prentice Hall.

VINCENT, L. (1989). Graphs and mathematical morphology. *Signal Processing* **16**, 365–388.

VINCENT, L. (1991). Exact euclidean distance function by chain propagation. In: *Proceedings of Computer Vision and Pattern Recognition*. Maui, Hawaii.

VINCENT, L., NACKEN, P., TOET, A. & HEIJMANS, H. J. A. M. (1992). Graph morphology. *Journal of Visual Communications and Image Representation* **3**(1), 24–38.

VINCENT, L. & SOILLE, P. (1991). Watersheds in digital spaces: An efficient algorithm based on immersion simulations. *IEEE Transactions on Pattern Analysis and Machine Intelligence.* **13**(6), 583–598.

VIOLA, P. & WELLS, W. M. (1997). Alignment by maximization of mutual information. *International Journal of Computer Vision* **24**(2), 137–154.

WALL, C. T. C. (1972). *A Geometric Introduction to Topology*. Dover.

WALLACE, T. P. & WINTZ, P. A. (1980). An efficient 3-dimensional aircraft recognition algorithm using normalized Fourier descriptors. *Computer Vision Graphics and Image Processing* **13**(2), 99–126.

WANDELL, B. A. (1995). *Foundations of Vision*. Sunderland, MA: Sinauer Associates.

WATSON, A. B. (1987). The cortex transform: Rapid computation of simulated neural images. *CVGIP: Image Understanding* **39**, 311–327.

WATTS, D. & STROGATZ, S. (1998). Collective dynamics of'small-world' networks. *Nature(London)* **393**(6684), 440–442.

WECHSLER, H. (1990). *Computational Vision*. Academic Press.

WEINBERG, S. (1994). *Dreams of a Final Theory*. Vintage Books.

WEISS, I. (1994). High-order differentiation filters that work. *IEEE Transactions on Pattern Analysis and Machine Intelligence* **16**(7), 734–739.

WEST, G. A. W. & ROSIN, P. L. (1991). Technique for segmenting image curves into meaningful descriptors. *Pattern Recognition* **24**(7), 643–652.

WEYL, H. (1980). *Symmetry*. Princeton University Press.

WIDDER, D. V. (1975). *The heat equation*. Academic Press.

WILLIAMSON, R. E. & TROTTER, H. F. (1996). *Multivariable Mathematics*. Prentice-Hall.

WITKIN, A. P. (1983). Scale-space filtering. In: *Proceedings of 8th International Joint Conference on Artificial Intelligence*.

WOOD, J. (1996). Invariant pattern recognition: A review. *Pattern Recognition* **29**(1), 1–17.

WORRING, M. & SMEULDERS, A. W. M. (1993). Digital curvature estimation. *Computer Vision Graphics and Image Processing* **58**, 366–382.

WRIGHT, M. W. (1995). *The Extended Euclidean Distance Transform*. Ph.D. thesis, University of Cambridge.

WU, K. & LEVINE, M. D. (1995). 2D shape segmentation: a new approach. Tech. rep., Centre for Intelligent Machines, McGill University, Canada. Technical Report TR-CIM-95-01.

WU, L. D. (1982). On the chain code of a line. *IEEE Transactions on Pattern Analysis and Machine Intelligence* **4**(3), 347–353.

WUESCHER, D. M. & BOYER, K. L. (1991). Robust contour decomposition using a constant curvature criterion. *IEEE Transactions Pattern Anaysis and Machine Intelligence* **13**(1), 41–51.

WYSZECKI, G. & STILES, W. (1982). *Color Science: Concepts and Methods, Quantitative Data and Formulae*. New York: Wiley, 2nd ed.

YANG, G. (1991). On the knowledge-based pattern recognition using syntactic approach. *Pattern Recognition* **24**(3), 185–193.

YANG, J. & WAIBEL, A. (1996). A real-time face tracker. In: *Proceedings of IEEE Workshop on Applications of Computer Vision*. Princeton, NJ.

YATES, R. C. (1952). *Curves and their Properties*. The National Council of Teachers of Mathematics.

YOU, K. C. & FU, K. S. (1979). A syntactic approach to shape recognition using attributed grammar. *IEEE Transactions on Systems, Man and Cybernetics* **9**(6), 334–345.

YOUNG, I. T., WALKER, J. E. & BOWIE, J. E. (1974). An analysis technique for biological shape - i. *Information and Control* **25**, 357–370.

YOUNG, T. Y. & CALVERT, T. W. (1974). *Classification, Estimation and Pattern Recognition*. American Elsevier.

ZAHN, C. T. & ROSKIES, R. Z. (1972). Fourier descriptors for plane closed curves. *IEEE Transactions on Computers* **21**, 269–281.

ZEKI, S. (2000). *Inner vision: an exploration of art and the brain*. Oxford University Press.

ZHOU, H. (2003). Distance, dissimilarity index, and network community structure. *Physical Review E* **67**(6), 61901.

ZHU, L., CHEN, Y. & YUILLE, A. (2006). Unsupervised Learning of a Probabilistic Grammar for Object Detection and Parsing. *Advances in Neural Information Processing Systems* **19**.

ZHU, P. & CHIRLIAN, P. M. (1995). On critical point detection of digital shapes. *IEEE Transactions on Pattern Analysis and Machine Intelligence* **17**(8), 737–748.

ZOIS, E. N. & ANASTASSOPOULOS, V. (2000). Morphological waveform coding for writer identification. *Pattern Recognition* **33**(3), 385–398.

ZUCKER, S. W. (1985). Early orientation selection: Tangent fields and the dimensionality of their support. *Computer Vision, Graphics and Image Processing* **32**, 74–103.

Index

T - #0967 - 101024 - C692 - 234/156/37 [39] - CB - 9780849379291 - Gloss Lamination